Welcome to the 69th International Symposium on Molecular Spectroscopy
June 16-20, 2014
Urbana-Champaign, IL

I0482620

On behalf of the Executive Committee, I extend a heartfelt welcome to all the attendees of the 69th Symposium and welcome you to the University of Illinois at Urbana-Champaign.

The Symposium presents research in fundamental molecular spectroscopy and a wide variety of related fields and applications. The continued vitality and significance of spectroscopy is annually re-affirmed by the number of talks, their variety, and the fact that many are given by students. These presentations are the heart of the meeting and are documented by this Abstract Book. Equally important is the information flowing from informal exchanges and discussions. As organizers, we strive to provide an environment that facilitates both kinds of interactions.

The essence of the meeting lies in the scientific discussions and your personal experiences this week independent of the number of times that you have attended this meeting. It is our sincere hope that you will find this meeting informative and enjoyable both scientifically and personally, whether it is your first or 50th meeting. If we can help to enhance your experience, please do not hesitate to ask the Symposium staff or the Executive Committee.

Ben McCall
Symposium Chair

VENUE AND SPONSOR INFORMATION FOLLOWS AUTHOR INDEX

69th INTERNATIONAL SYMPOSIUM ON MOLECULAR SPECTROSCOPY

International Advisory Committee
Allan Cheung, University of Hong Kong
Jose Domenech, CSIC
Brian Drouin, NASA Jet Propulsion Laboratory
Thomas Giesen, Universität Kassel
Michael Heaven*, Emory University, Chair
Mark Johnson, Yale University
Hideto Kanamori, Tokyo Institute of Technology
Isabelle Kleiner, CNRS
Sang Kuk Lee, Pusan National University
Terry Miller, Ohio State University
Nasser Moazzen-Ahmadi*, University of Calgary
Trevor Sears*, Brookhaven National Laboratory
Ned Sibert*, University of Wisconsin
David Skatrud, Army Research Office
Mary Ann Smith*, NASA Langley Research Center
Jonathan Tennyson, University College London
Jennifer van Wijngaarden, University of Manitoba
Jun Ye, JILA, University of Colorado and NIST
* steering committee member

Executive Committee
Ben McCall, Chair
Brian DeMarco
Dana Dlott
Gary Eden
Nick Glumac
Martin Gruebele
So Hirata
Leslie Looney
Dave Woon

Please send correspondence to
Ben McCall
International Symposium on Molecular Spectroscopy
Department of Chemistry
600 S. Mathews Avenue
Champaign IL 61801 USA
e-mail: chair@isms.illinois.edu
http://isms.illinois.edu

Special Sessions

For the 69th Symposium, three mini-symposia have been organized:

Astronomical Molecular Spectroscopy in the Age of ALMA

Organized by **Leslie Looney** (University of Illinois) and **Anthony Remijan** (NRAO). This mini-symposium will present new observational and experimental results that highlight the improvements in our understanding of molecular astronomy over the last few years. Invited Speakers: **Ted Bergin** (University of Michigan) and **Geoff Blake** (Caltech)

Beyond the Mass-to-Charge Ratio: Spectroscopic Probes of the Structures of Ions

Organized by **Mike Duncan** (University of Georgia) and **Tim Zwier** (Purdue). This mini-symposium will cover a variety of experimental and computational approaches to the spectroscopy and structure of molecular ions and their clusters. Invited Speakers: **Knut Asmis** (Universität Leipzig), **Yuan-Pern Lee** (National Chiao Tung University), **Cheuk Ng** (University of California Davis)

Spectroscopy in Kinetics and Dynamics

Organized by **Martin Gruebele** (University of Illinois) and **Ian Sims** (Université de Rennes 1). This mini-symposium embraces any research that uses spectroscopic techniques to elucidate the dynamics and kinetics of molecular systems. Invited Speakers: **Joel Bowman** (Emory University), **Ian Sims** (Université de Rennes 1), **Toshinori Suzuki** (Kyoto University)

Picnic

The Symposium picnic will be held on Wednesday evening, June 18, at Ikenberry Commons. The cost of the picnic is included in your registration (at below cost to students), so that all may attend the event. The **Coblentz Society** is the host for refreshments for one hour starting at 6:15 PM. Food will be served starting at 7:00 PM.

Sponsorship

We are pleased to acknowledge the many organizations that support the 69th Symposium. Principal funding comes from the **Army Office of Research** (ARO). We are most grateful to ARO for their long-standing support. We also acknowledge the many efforts and contributions of **The University of Illinois** in hosting the meeting, including financial contributions from the Office of the Vice Chancellor for Research and the Departments of Chemistry, Electrical and Computer Engineering, Astronomy, Physics, and Mechanical Science and Engineering.

Our Corporate Sponsors are **Bristol Instruments, Coherent, Elsevier/JMS, Ideal Vacuum Products, Journal of Physical Chemistry A, Newport/Spectra-Physics, Quantel, and Virginia Diodes**. Please see the back of this book for their advertisements.

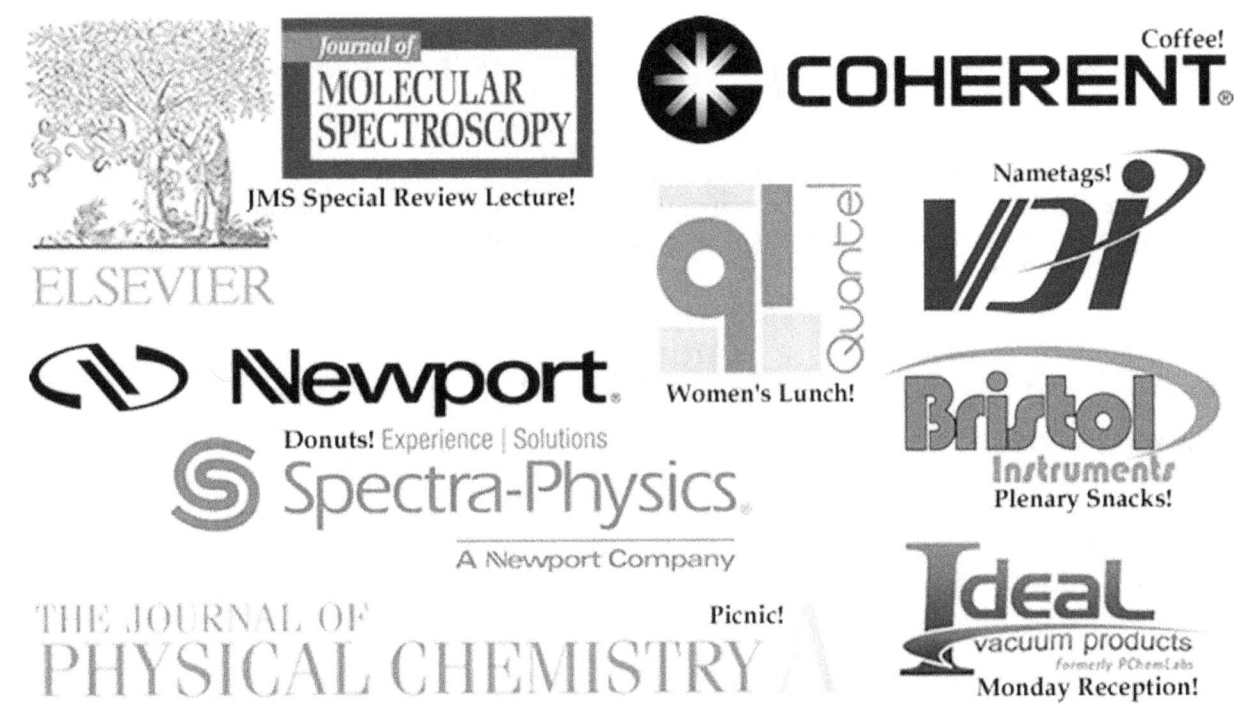

We are also pleased to acknowledge **Agilent Technologies, Block Engineering, Bruker Optics, Continuum, Daylight Solutions, IMRA, Litron Lasers, Lockheed Martin Aculight, Menlo Systems,** and **Princeton Instruments** as Contributing Sponsors.

IOS Press and **Canadian Light Source** have special inserts in your conferee packet. **Bruker Optics** will have a special Tuesday lunch presentation by Dr. Sergey Shilov entitled <u>Time-resolved FT-IR spectroscopy for fast reaction monitoring</u> (see your packet for a flyer). Our sponsors will have exhibits at the Symposium and we encourage you to visit their displays.

Rao Prize

The three Rao Prizes for the most outstanding student talks at the 2013 meeting will be presented. The winners are **Brett McGuire**, California Institute of Technology; **Gerard Dean O'Connor,** University of Sydney; and **Stefan Truppe**, Imperial College London. The Rao Prize was created by a group of spectroscopists who, as graduate students, benefited from the emphasis on graduate student participation, which has been a unique characteristic of the Symposium. This year three more Rao Prize winners will be selected.

The award is administered by a Prize Committee chaired by Yunjie Xu, University of Alberta, and comprised of Kevin Lehmann, University of Virginia; Brooks Pate, University of Virginia; Rebecca Peebles, Eastern Illinois University; Brenda Winnewisser, The Ohio State University and Tim Zwier, Purdue University. Any questions or suggestions about the Prize should be addressed to the Committee. Anyone (especially post-docs) willing to serve on a panel of judges should contact Yunjie Xu (yunjie.xu@ualberta.ca).

Miller Prize

The Miller Prize was created in honor of Professor Terry A. Miller, who served as chair of the International Symposium on Molecular Spectroscopy from 1992 to 2013. This prize will be awarded for the best presentation given by a recent PhD.

The Miller Prize winner and his or her co-authors will be invited to submit an article to the Journal of Molecular Spectroscopy based on the research in the prize-winning talk. After passing the normal review process, the article will appear in the Journal with a caption identifying the paper with the talk that received the Miller Prize.

Information

ACCOMMODATIONS

The check-in for dormitory accommodations is located in Bousfield Hall, 1214 South First Street, opens at noon on Sunday, June 15, and remains open 24 hours a day through the Symposium. Hotel information is listed on the ISMS website.

PARKING

Parking permits are for lot E14 (see map at end of book). If you did not purchase parking as part of your registration, and wish to purchase it after arrival, please contact the registration desk.

REGISTRATION

The registration desk is located in Room 165 Noyes Lab, and is open on Sunday from 4:00-6:00 PM, and Monday through Friday from 8:00 AM - 4:30 PM.

CHEMISTRY LIBRARY

The Chemistry Library will be open and available for your use during Symposium hours. The library has a number of computers, desks and tables to work at, and comfy chairs (and books!).

READY ROOM/STATION

We have set up Noyes Lab 164 as a "Ready Room" with computers that you can use to test your powerpoint presentation. If you have any problems, we will also have a staffed "Ready Station" in Noyes Lab 165 (right next to registration) where you can come for assistance.

COMPUTER LAB (VizLab)

Noyes Lab 151 is a small computer lab with Apple computers that is available for your use during the meeting. Please look in your packet for an access code to enter the room.

INTERNET ACCESS/Wi-Fi

Each attendee will receive a login and password to access campus WiFi (SSID: UIUCnet) as a guest. This access should work in most locations through campus. Please read the Internet Acceptable Use Policy below.

AUDIO/VIDEO INFORMATION

Each session room is equipped with a computer, onto which presentation files will be pre-loaded by Symposium staff. To submit your presentation file, you must go to the **Manage Presentations** link on our web site and follow the instructions. All files must be submitted by **11:59 PM CDT THE DAY BEFORE** your presentation session. All submitted files will be loaded onto the presentation computer one half-hour prior to the beginning of the session.

ACKNOWLEDGMENTS

The Symposium Chair wishes to acknowledge the hard work of numerous people who made this meeting possible. First and foremost is the Symposium Coordinator Birgit McCall, who has smoothly and single-handedly taken care of almost all of the electronic and logistical aspects of the meeting. Second are our student assistants, James Hodges, Michael Porambo, and Brad Gibson, who have handled innumerable important details to ensure the sessions and exhibitions go well. The other students in my group also play vital roles in monitoring the audiovisual systems and other aspects of the meeting. I wish to acknowledge the hospitality of the Chemistry Department and the School of Chemical Sciences (as well as the School of Molecular and Cell Biology) in tolerating our takeover of their buildings. Finally, I'd like to extend a special thanks to Becky Gregory and Terry Miller for their facilitation of a smooth transition to Illinois, and for answering countless questions along the way.

LIABILITY

The Symposium fees DO NOT include provisions for the insurance of participants against personal injuries, sickness, theft, or property damage. Participants and companions are advised to obtain whatever insurance they consider necessary. The Symposium organizing committee, its sponsors, and individual committee members DO NOT assume any responsibility for loss, injury, sickness, or damages to persons or belongings, however caused. The statements and opinions stated during oral presentations or in written abstracts are solely the author's responsibilities and do not necessarily reflect the opinions of the organizers.

DISCLAIMER

The views, opinions, and/or findings contained in this report are those of the authors and should not be construed as an official Department of the Army position, policy, or decision, unless so designated by other documentation.

INTERNET ACCEPTABLE USE POLICY

Each attendee will receive a login and password to access campus WiFi (SSID: UIUCnet) as a guest. Guest accounts are intended to support a broad range of communications. Professional and appropriate etiquette is required. Anonymous access and posting through guest accounts is forbidden. All users must accept that their identity may be associated with any content they provide while using the service. By accessing the campus WiFi network, you expressly acknowledge and agree to the following:

Use of the guest account service is at your sole risk and the entire risk as to satisfactory quality and performance is with you. You agree not to use the guest account intentionally or unintentionally to violate any applicable local, state, national or international law, including, but not limited to, any regulations having the force of law. To the extent not prohibited by law, in no event shall the university be liable for personal injury, or any incidental, special, indirect or consequential damages whatsoever, including, without limitation, damages for loss of profits, loss of data, business interruption or any other commercial damages or losses, arising out of or related to your use or inability to use the guest account, however caused, regardless of the theory of liability (contract, tort or otherwise) and even if the university has been advised of the possibility of such damages. The use of the guest account is subject, but not limited to, all University policies and regulations detailed at the Campus Administrative Manual (http://www.cam.illinois.edu). See the University's Web Privacy Notice (http://www.vpaa.uillinois.edu/policies/web_privacy.cfm) for all applicable laws and policies.

MA. Plenary
Monday, June 16, 2014 – 8:45 AM
Room: Theater Lincoln Hall

Chair: Gregory S. Girolami, University of Illinois, Urbana, IL, USA

Welcome   **8:45**
Peter E. Schiffer, Vice Chancellor for Research
University of Illinois at Urbana-Champaign

MA01   **9:00 – 9:40**
EXPLORING THE HIGH-RESOLUTION SPECTROSCOPY OF MOLECULES THAT CAN AFFECT THE QUALITY OF YOUR LIFE, Terry A. Miller

MA02   **9:45 – 10:25**
WELCOME TO RYDBERG-LAND, Yan Zhou, David Grimes, Tony Colombo, Ethan Klein, Timothy J Barnum, Robert W Field

Intermission

RAO AWARDS   **10:50**
Presentation of Awards by Yunjie Xu, University of Alberta

2013 Rao Award Winners
Brett A. McGuire, California Institute of Technology
Gerard Dean O'Connor, University of Sydney
Stefan Truppe, Imperial College London

COBLENTZ AWARD   **11:05**
Presentation of Award by Rohit Bhargava, Coblentz Society

MA03   *Coblentz Society Award Lecture*   **11:10 – 11:50**
SINGLE-MOLECULE MICROSCOPY OF NANOCATALYSIS, Peng Chen

MF. Astronomy
Monday, June 16, 2014 – 1:30 PM
Room: 116 Roger Adams Lab

Chair: Erika Gibb, University of Missouri - St. Louis, St. Louis, MO, USA

MF01   1:30 – 1:45
THE DISTRIBUTION, EXCITATION, AND ABUNDANCE OF CH^+ IN ORION KL, Harshal Gupta, Patrick Morris, Zsofia Nagy, John Pearson

MF02   1:47 – 2:02
MODELING LINEAR MOLECULES AS CARRIERS OF THE $\lambda 5797$ Å AND $\lambda 6613$ Å DIFFUSE INTERSTELLAR BANDS , Jane Huang, Takeshi Oka

MF03   2:04 – 2:19
SMALL AND LARGE MOLECULES IN THE DIFFUSE INTERSTELLAR MEDIUM, Takeshi Oka, Jane Huang

MF04   2:21 – 2:36
TEMPERATURE, DENSITY, IONIZATION RATE, AND MORPHOLOGY OF DIFFUSE GAS NEAR THE GALACTIC CENTER PROBED BY H_3^+, Takeshi Oka, Thomas R. Geballe, Miwa Goto, Tomonori Usuda

MF05   2:38 – 2:53
ESTIMATED SOFT X-RAY SPECTRUM AND IONIZATION OF MOLECULAR HYDROGEN IN THE CENTRAL MOLECULAR ZONE OF THE GALACTIC CENTER, Masahiro Notani, Takeshi Oka

MF06   2:55 – 3:05
MOLECULAR GAS NEAR UNUSUAL GALACTIC CENTER RADIO SOURCE N3, Dominic A. Ludovici, James Toomey, Cornelia Lang

MF07   3:07 – 3:22
SMALL CARBON CHAINS IN CIRCUMSTELLAR ENVELOPES, Robert J. Hargreaves, Kenneth Hinkle, Peter F. Bernath

MF08   3:24 – 3:39
ASTRONOMICAL MASERS: POLARIZATION PROPERTIES OF 22-GHZ WATER AND 6.7-GHZ METHANOL MASERS. , Gabriele Surcis, Wouter H.T. Vlemmings, Huib Jan van Langevelde

MF09   3:41 – 3:56
MODELING OF ASTROCHEMISTRY DURING STAR FORMATION, Ugo Hincelin, Eric Herbst, Qiang Chang, Tatiana Vasyunina, Yuri Aikawa, Kenji Furuya

Intermission

MF10   4:13 – 4:28
OSCILLATOR STRENGTHS AND PREDISSOCIATION RATES FOR $W - X$ BANDS AND THE $4P5P$ COMPLEX IN $^{13}C^{16}O$ AND $^{12}C^{18}O$, Michele Eidelsberg, Jean Louis Lemaire, Steven Federman, Glenn Stark, Alan Heays, Lisseth Gavilan, James R Lyons, Peter L Smith, Nelson de Oliveira, Denis Joyeux

MF11   4:30 – 4:40
STUDY OF THE PERTURBED $W\,^1\Pi(v = 1)$ STATE OF CO IN FIVE ISOTOPOLOGUES, Alan Heays, Michele Eidelsberg, Jean Louis Lemaire, Glenn Stark, Steven Federman, James R Lyons, Peter L Smith, Lisseth Gavilan, Nelson de Oliveira, Denis Joyeux

MF12   4:42 – 4:57
THE 3.1 μm INFRARED SPECTRA OF VIBRATIONALLY EXCITED C_3 IN A SUPERSONIC PLASMA JET, Dongfeng Zhao, Kirstin D Doney, Harold Linnartz

MF13   4:59 – 5:14
SUB-DOPPLER JET-COOLED INFRARED SPECTROSCOPY OF $ND_2H_2^+$ AND ND_3H^+ IN THE NH STRETCH FUNDAMENTAL MODES, Chih-Hsuan Chang, David Nesbitt

MF14   5:16 – 5:31
ELECTRONIC TRANSITION SPECTRA OF THIOPHENOXY AND PHENOXY RADICALS IN HOLLOW CATHODE DISCHARGES, Mitsunori Araki, Hiromichi Wako, Kei Niwayama, Koichi Tsukiyama

MF15   5:33 – 5:48
SYNCHROTRON-BASED HIGH RESOLUTION SPECTROSCOPY OF N-BEARING PAHS, Sébastien Gruet, Olivier Pirali, Manuel Goubet, Philippe Brechignac

MF16   *Post-Deadline Abstract*   5:50 – 6:05
THE HYPERFINE STRUCTURE OF ALUMINUM MONOXIDE, AlO, A. Breier, Thomas Büchling, Thomas Giesen, Jürgen Gauss

MG. Small molecules
Monday, June 16, 2014 – 1:30 PM
Room: 100 Noyes Laboratory

Chair: Gary Eden, University of Illinois, Urbana, IL, USA

MG01   1:30 – 1:45
ROUNDING AND UNCERTAINTIES IN PARAMETERS DETERMINED FROM FITS TO EXPERIMENTAL DATA,
or
A FAILURE TO ROUND DATA-ANALYSIS FIT PARAMETERS PROPERLY MAY MAKE THEM USELESS, Robert Le Roy

MG02   1:47 – 2:02
FULL EMPIRICAL POTENTIAL CURVES AND IMPROVED DISSOCIATION ENERGIES FOR THE $X\,^1\Sigma^+$ AND $A\,^1\Pi$ STATES OF CH^+, Young-Sang Cho, Robert Le Roy

MG03   2:04 – 2:19
ACCURATE ANALYTIC POTENTIAL FUNCTIONS FOR THE $A\,^3\Pi_1$ and $X\,^1\Sigma^+$ STATES OF IBr, Tokio Yukiya, Nobuo Nishimiya, Masao Suzuki, Robert Le Roy

MG04   2:21 – 2:36
OBSERVATION OF THE FORBIDDEN TRANSITIONS BETWEEN THE $A^1\Pi_u$ AND $b^3\Sigma_g^-$ STATES OF C_2, Wang Chen, Jian Tang, Kentarou Kawaguchi

MG05   2:38 – 2:53
ROTATIONAL ANALYSIS OF HIGH RESOLUTION F. T. SPECTRUM OF a' $^3\Sigma$- a $^3\Pi$ TRANSITION OF CS MOLECULE, Madhav Das Saksena, K Sunanda, M N Deo, Kentarou Kawaguchi

MG06   2:55 – 3:10
DEPERTURBATION STUDIES OF d $^3\Delta$ - a $^3\Pi$ TRANSITION OF CS MOLECULE, K Sunanda, M N Deo, Madhav Das Saksena, Kentarou Kawaguchi

MG07   3:12 – 3:27
CO^+ AND C_2 SPECTRA GENERATED BY CO_2 ATMOSPHERIC PRESSURE GLOW DISCHARGES IN MICROCHANNELS, Chul Shin, Zhen Dai, Thomas J. Houlahan, Jr., Sung-Jin Park, Gary Eden

MG08   3:29 – 3:44
STARK AND ZEEMAN EFFECT STUDY OF THE [18.6]3.5 - X(1)4.5 BAND OF URANIUM MONOFLUORIDE, UF., Colan Linton, Allan G. Adam, Timothy Steimle

Intermission

MG09   4:01 – 4:16
ANALYSIS OF $2\nu_3$ BAND OF HTO, Kaori Kobayashi, Hiroki Maki, Takuya Yamamoto, Masanori Hara, Yuji Hatano, Hiroyuki Ozeki

MG10   4:18 – 4:33
LIF SPECTROSCOPY OF JET COOLED MgOH, Masaru Fukushima, Takashi Ishiwata

MG11   4:35 – 4:50
ANALYSIS OF BOUND-FREE AND BOUND-BOUND EMISSION SPECTRA OF SCANDIUM MONOIODIDE PRODUCED BY THE PHOTODISSOCIATION OF ScI_3, Wenting Wendy Chen, Thomas C. Galvin, Thomas J. Houlahan, Jr., Gary Eden

MG12   4:52 – 5:07
MILLIMETER-WAVE SPECTROSCOPY OF OSSO, Marie-Aline Martin-Drumel, Jennifer van Wijngaarden, Oliver Zingsheim, Sven Thorwirth, Frank Lewen, Stephan Schlemmer

MG13   5:09 – 5:19
CALCULATION OF ANHARMONICITIES IN OVERTONE MODES AND SMALL-CLUSTER SHIFTS OF SPHERICAL-TOP MOLECULES, Javier D. Fuhr, Juan Fiol, Eduardo Cortizo, Pablo D. Fainstein, Daniel E Fregenal, Tomás Guozden, Enrique Kaúl, Pablo Knoblauch, Alberto Lamagna, Pablo Maceira, Guillermo Rozas, Martín Zarco

MG14   5:21 – 5:36
MM-WAVE ROTATIONAL SPECTRUM OF METHYL NITRATE, Jessica Thomas, Ivan Medvedev, David Dolson

MG15   5:38 – 5:53
PURE ROTATIONAL SPECTROSCOPY OF VINYL MERCAPTAN, Marie-Aline Martin-Drumel, Oliver Zingsheim, Sven Thorwirth, Holger S. P. Müller, Frank Lewen, Stephan Schlemmer

4

MH. Mini-symposium: Spectroscopy in Kinetics and Dynamics
Monday, June 16, 2014 – 1:30 PM
Room: B102 Chemical and Life Sciences

Chair: Mitchio Okumura, California Institute of Technology, Pasadena, CA, USA

MH01   *Journal of Molecular Spectroscopy Review Lecture*   1:30 – 2:00
ROAMING AND SPECTROSCOPY, Joel Bowman

MH02   2:05 – 2:20
THE SIGNATURE OF A ROAMING MECHANISM IN THE THERMAL DECOMPOSITION OF ETHYL NITRITE: CHIRPED-PULSE MILLIMETER-WAVE EXPERIMENT AND KINETIC MODELING, Kirill Prozument, Yury V. Suleimanov, Beat Buesser, William H. Green, Arthur Suits, Robert W Field

MH03   2:22 – 2:37
ROLES OF LARGE AMPLITUDE MOTIONS IN THE DYNAMICS OF THE PROTON TRANSFER REACTION $H_3^+ + H_2 \rightarrow H_5^+ \rightarrow H_3^+ + H_2$, Zhou Lin, Anne B McCoy

MH04   2:39 – 2:54
FULL-DIMENSIONAL FRANCK-CONDON FACTORS IN THE HARMONIC NORMAL MODE BASIS FOR THE $\tilde{A}\,^1A_u$—$\tilde{X}\,^1\Sigma_g^+$ TRANSITION OF ACETYLENE , Barratt Park, Robert W Field

MH05   2:56 – 3:11
CORRELATION BETWEEN THE SHAPE AND THE EXCITATION OF THE ANGULAR MOMENTUM FOR THE HCN/HNC MOLECULE, Georg Mellau, Robert W Field

MH06   3:13 – 3:28
DIRECT OBSERVATION OF b_2 VIBRATIONAL LEVELS IN THE $^1B_2$ $\tilde{C}$ STATE OF SO_2: PRECISE MEASUREMENT OF ν_3 LEVEL STAGGERINGS, Barratt Park, Jun Jiang, Carrie Womack, Peter Richter, Robert W Field, Andrew Richard Whitehill, Shuhei Ono

Intermission

MH07   3:45 – 4:00
VIBRATIONAL STATES AT THE DISSOCIATION LIMIT PROTECTED FROM VIBRATIONAL ENERGY FLOW, Martin Gruebele

MH08   4:02 – 4:17
TIME-RESOLVED FREQUENCY COMB SPECTROSCOPY OF TRANSIENT FREE RADICALS IN THE MID-INFRARED SPECTRAL REGION, Bryce J Bjork, Adam J. Fleisher, Bryan Changala, Thinh Quoc Bui, Kevin Cossel, Mitchio Okumura, Jun Ye

MH09   4:19 – 4:34
CHIRPED PULSE MICROWAVE SPECTROSCOPY IN PULSED UNIFORM SUPERSONIC FLOWS, Chamara Abeysekera, James Oldham, Kirill Prozument, Baptiste Joalland, Barratt Park, Robert W Field, Ian Sims, Arthur Suits

MH10   4:36 – 4:46
GAS-PHASE STRUCTURE DETERMINATION OF DIHYDROXYCARBENE, ONE OF THE SMALLEST STABLE SINGLET CARBENES, Carrie Womack, Kyle N Crabtree, Laura McCaslin, Oscar Martinez Jr., Robert W Field, John F. Stanton, Michael C McCarthy

MH11   4:48 – 5:03
IR-DRIVEN DYNAMICS OF THE 3-AMINOPHENOL-AMMONIA COMPLEX, Cornelia G Heid, W G Merrill, Amanda Case, Fleming Crim

MH12   5:05 – 5:20
VIBRATIONAL LEVELS AND RESONANCES ON A NEW POTENTIAL ENERGY SURFACE FOR THE GROUND ELECTRONIC STATE OF OZONE, Steve Alexandre Ndengue, Richard Dawes, Xiao-Gang Wang, Tucker Carrington

MH13   5:22 – 5:37
SHORT-LIVED ELECTRONICALLY-EXCITED DIATOMIC MOLECULES COOLED VIA SUPERSONIC EXPANSION FROM A PLASMA MICROJET, Thomas J. Houlahan, Jr., Rui Su, Gary Eden

MH14   5:39 – 5:54
CAVITY-ENHANCED ULTRAFAST TRANSIENT ABSORPTION SPECTROSCOPY, Yuning Chen, Melanie Roberts Reber, Kevin Keleher, Thomas K Allison

MI. Radicals
Monday, June 16, 2014 – 1:30 PM
Room: 112 Chemistry Annex

Chair: Dmitry G. Melnik, The Ohio State University, Columbus, OH, USA

MI01   1:30 – 1:45
VIBRONIC EMISSION SPECTROSCOPY OF BENZYL-TYPE RADICALS GENERATED BY CORONA DISCHARGE, Eun Hye Yi, Young Yoon, Sang Lee

MI02   *Post-Deadline Abstract*   1:47 – 2:02
HIGHER ELECTRONIC EXCITED STATES OF JET-COOLED AROMATIC HYDROCARBON RADICALS: 1-PHENYLPROPARGYL, 1-NAPHTHYLMETHYL, 2-NAPHTHYLMETHYL AND 9-ANTHRACENYLMETHYL, Gerard O'Connor, Gabrielle Victoria Grace Woodhouse, Tyler Troy, Klaas Nauta, Timothy Schmidt

MI03   2:04 – 2:19
INFRARED LASER SPECTROSCOPY OF THE HELIUM-SOLVATED ALLYL AND ALLYL PEROXY RADICALS, Chris Moradi, Christopher Leavitt, Brad Acrey, Gary Douberly

MI04   2:21 – 2:36
STUDY OF THE $CH_2I + O_2$ REACTION WITH A STEP-SCAN FOURIER-TRANSFORM INFRARED ABSORPTION SPECTROMETER: SPECTRA OF THE CRIEGEE INTERMEDIATE CH_2OO AND DIOXIRANE(?), Yu-Hsuan Huang, Yuan-Pern Lee

MI05   2:38 – 2:53
THE ELECTRONIC SPECTRUM OF BH_2 REVISITED, Mohammed Gharaibeh, Fumie Sunahori, Dennis Clouthier, Riccardo Tarroni

MI06   2:55 – 3:10
THE X_2BO and X_2BS (X = HYDROGEN OR HALOGEN) FREE RADICALS, Dennis Clouthier, Robert Grimminger, Bing Jin, Phillip Sheridan

MI07   3:12 – 3:27
OBSERVATION OF PURE ROTATIONAL SPECTRA OF SiCCN BY FOURIER-TRANSFORM MICROWAVE SPECTROSCOPY, Hiroya Umeki, Masakazu Nakajima, Yasuki Endo

MI08   3:29 – 3:44
PERTURBATION FACILITATED OPTICAL OPTICAL DOUBLE RESONANCE INVESTIGATION OF THE QUINTET MANIFOLD OF C_2 BY APPLYING TWO-COLOR FOUR-WAVE MIXING, Peter Bornhauser, Roberto Marquardt, Peter Radi

MI09   3:46 – 4:01
ANALYSIS OF BANDS OF THE 405 nm ELECTRONIC TRANSITION OF C_3Ar , Yen-Chu Hsu, Yi-Jen Wang, Anthony Merer

Intermission

MI10   4:18 – 4:33
VIBRONIC INTERACTION AND VIBRATIONAL ASSIGNMENT FOR NO_3 IN THE GROUND ELECTRONIC STATE, Eizi Hirota

MI11   4:35 – 4:50
VIBRONIC ANALYSIS OF THE $\tilde{A}^2E''$ STATE OF NO_3 RADICAL, Terrance Joseph Codd, John F. Stanton, Terry A. Miller

MI12   4:52 – 5:07
ROVIBRONIC ANALYSIS OF THE e' BANDS IN THE $\tilde{A}^2E''$ STATE OF NO_3 RADICAL, Henry Tran, Terrance Joseph Codd, Dmitry G. Melnik, Mourad Roudjane, Terry A. Miller

MI13   5:09 – 5:24
ROTATIONALLY-RESOLVED HIGH-RESOLUTION LASER SPECTROSCOPY OF THE $B\,^2E' \leftarrow X\,^2A_2'$ TRANSITION OF $^{14}NO_3$ RADICAL, Shunji Kasahara, Kohei Tada, Takashi Ishiwata, Eizi Hirota

MI14   5:26 – 5:41
ROTATIONALLY-RESOLVED HIGH-RESOLUTION LASER SPECTROSCOPY OF THE $B\,^2E' \leftarrow X\,^2A_2'$ TRANSITION OF $^{15}NO_3$ RADICAL, Kohei Tada, Shunji Kasahara, Takashi Ishiwata, Eizi Hirota

MI15   5:43 – 5:58
MULTISTATE VIBRONIC HAMILTONIAN FOR THE NITRATE RADICAL, John F. Stanton

MI16   6:00 – 6:15
AB INITIO CALCULATION FOR THE SPIN-ORBIT SPLITTINGS OF THE NITRATE RADICAL (NO_3), Lan Cheng, John F. Stanton

MI17   *Post-Deadline Abstract*   6:17 – 6:27
FOUR WAVE MIXING SPECTROSCOPY OF THE NO_3 $\tilde{B}\,^2E' - \tilde{X}\,^2A_2'$ transition , Masaru Fukushima, Takashi Ishiwata

MJ. Analytical, Combustion, Plasma
Monday, June 16, 2014 – 1:30 PM
Room: 274 Medical Sciences Building

Chair: Albert Ratner, The University of Iowa, Iowa City, IA, USA

MJ01   1:30 – 1:45
HEADSPACE ANALYSIS OF VOLATILE COMPOUNDS USING SEGEMENTED CHIRPED-PULSE FOURIER TRANSFORM MM-WAVE SPECTROSCOPY, Brent Harris, Amanda Steber, Brooks Pate

MJ02   1:47 – 2:02
HIGH SPEED, ULTRASENSITIVE TRACE GAS SENSING, David A. Long, Adam J. Fleisher, David F. Plusquellic, Joseph Hodges

MJ03   2:04 – 2:19
CHEMICAL ANALYSIS OF EXHALED HUMAN BREATH USING HIGH RESOLUTION MM-WAVE ROTATIONAL SPECTRA, Tianle Guo, Daniela Branco, Jessica Thomas, Ivan Medvedev, David Dolson, Hyun-Joo Nam, Kenneth O

MJ04   2:21 – 2:36
THZ/MM-WAVE SPECTROSCOPIC SENSORS, CATALOGS, AND UNCATALOGUED LINES , Ivan Medvedev, Christopher F. Neese, Frank C. De Lucia

MJ05   2:38 – 2:53
IDENTIFICATION OF FORGED BANK OF ENGLAND 20 GBP BANKNOTES USING IR SPECTROSCOPY, Emily Sonnex

MJ06   2:55 – 3:10
INTERFACIAL PROCESSES IN MODEL LITHIUM ION SYSTEMS PROBED WITH VIBRATIONAL SUM FREQUENCY GENERATION SPECTROSCOPY, Bruno G Nicolau, Natalia Garcia Rey, Dana Dlott

MJ07   3:12 – 3:27
DISTINCTIONS IN THE RAMAN SPECTROSCOPY FEATURES OF WO_3 MATERIALS WITH INCREASING TEMPERATURE, Raul F Garcia-Sanchez, Prabhakar Misra

Intermission

MJ08   3:44 – 3:59
PROBING THIN FILMS AND MONOLAYERS ON GOLD WITH LARGE AMPLITUDE TEMPERATURE JUMPS, Yuxiao Sun, Christopher M Berg, Dana Dlott

MJ09   4:01 – 4:16
USING LASER-DRIVEN FLYER-PLATES TO STUDY MECHANOCHEMICAL SHOCK WAVE ENERGY DISSIPATION, William L. Shaw, Dana Dlott

MJ10   4:18 – 4:33
SIMULTANEOUS QUANTIFICATION OF OH AND HO_2 IN DIMETHYL ETHER OXIDATION USING FARADAY ROTATION SPECTROSCOPY, Brian Brumfield, Xueliang Yang, Joseph Lefkowitz, Yiguang Ju, Gerard Wysocki

MJ11   4:35 – 4:50
CHIRPED PROBE PULSE FEMTOSECOND COHERENT ANTI-STOKES RAMAN SCATTERING FOR TURBULENT COMBUSTION DIAGNOSTICS, Claresta N. Fineman, Robert P. Lucht

MJ12   4:52 – 5:07
FTIR ANALYSIS OF FLOWING AFTERGLOW FROM A HIGH-FREQUENCY SPARK DISCHARGE, Allen White, Gary M Hieftje, Steve Ray, Kevin Pfeuffer

MJ13   5:09 – 5:24
EXPERIMENTAL EXAMINATION OF THE THERMOACOUSTIC INSTABILITY OF A LOW SWIRL FLAME WITH PLANAR LASER INDUCED FLUORESCENCE OF OH, Jianan Zhang, Kelsey Kaufman, Albert Ratner

MJ14   5:26 – 5:41
ABSORPTION SPECTROSCOPY IN THE 4.4-4.6 μm INFRARED WAVELENGTH RANGE FOR THE 10 KHZ HIGH-SPEED MEASUREMENT OF CO AND CO_2 CONCENTRATIONS IN COMBUSTING ENVIRONMENTS., Matthew L. Fotia, Brian C. Sell, John Hoke, Fred Schauer

MJ15   5:43 – 5:58
SPECTROSCOPIC STUDIES OF A LOW-TEMPERATURE ATMOSPHERIC PLASMOID ANALOGOUS TO BALL LIGHTNING, Scott E. Dubowsky, David M. Friday, Kevin C. Peters, Richard H. Perry, Zhangji Zhao, Bradley Deutsch, Rohit Bhargava, Jui-Nung Liu, Benjamin J. McCall

MK. Linelists, Lineshapes, Collisions
Monday, June 16, 2014 – 1:30 PM
Room: 217 Noyes Laboratory

Chair: Keeyoon Sung, Jet Propulsion Laboratory/Caltech, Pasadena, CA, USA

MK01 1:30 – 1:45
ARE YOUR SPECTROSCOPIC DATA BEING USED?, Iouli E Gordon, Laurence S Rothman, Jonas Wilzewski

MK02 1:47 – 2:02
ROVIBRATIONAL LINE LISTS FOR NINE ISOTOPOLOGUES OF CO SUITABLE FOR MODELLING AND INTER-PRETING SPECTRA AT VERY HIGH TEMPERATURES AND DIVERSE ENVIRONMENT, Gang Li, Iouli E Gordon, Laurence S Rothman, Yan Tan, Shui-Ming Hu, Samir Kassi, Alain Campargue

MK03 2:04 – 2:19
CH_4, C_2H_4, SF_6 AND CF_4 CALCULATED SPECTROSCOPIC DATABASES FOR THE VIRTUAL ATOMIC AND MOLECULAR DATA CENTRE, Vincent Boudon, Christian Wenger, Romain Surleau, Maud Louviot, Mbaye Faye, Maud Rotger, Ludovic Daumont, David A. Bonhommeau, Vladimir Tyuterev, Yaye Awa Ba, Marie-Lise Dubernet

MK04 2:21 – 2:36
HOT EXPERIMENTAL ABSORPTION SPECTRA OF CH_4 IN THE PENTAD AND OCTAD REGION, Robert J. Hargreaves, Michael Dulick, Peter F. Bernath

MK05 2:38 – 2:53
PREDICTING ROTATION-VIBRATION LEVELS OF ISOTOPICALLY SUBSTITUTED MOLECULES: WATER AS AN EXAMPLE, Oleg Polyansky, Aleksandra Kyuberis, Lorenzo Lodi, Jonathan Tennyson, Nikolay Fedorovich Zobov

MK06 2:55 – 3:10
NEW HIGH PRECISION LINELIST OF H_3^+, James N. Hodges, Adam J. Perry, Charles Markus, Paul A Jenkins II, G. Stephen Kocheril, Benjamin J. McCall

Intermission

MK07 3:27 – 3:37
PRECISE AND STABLE FREQUENCY SOURCE, AND MEASUREMENT OF $^{130}Te_2$ REFERENCE LINES FROM 443 TO 451 NM, James Coker, David La Mantia, John Furneaux, Jeffrey Gillean

MK08 3:39 – 3:49
ELECTRONIC TRANSITIONS ($BO_u^+ \leftarrow XO_g^+$) AND BANDHEAD FITTING FOR $^{130}Te_2$ IN THE INFRARED, David La Mantia, James Coker, John Furneaux, Jeffrey Gillean

MK09 3:51 – 4:06
NEW LINE LISTS FOR ROVIBRATIONAL AND ROTATIONAL TRANSITIONS WITHIN THE NH $X^3\Sigma^-$ AND OH $X^2\Pi$ GROUND STATES, James S.A. Brooke, Peter F. Bernath, Colin Western, Gang Li, Gerrit Groenenboom

MK10 4:08 – 4:23
DETERMINATION OF PRESSURE BROADENING AND SHIFTS FOR THE FIRST OVERTONE $2 \leftarrow 0$ OF HCL, Brian Drouin, Timothy J Crawford, Brennan M. Coffey

MK11 4:25 – 4:40
CONCENTRATION DEPENDENCE OF LINE SHAPES IN THE $\nu_1 + \nu_3$ BAND OF ACETYLENE, Matthew Cich, Damien Forthomme, Gregory Hall, C. McRaven, Trevor Sears

MK12 4:42 – 4:57
OBSERVATIONS OF DICKE NARROWING AND SPEED-DEPENDENCE IN CO2 LINESHAPES NEAR 2060NM, Thinh Quoc Bui, David A. Long, Cygan Agata, Vincent Sironneau, Daniel Hogan, Priyanka Milinda Rapusinghe, Mitchio Okumura

MK13 4:59 – 5:14
RAPID AND ACCURATE CALCULATION OF A SPEED DEPENDENT SPECTRAL LINE SHAPE, D. Reed Beverstock, Kendra Letchworth Weaver, D. Chris Benner

MK14 5:16 – 5:31
A THEORETICAL MODEL FOR WIDE-BAND INFRARED-ABSORPTION MOLECULAR SPECTRA AT ANY PRESSURE: FICTION OR REALITY?, Jeanna Buldyreva, Jean Vander Auwera

MK15 5:33 – 5:48
COLLISION-INDUCED SPECTRA: AN AVENUE TO INVESTIGATE MICROSCOPIC-SCALE DIFFUSION IN FLUIDS , Wouter A. Herrebout, Benjamin J. van der Veken, Alexander Kouzov

TA. Astronomy
Tuesday, June 17, 2014 – 8:30 AM
Room: 116 Roger Adams Lab

Chair: David E. Woon, University of Illinois, Urbana, IL, USA

TA01   8:30 – 8:45
TIME-DOMAIN TERAHERTZ SPECTROSCOPY (0.3 - 7.5 THz) OF MOLECULAR ICES OF SIMPLE ALCOHOLS, Brett A. McGuire, Sergio Ioppolo, Marco A. Allodi, Xander de Vries, Ian Finneran, Brandon Carroll, Geoffrey Blake

TA02   8:47 – 9:02
THz TIME-DOMAIN SPECTROSCOPY OF COMPLEX INTERSTELLAR ICE ANALOGS, Sergio Ioppolo, Brett A. McGuire, Marco A. Allodi, Xander de Vries, Ian Finneran, Brandon Carroll, Geoffrey Blake

TA03   9:04 – 9:19
TIME-DOMAIN TERAHERTZ SPECTRSOCOPY OF POLYCYCLIC AROMATIC HYDROCARBONS, Brandon Carroll, Marco A. Allodi, Sergio Ioppolo, Ian Finneran, Brett A. McGuire, Geoffrey Blake

TA04   9:21 – 9:36
PROBING GAS PHASE CHEMISTRY ABOVE ICE SURFACES WITH MILLIMETER/SUBMILLIMETER SPEC-TROSOCPY, AJ Mesko, Ian C Wagner, Stefanie N Milam, Susanna L. Widicus Weaver

TA05   9:38 – 9:53
MILLIMETER/SUBMILLIMETER SPECTROSCOPY OF PREBIOTIC MOLECULES FORMED FROM THE $O(^1D)$ IN-SERTION INTO METHYLAMINE, Brian Hays, Althea A. M. Roy, Susanna L. Widicus Weaver

TA06   9:55 – 10:10
ASTRONOMICAL APPLICATIONS OF NEW LINE LISTS FOR CN, C_2 AND THEIR ISOTOPOLOGUES, Peter F. Bernath, Chris Sneden, James S.A. Brooke, Ram Ram

TA07   10:12 – 10:27
MILLIMETER/SUBMILLIMETER STUDIES OF IONS AND RADICALS OF ASTROPHYSICAL INTEREST USING A HOLLOW CATHODE SPECTROMETER, Trevor Cross, Nadine Wehres, Mary Radhuber, Anne Carroll, Susanna L. Widicus Weaver

Intermission

TA08   10:44 – 10:59
THE MM-WAVE ROTATIONAL SPECTRUM OF GLYCOLIC ACID, Zbigniew Kisiel, Lech Pszczółkowski, Ewa Białkowska-Jaworska, Steven B Charnley

TA09   11:01 – 11:16
ROTATIONAL SPECTRA OF UREA IN ITS GROUND AND FIRST EXCITED VIBRATIONAL STATES, Jessica Thomas, Ivan Medvedev, Zbigniew Kisiel

TA10   11:18 – 11:33
THE LOWEST VIBRATIONAL STATES OF UREA FROM THE ROTATIONAL SPECTRUM, Zbigniew Kisiel, Jessica Thomas, Ivan Medvedev

TA11   11:35 – 11:50
HIGH RESOLUTION MEASUREMENTS AND ELECTRONIC STRUCTURE CALCULATIONS OF A DIAZANAPH-THALENE, Sébastien Gruet, Manuel Goubet, Olivier Pirali

TA12   11:52 – 12:07
Cis-METHYL VINYL ETHER: THE ROTATIONAL SPECTRUM UP TO 600 GHz, Lucie Kolesniková, Adam M Daly, José L. Alonso

TB. Instrument/Technique Demonstration
Tuesday, June 17, 2014 – 8:30 AM
Room: 100 Noyes Laboratory

Chair: Kevin Lehmann, The University of Virginia, Charlottesville, VA, USA

TB01   8:30–8:45
CHARACTERISATION AND CONTROL OF COLD CHIRAL COMPOUNDS, Chris Medcraft, Thomas Betz, V. Alvin Shubert, David Schmitz, Simon Merz, Melanie Schnell

TB02   8:47–9:02
MOLECULAR STRUCTURE AND CHIRALITY DETERMINATION FROM PULSED-JET FOURIER TRANSFORM MICROWAVE SPECTROSCOPY, Simon Lobsiger, Cristobal Perez, Luca Evangelisti, Nathan A Seifert, Brooks Pate, Kevin Lehmann

TB03   9:04–9:19
LIICG - A NEW METHOD FOR ROTATIONAL AND RO-VIBRATIONAL SPECTROSCOPY AT 4K, Lars Kluge, Alexander Stoffels, Sandra Bruenken, Oskar Asvany, Stephan Schlemmer

TB04   9:21–9:31
METHODS DEVELOPMENT FOR SPECTRAL SIMPLIFICATION OF ROOM-TEMPERATURE ROTATIONAL SPECTRA, Erin B Kent, Steven Shipman

TB05   9:33–9:48
DELIVERING MICROWAVE SPECTROSCOPY TO THE MASSES: A DESIGN OF A LOW-COST MICROWAVE SPECTROMETER OPERATING IN THE 18-26 GHZ FREQUENCY RANGE, Amanda Steber, Brooks Pate

Intermission

TB06   10:05–10:20
PRECISION FREQUENCY MEASUREMENT OF N_2O TRANSITIONS NEAR 4.5 μm AND ABOVE 150 μm , Wei-Jo Ting, Chun-Hung Chang, Shih-En Chen, Hsuan Chen Chen, Jow-Tsong Shy, Brian Drouin, Adam M Daly

TB07   10:22–10:37
SUBMILLIMETER SPECTROSCOPIC DIAGNOSTICS IN SEMICONDUCTOR PROCESSING PLASMAS, Yaser H. Helal, Christopher F. Neese, Frank C. De Lucia, Paul R. Ewing, Phillip J. Stout, Quentin Walker, Michael D. Armacost

TB08   10:39–10:54
INTERFERENCE EFFECTS IN NONLINEAR VIBRATIONAL SPECTROSCOPY FROM MULTILAYERED MATERIAL INTERFACES, Daniel B. O'Brien, Aaron M. Massari

TB09   10:56–11:11
MIXED POLARIZATION VIBRATIONAL SUM FREQUENCY GENERATION SPECTRA OF ORGANIC SEMICONDUCTING THIN FILMS , Patrick Kearns, Zahara Sohrabpour, Aaron M. Massari

TB10   11:13–11:28
SHOCK COMPRESSION INDUCED HOT SPOTS IN ENERGETIC MATERIAL DETECTED BY THERMAL IMAGING MICROSCOPY, Ming-Wei Chen, Dana Dlott

TB11   *Post-Deadline Abstract*   11:30–11:45
INFRARED SPECTROMETRIC DIAGNOSIS OF MALIGNANCY: TRENDS AND CONCERNS, Mohammadreza Khanmohammadi, Amir Bagheri Garmarudi

10

TC. Mini-symposium: Spectroscopy in Kinetics and Dynamics
Tuesday, June 17, 2014 – 8:30 AM
Room: B102 Chemical and Life Sciences

Chair: Ian Sims, Université de Rennes 1, Rennes, France

TC01   *INVITED TALK*   8:30 – 9:00
$O(^1D)$ REACTION WITH METHANE STUDIED BY STATE RESOLVED SCATTERING DISTRIBUTION MEASUREMENTS OF METHYL RADICALS, Toshinori Suzuki

TC02   9:05 – 9:20
COLLISION DYNAMICS OF EXCITED SODIUM MOLECULES, Burcin S Bayram, Phillip Arndt, Ceylan Guney, Jacob McFarland

TC03   9:22 – 9:37
SPECTROSCOPY AND DISSOCIATION DYNAMICS OF THE NO_3^+: A T-PEPICO STUDY, Kana Takematsu, Gustavo A. Garcia, John F. Stanton, Laurent Nahon, Mitchio Okumura

TC04   9:39 – 9:54
ULTRAVIOLET PHOTODISSOCIATION DYNAMICS OF THE CYCLOHEXYL RADICAL, Michael Lucas, Yanlin Liu, Jingsong Zhang

TC05   9:56 – 10:11
ULTRAVIOLET PHOTODISSOCIATION DYNAMICS OF THE 3-CYCLOHEXENYL RADICAL, Michael Lucas, Yanlin Liu, Raquel Bryant, Jasmine Minor, Jingsong Zhang

Intermission

TC06   10:28 – 10:43
THERMAL DECOMPOSITION OF BENZYL RADICAL VIA MULTIPLE ACTIVE PATHWAYS , Grant Buckingham, David Robichaud, Thomas Ormond, Mark R Nimlos, John W Daily, Barney Ellison

TC07   10:45 – 11:00
THE NICKEL ASSISTED DECOMPOSITION OF PENTANAL IN THE GAS PHASE AT VARIOUS INTERNAL ENERGIES, Adam Mansell, Darrin Bellert

TC08   11:02 – 11:17
STRONG-FIELD INDUCED DISSOCIATIVE IONIZATION OF VINYL BROMIDE PROBED BY FEMTOSECOND EXTREME ULTRAVIOLET (XUV) TRANSIENT ABSORPTION SPECTROSCOPY, Ming-Fu Lin, Daniel Neumark, Stephen R. Leone, Oliver Gessner

TC09   11:19 – 11:34
VELOCITY MAP IMAGING STUDIES OF NON-CONVENTIONAL METHANETHIOL PHOTOCHEMISTRY, Benjamin W. Toulson, Jonathan Alaniz, Craig Murray

TD. Biology, natural substances
Tuesday, June 17, 2014 – 8:30 AM
Room: 112 Chemistry Annex

Chair: Isabelle Kleiner, CNRS et Universités Paris Est et Paris Diderot, Créteil, France

TD01   8:30 – 8:45
CONFORMATION-SPECIFIC IR AND UV SPECTROSCOPY OF THE AMINO ACID GLUTAMINE: AMIDE-STACKING AND HYDROGEN BONDING IN AN IMPORTANT RESIDUE IN NEURODEGENERATIVE DISEASES, Patrick S. Walsh, Jacob C. Dean, Timothy S. Zwier

TD02   8:47 – 9:02
SURVEYING THE HYDROGEN BONDING LANDSCAPE OF AN ACHIRAL, α-AMINO ACID: CONFORMATION SPECIFIC IR AND UV SPECTROSCOPY OF 2-AMINOISOBUTYRIC ACID, Joseph R Gord, Daniel M. Hewett, Matthew A. Kubasik, Timothy S. Zwier

TD03   9:04 – 9:14
ROTATIONAL STUDY OF NATURAL AMINO ACID GLUTAMINE, Marcelino Varela, Carlos Cabezas, José L. Alonso

TD04   9:16 – 9:26
ROTATIONAL SPECTRUM OF TRYPTOPHAN, M. Eugenia Sanz, Carlos Cabezas, Santiago Mata, José L. Alonso

TD05   9:28 – 9:43
REMPI AND DOUBLE RESONANCE SPECTROSCOPY OF L-β-HOMOTRYPTOPHAN IN GAS PHASE, Hyuk Kang

Intermission

TD06   10:00 – 10:15
THE CONFORMATIONAL LANDSCAPE OF SERINOL, M. Eugenia Sanz, Donatella Loru, Isabel Peña, José L. Alonso

TD07   10:17 – 10:27
THE STRUCTURE OF PHENYLGLYCINOL, Alcides Simao, Isabel Peña, Carlos Cabezas, José L. Alonso

TD08   10:29 – 10:44
A NUCLEOSIDE UNDER OBSERVATION IN THE GAS PHASE: A ROTATIONAL STUDY OF URIDINE, Isabel Peña, José L. Alonso

TD09   10:46 – 11:01
THE CONFORMATIONAL BEHAVIOUR OF GLUCOSAMINE, Isabel Peña, Lucie Kolesniková, Carlos Cabezas, Celina Bermúdez, Matías Berdakin, Alcides Simao, José L. Alonso

TD10   11:03 – 11:13
THE DIPEPTIDE ALA-GLY IN THE GAS PHASE, Celina Bermúdez, Marcelino Varela, Carlos Cabezas, Isabel Peña, José L. Alonso

TD11   11:15 – 11:30
STRUCTURES AND ELECTRONIC SPECTRA OF THE CAFFEINE AND ITS HYDRATED CLUSTERS: A COMPUTATIONAL STUDY, Vipin Bahadur Singh

TD12   11:32 – 11:37
FLUORESCENCE SWITCH FOR SELECTIVELY SENSING COPPER AND HISTIDINE IN BOTH VITRO AND LIVING CELLS, Xiaojing Wang

TD13   11:39 – 11:49
BASED ON THE NATIONAL SCALE SOIL SPECTRAL DATABASE NITROGEN CONTENT INVERSION, Qianlong Wang

TD14   11:51 – 12:01
GREY INCIDENCE ANALYSIS (GIA): A NEW LOCAL METHOD FOR MODELLING CHINESE SOIL VIS-NIR SPECTRAL LIBRARY TO PREDICT SOIL TOTAL NITROGEN CONTENT, Qianlong Wang

TE. Structure determination
Tuesday, June 17, 2014 – 8:30 AM
Room: 274 Medical Sciences Building

Chair: Stewart E. Novick, Wesleyan University, Middletown, CT, USA

TE01   8:30 – 8:45
HIGH-RESOLUTION INFRARED SPECTROSCOPY SLIT-JET COOLED HYDROXYMETHYL RADICAL (CH_2OH): CH SYMMETRIC STRETCHING MODE, Fang Wang, Chih-Hsuan Chang, David Nesbitt

TE02   8:47 – 8:57
FOURIER-TRANSFORM MICROWAVE AND MILLIMETERWAVE SPECTROSCOPY OF CH_2IBr IN ITS GROUND VIBRATIONAL STATE, Kotomi Taniguchi, Shohei Sakai, Hiroyuki Ozeki, Toshiaki Okabayashi, William C. Bailey, Denis Duflot, Stephane Bailleux

TE03   8:59 – 9:14
GAS PHASE MICROWAVE MEASUREMENTS OF MONO-FLUOROBENZOIC ACIDS., Adam M Daly, Stephen G. Kukolich

TE04   9:16 – 9:31
THE COMBINED ORTHO / PARA HYDROGEN ASSIGNMENTS IN H_2 METAL CHLORIDES, Daniel A. Obenchain, G. S. Grubbs II, Derek S. Frank, Herbert M. Pickett, Stewart E. Novick

TE05   9:33 – 9:48
CHIRPED PULSE AND CAVITY FOURIER TRANSFORM MICROWAVE (CP-FTMW AND FTMW) INVESTIGATIONS INTO 3-BROMO-1,1,1,2,2-PENTAFLUOROPROPANE; A MOLECULE OF ATMOSPHERIC INTEREST, Nicholas Force, David Joseph Gillcrist, Cassandra C. Hurley, Frank E Marshall, Nicholas A. Payton, Thomas D. Persinger, N. E. Shreve, G. S. Grubbs II

Intermission

TE06   10:05 – 10:20
MICROWAVE SPECTRUM AND MOLECULAR STRUCTURE OF THE ARGON-(*E*)-1-CHLORO-1,2-DIFLUOROETHYLENE COMPLEX, Mark D. Marshall, Helen O. Leung, Hannah Tandon, Joseph P. Messinger, Eli Mlaver

TE07   10:22 – 10:37
THE EFFECT OF PROTIC ACID IDENTITY ON THE STRUCTURES OF COMPLEXES WITH VINYL CHLORIDE: FOURIER TRANSFORM MICROWAVE SPECTROSCOPY AND MOLECULAR STRUCTURE OF THE VINYL CHLORIDE-HYDROGEN CHLORIDE COMPLEX, Joseph P. Messinger, Helen O. Leung, Mark D. Marshall

TE08   10:39 – 10:54
MICROWAVE SPECTRUM OF THE HNO_3-HCOOH COMPLEX, Becca Mackenzie, Chris Dewberry, Ken Leopold

TE09   10:56 – 11:11
IMPACT FT-MW SPECTROSCOPY OF ORGANIC RINGS: INVESTIGATION OF THE CONFORMATIONAL LANDSCAPE, Dennis Wachsmuth, Jens-Uwe Grabow, Alberto Lesarri

TE10   11:13 – 11:28
MOLECULAR CHIRALITY: ENANTIOMER DIFFERENTIATION BY HIGH-RESOLUTION SPECTROSCOPY , Eizi Hirota

TE11   11:30 – 11:45
A PEPTIDE CO SOLVENT IN A CHIRALITY INDUCTION MODEL SYSTEM: BROADBAND ROTATIONAL SPECTROSCOPY OF THE 2,2,2-TRIFLUOROETHANOL- -PROPYL ENE OXIDE ADDUCT , Javix Thomas, Yunjie Xu

TE12   11:47 – 11:57
CONORMATONAL AND VIBRATIONAL ANALYSIS OF DIBENZO-18-CROWN-6.[a], A A El-Azhary, N. A. Al-jallal

TF. Mini-symposium: Astronomical Molecular Spectroscopy in the Age of ALMA
Tuesday, June 17, 2014 – 1:30 PM
Room: 116 Roger Adams Lab

Chair: Geoffrey Blake, California Institute of Technology, Pasadena, CA, USA

TF01   *INVITED TALK*   1:30 – 2:00
PHYSICS AND CHEMISTRY OF STAR AND PLANET FORMATION IN THE ALMA ERA, Edwin Bergin

TF02   2:05 – 2:20
UBIQUITOUS ARGONIUM, ArH^+, IN THE DIFFUSE INTERSTELLAR MEDIUM, P. Schilke, Holger S. P. Müller, C. Comito, A. Sanchez-Monge, D. A. Neufeld, Nick Indriolo, Edwin Bergin, D. C. Lis, Maryvonne Gerin, J. H. Black, M. G. Wolfire, John Pearson, Karl Menten, B. Winkel

TF03   2:22 – 2:37
ACCURATE LABORATORY MEASUREMENTS OF VIBRATION-ROTATION TRANSITIONS OF $^{36}ArH^+$ and $^{38}ArH^+$, Maite Cueto, Jose Cernicharo, Victor Jose Herrero, Isabel Tanarro, Jose Luis Domenech

TF04   2:39 – 2:54
THE ALMA ORION BAND 6 SCIENCE VERIFICATION DATA SPECTRAL LINE SURVEY - CONTENT AND DISCOVERY AVAILABLE FOR ALL SCIENCE, Anthony Remijan, Suzanna K. Randall, Catherine Vlahakis

TF05   2:56 – 3:11
THE ALMA SPECTRUM OF IRC+10216, Jose Cernicharo, Marcelino Agúndez, Fabien Daniel, Arancha Castro-Carrizo, Nuria Marcelino, Christine Joblin, Michel Guélin, Javier Goicoechea

TF06   3:13 – 3:28
^{13}C-METHYL FORMATE IN ORION-KL: ALMA OBSERVATIONS AND SPECTROSCOPIC CHARACTERIZATION, Cécile Favre, Miguel Carvajal, David Field, Edwin Bergin, Justin Neill, Nathan Crockett, Jes Jørgensen, Suzanne Bisschop, Nathalie Brouillet, Didier Despois, Alain Baudry, Isabelle Kleiner, L. Margulès, T. R. Huet, Jean Demaison

TF09   3:30 – 3:45
THE MOLECULAR COMPLEXITY OF G34.3+0.2, Douglas Friedel

Intermission

TF10   4:02 – 4:12
STAR FORMATION NEAR SGR A* AND THE ROLE OF COSMIC RAYS IN GALACTIC CENTER MOLECULAR CLOUDS , Farhad Yusef-Zadeh

TF11   4:14 – 4:24
MOLECULAR SPECTRAL LINES IN FILAMENTARY INFRARED DARK CLOUDS, Xing Lu, Qizhou Zhang, Hauyu Baobab Liu

TF12   4:26 – 4:41
THE CARMA LARGE-AREA STAR-FORMATION SURVEY: CLASSY , Leslie Looney, Manuel Fernandez-Lopez, Dominique M. Segura-Cox, Lee Mundy, Shaye Storm, Katherine Lee, Héctor Arce

TF13   4:43 – 4:58
TURBULENCE AND HEATING OF MOLECULAR CLOUDS IN THE GALACTIC CENTER, Natalie O Butterfield, Cornelia Lang, Betsy Mills, Dominic A. Ludovici

TF14   5:00 – 5:15
GROUND AND AIRBORNE OBSERVATIONS OF INTERSTELLAR HYDRIDES: NEW RESULTS FROM APEX AND SOFIA, Friedrich Wyrowski

TF15   5:17 – 5:27
MAGNETIC FIELDS AND STAR FORMATION - CN ZEEMAN MAPPING, Richard Crutcher

TF16   5:29 – 5:44
SURFACE CHEMISTRY UPDATE OF THE KOSMA-τ PDR CODE, Markus Röllig, Silke Andree-Labsch, Volker Ossenkopf

TF07   5:46 – 6:01
MOLECULES IN DUSTY IR-LUMINOUS GALAXIES WITH ALMA, Francesco Costagliola

TF08   6:03 – 6:18
HOT AMMONIA IN LUMINOUS HIGH-MASS STAR FORMING REGIONS IN THE GALAXY, Ciriaco Goddi

TG. Mini-symposium: Beyond the Mass-to-Charge Ratio: Spectroscopic Probes of the Structures of Ions

Tuesday, June 17, 2014 – 1:30 PM
Room: 100 Noyes Laboratory

Chair: Michael A Duncan, University of Georgia, Athens, GA, USA

TG01   ***INVITED TALK***   1:30 – 2:00
STATE-TO-STATE SPECTROSCOPY AND DYNAMICS OF IONS AND NEUTRALS BY PHOTOIONIZATION AND PHOTOELECTRON METHODS, Cheuk-Yiu Ng

TG02   2:05 – 2:20
ROVIBRONICALLY SELECTED AND RESOLVED LASER PHOTOIONIZATION AND PHOTOELECTRON STUDIES OF TRANSITION METAL CARBIDES, NITRIDES, AND OXIDES., Zhihong Luo, Yih-Chung Chang, Huang Huang, Cheuk-Yiu Ng

TG03   2:22 – 2:37
MULTIPLE ISOMERS OF $La(C_4H_6)$ FORMED IN REACTIONS OF La ATOM WITH SMALL HYDROCARBONS, Wenjin Cao, Dilrukshi Hewage, Dong-Sheng Yang

TG05   2:39 – 2:54
LANTHANUM ATOM-MEDIATED BOND ACTIVATION, COUPLING, AND CYCLIZATION OF 1,3-BUTADIENE PROBED BY MASS-ANALYZED THRESHOLD IONIZATION SPECTROSCOPY, Dilrukshi Hewage, Dong-Sheng Yang

TG06   2:56 – 3:11
CHARACTERIZATION OF THE RETINAL CHROMOPHORE IN THE GAS-PHASE VIA PHOTOISOMERIZATION ACTION SPECTROSCOPY, Katherine Jean Catani, Neville J Coughlan, Brian D Adamson, Evan Bieske

TG07   3:13 – 3:28
VIBRATIONAL SPECTROSCOPY OF TRANSIENT DIPOLAR RADICALS VIA AUTODETACHMENT OF DIPOLE-BOUND STATES OF COLD ANIONS, Dao-Ling Huang, Hong-Tao Liu, Phuong Diem Dau, Lai-Sheng Wang

TG08   3:30 – 3:45
UV PHOTOFRAGMENTATION SPECTROSCOPY OF MODEL LIGNIN-ALKALI ION COMPLEXES: EXTENDING LIGNOMICS INTO THE SPECTROSCOPIC REGIME, Jacob C. Dean, Nicole L Burke, John R. Hopkins, James Redwine, Bidyut Biswas, P. V. Ramachandran, Scott A McLuckey, Timothy S. Zwier

TG09   3:47 – 4:02
ISOMERIC EFFECTS ON FRAGMENTATIONS OF CROTONALDEHYDE AND METHACROLEIN IN LOW-ENERGY ELECTRON–MOLECULE COLLISIONS, Arup K. Ghosh, Aparajeo Chattopadhyay, Tapas Chakraborty

Intermission

TG10   4:19 – 4:34
STRUCTURAL ISOMERIZATION OF THE GAS PHASE 2-NORBORNYL CATION REVEALED WITH INFRARED SPECTROSCOPY AND COMPUTATIONAL CHEMISTRY, Daniel Mauney, Jonathan Mosley, Michael A Duncan

TG11   4:36 – 4:51
COLD, GAS-PHASE UV AND IR SPECTROSCOPY OF PROTONATED LEUCINE ENKEPHALIN AND ITS ANALOGUES , Nicole L Burke, James Redwine, Jacob C. Dean, Scott A McLuckey, Timothy S. Zwier

TG12   4:53 – 5:03
INTRAMOLECULAR HYDROGEN BONDING MOTIFS IN DEPROTONATED GLYCINE PEPTIDES REVEALED BY CRYOGENIC ION VIBRATIONAL SPECTROSCOPY, Brett Marsh, Erin M. Duffy, Michael Soukup, Jia Zhou, Etienne Garand

TG13   5:05 – 5:20
A GAS PHASE INVESTIGATION OF $CuOH(H_2O)^+$ AND Cu(II) OLIGOGLYCINE WATER OXIDATION CATALYSTS, Brett Marsh, Jia Zhou, Etienne Garand

TG14   5:22 – 5:32
GAS PHASE SPECTRA AND STRUCTURAL DETERMINATION OF GLUCOSE 6 PHOSPHATE USING CRYOGENIC ION VIBRATIONAL SPECTROSCOPY, Steven J. Kregel, Jonathan Voss, Brett Marsh, Etienne Garand

TG15   5:34 – 5:49
EXTENSIVE FREQUENCY COMB VELOCITY MODULATION SPECTROSCOPY OF ThF^+ FOR USE IN THE JILA ELECTRON EDM EXPERIMENT, Dan Gresh, Kevin Cossel, Jun Ye, Eric Cornell

TG16   5:51 – 6:06
MONITORING THERMODYNAMIC EQUILIBRIUM PROCESSES AT 10 K: CONFORMATIONAL ISOMERIZATION AND PHOTOCHROMISM OF O_4^+ IN ARGON MATRICES., Ryan M Ludwig, David T Moore

TG04   6:08 – 6:23
PULSED-FIELD IONIZATION AND ELECTRONIC STATES OF $La(C_6H_5X)$ (X = H, CH_3, AND OH) COMPLEXES, Tao Jin, Dong-Sheng Yang

TH. Mini-symposium: Spectroscopy in Kinetics and Dynamics
Tuesday, June 17, 2014 – 1:30 PM
Room: B102 Chemical and Life Sciences

Chair: Ahmed Heikal, University of Minnesota Duluth, Duluth, MN, USA

TH01   1:30 – 1:45
FEMTOSECOND TIME AND ANGLE-RESOLVED PHOTOELECTRON SPECTROSCOPY OF AQUEOUS SOLUTIONS, Toshinori Suzuki

TH02   1:47 – 2:02
EXCITED STATE STUDY OF $CuCl_4{}^{2-}$ COMPLEX IN ACETONITRILE, Andrey S Mereshchenko, Alexander Tarnovsky

TH03   2:04 – 2:19
DETECTION OF INTRAMOLECULAR CHARGE TRANSFER AND DYNAMIC SOLVATION IN EOSIN B BY FEMTOSECOND TWO-DIMENSIONAL ELECTRONIC SPECTROSCOPY, Soumen Ghosh, Jerome D. Roscioli, Warren F. Beck

TH04   2:21 – 2:36
COLLECTIVE VIBRATIONS OF WATER-SOLVATED HYDROXIDE IONS INVESTIGATED WITH BROADBAND 2DIR SPECTROSCOPY, Aritra Mandal, Krupa Ramasesha, Luigi De Marco, Martin Thämer, Andrei Tokmakoff

TH05   2:38 – 2:53
DYNAMICS OF MODEL HYDRAULIC FRACTURING LIQUID STUDIED BY TWO-DIMENSIONAL INFRARED SPECTROSCOPY, Kim Daley, Kevin J Kubarych

TH06   2:55 – 3:10
DYNAMIC STUDIES OF BOTH NON-EQUILIBRIA AND EQUILIBRIA PHENOMENA IN SILICA SOL-GEL MATERIALS, Christopher Jerald Huber, Aaron M. Massari

TH07   3:12 – 3:22
ENERGY TRANSFER IN A SYNTHETIC DENDRON-BASED LIGHT HARVESTING SYSTEM, Lea Nienhaus, Martin Gruebele

Intermission

TH08   3:39 – 3:44
EXCITED STATE DYNAMICS OF SUBSTITUTED TETRAPHENYLPORPHYRINS AND THEIR ADDUCTS WITH FULLERENE, Andrey S Mereshchenko, Alexey V Povolotskiy, Alexander S Konev, Yuriy S Tver'yanovich, Alexander F Khlebnikov

TH09   3:46 – 4:01
IDENTIFICATION OF A CRITICAL INTERMEDIATE IN GALVANIC EXCHANGE REACTIONS BY SINGLE-NANOPARTICLE RESOLVED KINETICS, Jeremy George Smith, Prashant Jain

TH10   4:03 – 4:18
ELECTRONIC GROUND AND EXCITED STATE SPECTRAL DIFFUSION OF A PHOTOCATALYST, Laura M. Kiefer, John T. King, Kevin J Kubarych

TH11   4:20 – 4:35
FEMTOSECOND NONLINEAR OPTICAL STUDIES OF RADIATIONLESS DECAY IN CAROTENOIDS AND IN THE PERIDININ–CHLOROPHYLL A PROTEIN, Jerome D. Roscioli, Soumen Ghosh, Michael M Bishop, Warren F. Beck, Harry A. Frank

TH12   4:37 – 4:52
THE EFFECT OF SULFUR SUBSTITUTION ON THE EXCITED-STATE DYNAMICS OF DNA AND RNA BASE DERIVATIVES, Marvin Pollum, Carlos E. Crespo-Hernández

TH13   4:54 – 5:09
PROTEIN DYNAMICS AND CONFORMATIONAL HETEROGENEITY CHARACTERIZED WITH TWO-DIMENSIONAL INFRARED SPECTROSCOPY, Megan Thielges, Edward Basom, James Spearman

TH14   5:11 – 5:26
KINETIC TRAPPING OF METASTABLE AMINO ACID POLYMORPHS, Garth Simpson

TH15   5:28 – 5:43
BIOENERGETICS AND DIFFUSION IN THE CROWDED MILIEU OF LIVING CELLS, Ahmed Heikal

TI. Atmospheric science
Tuesday, June 17, 2014 – 1:30 PM
Room: 112 Chemistry Annex

Chair: Iouli E Gordon, Harvard-Smithsonian Center for Astrophysics, Cambridge, MA, USA

TI01   1:30–1:45
METHACROLEIN IN THE IR ATMOSPHERIC WINDOW: MM-WAVE AND FTIR SPECTROSCOPIES COMPLEMENTED BY QUANTUM CALCULATIONS, Olena Zakharenko, Juan-Ramon Aviles Moreno, Haykal Imane, R. A. Motiyenko, T. R. Huet, Olivier Pirali

TI02   1:47–2:02
IR SPECTROSCOPY OF SELECTED ATMOSPHERIC MONOTERPENES AND OXYDATION PRODUCTS, Juan-Ramon Aviles Moreno, T. R. Huet, Manuel Goubet, Pascale Soulard, Pierre Asselin, Robert Georges, Olivier Pirali, P. Roy

TI03   2:04–2:19
HIGH-RESOLUTION INFRARED SPECTROSCOPY SLIT-JET COOLED HYDROXYMETHYL RADICAL (CH_2OH): OH STRETCHING MODE, Fang Wang, Chih-Hsuan Chang, David Nesbitt

TI04   2:21–2:36
VIBRONIC COUPLING OF $\tilde{B}^2 A'$ ELECTRONIC STATE WITH THE $\tilde{X}^2 A'$, $\tilde{A}^2 A''$ TWOFOLD OF ISOPROPOXY RADICAL., Mourad Roudjane, Rabi Chhantyal-Pun, Dmitry G. Melnik, Terry A. Miller, Jinjun Liu

TI05   2:38–2:53
DEVELOPMENT OF A NEAR-IR CAVITY ENHANCED ABSORPTION SPECTROMETER FOR THE DETECTION OF ATMOSPHERIC OXIDATION PRODUCTS AND ORGANOAMINES, Nathan C Eddingsaas, Breanna Jewell, Emily Thurnherr

TI06   2:55–3:10
INFRARED SPECTROSCOPY OF HALOGENATED SPECIES FOR ATMOSPHERIC REMOTE SENSING, Jeremy J. Harrison

TI07   3:12–3:27
PERFORMANCE OF A CRYOGENIC MULTIPATH HERRIOTT CELL VACUUM-COUPLED TO A BRUKER IFS-125HR SYSTEM, Arlan Mantz, Keeyoon Sung, Timothy J Crawford, Linda Brown, Mary Ann H. Smith

TI08   3:29–3:44
MEASUREMENTS AND MODELING OF $^{16}O^{12}C^{17}O$ SPECTROSCOPIC PARAMETERS AT 2 μm, David Jacquemart, Keeyoon Sung, Linda Brown, Max Coleman, Arlan Mantz, Mary Ann H. Smith

TI09   3:46–4:01
BROADBAND SPECTROSCOPY OF CO_2 BANDS NEAR 2 μm USING A FEMTOSECOND MODE-LOCKED LASER, Andrew Klose, Daniel L. Maser, Gabriel Ycas, Scott Diddams, Nathan R. Newbury, Ian Coddington

Intermission

TI10   4:18–4:33
FTS STUDIES OF THE ^{17}O ENRICHED ISOTOPOLOGUES OF CO_2 TOWARD CREATING A COMPLETE AND HIGHLY ACCURATE REFERENCE STANDARD, Ben Elliott, Keeyoon Sung, Linda Brown, Charles Miller

TI11   4:35–4:50
WATER VAPOR SELF-CONTINUUM BY CAVITY RING DOWN SPECTROSCOPY IN THE 1.6 MICRON TRANSPARENCY WINDOW , Alain Campargue, Samir Kassi, Didier Mondelain

TI12   4:52–5:07
AN ACCURATE AND COMPLETE EMPIRICAL LINE LIST FOR WATER VAPOR BETWEEN 5850 AND 7920 CM^{-1} , Semen Mikhailenko, Didier Mondelain, Samir Kassi, Alain Campargue

TI13   5:09–5:24
THE ROTATIONAL SPECTRUM OF HDO: ACCURATE SPECTROSCOPIC AND HYPERFINE PARAMETERS, Gabriele Cazzoli, Valerio Lattanzi, Cristina Puzzarini, Jürgen Gauss

TI14   5:26–5:41
O_2 ENERGY LEVELS, BAND CONSTANTS, POTENTIALS, FRANCK-CONDON FACTORS AND LINELISTS INVOLVING THE $X^3\Sigma_g^-$, $a^1\Delta_g$ AND $b^1\Sigma_g^+$ STATES, Shanshan Yu, Brian Drouin, Charles Miller, Iouli E Gordon

TI15   5:43–5:58
THE OXYGEN A BAND, D. Chris Benner, V. Malathy Devi, Jiajun Hoo, Joseph Hodges, David A. Long, Keeyoon Sung, Brian Drouin, Mitchio Okumura, Thinh Quoc Bui, Priyanka Rupasinghe

TI16   6:00–6:15
SELF- AND AIR-BROADENED LINE SHAPE PARAMETERS OF $^{12}CH_4$: 4500-4620 CM^{-1} , V. Malathy Devi, D. Chris Benner, Keeyoon Sung, Linda Brown, Timothy J Crawford, Mary Ann H. Smith, Arlan Mantz, Adriana Predoi-Cross

TJ. Instrument/Technique Demonstration
Tuesday, June 17, 2014 – 1:30 PM
Room: 274 Medical Sciences Building

Chair: Terry A. Miller, The Ohio State University, Columbus, OH, USA

TJ01   1:30 – 1:45
FAR-INFRARED BEAMLINE AT THE CANADIAN LIGHT SOURCE, Brant E Billinghurst, Tim E May

TJ02   1:47 – 2:02
COHERENT GENERATION OF BROADBAND PULSED LIGHT IN THE SWIR AND MWIR USING AN ALL POLARIZATION-MAINTAINING FIBER FREQUENCY COMB SOURCE, H. Hoogland, M. Engelbrecht, C. McRaven, R. Holzwarth, A. Thai, D. Sánchez, S. L. Cousin, M. Hemmer, M. Baudisch, K. Zawilski, P. G. Schunemann, J. Biegert

TJ03   2:04 – 2:19
DUAL-COMB SPECTROSCOPY OF C_2H_2, CH_4 AND H_2O OVER 1.0 - 1.7 μm, Kana Iwakuni, Sho Okubo, Hajime Inaba, Kazumoto Hosaka, Atsushi Onae, Hiroyuki Sasada, Feng-Lei Hong

TJ04   2:21 – 2:36
MULTIPLEXED DETECTION OF CO_2 USING A NOVEL DUAL COMB SPECTROMETER, Adam J. Fleisher, David A. Long, Joseph Hodges, David F. Plusquellic

TJ05   2:38 – 2:53
DOPPLER-FREE TWO-PHOTON ABSORPTION SPECTROSCOPY OF NAPHTHALENE ASSISTED BY AN OPTICAL FREQUENCY COMB, Akiko Nishiyama, Ayumi Matsuba, Masatoshi Misono

TJ06   2:55 – 3:10
MULTIPLEXED CHIRPED PULSE QUANTUM CASCADE LASER MEASUREMENTS OF AMMONIA AND OTHER SMALL MOLECULES, Craig Picken, Nigel Langford, Geoffrey Duxbury

Intermission

TJ07   3:27 – 3:42
BROADBAND HIGH-RESOLUTION SPECTROSCOPY WITH FABRY-PEROT QUANTUM CASCADE LASERS, Yin Wang, Gerard Wysocki

TJ08   3:44 – 3:59
DEVELOPMENT OF A FREQUENCY-STABILIZED MID-INFRARED EXTERNAL CAVITY-QCL CAVITY RING-DOWN SPECTROMETER, Bradley M. Gibson, Benjamin J. McCall

TJ09   4:01 – 4:16
HIGH-RESOLUTION SPECTROSCOPY OF THE ν_{16} BAND OF 1,3,5-TRIOXANE, Bradley M. Gibson, Nicole Koeppen, Benjamin J. McCall

TJ10   4:18 – 4:33
DUAL WAVELENGTH CAVITY RINGDOWN SPECTROSCOPY FOR HIGH PRECISION METHANE ISOTOPE RATIO MEASUREMENTS, Thinh Quoc Bui, Linhan Shen, Daniel Hogan, Pin Chen, Mitchio Okumura

TJ11   4:35 – 4:50
RECENT PROGRESS IN DEVELOPING A COMMERCIAL FIBER-LOOP CAVITY RINGDOWN SYSTEM, Brian Siller, Ryan Matz, Helen Waechter

TJ12   4:52 – 5:07
DOPPLER BROADENING THERMOMETRY BASED ON CAVITY RING-DOWN SPECTROSCOPY, Shui-Ming Hu, Cunfeng Cheng, Jin Wang, Yan Tan, Yu Robert Sun, An-Wen Liu, Jin-Tao Zhang

TJ13   5:09 – 5:24
MID-IR SUB-DOPPLER ERESOLUTION SPECTROMETER USING AN ENHANCED-CAVITY ABSORPTION CELL COUPLED WITH A WIDE BEAM, Masashi Abe, Kana Iwakuni, Sho Okubo, Hiroyuki Sasada

TJ14   5:26 – 5:41
QUANTITATIVE ABSORPTION AND KINETIC STUDIES OF TRANSIENT SPECIES USING GAS PHASE OPTICAL CALORIMETRY, Dmitry G. Melnik

TJ15   5:43 – 5:58
COLLINEAR TWO-COLOR SATURATION SPECTROSCOPY IN CN A-X (1-0) AND (2-0) BANDS, Damien Forthomme, C. McRaven, Trevor Sears, Gregory Hall

TK. Metal containing
Tuesday, June 17, 2014 – 1:30 PM
Room: B102 Chemical and Life Sciences

Chair: Manori Perera, Illinois Wesleyan University, Bloomington, IL, USA

TK01   1:30 – 1:45
LASER SPECTROSCOPY OF THE $C^2\Pi$ (41242 cm^{-1}) AND $^2\Delta$ (42192 cm^{-1}) STATES OF MAGNESIUM HYDRIDE, Nicholas Caron, Dennis Tokaryk, Allan G. Adam

TK02   1:47 – 2:02
VIBRONIC PERTURBATIONS IN THE ELECTRONIC SPECTRUM OF MgC, Phalgun Lolur, Richard Dawes, Michael Heaven

TK03   2:04 – 2:19
ELECTRONIC TRANSITIONS OF SCANDIUM MONOXIDE , Na Wang, Yuk Wai Ng, Allan S.C. Cheung

TK04   2:21 – 2:36
ELECTRONIC TRANSITIONS OF SCANDIUM MONOPHOSPHIDE, Na Wang, Kiu Fung Ng, Allan S.C. Cheung

TK05   2:38 – 2:48
PERTURBATION ANALYSIS OF THE (0,0) BAND OF THE $A^2\Pi_{3/2}$ - $X^2\Sigma^+$ TRANSITION IN ZrN, Kaitlin A Womack, Taylor Dahms, Leah C O'Brien, James J O'Brien

TK06   2:50 – 3:05
ANALYSIS OF A NEW ELECTRONIC TRANSITION OF MoO IN THE NEAR-INFRARED, Jack C Harms, Kaitlin A Womack, Leah C O'Brien

TK07   3:07 – 3:22
HIGH RESOLUTION LASER SPECTROSCOPY OF RHENIUM CARBIDE, Allan G. Adam, Ryan M. Hall, Colan Linton, Dennis Tokaryk

TK08   3:24 – 3:39
OBSERVATION OF A NEW $^2\Sigma^+$ - $^2\Sigma^+$ TRANSITION OF PtF BY INTRACAVITY LASER ABSORPTION SPECTROSCOPY, Taylor Dahms, Leah C O'Brien, Kaitlin A Womack, James J O'Brien

Intermission

TK09   3:56 – 4:11
OPTICAL STARK SPECTROSCOPY OF GOLD CHROLRIDE, Ruohan Zhang, Timothy Steimle

TK10   4:13 – 4:28
OBSERVATION OF TWO $\Omega=0^+$ EXCITED ELECTRONIC STATES IN JET-COOLED LaH, Suresh Yarlagadda, Sheo Mukund, Soumen Bhattacharyya, Sanjay G. Nakhate

TK11   4:30 – 4:45
THE SUBMILLIMETER SPECTRUM OF UO, Jennifer Holt, Christopher F. Neese, Frank C. De Lucia, Ivan Medvedev, Michael Heaven

TK12   4:47 – 5:02
APPLICATION OF TWO DIMENSIONAL FLOURESCENCE SPECTROSCOPY TO TRANSITION METAL CLUSTERS., Damian L Kokkin, Timothy Steimle

TK13   5:04 – 5:19
FOURIER TRANSFORM INFRARED EMISSION SPECTRA OF MgF_2, Daniel J. Frohman, Peter F. Bernath, Jacek Koput

TK14   5:21 – 5:26
MICROWAVE FREQUENCY TRANSITIONS REQUIRING LASER ABLATED URANIUM METAL DISCOVERED USING CHIRP-PULSE FOURIER TRANSFORM SPECTROSCOPY, B. E. Long, S. A. Cooke

TK15   5:28 – 5:43
ANION PHOTOELECTRON SPECTROSCOPIC STUDIES OF THE $NbC_4H_4^-$, $NbC_6H_6^-$ AND $NbC_6H_4^-$ PRODUCTS OF FLOW TUBE REACTIONS OF NIOBIUM WITH BUTADIENE, Melissa A. Baudhuin, Praveenkumar Boopalachandran, D. Alex Schnepper, Doreen Leopold, Stephen R Miller

TK16   5:45 – 5:55
THEORETICAL STUDY OF M$^+$—RG$_2$ (M$^+$ = CA — RA; RG = HE — RN), Anna Andrejeva, Adrian Gardner, Jack B Graneek, Richard Plowright, William Breckenridge, Tim Wright

WA. Plenary
Wednesday, June 18, 2014 – 8:30 AM
Room: Theater Lincoln Hall

Chair: Dale Van Harlingen, University of Illinois, Urbana, IL, USA

WA01   8:30 – 9:10
BROADBAND ROTATIONAL SPECTROSCOPY, Brooks Pate

WA02   9:15 – 9:55
DECELERATION AND TRAPPING OF COLD FREE RADICALS BY PULSED MAGNETIC FIELDS, Takamasa Momose

Intermission

WA03   10:15 – 10:55
CHIRAL MOLECULES REVISITED BY BROADBAND MICROWAVE SPECTROSCOPY, Melanie Schnell

WA04   11:00 – 11:40
HIGH-RESOLUTION SPECTROSCOPIC STUDIES OF REACTION INTERMEDIATES RELEVANT TO ATMO-SPHERIC CHEMISTRY , Yasuki Endo

WF. Mini-symposium: Astronomical Molecular Spectroscopy in the Age of ALMA
Wednesday, June 18, 2014 – 1:30 PM
Room: 116 Roger Adams Lab

Chair: Cécile Favre, University of Michigan, Ann Arbor, MI, USA

WF01   1:30 – 1:45
THE GBT PRIMOS PROGRAM: 7 YEARS OF ASTRONOMICAL DISCOVERY , Joanna F. Corby, Brett A. McGuire, Mike Hollis, Frank J Lovas, Philip Jewell, Anthony Remijan

WF02   1:47 – 2:02
A LOOK AT NITRILE CHEMISTRY IN SGR B2(N) USING THE COMBINED POWER OF THE GBT AND THE VLA , Amanda Steber, Daniel P. Zaleski, Nathan A Seifert, Justin Neill, Matt Muckle, Brooks Pate, Joanna F. Corby, Anthony Remijan

WF03   2:04 – 2:19
METHANIMINE AT HIGH SPATIAL RESOLUTION IN SGR B2: IMPLICATIONS FOR THE FORMATION OF CYANOMETHANIMINE, Joanna F. Corby, Amanda Steber, Nathan A Seifert, Cristobal Perez, Anthony Remijan, Brooks Pate

WF04   2:21 – 2:36
CH^+ AND SH^+ ABSORPTION SPECTROSCOPY WITH HERSCHEL: PROBING THE TURBULENT DISSIPATION IN THE DIFFUSE ISM. , Benjamin Godard, Edith G. Falgarone, Guillaume Pineau des Forêts, Maryvonne Gerin, Pierre Lesaffre, D. A. Neufeld, François Levrier

WF05   2:38 – 2:53
MEASUREMENT OF THE LOWEST MILLIMETER-WAVE TRANSITION FREQUENCY OF THE CH RADICAL, Stefan Truppe, Richard James Hendricks, Ed Hinds, Michael Tarbutt

WF06   2:55 – 3:05
LABORATORY CHARACTERIZATION AND ASTRONOMICAL DETECTION OF THE NITROSYLIUM ION, NO^+, Stephane Bailleux, E. A. Alekseev, Jose Cernicharo, Belén Tercero, Asuncion Fuente, Rafael Bachiller, Evelyne Roueff, Maryvonne Gerin, Sandra Treviño-Morales, Nuria Marcelino, Bertrand Lefloch

WF07   3:07 – 3:22
THE SEARCH FOR l-C_3H^+ (B11244) IN MORE THAN 40 ASTRONOMICAL SOURCES, Brett A. McGuire, Brandon Carroll, James Sanders, Susanna L. Widicus Weaver, Geoffrey Blake, Anthony Remijan

WF08   3:24 – 3:39
THE DISTRIBUTION OF ASTRONOMICAL ALDEHYDES - THE CASE FOR EXTENDED EMISSION OF ACETALDEHYDE (CH_3CHO)., Andrew Burkhardt, Ryan Loomis, Niklaus Dollhopf, Joanna F. Corby, Anthony Remijan

WF09   3:41 – 3:56
THE SEARCH FOR A COMPLEX MOLECULE IN A SELECTED HOT CORE REGION: A RIGOROUS ATTEMPT TO CONFIRM TRANS-ETHYL METHYL ETHER TOWARD W51 E1/E2, Brandon Carroll, Brett A. McGuire, Aldo J. Apponi, Lucy Ziurys, Geoffrey Blake, Anthony Remijan

Intermission

WF10   4:13 – 4:28
SUBMILLIMETER WAVE SPECTROSCOPY OF ACETYL ISOCYANATE : $CH_3C(O)NCO$, L. Margulès, R. A. Motiyenko, J.-C. Guillemin, Belén Tercero, Jose Cernicharo, Atef Jabri, Isabelle Kleiner, V. Ilyushin

WF11   4:30 – 4:45
LABORATORY CHARACTERIZATION AND ASTROPHYSICAL DETECTION IN ORION KL OF HIGHER EXCITED VIBRATIONAL STATES OF VINYL CYANIDE, Alicia López, Belén Tercero, Jose Cernicharo, Zbigniew Kisiel, Lech Pszczółkowski, Celina Bermúdez, José L. Alonso, Ivan Medvedev, Christopher F. Neese, Brian Drouin, Adam M Daly, Nuria Marcelino, Serena Viti, Hannah Calcutt

WF12   4:47 – 5:02
LABORATORY AND ASTRONOMICAL DISCOVERY OF HYDROMAGNESIUM ISOCYANIDE, Carlos Cabezas, Isabel Peña, Santiago Mata, José L. Alonso, Jose Cernicharo, Marcelino Agúndez, Michel Guélin

WF13   5:04 – 5:19
SPECTROSCOPIC CHARACTERIZATION AND DETECTION OF ETHYL MERCAPTAN IN ORION, Lucie Kolesniková, Adam M Daly, José L. Alonso, Belén Tercero, Jose Cernicharo, Bri Gordon, Steven Shipman

WF14   5:21 – 5:31
METHODS FOR DETECTION OF FAMILIES OF MOLECULES IN THE INTERSTELLAR MEDIUM, Glen Langston

WF15   5:33 – 5:48
INVESTIGATING THE "MINIMUM ENERGY PRINCIPLE" IN SEARCHES FOR NEW MOLECULAR SPECIES - THE CASE OF H_2C_3O ISOMERS, Ryan Loomis, Amy Robertson, Chelen Johnson, Samantha Blair, Anthony Remijan

WF16   5:50 – 6:05
OBSERVING ORGANIC MOLECULES IN INTERSTELLAR GASES: NON EQUILIBRIUM EXCITATION., Laurent Wiesenfeld, Alexandre Faure, Anthony Remijan, Krzysztof Szalewicz

WG. Mini-symposium: Beyond the Mass-to-Charge Ratio: Spectroscopic Probes of the Structures of Ions
Wednesday, June 18, 2014 – 1:30 PM
Room: 100 Noyes Laboratory

Chair: Jaime A. Stearns, Air Force Research Laboratory, Kirtland AFB, NM, USA

WG01   *INVITED TALK*   1:30 – 2:00
INFRARED SPECTRA OF PROTONATED AROMATIC HYDROCARBONS AND THEIR NEUTRAL COUNTERPARTS IN SOLID *PARA*-HYDROGEN, Mohammed Bahou, Yu-Jong Wu, <u>Yuan-Pern Lee</u>

WG02   2:05 – 2:20
VISIBLE PHOTODISSOCIATION SPECTRA OF THE 1-METHYL AND 2-METHYLNAPHTHALENE CATIONS: LASER SPECTROSCOPY AND THEORETICAL SIMULATIONS , Hela Friha, Geraldine Feraud, Cyril Falvo, Pascal Parneix, Thomas Pino, <u>Philippe Brechignac</u>, Tyler Troy, Timothy Schmidt, Zoubeida Dhaouadi

WG03   2:22 – 2:37
STRUCTURE AND ELECTRONIC PROPERTIES OF IONIZED PAH CLUSTERS , <u>Christine Joblin</u>, Damian L Kokkin, Hassan Sabbah, Anthony Bonnamy, Leo Dontot, Mathias Rapacioli, Aude Simon, Fernand Spiegelman, Pascal Parneix, Thomas Pino, Olivier Pirali, Cyril Falvo, Antonio Gamboa, Philippe Brechignac, Gustavo A. Garcia, Laurent Nahon

WG04   2:39 – 2:54
ABSORPTIONS IN THE VISIBLE OF PROTONATED PYRENE COLLISIONALLY COOLED TO 15 K, <u>C. A. Rice</u>, Francois Xavier Hardy, Oliver Gause, John P. Maier

WG05   2:56 – 3:11
ULTRAVIOLET PHOTODISSOCIATION ACTION SPECTROSCOPY OF PROTONATED AZABENZENES, <u>Christopher S. Hansen</u>, Stephen J. Blanksby, Evan Bieske, Jeffrey R. Reimers, Adam J. Trevitt

WG06   3:13 – 3:28
SIMULTANEOUS COUNTER-ION CO-DEPOSITION: A TECHNIQUE ENABLING MATRIX ISOLATION SPECTROSCOPY STUDIES USING LOW-ENERGY BEAMS OF MASS-SELECTED IONS, <u>Ryan M Ludwig</u>, David T Moore

WG07   3:30 – 3:45
CONTROLLED NEUTRALIZATION OF ANIONS IN CRYOGENIC MATRICES BY NEAR-THRESHOLD PHOTODETACHMENT, <u>Ryan M Ludwig</u>, David T Moore

Intermission

WG08   4:02 – 4:17
ANOMALOUS BEHAVIOR OBSERVED UPON ANNEALING AND PHOTODETACHMENT OF ANIONIC COPPER CARBONYL CLUSTERS IN ARGON MATRICES, Ryan M Ludwig, <u>David T Moore</u>

WG09   4:19 – 4:34
CARBON DIOXIDE CLUSTERS AND COPPER COMPLEXES FORMED IN ARGON MATRICES, <u>Michael E. Goodrich</u>, David T Moore

WG10   4:36 – 4:51
THEORETICAL HIGH-RESOLUTION SPECTROSCOPY BEYOND CCSD(T): THE INTERSTELLAR ANIONS CN^-, CCH^-, C_3N^-, AND C_4H^-, <u>Peter Botschwina</u>, Benjamin Schröder, Peter Sebald, Rainer Oswald

WG11   4:53 – 5:08
HIGH-J ROTATIONAL LINES OF HCO^+ AND ITS ISOTOPOLOGUES MEASURED BY USING EVENSON-TYPE TUNABLE FIR SPECTROMETER , Ryo Oishi, Tatsuya Miyamoto, Mari Suzuki, Yoshiki Moriwaki, Fusakazu Matsushima, <u>Takayoshi Amano</u>

WG12   5:10 – 5:25
MID-INFRARED NICE-OHMS SPECTROMETER FOR THE STUDY OF COLD MOLECULAR IONS, <u>Michael Porambo</u>, Jessica Pearson, Courtney Talicska, Benjamin J. McCall

WG13   5:27 – 5:42
RIGIDITY OF THE MOLECULAR ION H_5^+, Csaba Fábri, János Sarka, <u>Attila Császár</u>

WG14   5:44 – 5:59
FULL DIMENSIONAL POTENTIALS, DIPOLE MOMENT SURFACES AND (RO)VIBRATIONAL CALCULATIONS FOR H_5^+, H_7^+ AND HOCO, <u>Joel Bowman</u>, Stuart Carter, Yimin Wang

WH. Mini-symposium: Spectroscopy in Kinetics and Dynamics
Wednesday, June 18, 2014 – 1:30 PM
Room: B102 Chemical and Life Sciences

Chair: Toshinori Suzuki, Kyoto University, Kyoto, Japan

WH01   *INVITED TALK*   1:30 – 2:00
RADICALLY DIFFERENT KINETICS AT LOW TEMPERATURES, Ian Sims

WH02   2:05 – 2:20
H-ATOM REACTION KINETICS IN SOLID PARAHYDROGEN FOLLOWED BY RAPID SCAN FTIR, David T. Anderson

WH03   2:22 – 2:37
TO TUNNEL OR NOT TO TUNNEL, PROTON TRANSFER IS THE QUESTION., Kathryn Chew, Deacon Nemchick, Patrick Vaccaro

WH04   2:39 – 2:54
SPECTROSCOPIC AND KINETIC STUDIES OF ATMOSPHERIC FREE RADICALS, Elizabeth Foreman, YiTien Jou, Kara Kapnas, Craig Murray

WH05   2:56 – 3:11
RADICAL INTERMEDIATES IN THE ADDITION OF OH TO PROPENE: PHOTOLYTIC PRECURSORS AND ANGULAR MOMENTUM EFFECTS , Matthew D Brynteson, Carrie C Womack, Ryan S Booth, Shih -H Lee, Jim J Lin, Laurie Butler

WH06   3:13 – 3:28
STATE-RESOLVED AND STATE-TO-STATE PHOTODISSOCIATION STUDY OF CO_2 BY TWO-COLOR VUV-VUV LASER PUMP-PROBE METHOD , Zhou Lu, Yih-Chung Chang, William M. Jackson, Cheuk-Yiu Ng

Intermission

WH07   3:45 – 4:00
SUBMILLIMETER MEASUREMENTS OF THE CRIEGEE INTERMEDIATE, CH_2OO, IN THE GAS PHASE, Adam M Daly, Brian Drouin, Shanshan Yu

WH08   4:02 – 4:17
MEASURING RATE CONSTANTS FOR REACTIONS OF THE SIMPLEST CRIEGEE INTERMEDIATE CH_2OO BY MONITORING THE OH RADICAL, Yingdi Liu, Kyle D Bayes, Stanley P. Sander

WH09   4:19 – 4:34
EFFECTS OF REACTANT ROTATIONAL EXCITATION ON $Cl + CH_4 / CHD_3$ REACTIONS, Huilin Pan, Fengyan Wang, Yuan Cheng, Jui-San Lin, Kopin Liu

WH10   4:36 – 4:51
ROTATIONAL ENERGY TRANSFER AND DEPOLARIZATION IN RARE GAS + CN ($\tilde{X}$, v=0) COLLISIONS, Gregory Hall, Damien Forthomme, Trevor Sears

WH11   4:53 – 5:08
THE VIBRATIONALLY DRIVEN H-ATOM ABSTRACTION FROM METHANE BY BROMINE RADICALS, Ethan Volpa, Andrew Berke, Fleming Crim

WH12   5:10 – 5:25
STUDYING OZONOLYSIS REACTIONS OF 2-BUTENES USING CAVITY RING-DOWN SPECTROSCOPY, Liming Wang, Yingdi Liu, Mixtli Campos-Pineda, Chad Priest, Jingsong Zhang

WH13   5:27 – 5:37
COLLISIONALLY-MEDIATED SINGLET-TRIPLET CROSSING IN $\tilde{a}^1A_1$ CH_2 REVISITED: (010) COUPLING , Anh T. Le, Gregory Hall, Trevor Sears

WH14   5:39 – 5:54
THEORETICAL INVESTIGATION OF THE UV-VIS PHOTODISSOCIATION DYNAMICS OF $Ar_n(BRCN^-)$, Bernice Opoku-Agyeman, Julia H Lehman, Amanda Case, Carl Lineberger, Anne B McCoy

WI. Comparing theory and experiment
Wednesday, June 18, 2014 – 1:30 PM
Room: 112 Chemistry Annex

Chair: Richard Dawes, University of Missouri - Rolla, Rolla, MO, USA

WI01 1:30 – 1:45
ROVIBRATIONAL STATES OF HBF^+ AND HCO^+ ISOTOPOLOGUES UP TO HIGH J: THEORY AND EXPERIMENT, Peter Botschwina, Peter Sebald, Benjamin Schröder, Kentarou Kawaguchi, Takayoshi Amano

WI02 1:47 – 2:02
A COMPARISON OF THE METHODS OF STUDYING THE SPECTRA OF THE AsH_2 RADICAL, Geoffrey Duxbury, Alexander Alijah

WI03 2:04 – 2:19
EFFECTS OF SPIN-ORBIT COUPLING ON THE SPIN-ROTATION INTERACTION IN THE AsH_2 RADICAL, Geoffrey Duxbury, Alexander Alijah

WI04 2:21 – 2:36
AB INITIO STUDY OF ION-PAIR STATES OF THE IODINE MOLECULE, Vadim A Alekseev

WI05 2:38 – 2:53
EXPERIMENTAL AND THEORETICAL STUDIES ON THE ELECTRONIC ABSORPTION SPECTRA OF QUINOLINE CARBOXALDEHYDES , Mustafa Kumru, Mustafa Kocademir, Haidar M Alfanda

WI06 2:55 – 3:10
POLARIZED MATRIX INFRARED SPECTRA OF CYCLOPENTADIENONE - AN IMPORTANT REACTIVE INTERMEDIATE IN COMBUSTION AND BIOMASS PYROLYSIS, Thomas Ormond, Barney Ellison, John F. Stanton

Intermission

WI07 3:27 – 3:42
CONSISTENT ASSIGNMENT OF THE VIBRATIONS OF MONOHALOSUBSTITUTED BENZENES , Joe Harris, Anna Andrejeva, William Duncan Tuttle, Igor Pugliesi, Christian Schriever, Tim Wright

WI08 3:44 – 3:54
CONSISTENT ASSIGNMENT OF THE VIBRATIONS OF CHLORO- AND BROMOBENZENE MOLECULES AND THEIR DEUTERATED ANALOGUES, Anna Andrejeva, Joe Harris, William Duncan Tuttle, Tim Wright

WI09 3:56 – 4:11
ACCURATE ANHARMONIC IR SPECTRA FROM INTEGRATED CC/DFT APPROACH , Vincenzo Barone, Malgorzata Biczysko, Julien Bloino, Ivan Carnimeo, Cristina Puzzarini

WI10 4:13 – 4:28
A FIRST-PRINCIPLES MODEL OF FERMI RESONANCE IN THE ALKYL CH STRETCH REGION: APPLICATION TO HYDRONAPHTHALENES, INDANES, AND CYCLOHEXANE, Edwin Sibert, Nathanael Kidwell, Timothy S. Zwier

WI11 4:30 – 4:45
THE ROLE OF ELECTRICAL ANHARMONICITY IN DETERMINING INTENSITY IN THE 2100 cm^{-1} REGION OF THE WATER SPECTURM, Anne B McCoy

WI12 4:47 – 4:57
A PADE APPROXIMATION OF THE TRANSFORMED TRANSITION MOMENT OPERATOR OF THE ν_2 BAND OF WATER VAPOR, Oleg V. Egorov

WI13 4:59 – 5:14
VIBRATIONAL SPECTROSCOPY AND GAS-PHASE THERMOCHEMISTRY OF THE MODEL DIPEPTIDE N-ACETYL GLYCINE METHYL AMIDE, Christopher Leavitt, Paul Raston, Grant Moody, Caitlyne Shirley, Gary Douberly

WI14 5:16 – 5:31
EXPERIMENTAL AND THEORETICAL STUDIES OF THE PURE ROTATIONAL SPECTRA OF LEAD HALIDES: PbF AND PbCl, Spencer Norman, Richard Dawes, G. S. Grubbs II, S. A. Cooke, B. E. Long, Chris Dewberry

WJ. Structure determination
Wednesday, June 18, 2014 – 1:30 PM
Room: 274 Medical Sciences Building

Chair: Isabel Peña, Universidad de Valladolid, Valladolid, Spain

WJ01   1:30 – 1:45
MILLIMETER-WAVE STUDIES OF THE ISOTOPOLOGUES OF IZnCH$_3$(X^{1}A$_1$): GEOMETRIC PARAMETERS AND EVIDENCE FOR ZINC INSERTION, Matthew Bucchino, Justin Young, Phillip Sheridan, Lucy Ziurys

WJ02   1:47 – 2:02
LABORATORY DETECTION OF ClZnCH$_3$ (X^{1}A$_1$): FURTHER EVIDENCE FOR ZINC INSERTION, Matthew Bucchino, Lucy Ziurys

WJ03   2:04 – 2:19
THE SIMPLEST CRIEGEE INTERMEDIATE (H$_2$C = O–O): EQUILIBRIUM STRUCTURE AND POSSIBLE FORMATION FROM ATMOSPHERIC LIGHTNING, Michael C McCarthy, Lan Cheng, Kyle N Crabtree, Oscar Martinez Jr., Thanh Lam Nguyen, Carrie Womack, John F. Stanton

WJ04   2:21 – 2:36
MILLIMETER AND SUBMILLIMETER SPECTROSCOPIC STUDIES OF HO$_3$, Luyao Zou, Susanna L. Widicus Weaver

WJ05   2:38 – 2:53
MILLIMETER-WAVE SPECTROSCOPY OF HYDRAZOIC ACID (HN$_3$), Brent K. Amberger, Brian J. Esselman, R. Claude Woods, Robert J. McMahon

WJ06   2:55 – 3:10
PHOSPHORUS AND SILICON ANALOGS OF ISOCYANIC ACID:
FOURIER-TRANSFORM MICROWAVE SPECTROSCOPY OF HPCO AND HNSiO, Sven Thorwirth, Valerio Lattanzi, Michael C McCarthy

Intermission

WJ07   3:27 – 3:42
BOND ANGLES AROUND A TETRAVALENT CENTRAL ATOM, Robert Karl Bohn

WJ08   3:44 – 3:59
THE CHIRPED PULSE AND CAVITY FOURIER TRANSFORM MICROWAVE (CP-FTMW AND FTMW) SPECTRUM OF BROMOPERFLUOROACETONE, Nicholas Force, David Joseph Gillcrist, Cassandra C. Hurley, Frank E Marshall, Nicholas A. Payton, Thomas D. Persinger, N. E. Shreve, G. S. Grubbs II

WJ09   4:01 – 4:16
PERFLUOROBUTYRIC ACID AND ITS MONOHYDRATE: A CHIRPED PULSE AND CAVITY BASED FOURIER TRANSFORM MICROWAVE SPECTROSCOPIC STUDY , Javix Thomas, Agapito Serrato III, Wei Lin, Wolfgang Jaeger, Yunjie Xu

WJ10   4:18 – 4:33
ANALYSIS OF THE ROTATIONAL STRUCTURE IN THE HIGH-RESOLUTION IR SPECTRUM OF *trans*-HEXATRIENE-3-d_1, Norman C. Craig, Yihui Chen, Thomas Blake

WJ11   4:35 – 4:50
MOLECULAR STRUCTURE OF THE PHENYL RADICAL (C$_6$H$_5$), Kyle N Crabtree, Oscar Martinez Jr., John F. Stanton, Michael C McCarthy

WJ12   4:52 – 5:07
EQUILIBRIUM STRUCTURE OF PIPERIDINE, Jean Demaison, Heinz Dieter Rudolph, Norman C. Craig, Patricia Ecija, Emilio J. Cocinero, Alberto Lesarri

WJ13   5:09 – 5:24
MICROWAVE SPECTRA AND MOLECULAR GEOMETRIES OF BENZONITRILE AND PENTAFLUOROBENZONITRILE, Mahdi Kamaee, Jennifer van Wijngaarden

WJ14   5:26 – 5:41
MICROWAVE SPECTROSCOPY OF MONOTERPENES OF ATMOSPHERIC INTEREST: α-PINENE, β-PINENE, AND NOPINONE, Juan-Ramon Aviles Moreno, Elias Neeman, T. R. Huet

RA. Mini-symposium: Astronomical Molecular Spectroscopy in the Age of ALMA
Thursday, June 19, 2014 – 8:30 AM
Room: 116 Roger Adams Lab

Chair: Anthony Remijan, NRAO, Charlottesville, VA, USA

RA01   8:30 – 8:45
IMPROVED INFRASTUCTURE FOR CDMS AND JPL MOLECULAR SPECTROSCOPY CATALOGUES ,
Christian Endres, Stephan Schlemmer, Brian Drouin, John Pearson, Holger S. P. Müller, P. Schilke, Jürgen Stutzki

RA02   8:47 – 9:02
HIGHLY ACCURATE QUANTUM-CHEMICAL CALCULATIONS FOR THE INTERSTELLAR MOLECULES C_3 AND
l-C_3H^+, Peter Botschwina, Benjamin Schröder, Christopher Stein, Peter Sebald, Rainer Oswald

RA03   9:04 – 9:19
TERAHERTZ MEASUREMENTS OF HOT HYDRONIUM IONS (H_3O^+) WITH AN EXTENDED NEGATIVE GLOW
DISCHARGE, Shanshan Yu, John Pearson

RA04   9:21 – 9:36
TORSION-ROTATION-VIBRATION EFFECTS IN THE v_{20}, $2v_{21}$, $2v_{13}$ AND $v_{21}+v_{13}$ STATES OF CH_3CH_2CN,
Adam M Daly, John Pearson, Shanshan Yu, Brian Drouin, Celina Bermúdez, José L. Alonso

RA05   9:38 – 9:53
LABORATORY MEASUREMENTS IN SUPPORT OF ASTRONOMICAL OBSERVATIONS: ROTATIONAL SPEC-
TROSCOPY UP THE THz REGION, Gabriele Cazzoli, Cristina Puzzarini

RA06   9:55 – 10:10
THE COMPLETE, TEMPERATURE RESOLVED SPECTRUM OF METHANOL BETWEEN 214 AND 265 GHZ,
James McMillan, Christopher F. Neese, Sarah Fortman, Frank C. De Lucia

Intermission

RA07   10:27 – 10:42
AN ANALYSIS OF THE ROTATIONAL SPECTRUM OF ACETONITRILE (CH_3CN) IN EXCITED VIBRATIONAL
STATES, Christopher F. Neese, James McMillan, Sarah Fortman, Frank C. De Lucia

RA08   10:44 – 10:59
THE MILLIMETER- AND SUBMILLIMETER-WAVE SPECTRUM OF PROPENAL, Adam M Daly, Celina Bermúdez,
Lucie Kolesniková, José L. Alonso

RA09   11:01 – 11:16
A COMPREHENSIVE INTENSITY STUDY OF THE ν_4 TORSIONAL BAND OF ETHANE, Jalal Norooz Oliaee,
Nasser Moazzen-Ahmadi, Irving Ozier, Keeyoon Sung, Timothy J Crawford, Linda Brown, Edward H Wishnow, V. Malathy
Devi

RA10   11:18 – 11:33
ISOTOPE SELECTIVE PHOTODISSOCIATION OF N_2 BY THE INTERSTELLAR RADIATION FIELD AND COSMIC
RAYS, Alan Heays, Ewine van Dishoeck, Ruud Visser, Roland Gredel, Wim Ubachs, Brenton R Lewis, Stephen Gibson

RA11   11:35 – 11:50
THEORETICAL STUDY ON VIBRONIC INTERACTIONS AND PHOTOPHYSICS OF LOW-LYING EXCITED ELEC-
TRONIC STATES OF POLYCYCLIC AROMATIC HYDROCARBONS, Nagaprasad Reddy Samala, S. Mahapatra

RB. Clusters/Complexes
Thursday, June 19, 2014 – 8:30 AM
Room: 100 Noyes Laboratory

Chair: Bob McKellar, National Research Council of Canada, Ottawa, ON, Canada

RB01　　　　　　　　　　　　　　　　　　　　　　　　　　　　　　　　　　　8:30 – 8:45
INFRARED SPECTROSCOPY OF $C_6D_6 - Rg_n$(n=1,2) , Jobin George, Mahdi Yousefi, Mojtaba Rezaei, Bob McKellar, Nasser Moazzen-Ahmadi

RB02　　　　　　　　　　　　　　　　　　　　　　　　　　　　　　　　　　　8:47 – 9:02
SUB-DOPPLER ELECTRONIC SPECTRUM OF THE BENZENE–D_2 COMPLEX, Masato Hayashi, Yasuhiro Ohshima

RB03　　　　　　　　　　　　　　　　　　　　　　　　　　　　　　　　　　　9:04 – 9:19
ISOMER-SPECIFIC IR SPECTROSCOPY OF BENZENE-(WATER)$_N$ CLUSTERS WITH N=1-8: NEW INSIGHTS FROM THE WATER BEND FUNDAMENTALS AND ISOTOPICALLY SUBSTITUTED CLUSTERS, Ryoji Kusaka, Patrick S. Walsh, Timothy S. Zwier

RB04　　　　　　　　　　　　　　　　　　　　　　　　　　　　　　　　　　　9:21 – 9:36
MICROWAVE SPECTRUM, STRUCTURE, AND INTERNAL DYNAMICS OF THE PYRIDINE - ACETYLENE WEAKLY BOUND COMPLEX, Becca Mackenzie, Chris Dewberry, Emma Jarrett, Anthony Legon, Ken Leopold

RB05　　　　　　　　　　　　　　　　　　　　　　　　　　　　　　　　　　　9:38 – 9:53
ROTATIONAL SPECTRA OF ADDUCTS OF PYRIDINE WITH METHANE AND ITS HALIDES, Qian Gou, Lorenzo Spada, Montserrat Vallejo-López, Alberto Lesarri, Emilio J. Cocinero, Walther Caminati

Intermission

RB06　　　　　　　　　　　　　　　　　　　　　　　　　　　　　　　　　　10:10 – 10:25
REACTIVE PATHWAYS IN THE CHLOROBENZENE-AMMONIA DIMER CATION RADICAL: NEW INSIGHTS FROM EXPERIMENT AND THEORY, Silver Nyambo, Brandon Uhler, Aimable Kalume, Lloyd Muzangwa, Scott Reid

RB07　　　　　　　　　　　　　　　　　　　　　　　　　　　　　　　　　　10:27 – 10:42
PHOTOIONIZATION INDUCED INTERMOLECULAR PROTON TRANSFER IN THE CH...O HYDROGEN BONDED CYCLOPENTANONE DIMER IN THE GAS PHASE, Arup K. Ghosh, Tapas Chakraborty

RB08　　　　　　　　　　　　　　　　　　　　　　　　　　　　　　　　　　10:44 – 10:59
MICROWAVE SPECTRUM OF HYDROGEN BONDED HEXAFLUOROISOPROPANOL•••WATER COMPLEX, Abhishek Shahi, Elangannan Arunan

RB09　　　　　　　　　　　　　　　　　　　　　　　　　　　　　　　　　　11:01 – 11:16
HYDROGEN BOUND COMPLEXES WITH TROPOLONE: BINDING MOTIFS, BARRIER HEIGHTS, AND THE SEARCH FOR BIFURCATING SYSTEMS, Deacon Nemchick, Kathryn Chew, Patrick Vaccaro

RB10　　　　　　　　　　　　　　　　　　　　　　　　　　　　　　　　　　11:18 – 11:33
INTERMOLECULAR INTERACTIONS BETWEEN FORMALDEHYDE AND DIMETHYL ETHER AND BETWEEN FORMALDEHYDE ABD DIMETHYL SULFIDE IN THE COMPLEX, INVESTIGATED BY FOURIER TRANSFORM MICROWAVE SPECTROSCOPY AND AB INITIO CALCULATIONS , Yoshiyuki Kawashima, Yoshio Tatamitani, Yoshi-hiro Osamura, Eizi Hirota

RB11　　　　　　　　　　　　　　　　　　　　　　　　　　　　　　　　　　11:35 – 11:50
A VIBRATIONAL MODEL FOR ACCURATE DETERMINATION OF OSCILLATOR STRENGTHS IN HYDROGEN BONDED COMPLEXES, Kasper Mackeprang, Henrik Kjærgaard

RB12　　　　　　　　　　　　　　　　　　　　　　　　　　　　　　　　　　11:52 – 12:07
EFFECT OF SUBSTITUENTS IN ALCOHOL-AMINE COMPLEXES, Anne Schou Hansen, Lin Du, Henrik Kjærgaard

RC. Theory and Computation
Thursday, June 19, 2014 – 8:30 AM
Room: B102 Chemical and Life Sciences

Chair: So Hirata, University of Illinois, Urbana, IL, USA

RC01   8:30 – 8:45
SIMULTANEOUS EVALUATION OF MULTIPLE ROTATIONALLY EXCITED STATES OF FLOPPY MOLECULES USING DIFFUSION MONTE CARLO, Anne B McCoy, Jason E. Ford, Melanie L. Marlett, Andrew S. Petit

RC02   8:47 – 9:02
NUMERICALLY EXACT CALCULATION OF ROVIBRATIONAL LEVELS OF $Cl^- H_2O$, Xiao-Gang Wang, Tucker Carrington

RC03   9:04 – 9:19
AUTOMATIC GENERATION OF ANALYTIC EQUATIONS FOR VIBRATIONAL AND ROVIBRATIONAL CONSTANTS FROM FOURTH-ORDER VIBRATIONAL PERTURBATION THEORY, Devin A. Matthews, Justin Z Gong, John F. Stanton

RC04   9:21 – 9:36
ROVIBRATIONAL CONSTANTS FROM FOURTH-ORDER PERTURBATION THEORY AND THE RELATIONSHIP TO THE CONTACT TRANSFORMATION APPROACH, Devin A. Matthews, Justin Z Gong, John F. Stanton

RC05   9:38 – 9:53
APPLICATION OF FOURTH ORDER VIBRATIONAL PERTURBATION THEORY WITH ANALYTIC HARTREE-FOCK FORCE FIELDS, Justin Z Gong, Devin A. Matthews, John F. Stanton

RC06   9:55 – 10:05
ROTATIONAL SPECTRUM OF SO_3 AND THEORETICAL EVIDENCE FOR THE FORMATION OF ROTATIONAL ENERGY LEVEL CLUSTERS IN ITS VIBRATIONAL GROUND STATE, Daniel S Underwood, Sergei N. Yurchenko, Jonathan Tennyson, Per Jensen

RC07   10:07 – 10:22
CONSISTENT ASSIGNMENT OF THE VIBRATIONS OF PARA DISUBSTITUTED BENZENE MOLECULES, Anna Andrejeva, Alison Jasmin Lee, Tim Wright

Intermission

RC08   10:39 – 10:54
ANHARMONIC VIBRATIONAL SPECTROSCOPY ON METAL TRANSITION COMPLEXES, Camille Latouche, Julien Bloino, Vincenzo Barone

RC09   10:56 – 11:11
FREE O-H ANHARMONIC STRETCHING MOTIONS IN $H^+(CH_3OH)_{1-3}$ WITH/WITHOUT ATTACHED ARGON., Hsiao-Han Chuang, Jer-Lai Kuo, Kaito Takahashi, Asuka Fujii

RC10   11:13 – 11:28
UNFOLDING THE QUANTUM NATURE OF PROTON BOUND SYMMETRIC DIMERS OF $(MeOH)_2H^+$ AND $(Me_2O)_2H^+$: A THEORETICAL STUDY , Jake Acedera Tan, Jer-Lai Kuo

RC11   11:30 – 11:45
ANHARMONIC IR SPECTRA OF BIOMOLECULES: NUCLEOBASES AND THEIR OLIGOMERS, Vincenzo Barone, Malgorzata Biczysko, Julien Bloino, Ivan Carnimeo, Teresa Fornaro

RC12   11:47 – 12:02
ACCURATE CHARACTERIZATION OF THE PEPTIDE LINKAGE IN THE GAS PHASE: A JOINT QUANTUM-CHEMICAL AND ROTATIONAL SPECTROSCOPY STUDY OF THE GLYCINE DIPEPTIDE ANALOGUE, Cristina Puzzarini, Malgorzata Biczysko, Vincenzo Barone, Laura Largo, Isabel Peña, Carlos Cabezas, José L. Alonso

RD. Fundamental interest
Thursday, June 19, 2014 – 8:30 AM
Room: 112 Chemistry Annex

Chair: Ming-Wei Chen, University of Illinois, Urbana, IL, USA

RD01   8:30 – 8:45
HALO NUCLEIC MOLECULES: MOLECULES FORMED FROM AT LEAST ONE ATOM WITH A HALO NUCLEUS. EMPHASIS ON $^{11,11}Li_2$ ALONG WITH OTHER EXOTIC ISOTOPOLOGUES. , Nikesh S. Dattani, Staszek Welsh

RD02   8:47 – 9:02
QED CORRECTION FOR H_3^+, Lorenzo Lodi, Oleg Polyansky, Jonathan Tennyson, Alexander Alijah, Nikolay Fedorovich Zobov

RD03   9:04 – 9:19
CALCULATING POTENTIAL ENERGY CURVES WITH QUANTUM MONTE CARLO, Andrew D Powell, Richard Dawes

RD04   9:21 – 9:36
MILLIMETER-WAVE SPECTROSCOPY OF S_2Cl_2: A CANDIDATE MOLECULE FOR THE DETECTION OF *ORTHO-PARA* TRANSITION, Zeinab Tafti Dehghani, Asao Mizoguchi, Hideto Kanamori

RD05   9:38 – 9:53
HIGH-RESOLUTION INFRARED SPECTROSCOPY OF CUBANE, C_8H_8, Vincent Boudon, Olivier Pirali, Sébastien Gruet, Lucia D'accolti, Caterina Fusco, Cosimo Annese

Intermission

RD06   10:10 – 10:25
MICROWAVE SPECTROSCOPY AND MOLECULAR STRUCTURE OF ISONITROSYL HYDROXIDE (HOON), Kyle N Crabtree, Marat R Talipov, Gerard O'Connor, Oscar Martinez Jr., Sergey L Khursan, Michael C McCarthy

RD07   10:27 – 10:37
HIGH RESOLUTION JET-COOLED INFRARED ABSORPTION SPECTRA OF FORMIC ACID DIMER: A REINVESTIGATION OF THE FERMI-TRIAD SYSTEM IN THE C-O STRETCHING REGION, Chuanxi Duan

RD08   10:39 – 10:54
BROADBAND MID-INFRARED COMB-RESOLVED FOURIER TRANSFORM SPECTROSCOPY, Andrew Mills, Kevin Lee, Christian Mohr, Jie Jiang, Martin Fermann, Piotr Maslowski

RD09   10:56 – 11:06
SINGLE MOLECULE RAMAN SPECTROSCOPY UNDER HIGH PRESSURE, Yuanxi Fu, Dana Dlott

RD10   11:08 – 11:18
THERMAL EXPANSION INVESTIGATIONS OF SINGLE WALLED CARBON NANOTUBES BY RAMAN SPECTROSCOPY AND MOLECULAR DYNAMICS SIMULATIONS, Daniel Casimir, Raul F Garcia-Sanchez, Prabhakar Misra

RD11   11:20 – 11:25
SPECTROSCOPIC INVESTIGATION OF THE EFFECTS OF ENVIRONMENT ON NEWLY DEVELOPED EMISSIVE MATERIALS, Louis E. McNamara, Nathan I Hammer, Hemali Rathnayake, Kieth Hollis, Jared Delcamp

RD12   11:27 – 11:37
NQR STUDY OF LATTICS DYNAMICS in $RbIO_3$ and $KBrO_3$, Igor Vertegel, Eugeny Chesnokov, Alexander Barabash, Alexander Ovcharenko, Oleg Ponkratenko, Ivan Vertegel

RD13   11:39 – 11:49
INVESTIGATION OF THE PECULIARITIES OF THE ABSORBED WATER ON TiO_2 BY NMR-RELAXATION METHOD, Igor Vertegel, Eugeny Chesnokov, Alexander Ovcharenko, Vladimir Gaivoronsky, Ivan Vertegel

RE. Chirped pulse
Thursday, June 19, 2014 – 8:30 AM
Room: 274 Medical Sciences Building

Chair: Brooks Pate, The University of Virginia, Charlottesville, VA, USA

RE01   8:30 – 8:45
A LOW-COST CHIRPED-PULSE FOURIER TRANSFORM MICROWAVE SPECTROMETER FOR UNDERGRADUATE PHYSICAL CHEMISTRY LAB, Brandon Carroll, Ian Finneran, Geoffrey Blake

RE02   8:47 – 8:57
A CHIRPED-PULSE FOURIER TRANSFORM SPECTROMETER OPERATING FROM 110 TO 170 GHZ, Lauren E. Bernier, Steven Shipman

RE03   8:59 – 9:14
SPECTRAL TAXONOMY: A SEMI-AUTOMATED COMBINATION OF CHIRPED-PULSE AND CAVITY FOURIER TRANSFORM MICROWAVE SPECTROSCOPY, Kyle N Crabtree, Michael C McCarthy

RE04   9:16 – 9:31
BUFFER GAS COOLED MOLECULE SOURCE FOR CPMMW SPECTROSCOPY, Yan Zhou, David Grimes, Timothy J Barnum, Ethan Klein, Robert W Field

RE05   9:33 – 9:48
DIRECT OBSERVATION OF RYDBERG-RYDBERG TRANSITIONS VIA CPMMW SPECTROSCOPY, Yan Zhou, David Grimes, Ethan Klein, Timothy J Barnum, Robert W Field

RE06   9:50 – 10:05
TOWARD THE USE OF RYDBERG STATES FOR STATE-SELECTIVE PRODUCTION OF MOLECULAR IONS, David Grimes, Timothy J Barnum, Stephen Coy, Robert W Field

RE07   10:07 – 10:22
SELECTIVE POPULATION OF MOLECULAR CORE NONPENETRATING RYDBERG STATES, David Grimes, Yan Zhou, Timothy J Barnum, Ethan Klein, Robert W Field

Intermission

RE08   10:39 – 10:54
THE PURE ROTATIONAL SPECTRUM OF PbI FROM BROADBAND ROTATIONAL SPECTROSCOPY, Daniel P. Zaleski, Hansjochen Köckert, Susanna Louise Stephens, Nick Walker, Lisa-Maria Dickens, Corey Evans

RE09   10:56 – 11:06
WAVEGUIDE CHIRPED-PULSE FOURIER TRANSFORM MICROWAVE SPECTROSCOPY OF ALLYL BROMIDE, Morgan N McCabe, Steven Shipman

RE10   11:08 – 11:18
CHIRPED-PULSE FOURIER TRANSFORM MICROWAVE SPECTROSCOPY OF 3-METHOXYPROPYLAMINE, Morgan N McCabe, Steven Shipman, Sean Arnold, J. Chase Chewning, Miranda Smith, Gordon Brown

RE11   11:20 – 11:35
THE ROTATIONAL SPECTRA, STRUCTURES, AND CHLORINE NUCLEAR ELECTRIC QUADRUPOLE COUPLING CONSTANTS FOR A FAMILY OF THREE HALOGENATED CYCLIC ALKENES, $C_nF_{2n-4}Cl_2$: n = 4, 5, AND 6, B. E. Long, E. A. Arsenault, Lucas Hansen, S. A. Cooke

RE12   11:37 – 11:52
CHIRPED-PULSE FOURIER TRANSFORM MICROWAVE SPECTROSCOPY OF 2-CHLORO-3-FLUOROPYRIDINE AND 2-CHLORO-6-FLUOROPYRIDINE, Sean Arnold, J. Chase Chewning, Gordon Brown

RF. Mini-symposium: Astronomical Molecular Spectroscopy in the Age of ALMA
Thursday, June 19, 2014 – 1:30 PM
Room: 116 Roger Adams Lab

Chair: Leslie Looney, University of Illinois, Urbana, IL, USA

RF01   *INVITED TALK*   1:30 – 2:00

OBSERVATIONS OF VOLATILE SPECIES IN PROTOPLANETARY DISKS, Geoffrey Blake

RF02   2:05 – 2:15

THE FIRST UNBIASED RADIO EMISSION LINE SURVEY OF THE PROTOPLANETARY DISK ORBITING LKCA 15, Kristina Marie Punzi, Joel H Kastner, Pierre Hily-Blant, Thierry Forveille, G G Sacco

RF03   2:17 – 2:32

NEAR-INFRARED SPECTROSCOPY OF SIMPLE ORGANIC MOLECULES IN THE GV TAU N PROTOPLANETARY DISK, Erika Gibb

RF04   2:34 – 2:49

NEAR-INFRARED SPECTROSCOPIC STUDY OF AA TAU: WATER AND OH OBSERVATIONS, Logan Ryan Brown, Erika Gibb

RF05   2:51 – 3:06

PHYSICS AND CHEMISTRY IN UV ILLUMINATED REGIONS: THE HORSEHEAD CASE, Viviana V. Guzman, Jérôme Pety, Pierre Gratier, Javier Goicoechea, Maryvonne Gerin, Evelyne Roueff

RF06   3:08 – 3:23

WHAT MOLECULAR ABUNDANCES CAN TELL US ABOUT THE DYNAMICS OF STAR FORMATION , Konstantinos Tassis, Karen Willacy, Harold W Yorke, Neal J Turner

RF07   3:25 – 3:40

PROBING THE CHEMISTRY AND DYNAMICS OF HOT MOLECULAR CORES USING HIGHLY EXCITED CYANOPOLYYNIC TRANSITIONS, Robert John Loughnane, François Lique, Navtej Singh, Stan Kurtz

RF08   3:42 – 3:57

THE CO AND SIO PROTOSTELLAR OUTFLOWS OF 30 PROTOSTARS, Dominique M. Segura-Cox, Leslie Looney

Intermission

RF09   4:14 – 4:29

SHOCKS AND MOLECULES IN PROTOSTELLAR OUTFLOWS, Héctor Arce

RF10   4:31 – 4:46

DETECTION, IDENTIFICATION AND CORRELATION OF COMPLEX ORGANIC MOLECULES IN 32 INTERSTELLAR CLOUDS USING SUBMM OBSERVATIONS, Nadine Wehres, Shiya Wang, Mary Radhuber, James Sanders, Jay A Kroll, Jacob Laas, Brian Hays, Trevor Cross, D. C. Lis, Eric Herbst, Susanna L. Widicus Weaver

RF11   4:48 – 5:03

LIGHTING THE DARK MOLECULAR GAS USING THE MID INFRARED H_2 ROTATIONAL LINES, Aditya Togi, JD Smith

RF12   5:05 – 5:20

COMPARATIVE CHEMISTRY IN PLANETARY NEBULAE: THE ROLE OF THE CARBON TO OXYGEN RATIO, Jessica L Edwards, Lucy Ziurys

RF13   5:22 – 5:37

MOLECULAR ABUNDANCES IN THE CIRCUMSTELLAR ENVELOPE OF OXYGEN-RICH SUPERGIANT VY CANIS MAJORIS, Jessica L Edwards, Lucy Ziurys

RF14   5:39 – 5:49

ROTATIONALLY EXCITED H_2 IN THE MAGELLANIC CLOUDS, Rui Xue, Tony Wong

RF15   5:51 – 6:06

DOES PLASMA STRUCTURE INFLUENCE MOLECULE FORMATION AND RADIATION CHARACTERISTICS?, Daan C Schram

RG. Mini-symposium: Beyond the Mass-to-Charge Ratio: Spectroscopic Probes of the Structures of Ions
Thursday, June 19, 2014 – 1:30 PM
Room: 100 Noyes Laboratory

Chair: Etienne Garand, University of Wisconsin, Madison, WI, USA

RG01   *INVITED TALK*   1:30 – 2:00
ISOMER-SPECIFIC IR^{2}MS^2 SPECTROSCOPY OF PROTONATED WATER CLUSTERS, Knut R. Asmis

RG02   2:05 – 2:20
A THEORETICAL STUDY ON THE STRUCTURAL EVOLUTION OF IONIZED WATER CLUSTERS, $(H_2O)_n^+$, n$\leq$ 3 ~ 8, En-Ping Lu, Ying-Cheng Li, Jer-Lai Kuo, Ming-Kang Tsai

RG03   2:22 – 2:37
MULTIDIMENTIONAL NORMAL MODE CALCULATIONS FOR THE OH VIBRATIONAL SPECTRA OF $(H_2O)_3^+$, $(H_2O)_3^+$Ar, $H^+(H_2O)_3$, AND $H^+(H_2O)_3$Ar, Ying-Cheng Li, Hsiao-Han Chuang, Jake Acedera Tan, Kaito Takahashi, Jer-Lai Kuo

RG04   2:39 – 2:54
COMPUTATIONAL FRAMEWORK FOR ANALYSIS OF HYDROGEN BONDING IN THE OH STRETCH REGION, Laura C. Dzugan, Anne B McCoy

RG05   2:56 – 3:11
ADDING THE TEMPERATURE DIMENSION TO SIZE-SELECTED ION VIBRATIONAL PREDISSOCIATION SPECTROSCOPY: OBSERVATION OF "MELTING" IN THE PRIMARY SOLVATION SHELL OF MICROHYDRATED IODIDE CLUSTERS, Olga Gorlova, Conrad Wolke, Mark Johnson

Intermission

RG06   3:28 – 3:43
STRUCTURES OF HYDRATED ALKALI METAL CATIONS, $M^+(H_2O)_n$Ar (M = Li, Na, K, Rb and Cs, n = 3-5), USING INFRARED PHOTODISSOCIATION SPECTROSCOPY AND THERMODYNAMIC ANALYSIS, Haochen Ke, Christian van der Linde, James M. Lisy

RG07   3:45 – 4:00
UNRAVELING THE ROLES OF HYDROGEN BONDING, ELECTROSTATICS, AND FERMI RESONANCES IN THE IONIC LIQUID [EMIM][BF$_4$] THROUGH CRYOGENIC ION VIBRATIONAL SPECTROSCOPY, Joseph Fournier, Conrad Wolke, Christopher Johnson, Mark Johnson

RG08   4:02 – 4:17
ION PAIR STRUCTURE AND PHOTODISSOCIATION DYNAMICS OF IONIC LIQUID [EMIM][TF$_2$N], Jaime A. Stearns, Russell Cooper, David Sporleder, Alexander M. Zolot, Jerry Boatz

RG09   4:19 – 4:34
INFRARED SPECTRA OF ANIONIC COBALT-CARBON DIOXIDE CLUSTERS, Benjamin Knurr, J. Mathias Weber

RG10   4:36 – 4:51
INFRARED PHOTODISSOCIATION SPECTROSCOPY OF DOUBLY CHARGED M(CO)$_n$ CATIONS, Jon Maner, Antonio Brathwaite, Nicki Reishus, Michael A Duncan

RG11   4:53 – 5:08
VIBRATIONAL SPECTROSCOPY AND THEORY OF $Cu^+(CH_4)_n$ AND $Ag^+(CH_4)_n$ (n = 1 - 6), Abdulkadir Kocak, Muhammad Affawn Ashraf, Ricardo B. Metz

RG12   5:10 – 5:25
STRUCTURES, ENERGETICS, AND VIBRATIONS OF SMALL TRANSITION METAL OXIDE CLUSTERS BY HIGH-RESOLUTION ANION PHOTOELECTRON SPECTROSCOPY, Jongjin B. Kim, Marissa L. Weichman, Daniel Neumark

RG13   5:27 – 5:42
IR SPECTROSCOPY OF FIRST-ROW TRANSITION METAL CLUSTERS AND THEIR COMPLEXES WITH SIMPLE MOLECULES , D. M. Kiawi, J. Bakker, J. Oomens, W. J. Buma, L.B.F.M Waters

RG14   5:44 – 5:59
DETERMINATION OF THE STRUCTURES OF SILICON AND METAL DOPED SILICON CLUSTERS , Jonathan T Lyon, Andre Fielicke, Ewald Janssens, Peter Lievens

RG15   6:01 – 6:16
PHOTOELECTRON SPECTROSCOPY STUDY OF [Ta$_2$B$_6$]$^-$: A HEXAGONAL BIPYRAMDIAL CLUSTER, Tian Jian, Weili Li, Constantin Romanescu, Lai-Sheng Wang

RH. Cold/Ultra-cold/Physics
Thursday, June 19, 2014 – 1:30 PM
Room: B102 Chemical and Life Sciences

Chair: Brian DeMarco, University of Illinois, Urbana, IL, USA

RH01   1:30 – 1:45
A NOVEL METHOD TO MEASURE SPECTRA OF COLD MOLECULAR IONS, Satrajit Chakrabarty, Mathias Holz, Ewen Campbell, Agniva Banerjee, Dieter Gerlich, John P. Maier

RH02   1:47 – 2:02
INFRARED LASER STARK SPECTROSCOPY AND AB INITIO COMPUTATIONS OF THE OH$\cdots$CO COMPLEX, Tao Liang, Paul Raston, Gary Douberly

RH03   2:04 – 2:19
MOLECULAR BEAM OPTICAL STARK SPECTROSCOPY OF MAGNESIUM DEUTERIDE , Timothy Steimle, Ruohan Zhang, Hailing Wang

RH04   2:21 – 2:36
MM-WAVE SPECTROSCOPY AND DETERMINATION OF THE RADIATIVE BRANCHING RATIOS OF ^{11}BH FOR LASER COOLING EXPERIMENTS, Stefan Truppe, Darren Holland, Richard James Hendricks, Ed Hinds, Michael Tarbutt

RH05   2:38 – 2:53
CHARACTERIZATION OF CaO$^+$ AND BaO$^+$ BY TWO-PHOTON IONIZATION SPECTROSCOPY, Joshua Bartlett, Robert A. VanGundy, Michael Heaven

RH06   2:55 – 3:10
RYDBERG SPECTROSCOPY OF ZEEMAN-DECELERATED BEAMS OF METASTABLE HELIUM MOLECULES, Paul Jansen, Michael Motsch, Daniel Sprecher, Frederic Merkt

RH07   3:12 – 3:27
LASER COOLING THE DIATOMIC MOLECULE CaH, Joe Velasquez, III, Michael Di Rosa

Intermission

RH08   3:44 – 3:59
VIBRATIONAL SPECTROSCOPY ON TRAPPED COLD MOLECULAR IONS, Ncamiso B Khanyile, Kenneth R Brown

RH09   4:01 – 4:16
SYMPATHETIC SIDEBAND COOLING OF CaH$^+$, Rene Rugango, Kenneth R Brown

RH10   4:18 – 4:33
BROADBAND OPTICAL COOLING OF AlH$^+$ TO THE ROTATIONAL GROUND STATE, Christopher M. Seck, Chien-Yu Lien, Brian C. Odom

RH11   *Post-Deadline Abstract*   4:35 – 4:50
PHYSICS WITH COLD MOLECULES USING BUFFER GAS COOLING: PRECISION MEASUREMENT, COLLISIONS, AND LASER COOLING, Nicholas R Hutzler, John M. Doyle

RH12   4:52 – 5:07
THE [18.1], [18.6] and [18.7] EXCITED STATES OF YTTERBIUM FLUORIDE, Timothy Steimle, Fang Wang, Joe Smallman

RH13   5:09 – 5:24
HIGH-RESOLUTION MOLECULAR SPECTROSCOPY OF H$_2$ AT 10% THE AGE OF THE UNIVERSE; TESTING THE CONSTANCY OF PHYSICAL LAW, Wim Ubachs, Julija Bagdonaite, Mario Dapra, Michael T Murphy, Lex Kaper

RH14   5:26 – 5:41
SEARCH FOR A VARIATION OF THE PROTON-ELECTRON MASS RATIO FROM METHANOL OBSERVATIONS, Wim Ubachs, Julija Bagdonaite, Mario Dapra, Hendrick Bethlem, Nissim Kanekar, Sebastien Muller, Christian Henkel, Karl Menten

RH15   5:43 – 5:58
"SIMPLEST MOLECULE" CLARIFIES MODERN PHYSICS I. CW LASER SPACE-TIME FRAME DYNAMICS, T.C. Reimer, W. G. Harter

RH16   6:00 – 6:15
"SIMPLEST MOLECULE" CLARIFIES MODERN PHYSICS II. RELATIVISTIC QUANTUM MECHANICS, T.C. Reimer, W. G. Harter

RI. Matrix isolation (and droplets)
Thursday, June 19, 2014 – 1:30 PM
Room: 112 Chemistry Annex

Chair: Takamasa Momose, University of British Columbia, Vancouver, BC, Canada

RI01   1:30 – 1:45
MARILYN JACOX: HER CONTRIBUTIONS TO MATRIX ISOLATION SPECTROSCOPY AND BEYOND, Barbara Miller, Terry A. Miller

RI02   1:47 – 2:02
INFRARED ABSORPTION OF CH_3O/CD_3O RADICALS PRODUCED UPON PHOTOLYSIS OF CH_3ONO/CD_3ONO IN A p-H_2 MATRIX , Yu-Fang Lee, Wei-Te Chou, Britta Johnson, Edwin Sibert, Yuan-Pern Lee

RI03   2:04 – 2:19
ELECTRONIC SPECTROSCOPY OF MASS-SELECTED $C_7H_3^+$ AND C_7H_3 ISOMERS IN 6 K NEON MATRICES, John P. Maier, Arghya Chakraborty, Jan Fulara

RI04   2:21 – 2:36
ASSIGNING STATES IN THE JAHN-TELLER COUPLED INFRARED SPECTRA OF CH_3O AND CD_3O, Britta Johnson, Edwin Sibert

RI05   2:38 – 2:53
FTIR STUDIES OF THE PHOTOCHEMISTRY OF DEUTERATED FORMIC ACID IN A PARAHYDROGEN MATRIX, David T. Anderson

RI06   2:55 – 3:10
STEERING H-ATOM DIFFUSION THROUGH IMPURITY-DOPED SOLID PARAHYDROGEN: THE ROLE OF DIFFERENTIAL SOLVATION ENERGIES, Robert Hinde

RI07   3:12 – 3:27
INFRARED SPECTRA AND CALCULATED BINDING ENERGIES OF γ-BUTYROLACTONE DIMERS AND TRIMERS, Eric Willis, Chris Baumann

Intermission

RI08   3:44 – 3:59
OBSERVATION OF TRANSIENT SURFACE-BOUND INTERMEDIATES BY INTERFACIAL MATRIX STABILIZATION SPECTROSCOPY (IMSS), Nina K Jarrah, David T Moore

RI09   4:01 – 4:16
TEMPERATURE DEPENDENT ANISOTROPIC LINE SPLITTING AND LINEWIDTH IN MOTIONALLY AVERAGED EPR SPECTRA OF CH_3 RADICALS IN SOLID Ar, Yurij Dmitriev, Nikolas Ploutarch Benetis

RI10   4:18 – 4:33
INFRARED SPECTROSCOPY OF ALANINE IN SOLID PARAHYDROGEN , Shin Yi Toh, Ying-Tung Angel Wong, Pavle Djuricanin, Takamasa Momose

RI11   4:35 – 4:50
ELECTRONIC ABSORPTION SPECTROSCOPY AND FRANCK-CONDON SIMULATIONS FOR HC_7H and MeC_7H, Benjamin C. Haenni, Christopher J. Shaffer, John F. Stanton, Robert J. McMahon

RI12   4:52 – 5:02
CONFORMATIONAL ANALYSIS OF R-(+)-3-METHYLCYCLOPENTANONE BY IR SPECTROSCOPY IN PARAHYDROGEN CRYSTAL , Watheq Al-Basheer, Shin Yi Toh, Jun Miyazaki, Takamasa Momose

RI13   5:04 – 5:19
THE CHARACTERIZATION OF GeH_2 AND GeH USING MATRIX ISOLATION INFRARED SPECTROSCOPY, Jay Amicangelo, Christopher Bailey, Madelyn Hoover, Bruce Huffman

RI14   5:21 – 5:36
ELECTRONIC RELAXATION PROCESSES OF TRANSITION METAL ATOMS IN HELIUM NANODROPLETS, Andreas Kautsch, Friedrich Lindebner, Markus Koch, Wolfgang E. Ernst

RI15   5:38 – 5:53
MOLECULAR SPECTRA OF RbSr: HELIUM DROPLET ASSISTED PREPARATION OF A DIATOMIC MOLECULE, Florian Lackner, Günter Krois, Thomas Buchsteiner, Johann V. Pototschnig, Wolfgang E. Ernst

RI16   5:55 – 6:10
HELIUM NANODROPLET ISOLATION AND ROVIBRATIONAL SPECTROSCOPY OF HYDROXYMETHYLENE, Christopher Leavitt, Chris Moradi, John F. Stanton, Gary Douberly

RJ. Large amplitude motions, internal rotation
Thursday, June 19, 2014 – 1:30 PM
Room: 274 Medical Sciences Building

Chair: Marie-Aline Martin-Drumel, Universität zu Köln, Köln, Germany

RJ01   1:30 – 1:45
SPECTRAL ASSIGNMENTS AND ANALYSIS OF THE GROUND STATE OF NITROMETHANE IN HIGH-RESOLUTION FTIR SYNCHROTRON SPECTRA, Sylvestre Twagirayezu, Brant E Billinghurst, Tim E May, Mahesh B. Dawadi, David S. Perry

RJ02   1:47 – 1:57
ASSIGNMENT AND ANALYSIS OF THE NO_2 IN-PLANE ROCK BAND OF NITROMETHANE RECORDED BY HIGH-RESOLUTION FTIR SYNCHROTRON SPECTROSCOPY , Mahesh B. Dawadi, David S. Perry, Sylvestre Twagirayezu, Brant E Billinghurst

RJ03   1:59 – 2:09
A MICROWAVE SPECTROSCOPIC STUDY OF METHYLATED INDOLES: INTERNAL ROTATION AND NUCLEAR QUADRUPOLE COUPLING, Ranil Gurusinghe, Michael Tubergen

RJ04   2:11 – 2:26
MILLIMETER AND SUBMILLIMETER WAVE SPECTRA OF N-METHYLFORMAMIDE AND PROPIONAMIDE, A. A. Mescheryakov, E. A. Alekseev, V. Ilyushin, R. A. Motiyenko, L. Margulès

RJ05   2:28 – 2:43
THE MILLIMETER-WAVE SPECTRUM OF VINYL ACETATE, Lucie Kolesniková, Isabel Peña, José L. Alonso, Jose Cernicharo, Isabelle Kleiner

RJ06   2:45 – 3:00
COMPARISON OF INDEPENDENTLY CALCULATED AB-INITIO NORMAL-MODE DISPLACEMENTS FOR THE THREE C-H STRETCHING VIBRATIONS OF METHANOL ALONG THE INTERNAL ROTATION PATH, Li-Hong Xu, Ronald Lees, Jon T. Hougen, Joel Bowman, Xinchuan Huang, Stuart Carter

RJ07   3:02 – 3:17
STUDIES ON THE CONFORMATIONAL LANDSCAPE OF TERT-BUTYL ACETATE USING MICROWAVE SPECTROSCOPY AND QUANTUM CHEMICAL CALCULATIONS, YueYue Zhao, Halima Mouhib, Guohua Li, Wolfgang Stahl, Isabelle Kleiner

RJ08   3:19 – 3:34
TORSION - VIBRATION COUPLING IN THE METHYL ROTOR SYSTEMS, Meng Huang, Anne B McCoy, Terry A. Miller

RJ09   3:36 – 3:51
ANALYSIS OF THE FAR IR SPECTRUM OF TRIMETHYLENE SULFIDE USING EVOLUTIONARY ALGORITHMS, Jennifer van Wijngaarden, Durell Desmond, W. Leo Meerts

Intermission

RJ10   4:08 – 4:23
FIRST OBSERVATION OF THE SPIN ROTATIONAL STRUCTURE OF THE HYDROXYMETHYL RADICAL (H_2COH) IN THE CH_2 ASYMMETRIC MODE, Chih-Hsuan Chang, Fang Wang, David Nesbitt

RJ11   4:25 – 4:40
EFFECTS OF MULTIPLE ARGON TAGGING IN ALKALI METAL $M^+H_2OAr_n$ AND $M^+D_2OAr_n$ STUDIED BY IRPD SPECTROSCOPY, Christian van der Linde, Haochen Ke, James M. Lisy

RJ12   4:42 – 4:57
LARGE-AMPLITUDE TUNNELING DYNAMICS IN HYDROXYMETHYL RADICAL , David Nesbitt, Chih-Hsuan Chang, Fang Wang

RJ13   4:59 – 5:14
SYNCHROTRON RADIATION AND THE FAR-INFRARED AND MID-INFRARED SPECTRA OF NCNCS, Manfred Winnewisser, Brenda P. Winnewisser, Frank C. De Lucia, Dennis Tokaryk, Stephen Cary Ross, Brant E Billinghurst

RJ14   5:16 – 5:31
SPECTROSCOPY OF NCNCS AT THE CANADIAN LIGHT SOURCE: THE FAR-INFRARED SPECTRUM OF THE ν_7 REGION FROM 60-140 cm^{-1}, Dennis Tokaryk, Stephen Cary Ross, Brenda P. Winnewisser, Manfred Winnewisser, Frank C. De Lucia, Brant E Billinghurst

RJ15   5:33 – 5:48
FITTING THE HIGH-RESOLUTION SPECTROSCOPIC DATA FOR NCNCS, Zbigniew Kisiel, Brenda P. Winnewisser, Manfred Winnewisser, Frank C. De Lucia, Dennis Tokaryk, Stephen Cary Ross, Brant E Billinghurst

RJ16   5:50 – 6:05
NON COVALENT INTERACTIONS AND INTERNAL DYNAMICS IN ADDUCTS OF FREONS, Walther Caminati, Qian Gou, Luca Evangelisti, Gang Feng, Lorenzo Spada, Montserrat Vallejo-López, Alberto Lesarri, Emilio J. Cocinero

FA. Mini-symposium: Astronomical Molecular Spectroscopy in the Age of ALMA
Friday, June 20, 2014 – 8:30 AM
Room: 116 Roger Adams Lab

Chair: Susanna L. Widicus Weaver, Emory University, Atlanta, GA, USA

FA01　　　　　　　　　　　　　　　　　　　　　　　　　　　　　　　　　　**8:30 – 8:45**
HIGH-PRECISION SUB-DOPPLER INFRARED SPECTROSCOPY OF HeH^+, Adam J. Perry, James N. Hodges, Charles Markus, G. Stephen Kocheril, Paul A Jenkins II, Benjamin J. McCall

FA02　　　　　　　　　　　　　　　　　　　　　　　　　　　　　　　　　　**8:47 – 9:02**
HeH^+, HeD^+, AND HeT^+ POTENTIALS THAT REPRODUCE ALL MEASURED ENERGY TRANSITIONS, AND ARE ACCURATE UP TO THE INCLUSION OF RELATIVISTIC AND LEADING FOURTH ORDER QED EFFECTS IN THE LONG-RANGE REGION BEYOND AVAILABLE EXPERIMENTS, Staszek Welsh, Mariusz Puchalski, Grzegorz Lach, Tung Wei-Cheng, Ludwik Adamowicz, Nikesh S. Dattani

FA03　　　　　　　　　　　　　　　　　　　　　　　　　　　　　　　　　　**9:04 – 9:19**
ACCURATE, ANALYTIC, EMPIRICAL POTENTIALS AND BORN-OPPENHEIMER BREAKDOWN FUNCTIONS FOR THE $X(1^1\Sigma)$-STATES OF BeH, BeD, and BeT, Nikesh S. Dattani, Staszek Welsh

FA04　　　　　　　　　　　　　　　　　　　　　　　　　　　　　　　　　　**9:21 – 9:36**
RELIABLE IR LINE LISTS FOR 13 CO_2 ISOTOPOLOGUES, UP TO 1000K, E'=18,000 CM^{-1} AND WITH LINE SHAPE PARAMETERS INCLUDED, Xinchuan Huang, Robert R. Gamache, Richard S Freedman, David Schwenke, Timothy Lee

FA05　　　　　　　　　　　　　　　　　　　　　　　　　　　　　　　　　　**9:38 – 9:53**
VIBRATIONAL AVERAGING OF THE ISOTROPIC HYPERFINE COUPLING CONSTANTS FOR THE METHYL RADICAL, Ahmad Adam, Per Jensen, Andrey Yachmenev, Sergei N. Yurchenko

Intermission

FA06　　　　　　　　　　　　　　　　　　　　　　　　　　　　　　　　　**10:10 – 10:25**
THE TORSIONAL SPECTRUM OF DOUBLY DEUTERATED METHANOL CHD_2OH, M. Ndao, L. H. Coudert, F. Kwabia Tchana, J. Barros, L. Margulès, Laurent Manceron, P. Roy

FA07　　　　　　　　　　　　　　　　　　　　　　　　　　　　　　　　　**10:27 – 10:42**
THE MICROWAVE SPECTRUM OF MONODEUTERATED ACETAMIDE $CH_2DC(=O)NH_2$, I. A. Konov, L. H. Coudert, C. Gutle, T. R. Huet, L. Margulès, R. A. Motiyenko, H. Møllendal, J.-C. Guillemin

FA08　　　　　　　　　　　　　　　　　　　　　　　　　　　　　　　　　**10:44 – 10:59**
TERAHERTZ SPECTROSCOPY OF DEUTERATED ACETALDEHYDE: CH_2DCHO, L. Margulès, R. A. Motiyenko, L. H. Coudert, J.-C. Guillemin

FA09　　　　　　　　　　　　　　　　　　　　　　　　　　　　　　　　　**11:01 – 11:16**
HIGH RESOLUTION SPECTROSCOPY OF THE TWO LOWEST VIBRATIONAL STATES OF QUINOLINE C_9H_7N, Olivier Pirali, Zbigniew Kisiel, Manuel Goubet, Sébastien Gruet, Marie-Aline Martin-Drumel, Arnaud Cuisset, Gael Mouret

FA10　　　　　　　　　　　　　　　　　　　　　　　　　　　　　　　　　**11:18 – 11:33**
CALCULATED DIPOLE MOMENTS AND DIPOLE POLARIZABILITIES OF OBSERVED AND CANDIDATE ASTRO-MOLECULES CONTAINING SILICON AND PHOSPHORUS, David E. Woon, Holger S. P. Müller

FA11　　　　　　　　　　　　　　　　　　　　　　　　　　　　　　　　　**11:35 – 11:50**
ANALYTIC EMPIRICAL POTENTIALS FOR ALL STABLE ISOTOPOLOGUES OF THE GROUND $X(^1\Sigma^+)$ STATE OF ZnO FROM PURELY ROTATIONAL MEASUREMENTS, Nikesh S. Dattani, Lindsay Zack, Ming Sun, Erin R Johnson, Robert Le Roy, Lucy Ziurys

FB. Clusters/Complexes
Friday, June 20, 2014 – 8:30 AM
Room: 100 Noyes Laboratory

Chair: Michael Heaven, Emory University, Atlanta, GA, USA

FB01   8:30 – 8:45
DEVELOPMENT OF A CAVITY RINGDOWN SPECTROMETER FOR MEASURING ELECTRONIC STATES OF Be CLUSTERS, Jacob Stewart, Michael Sullivan, Michael Heaven

FB02   8:47 – 9:02
A THEORETICAL SEARCH FOR AN ELECTRONIC SPECTRUM OF THE He-BeO COMPLEX , Adrian Gardner, Michael Heaven

FB03   9:04 – 9:19
A UNIFIED PERSPECTIVE ON THE NATURE OF PAIRWISE INTERATOMIC INTERACTIONS FROM Ar_2 TO CARBON MONOXIDE., Charles K. Rosales, Luis A. Rivera-Rivera, Blake A. McElmurry, Robert R. Lucchese, John W. Bevan, Jay R. Walton

FB04   9:21 – 9:36
MID-INFRARED SPECTRUM OF THE ATMOSPHERICALLY SIGNIFICANT N_2-H_2O COMPLEX, Sean D. Springer, Blake A. McElmurry, Robert R. Lucchese, John W. Bevan, L. H. Coudert

FB05   9:38 – 9:53
HIGH RESOLUTION INFRARED SPECTROSCOPY OF N_2O-DIACETYLENE AND $CS_2 - C_2D_2$ DIMERS, Mahdi Yousefi, S. Sheybani-Deloui, Jalal Norooz Oliaee, Bob McKellar, Nasser Moazzen-Ahmadi

FB06   9:55 – 10:10
FUNDAMENTAL AND COMBINATION BANDS OF CO_2-C_2H_2 AND CO_2-C_2D_2 IN THE MID-INFRARED REGION, Mojtaba Rezaei, Jobin George, Luis Welbanks, Nasser Moazzen-Ahmadi

Intermission

FB07   10:27 – 10:42
QUANTUM MONTE CARLO SIMULATION OF VIBRATIONAL FREQUENCY SHIFTS OF CO IN SOLID para-HYDROGEN, Lecheng Wang, Robert Le Roy, Pierre-Nicholas Roy

FB08   10:44 – 10:59
PROGRESS IN UNDERSTANDING THE INFRARED SPECTRA OF He- AND Ne-C_2D_2, Nasser Moazzen-Ahmadi, Bob McKellar

FB09   11:01 – 11:16
FIRST OBSERVATION AND ANALYSIS OF $OCS - C_4H_2$ DIMER AND $(OCS)_2 - C_4H_2$ TRIMER, S. Sheybani-Deloui, Mahdi Yousefi, Jalal Norooz Oliaee, Bob McKellar, Nasser Moazzen-Ahmadi

FB10   11:18 – 11:33
CHIRPED-PULSE BROADBAND MICROWAVE SPECTRA AND STRUCTURES OF THE OCS TRIMER AND TETRAMER, Luca Evangelisti, Cristobal Perez, Nathan A Seifert, Brooks Pate, Mehdi Dehghany, Nasser Moazzen-Ahmadi, Bob McKellar

FB11   11:35 – 11:50
OCS TRIMER AND TETRAMER: CALCULATED STRUCTURES AND INFRARED SPECTRA, Mehdi Dehghany, Nasser Moazzen-Ahmadi, Bob McKellar

FB12   11:52 – 11:57
SPECTROSCOPIC AND COMPUTATIONAL CHARACTERIZATION OF HYDRATED PYRIMIDINE ANIONS, John T. Kelly, Nathan I Hammer

FC. Theory and Computation
Friday, June 20, 2014 – 8:30 AM
Room: B102 Chemical and Life Sciences

Chair: Anne B McCoy, The Ohio State University, Columbus, OH, USA

FC01   8:30 – 8:45
THE THREE-DIMENSIONAL POTENTIAL ENERGY SURFACE OF Ar-CO, Yoshihiro Sumiyoshi, Yasuki Endo

FC02   8:47 – 8:57
AB INITIO CALCULATIONS OF THE GROUND ELECTRONIC STATES OF THE C_3-Ar AND C_3-Ne COMPLEXES, Yi-Ren Chen, Yi-Jen Wang, Yen-Chu Hsu

FC03   8:59 – 9:14
VIBRATIONAL LEVEL STRUCTURES OF THE GROUND ELECTRONIC STATES OF THE C_3-Ar and C_3-Ne COMPLEXES, Yi-Ren Chen, Yen-Chu Hsu

FC04   9:16 – 9:31
THEORETICAL STUDY OF THE VIBRATIONAL SPECTROSCOPY OF THE ETHYL RADICAL, Daniel P. Tabor, Edwin Sibert

FC05   9:33 – 9:48
SIMULATION OF ACCURATE VIBRATIONALLY RESOLVED ELECTRONIC SPECTRA: THE INTEGRATED TIME-DEPENDENT AND TIME-INDEPENDENT FRAMEWORK, Alberto Baiardi, Vincenzo Barone, Malgorzata Biczysko, Julien Bloino

FC06   9:50 – 10:05
HIGH ACCURACY AB INITIO CALCULATION OF METAL QUADRUPOLE-COUPLING PARAMETERS, Lan Cheng, John F. Stanton, Jürgen Gauss

FC07   10:07 – 10:22
ELECTRONIC EXCITATIONS OF ALKALI-ALKALINE EARTH DIATOMIC MOLECULES - RESULTS FROM AB INITIO CALCULATIONS, Johann V. Pototschnig, Günter Krois, Florian Lackner, Wolfgang E. Ernst

Intermission

FC08   10:39 – 10:54
CONICAL INTERSECTIONS BETWEEN VIBRATIONALLY ADIABATIC SURFACES IN METHANOL, Mahesh B. Dawadi, David S. Perry

FC09   10:56 – 11:11
NONADIABATIC PHOTO-PROCESS INVOLVING THE $\pi\sigma*$ STATE IN INTRAMOLECULAR CHARGE TRANSFER: A CONCERTED SPECTROSCOPIC AND COMPUTATIONAL STUDY ON 4-(DIMETHYLAMINO)BENZETHYNE AND 4-(DIMETHYLAMINO)BENZONITRILE., Takashige Fujiwara, Javier Segarra-Martí, Pedro B. Coto

FC10   11:13 – 11:28
A COMPUTATIONAL TDDFT STUDY ON INTRAMOLECULAR CHARGE TRANSFER IN DI-*TERT*-BUTYLAMINOBENZONITRILES AND 2,4,6-TRICYANOANILINES., Takashige Fujiwara, Marek Z. Zgierski

FC11   11:30 – 11:45
FULL DIMENSIONAL VIBRATIONAL CALCULATIONS FOR METHANE USING AN ACCURATE NEW AB INITIO BASED POTENTIAL ENERGY SURFACE, Moumita Majumder, Richard Dawes, Xiao-Gang Wang, Tucker Carrington, Jun Li, Hua Guo, Sergei Manzhos

FD. Chirped pulse
Friday, June 20, 2014 – 8:30 AM
Room: 112 Chemistry Annex

Chair: Steven Shipman, New College of Florida, Sarasota, FL, USA

FD01   8:30 – 8:45
MONITORING THE REACTION PRODUCTS OF PERFLUOROPROPIONIC ACID AND ALLYL PHENYL ETHER USING CHIRPED-PULSE FOURIER TRANSFORM MICROWAVE (CP-FTMW) SPECTROSCOPY, Derek S. Frank, Daniel A. Obenchain, Wei Lin, Stewart E. Novick, S. A. Cooke, G. S. Grubbs II

FD02   8:47 – 9:02
ENANTIOMER IDENTIFICATION IN CHIRAL MIXTURES WITH BROADBAND MICROWAVE SPECTROSCOPY, V. Alvin Shubert, David Schmitz, Chris Medcraft, David Patterson, John M. Doyle, Melanie Schnell

FD03   9:04 – 9:19
BROADBAND MICROWAVE SPECTRUM AND STRUCTURE OF CYCLOPROPYL CYANOSILANE, Nathan A Seifert, Simon Lobsiger, Brooks Pate, Gamil A Guirgis, Jason S Overby, James R. Durig, Peter Groner

FD04   9:21 – 9:36
BROADBAND MICROWAVE SPECTROSCOPY AND AUTOMATED ANALYSIS OF 12 CONFORMERS OF 1-HEXANAL , Nathan A Seifert, Cristobal Perez, Daniel P. Zaleski, Justin Neill, Amanda Steber, Richard D. Suenram, Brooks Pate, Steven Shipman, Ian Finneran, Alberto Lesarri

FD05   9:38 – 9:53
CONFORMATIONAL ANALYSIS OF IBUPROFEN USING BROADBAND MICROWAVE SPECTROSCOPY, Sabrina Zinn, Thomas Betz, Melanie Schnell

FD06   9:55 – 10:10
ROTATIONAL SPECTRA OF HYDROGEN BONDED NETWORKS OF AMINO ALCOHOLS, Di Zhang, Timothy S. Zwier

Intermission

FD07   10:27 – 10:42
PREFERENCE FOR TOP- VS. SIDE-BINDING IN FLUORINATED ETHYLENE$\cdots$CO$_2$ COMPLEXES, Rebecca A. Peebles, Ashley M. Anderton, Cori L. Christenholz, Rachel E. Dorris, Sean A. Peebles

FD08   10:44 – 10:59
CHIRPED-PULSE FOURIER-TRANSFORM MICROWAVE SPECTROSCOPY OF THE PROTOTYPICAL C–H$\cdots\pi$ INTERACTION: THE BENZENE$\cdots$ACETYLENE WEAKLY BOUND DIMER, Nathan Ulrich, Nathan A Seifert, Rachel E. Dorris, Rebecca A. Peebles, Sean A. Peebles, Brooks Pate

FD09   11:01 – 11:16
EVIDENCE FROM BROADBAND ROTATIONAL SPECTROSCOPY FOR A COMPLEX BETWEEN AgCCH AND C$_6$H$_6$, Daniel P. Zaleski, Susanna Louise Stephens, Nick Walker, Anthony Legon

FD10   11:18 – 11:33
MICROWAVE SPECTRUM AND GEOMETRY OF CF$_3$I $\cdots$ PH$_3$, Susanna Louise Stephens, Nick Walker, Anthony Legon

FD11   11:35 – 11:50
MICROWAVE SPECTRA AND GEOMETRIES OF C$_2$H$_2$ $\cdots$ AuI and C$_2$H$_4$ $\cdots$ AuI, Susanna Louise Stephens, John Connor Mullaney, Matt John Sprawling, David Peter Tew, Nick Walker, Anthony Legon

FD12   11:52 – 12:02
DEUTERATED WATER HEXAMER OBSERVED BY CHIRPED-PULSE ROTATIONAL SPECTROSCOPY, Luca Evangelisti, Cristobal Perez, Simon Lobsiger, Nathan A Seifert, Daniel P. Zaleski, Brooks Pate, Zbigniew Kisiel, Berhane Temelso, George C. Shields

FE. Planetary atmospheres
Friday, June 20, 2014 – 8:30 AM
Room: 274 Medical Sciences Building

Chair: Vincent Boudon, CNRS / Université de Bourgogne, Dijon, France

FE01 8:30 – 8:45
PHOTON AND WATER MEDIATED SULFUR OXIDE AND ACID CHEMISTRY IN THE ATMOSPHERE OF VENUS, Jay A Kroll, Veronica Vaida

FE02 8:47 – 9:02
LABORATORY SIMULATIONS OF TITAN'S SURFACE COMPOSITION AND ITS RELATION TO ATMOSPHERIC HAZE LAYERS, Joshua A Sebree, Angela M Schmitt, Melissa G Trainer, Xiang Li, Veronica T Pinnick, Stephanie A Getty, Mark Loeffler, Carrie M Anderson, William B Brinckerhoff

FE03 9:04 – 9:19
THE HIGH-RESOLUTION EXTREME-ULTRAVIOLET SPECTRUM OF N_2 BY ELECTRON IMPACT, Alan Heays, Joe M Ajello, Alejandro Aguilar, Brenton R Lewis, Stephen Gibson

FE04 9:21 – 9:36
FORMATION OF HYDROXYLAMINE FROM AMMONIA AND HYDROXYL RADICALS , Lahouari Krim, Emilie-Laure Zins

FE05 9:38 – 9:53
THE HIGH J SPECTRUM OF THE $2\nu_2$ and ν_4 STATES OF AMMONIA, John Pearson, Shanshan Yu

Intermission

FE06 10:10 – 10:25
HIGH-RESOLUTION ABSORPTION CROSS SECTIONS OF ETHANE AT LOW TEMPERATURES, Robert J. Hargreaves, Dominique Appadoo, Peter F. Bernath

FE07 10:27 – 10:42
THZ SPECTROSCOPY OF 1d-ETHANE: Assignment of ν_{18} , Adam M Daly, Brian Drouin, Linda Brown, Peter Groner

FE08 10:44 – 10:59
LINE POSITIONS AND INTENSITIES FOR THE ν_{12} BAND OF $^{13}C^{12}CH_6$, V. Malathy Devi, D. Chris Benner, Keeyoon Sung, Timothy J Crawford, Arlan Mantz, Mary Ann H. Smith

FE09 11:01 – 11:11
HIGH-RESOLUTION INFRARED SPECTRUM OF THE $\nu_3+\nu_8$ COMBINATION BAND OF JET-COOLED PROPYNE, Dongfeng Zhao, Harold Linnartz

FE10 11:13 – 11:28
HIGH-RESOLUTION INFRARED SPECTRA OF THE ν_1 FUNDAMENTAL BANDS OF ^{13}C MONO-SUBSTITUTED PROPYNE IN A SUPERSONIC SLIT JET, Dongfeng Zhao, Kirstin D Doney, Harold Linnartz

FE11 11:30 – 11:45
FT-IR MEASUREMENTS OF COLD CROSS SECTIONS OF BENZENE (C_6H_6) FOR CASSINI/CIRS, Keeyoon Sung, Linda Brown, Geoffrey C. Toon

FE12 11:47 – 12:02
SPECTROSCOPIC INVESTIGATION OF O-,M-, AND P-CYANOSTYRENES, Joseph A. Korn, Stephanie N. Knezz, Robert J. McMahon, Timothy S. Zwier

MA. Plenary
Monday, June 16, 2014 – 8:45 AM
Room: Theater Lincoln Hall

Chair: Gregory S. Girolami, University of Illinois, Urbana, IL, USA

Welcome   8:45
Peter E. Schiffer, Vice Chancellor for Research
University of Illinois at Urbana-Champaign

MA01   9:00 – 9:40

EXPLORING THE HIGH-RESOLUTION SPECTROSCOPY OF MOLECULES THAT CAN AFFECT THE QUALITY OF YOUR LIFE

TERRY A. MILLER, *Department of Chemistry and Biochemistry, The Ohio State University, Columbus, OH, USA.*

Few things affect your quality of life more than the air you breathe and the temperature of your immediate environment. Since more than 80% of the energy used in the industrialized world today is still derived from fossil fuels, these two quantities are not unrelated. Most organic molecules injected into the troposphere are degraded via oxidative processes involving free radical intermediates, and many of these intermediates are the same as the ones involved in the combustion of fossil fuels. Key oxidizing intermediates are hydroxyl, OH (day), and nitrate, NO_3 (night), and early intermediates of oxidized organic compounds include the alkoxy (RO) and peroxy (RO_2) families of radicals. Recently we have explored the spectroscopy of RO, RO_2, and NO_3 radicals both for diagnostic purposes and to characterize their molecular properties and benchmark quantum chemistry calculations.

We have utilized moderate resolution cavity ringdown spectroscopy (CRDS) to study ambient temperature radicals and high resolution CRDS and laser induced fluorescence (LIF) to study jet-cooled radicals. Peroxy radicals and NO_3 have weak $\tilde{A} - \tilde{X}$ electronic transitions in the near infrared which we have studied with CRDS. Comparable LIF measurements have been made for the alkoxy species in the UV. Both vibrational and rotational resolution of the electronic spectra is observed. Data obtained from the spectral observations provide information about both the geometric and electronic structure of these radicals as well as their dynamics and also provide the capability for unambiguous diagnostics of their concentrations and reactions.

MA02   9:45 – 10:25

WELCOME TO RYDBERG-LAND

YAN ZHOU, DAVID GRIMES, *Department of Chemistry, MIT, Cambridge, MA, USA*; TONY COLOMBO, *Physical Chemistry, Sandia National Laboratories, Albuquerque, NM, USA*; ETHAN KLEIN, TIMOTHY J BARNUM, ROBERT W FIELD, *Department of Chemistry, MIT, Cambridge, MA, USA.*

Rydberg-Rydberg electronic transitions provide information about the electronic structure of the ion-core and each of the fundamental mechanisms by which a light electron exchanges energy and angular momentum with heavy nuclei. Normally, Rydberg electronic states have been indirectly observed via a sequence of laser-excitation steps, for which detection of transitions is accomplished by either fluorescence- or ionization-based schemes. Electronic transitions of $|\Delta n^*| < 1$ between Rydberg states (n* is the effective principal quantum number) have kilo-Debye electric dipole transition moments when $n^* > 30$. Such enormous transition moments render Rydberg-Rydberg electronic transitions directly observable. A chirped millimeter wave pulse can simultaneously polarize a 23 GHz chunk of two-level systems. In our spectra of Ca atoms (10^4 Rydberg atoms/cm^3 in a volume of 100 cm^3), the resultant Free Induction Decay (FID) from each of these two level systems is down-converted and heterodyne detected at <500 kHz resolution (at 3:1 S:N in a single chirp). Willis Flygare and Brooks Pate are to be thanked! But there is more, especially for molecules!

Recently, the Doyle and DeMille research groups have developed a cryogenic buffer gas cooled ablation source, our version of which produces beams of alkaline earth monohalide molecules that are >100x brighter and 10x slower than those produced by our Smalley type supersonic jet ablation source. Our 20 K Neon buffer gas cooled ablation source, in combination with redesign of the resonance region (300 cm^3, mm-wave radiation on-axis with the molecular beam) of our CPmmW spectrometer, has resulted in a 1000x increase in brightness of a BaF molecular beam (10^8 Rydberg molecules/cm^3 in a single quantum state) and a 10x improvement in resolution (50 kHz @ 100 GHz).

When buffer gas cooled ablation sources are combined with direct detection of FID, a new domain of high resolution molecular spectroscopy begs for exploration!

Intermission

RAO AWARDS **10:50**
Presentation of Awards by Yunjie Xu, University of Alberta

2013 Rao Award Winners
Brett A. McGuire, California Institute of Technology
Gerard Dean O'Connor, University of Sydney
Stefan Truppe, Imperial College London

COBLENTZ AWARD **11:05**
Presentation of Award by Rohit Bhargava, Coblentz Society

MA03 *Coblentz Society Award Lecture* **11:10 – 11:50**

SINGLE-MOLECULE MICROSCOPY OF NANOCATALYSIS

PENG CHEN, *Department of Chemistry and Chemical Biology, Cornell University, Ithaca, NY, USA.*

Nanoparticles are important catalysts. Understanding their structure-activity correlation is paramount for developing better catalysts, but hampered by their inherent inhomogeneity: individual nanoparticles differ from one to another, and for every nanoparticle, it can change from time to time, especially during catalysis. Furthermore, each nanoparticle presents on its surface various types of sites, which are often unequal in catalytic activity. I will present our work of using single-molecule fluorescence microscopy to overcome these challenges and study single-nanoparticle catalysis at the single-turnover resolution and nanometer precision. I will present how we interrogate the catalytic activity and dynamics of individual metal nanoparticles, map the reactivity of different surface sites, and uncover surprising spatial reactivity patterns within single facets at the nanoscale. This spatiotemporally resolved catalysis mapping also enables us to probe the communication between catalytic reactions at different locations on a single nanocatalyst, in much relation to allosteric effects in enzymes.

MF. Astronomy

Monday, June 16, 2014 – 1:30 PM

Room: 116 Roger Adams Lab

Chair: Erika Gibb, University of Missouri - St. Louis, St. Louis, MO, USA

MF01 1:30 – 1:45

THE DISTRIBUTION, EXCITATION, AND ABUNDANCE OF CH^+ IN ORION KL

HARSHAL GUPTA, PATRICK MORRIS, *Infrared Processing and Analysis Center, California Institute of Technology, Pasadena, CA, USA*; ZSOFIA NAGY, *I. Physikalisches Institut, University of Cologne, Cologne, Germany*; JOHN PEARSON, *Jet Propulsion Laboratory, California Institute of Technology, Pasadena, CA, USA.*

The CH^+ ion was one of the first molecules identified in the interstellar gas more than 75 years ago, but the high observed abundances of CH^+ remain a puzzle, because the main reaction proposed for the formation of CH^+, *viz.*, $C^+ + H_2 \rightarrow CH^+ + H$, is so endothermic (4640 K), that it is unlikely to proceed at the typical temperatures of molecular clouds. One way in which the high endothermicity may be overcome, is if a significant fraction of the H_2 is vibrationally excited, as is the case in dense molecular gas exposed to intense far-ultraviolet radiation fields. Elucidating the formation of CH^+ in molecular clouds requires characterization of its spatial distribution, as well as that of the key reactants in the chemical pathways yielding CH^+. Here we present high-resolution spectral maps of the two lowest rotational transitions of CH^+ and the fine structure transition of C^+ in a $\sim 3' \times 3'$ region around the Orion Kleinmann-Low (KL) nebula, obtained with the *Herschel Space Observatory's* Heterodyne Instrument for the Far-Infrared (HIFI).[a] We compare these maps to those of CH^+ and C^+ in the Orion Bar photodissociation region (PDR), and discuss the excitation and abundance of CH^+ toward Orion KL in the context of chemical and radiative transfer models, which have recently been successfully applied to the Orion Bar PDR.[b]

---

[a] These observations were done as part of the Herschel observations of EXtraordinary sources: the Orion and Sagittarius star-forming regions (HEXOS) Key Programme, led by E. A. Bergin at the University of Michigan, Ann Arbor, MI.

[b] Nagy, Z. et al. 2013, A&A 550, A96

MF02 1:47 – 2:02

MODELING LINEAR MOLECULES AS CARRIERS OF THE $\lambda 5797$ Å AND $\lambda 6613$ Å DIFFUSE INTERSTELLAR BANDS

JANE HUANG, *Department of Chemistry, The University of Chicago, Chicago, IL, USA*; TAKESHI OKA, *Department of Astronomy and Astrophysics, Chemistry, The University of Chicago, Chicago, IL, USA.*

Electronic transitions of polar linear molecules have been modeled and compared to archival high resolution spectra of the diffuse interstellar bands (DIBs) at 5797 and 6613 Å. These two bands are notable for fine structure that has most commonly been attributed to the rotational structure of electronic transitions of gas-phase molecules.[a] Most strikingly, the 5797 DIB has a sharp, narrow center peak that is characteristic of the Q branch of parallel transitions with non-zero Λ. This work is also motivated by Oka et al.'s analysis of the anomalously extended redward tails seen in certain DIBs toward Herschel 36, which are reminiscent of electronic transitions of polar linear molecules at high radiative temperatures.[b] The determination of rotational distributions, which includes radiative and collisional effects, is based on the model presented in the earlier work. Thus far, the most promising models are a $^2\Pi \leftarrow \,^2\Pi$ transition for the 5797 DIB and a $^2\Delta \leftarrow \,^2\Pi$ transition for the 6613 DIB, with the effects of spin-orbit coupling examined in each case. The degree of consistency of these transitions with respect to the anomalous DIBs toward Herschel 36 is also discussed.

---

[a] Sarre, P. J., Miles, J. R., Kerr, T. H. et al. 1995, MNRAS, **177L** 41

[b] Oka, T., Welty, D. E., Johnson, S. et al. 2013, ApJ, **773** 42

SMALL AND LARGE MOLECULES IN THE DIFFUSE INTERSTELLAR MEDIUM

TAKESHI OKA, *Department of Astronomy and Astrophysics, Chemistry, The University of Chicago, Chicago, IL, USA*; JANE HUANG, *Department of Chemistry, The University of Chicago, Chicago, IL, USA*.

Although molecules with a wide range of sizes exist in dense clouds (e.g. $H(C\equiv C)_n C\equiv N$ with $n = 0 - 5$), molecules identified in diffuse clouds are all small ones. Since the initial discovery of CH, CN, and CH^+, all molecules detected in the optical region are diatomics except for H_3^+ in the infrared and C_3 in the visible. Radio observations have been limited up to triatomic molecules except for H_2CO and the ubiquitous C_3H_2.[a] The column densities of all molecules are less than 10^{14} cm^{-2} with the two exceptions of CO and H_3^+ as well as CH and C_2 in a few special sightlines. Larger molecules with many carbon atoms have been searched for but have not been detected.

On the other hand, the observations of a great many diffuse interstellar bands (380 toward HD 204827[b] and 414 toward HD 183143[c]) with equivalent widths from 1 to 5700 mÅ indicate high column densities of many heavy molecules. If an electronic transition dipole moment of 1 Debye is assumed, the observed equivalent widths translate to column densities from 5×10^{11} cm^{-2} to 3×10^{15} cm^{-2}. It seems impossible that these large molecules are formed from chemical reactions in space from small molecules. It is more likely that they are fragments of aggregates, perhaps mixed aromatic/aliphatic organic nanoparticles (MAONS).[d] MAONS and their large fragment molecules are stable against photodissociation in the diffuse ISM because the energy of absorbed photons is divided into statistical distributions of vibrational energy and emitted in the infrared rather than breaking a chemical bond. We use a simple Rice-Ramsperger-Kassel-Marcus theory[e] to estimate the molecular size required for the stabilization.

[a] Snow, T. P. & McCall, B. J. 2006, ARA&A, **44** 367

[b] Hobbs, L. M., York, D. G., Snow, T. P., Oka, T., Thorburn, J. A., et al. 2008, ApJ, **680** 1256

[c] Hobbs, L. M., York, D. G., Thorburn, J. A., Snow, T. P., Bishof, M., et al. 2009, ApJ, **705** 32

[d] Kwok, S. & Zhang, S. 2013, ApJ, **771** 5

[e] Freed, K. F., Oka, T., & Suzuki, H. 1982, ApJ, **263** 718

TEMPERATURE, DENSITY, IONIZATION RATE, AND MORPHOLOGY OF DIFFUSE GAS NEAR THE GALACTIC CENTER PROBED BY H_3^+

TAKESHI OKA, *Department of Astronomy and Astrophysics, Chemistry, The University of Chicago, Chicago, IL, USA*; THOMAS R. GEBALLE, *Gemini Observatory, Hilo, HI, USA*; MIWA GOTO, *University Observatory Munich, Munich, Germany*; TOMONORI USUDA, *National Astronomical Observatory of Japan, Tokyo, Japan*.

Since last year, infrared spectra of H_3^+ and CO have been obtained toward nine stars (designated by us $\alpha+$, β, γ, $\gamma-$, δ, θ, κ, λ, and $\lambda-$) along the Galactic plane from 138 pc to the west of Sgr A* to 115 pc east, using IRCS of the Subaru Telescope and GNIRS of the Gemini North Observatory. All of the objects lie within the Central Molecular Zone (CMZ), a region of radius $\sim$150 pc at the center of the Galaxy. All sightlines except that toward λ (a red giant not suitable for H_3^+ spectroscopy) have high H_3^+ column densities on the order of a few times 10^{15} cm^{-2}. The metastable $R(3,3)^l$ absorption line was sought on seven sightlines ($\alpha+$, β, γ, $\gamma-$, δ, θ, κ), each of which showed significant signal except κ for which detection of this line was inconclusive. These results indicate that the long (at least several tens of parsecs) columns of warm ($T \sim 250$ K) and diffuse ($n \leq 100$ cm^{-3}) gas in which a high ionization rate of ζ of a few times 10^{-15} s^{-1} exists, found earlier by us on sightlines passing through the central 30 pc of the CMZ[ab] are present over nearly the entire CMZ.

The velocity profiles of the H_3^+ absorption lines provide information on the morphology of the diffuse gas in the CMZ. The velocity profile toward star $\lambda-$ (2MASS J17482472-2824313) observed by GNIRS is particularly noteworthy. The sightline toward this star, located 115 pc to the east of Sgr A*, shows the presence of warm diffuse gas near 0 radial velocity and complements an identical result at the west end (on sightlines toward $\alpha+$ and previously observed sources α and β). Stars nearer to the center of the CMZ show the warm diffuse gas at negative velocities only. Although many more stars need to be observed, the results to date suggest the existence of an expanding molecular ring of diffuse gas which is, unlike previously reported, not rotating but purely expanding.[c]

[a] Oka, T., Geballe, T. R., Goto, M., Usuda, T., and McCall, B. J. 2005, ApJ, **632** 882

[b] Goto, M., Usuda, T., Nagata, T. et al. 2008, ApJ, **688** 306

[c] Oka, T. 2013, Chem. Rev. **113** 8738

MF05   2:38 – 2:53

ESTIMATED SOFT X-RAY SPECTRUM AND IONIZATION OF MOLECULAR HYDROGEN IN THE CENTRAL MOLECULAR ZONE OF THE GALACTIC CENTER

MASAHIRO NOTANI, TAKESHI OKA, *Department of Astronomy and Astrophysics, Chemistry, The University of Chicago, Chicago, IL, USA.*

From observed high H_3^+ column densities in the Central Molecular Zone (CMZ), a region with a radius of $\sim$150 pc at the center of our Milky Way Galaxy, H_2 ionization rates of $\zeta \sim 3 \times 10^{-15}$ s^{-1} have been reported.[ab] This ionization rate which is higher than those in dense clouds and diffuse clouds in the Galactic disk by $\sim$100 and $\sim$10, respectively, have been ascribed to high flux of cosmic rays due to the high density of supernova remnants in the region. We are studying the ionization rate due to X-rays intensely observed in the CMZ as a possible competing process. Last year we reported the estimated ionization rate due to observable X-rays with energy 1 - 10 keV as negligible compared to the observed ζ.

However, just like cosmic ray ionization is dominated by low energy ($E \leq 100$ MeV) cosmic rays that are not directly observable because of deflection by solar magnetic field, the X-ray ionization is dominated by soft X-rays ($E \leq 1$ keV) that are not observable due to optical depth of the foreground gas. Our task therefore resembles those by Hayakawa et al. (1961)[c] and Spitzer and Tomasko (1968)[d] who estimated the cosmic ray ionization rate ζ based on high energy (> 1 GeV) cosmic ray observations.

We use theoretical X-ray spectrum and interpolate the observed X-rays at 4 keV, which are observable nearly un-attenuated from the CMZ, to the low energy region. Two theoretical spectra are presented, one due to Bremsstrahlung with variable temperature and proper cut off and the other the advection-dominated accretion flow (ADAF) model.[e] Discussion of the calculations and the results will be presented.

[a]Oka, T., Geballe, T. R., Goto, M., Usuda, T., and McCall, B. J. 2005, ApJ, **632** 882

[b]Geballe, T. R., and Oka, T. 2010, ApJ, **709** L70.

[c]Hayakawa,S.,Nishimura, S., Takayanagi, K. 1961, PASJ, **18** 184

[d]Spitzer,L. T., Tomasko, M. G. 1968, ApJ, **152** 971

[e]Yuan, F., Quataert, E., Narayan, R. 2003, JPJ, **598** 301.

MF06   2:55 – 3:05

MOLECULAR GAS NEAR UNUSUAL GALACTIC CENTER RADIO SOURCE N3

DOMINIC A. LUDOVICI, JAMES TOOMEY, CORNELIA LANG, *Department of Physics and Astronomy, University of Iowa, Iowa City, IA, USA.*

Near the Galactic center, the presence of the supermassive black hole, high dust temperatures, and large densities produce an unique environment within our galaxy. Unfortunately, optical light in this region is obscured by dust, making optical observations difficult or impossible. By utilizing radio telescopes, we can peer through the dust to examine the inner workings of this fascinating region. Using the Very Large Array, we examined the Galactic center radio source N3, a point source located within the Radio Arc. Our observations examined both molecular line and continuum emission from the region at frequencies from 2 GHz to 49 GHz. Several molecular species are detected around N3. Using molecular line analysis, N3's spectral index, physical size limits, and possible interactions between N3 and the Radio Arc, we examine the physical nature of N3 and outline further work needed to complete the analysis of this interesting source.

MF07 3:07 – 3:22

SMALL CARBON CHAINS IN CIRCUMSTELLAR ENVELOPES

ROBERT J. HARGREAVES, *Department of Chemistry and Biochemistry, Old Dominion University, Norfolk, VA, USA*; KENNETH HINKLE, *950 North Cherry Avenue, NOAO, Tucson, AZ, USA*; PETER F. BERNATH, *Department of Chemistry and Biochemistry, Old Dominion University, Norfolk, VA, USA.*

Observations were made of a number of carbon-rich circumstellar envelopes using the Phoenix spectrograph on the Gemini South telescope to determine the presence of small carbon chain molecules. The circumstellar envelope of IRC+10216 (CRL 1381) has been extensively studied, due to its brightness in the infrared, and C_3 and C_5 have previously been observed[a][b]. Vibration-rotation lines of the ν_3 antisymmetric stretch of C_3 near 2040 cm^{-1} have been used to determine the column density of C_3 in three new circumstellar envelopes: CRL 865, CRL 1922 and CRL 2023. Our new observations support the column density determined from CRL 1381 and also demonstrate that C_3 is common in carbon-rich circumstellar shells. We additionally determine upper limits for the small carbon chains, C_5 and C_7.

[a]Hinkle, K.W., Keady, J.J., & Bernath, P.F. 1988, *Science*, 241, 1319

[b]Bernath, P.F., Hinkle, K.H., & Keady, J.J. 1989, *Science*, 244, 562

MF08 3:24 – 3:39

ASTRONOMICAL MASERS: POLARIZATION PROPERTIES OF 22-GHZ WATER AND 6.7-GHZ METHANOL MASERS.

GABRIELE SURCIS, *Joint Institute for VLBI in Europe, Dwingeloo, Netherlands*; WOUTER H.T. VLEM-MINGS, *Onsala Space Observatory, Chalmers University of Technology, Onsala, Sweden*; HUIB JAN VAN LANGEVELDE, *Joint Institute for VLBI in Europe, Dwingeloo, Netherlands.*

By observing the astronomical masers in the Milky Way we can determine for instance high-accurate distances of the hosting Galactic sources (e.g., Galactic star-forming regions) and the kinematic of the gas where the masers arise (e.g., the kinematic of Keplerian accretion disks and outflows in massive star-forming regions). In addition, the bright and narrow spectral line emissions of water and methanol masers are ideal for measuring the Zeeman splitting as well as for determining the orientation of the magnetic field in 3-dimensions around massive young stellar objects (YSOs). Therefore, water and methanol maser species can help us to answer several crucial questions about massive star-formation. For instance, one of the most debated question is whether magnetic fields are important in the formation of high-mass stars ($M > 8$ M_{sun}). The main difficulty in answering this question is related to the fast evolution of the high-mass stars that makes the massive YSOs rare. Furthermore, they are typically found at fairly large distance. Hence, it is very difficult to measure the magnetic fields at distances < 100 Astronomical Units from the central protostar by using dust polarized emissions. But fortunately, the direct measurement of magnetic fields at small scale (10-100 Astronomical Units) around massive YSOs is possible by observing the polarized emission of masers.

In my oral contribution, besides showing the polarization properties of 22-GHz water and 6.7-GHz methanol masers, I will show our most interesting results about the determination of the orientation and of the strength of magnetic fields around massive YSOs. We have also started a systematic study for determining if there exists a real alignment between magnetic fields and the large scale outflows that are launched from the central protostar, which is important to constrain future simulations. Furthermore, we are involved in laboratory and modelling efforts to calibrate the magnitude of the Zeeman effect for methanol masers.

MF09 3:41 – 3:56

MODELING OF ASTROCHEMISTRY DURING STAR FORMATION

UGO HINCELIN, ERIC HERBST, *Department of Chemistry, The University of Virginia, Charlottesville, VA, USA*; QIANG CHANG, *Radio Astronomy, Xinjiang Astronomical Observatory, Xinjiang, China*; TATIANA VASYUNINA, *Millimeter- und Submillimeter-Astronomie, Max-Planck-Institut für Radioastronomie, Bonn, NRW, Germany*; YURI AIKAWA, KENJI FURUYA, *Graduate School of Science, Kobe University, Kobe, Japan.*

Interstellar matter is not inert, but is constantly evolving. On the one hand, its physical characteristics such as its density and its temperature, and on the other hand, its chemical characteristics such as the abundances of the species and their distribution, can change drastically. The phases of this evolution spread over different timescales, and this matter evolves to create very different objects such as molecular clouds (T $\sim$ 10 K, n $\sim 10^4$ cm^{-3}, t $\sim 10^6$ years), collapsing prestellar cores (inner core : T $\sim$ 1000 K, n $\sim 10^{16}$ cm^{-3}, t $\sim 10^4$ years), protostellar cores (inner core : T $\sim 10^5$ K, n $\sim 10^{24}$ cm^{-3}, t $\sim 10^6$ years), or protoplanetary disks (T $\sim$ 10 $-$ 1000 K, n $\sim 10^9 - 10^{12}$ cm^{-3}, t $\sim 10^7$ years). These objects are the stages of the star formation process. Starting from the diffuse cloud, matter evolves to form molecular clouds. Then, matter can condense to form prestellar cores, which can collapse to form a protostar surrounded by a protoplanetary disk. The protostar can evolve in a star, and planets and comets can be formed in the disk. Thus, modeling of astrochemistry during star formation should consider chemical and physical evolution in parallel.

We present a new gas-grain chemical network involving deuterated species, which takes into account ortho, para, and meta states of H_2, D_2, H_3^+, H_2D^+, D_2H^+, and D_3^+. It includes high temperature gas phase reactions, and some ternary reactions for high density, so that it should be able to simulate media with temperature equal to [10; 800] K and density equal to [$\sim 10^4$; $\sim 10^{12}$] cm^{-3}. We apply this network to the modeling of low-mass and high-mass star formation, using a gas-grain chemical code coupled to a time dependent physical structure. Comparisons with observational constraints, such as the HDO/H_2O ratio in high mass star forming region, give good agreement which is promising. Besides, high density conditions have highlighted some limitations of our grain surface modeling. We present a numerical technique to model in a more realistic way H_2 diffusion and desorption in high density conditions.

Intermission

MF10 4:13 – 4:28

OSCILLATOR STRENGTHS AND PREDISSOCIATION RATES FOR $W - X$ BANDS AND THE $4P5P$ COMPLEX IN $^{13}C^{16}O$ AND $^{12}C^{18}O$

MICHELE EIDELSBERG, JEAN LOUIS LEMAIRE, *Meudon, Observatoire de Paris, Paris, France*; STEVEN FEDERMAN, *Physics and Astronomy, University of Toledo, Toledo, OH, USA*; GLENN STARK, *Department of Physics, Wellesley College, Wellesley, MA, USA*; ALAN HEAYS, *Leiden Observatory, University of Leiden, Leiden, Netherlands*; LISSETH GAVILAN, *Meudon, Observatoire de Paris, Paris, France*; JAMES R LYONS, *School of Earth and Space Exploration, Arizona State University, Tempe, AZ, USA*; PETER L SMITH, *93 Pleasant St., 93 Pleasant St., Watertown, MA, USA*; NELSON DE OLIVEIRA, DENIS JOYEUX, *DESIRS Beamline, Synchrotron SOLEIL, Saint Aubin, France.*

We are conducting experiments on the DESIRS beam-line at the SOLEIL Synchrotron to acquire the necessary data on oscillator strengths and predissociation rates for modeling CO photochemistry in astronomical environments. A VUV Fourier Transform Spectrometer provides a resolving power of about 350,000, allowing us to discern individual lines in electronic transitions. Here we focus on results obtained from absorption spectra of $^{13}C^{16}O$ and $^{12}C^{18}O$ for the $W\ ^1\Pi - X\ ^1\Sigma^+$ bands with $v' = 0 - 3$ and $v'' = 0$ and the four overlapping bands (three resolved and one diffuse) observed between 92.97 and 93.35 nm. The three resolved bands are transitions to the upper levels $4p\pi(2)$, $5p\pi(0)$, and $5p\sigma(0)$ of the $4p(2)$ and $5p(0)$ complexes, and the diffuse band is associated with a non Rydberg level I $^1\Pi$; weak features in $^{13}C^{16}O$ are likely associated with absorption to the $4p\sigma(2)$ and II $^1\Pi$ levels. Several perturbations are also revealed in the high-resolution spectra. We compare our results with earlier determinations for these isotopologues of CO, as well as our SOLEIL measurements on $^{12}C^{16}O$.

MF11

STUDY OF THE PERTURBED $W\,^1\Pi(v=1)$ STATE OF CO IN FIVE ISOTOPOLOGUES

ALAN HEAYS, *Leiden Observatory, University of Leiden, Leiden, Netherlands*; MICHELE EIDELSBERG, JEAN LOUIS LEMAIRE, *Meudon, Observatoire de Paris, Paris, France*; GLENN STARK, *Department of Physics, Wellesley College, Wellesley, MA, USA*; STEVEN FEDERMAN, *Physics and Astronomy, University of Toledo, Toledo, OH, USA*; JAMES R LYONS, *School of Earth and Space Exploration, Arizona State University, Tempe, AZ, USA*; PETER L SMITH, *93 Pleasant St., 93 Pleasant St., Watertown, MA, USA*; LISSETH GAV-ILAN, *Meudon, Observatoire de Paris, Paris, France*; NELSON DE OLIVEIRA, DENIS JOYEUX, *DESIRS Beamline, Synchrotron SOLEIL, Saint Aubin, France.*

As part of a series of new photoabsorption measurements of CO isotopologues in the vacuum-ultraviolet at the SOLEIL synchrotron, we have studied the $v=1$ level of the $W\,^1\Pi$ Rydberg state. This state is crossed by an unknown perturber that strongly influences the predissociation rate of $W\,^1\Pi$ rotational levels. A detailed multi-isotopologue study of this interaction reveals the molecular constants of the interacting species and gives clues to its identity.

Our measurement program has astrophysical applications in mind, with CO photodissociation being a critical step in the photochemistry of interstellar clouds, protoplanetary disks, and (exo)planetary atmospheres. The careful analysis of perturbations such as this and other weak features in the spectrum of CO give critical information regarding the electronic structure of the molecule. Illuminating this structure will help translate laboratory measurements to astrophysically-relevant cross sections and predissociation rates, and is itself an interesting problem in molecular physics.

MF12

THE 3.1 μm INFRARED SPECTRA OF VIBRATIONALLY EXCITED C_3 IN A SUPERSONIC PLASMA JET

DONGFENG ZHAO, KIRSTIN D DONEY, HAROLD LINNARTZ, *Leiden Observatory, Sackler Laboratory for Astrophysics, Universiteit Leiden, Leiden, Netherlands.*

The linear triatomic carbon (C_3), one of the most important molecules that have been identified in both dense and diffuse interstellar environments, has attracted great interest to astronomers and astrochemists. It is also of fundamental interest as it serves as a benchmark system for quantum chemistry. In this presentation, we report the high-resolution infrared spectra of C_3 in the 3.1 μm region. The C_3 molecules are produced in a supersonic pulsed planar plasma by discharging a propyne/helium/argon gas mixture. Continuous-wave cavity ringdown spectroscopy is used to record the infrared absorption spectra of C_3. In total, eighteen vibrational bands are observed in the 3110 - 3290 cm^{-1} range, and sixteen of them are reported for the first time. It is found that, the vibrational temperatures for the two CC stretch modes of C_3 are up to 8000 K in our plasma source, allowing to experimentally determine the ro-vibrational levels of C_3 to the 10 000 cm^{-1} region. Accurate spectroscopic parameters are obtained from the detailed analysis of our spectra. The molecular data reported here are used to test the very recent theoretical work beyond the 'gold standard' [a] for a comprehensive understanding of the ground-state potential energy surface of C_3.

---

[a]P. Botschiwina, private communication.

MF13 4:59–5:14

SUB-DOPPLER JET-COOLED INFRARED SPECTROSCOPY OF $ND_2H_2^+$ AND ND_3H^+ IN THE NH STRETCH FUNDAMENTAL MODES

CHIH-HSUAN CHANG, *JILA, The University of Colorado, Boulder, CO, USA*; DAVID NESBITT, *Department of Chemistry, JILA CU-NIST, Boulder, CO, USA.*

Sub-Doppler jet-cooled rovibrational spectra of ND_3H^+, $ND_2H_2^+$, and NDH_3^+ ions in various fundamental NH modes were observed and analyzed using difference frequency generation infrared spectroscopy. The ions were generated in the concentration-modulation slit-jet expansion via a H_3^+ proton transfer mechanism in a discharge mixture of ND_3/H_2O and H_2 gases. NH mode excitation in ND_3H^+ ion yielded a prominent Q branch feature and parallel band rotational structure. Rotational transitions were confirmed unambiguously by four-line ground state combination differences within frequency measurement accuracy (10 MHz). The band origin was determined to be 3316.8413(9) cm^{-1}. Perturbation in the upper state was observed from analysis of residuals. In the case of $ND_2H_2^+$, both NH symmetric (b-type) and anti-symmetric (c-type) modes were observed and assigned for the first time, yielding band origins of 3297.5440(1) and 3337.9050(1) cm^{-1}, respectively. The intensity for the two fundamental bands was interpreted with simple context of a bond-dipole model. The present study provided high precision ground state rotational constants ($A'' = 4.85675(4)$, $B'' = 3.96829(4)$, $C'' = 3.44667(6)$ cm^{-1}), which should facilitate microwave searches for isotope-substituted ammonium ions in the regions of interstellar medium, such as dense molecular clouds or younger stellar objects.

MF14 5:16–5:31

ELECTRONIC TRANSITION SPECTRA OF THIOPHENOXY AND PHENOXY RADICALS IN HOLLOW CATHODE DISCHARGES

MITSUNORI ARAKI, HIROMICHI WAKO, KEI NIWAYAMA, KOICHI TSUKIYAMA, *Faculty of Science Division I, Tokyo University of Science, Shinjuku-ku, Tokyo, Japan.*

Diffuse interstellar bands (DIBs) still remain the longest standing unsolved problem in spectroscopy and astrochemistry, although several hundreds of DIBs have been already detected. It is expected that identifications of DIBs can give us crucial information for extraterrestrial organic molecule. One of the best approaches to identify carrier molecules of DIBs is a measurement of DIB candidate molecule produced in the laboratory to compare their absorption spectra with astronomically observed DIB spectra.

Radical in a gas phase is a potential DIB candidate molecule. The electronic transitions of polyaromatic hydrocarbon radicals result in optical absorption. However, because radicals are unstable, their electronic transitions are difficult to observe using a laboratory spectrometer system. To solve this difficulty, we have developed a glow-discharge cell using a hollow cathode in which radicals can be effectively produced as a high-density plasma. The radicals produced were measured by using the cavity ringdown (CRD) spectrometer and the discharge emission spectrometer.

The CRD spectrometer, which consists of a tunable pulse laser system, an optical cavity and a discharge device, is an apparatus to observe an high-resolution optical absorption spectrum. The electronic transition of the thiophenoxy radical C_6H_5OS was observed in the discharge emission of thiophenol C_6H_5OH. The electronic transition frequency of the thiophenoxy radical was measured.

A optical discharge emission was examined by using a HORIBA Jobin Yvon iHR320 monochromator. We detected the phenoxy radical C_6H_5O in the discharge of phenol C_6H_5OH. The band observed at 6107 Å in the discharge was assigned to the electronic transition of the phenoxy radical on the basis of the sample gas dependences and the reported low resolution spectrum.[a] The electronic transition frequency of the phenoxy radical was measured.

Comparison studies of the thiophenoxy and phenoxy radicals were made with known DIB spectra.

---

[a]B. Ward, Spectrochimica Acta, 24A, 813, 1968

MF15   **5:33 – 5:48**

SYNCHROTRON-BASED HIGH RESOLUTION SPECTROSCOPY OF N-BEARING PAHS

SÉBASTIEN GRUET[a], *AILES beamline, Synchrotron SOLEIL, Saint Aubin, France*; OLIVIER PIRALI, *Institut des Sciences Moléculaires d'Orsay, Université Paris-Sud, Orsay, France*; MANUEL GOUBET, *Laboratoire PhLAM, Université de Lille 1, Villeneuve de Ascq, France*; PHILIPPE BRECHIGNAC, *Institut des Sciences Moléculaires d'Orsay, Université Paris-Sud, Orsay, France.*

For thirty years,[b] the Polycyclic Aromatic Hydrocarbons (PAHs) have been suspected to give rise to the numerous Unidentified Infrared Bands (UIBs) observed in most astrophysical objects. Pure carbon molecules as well as derivatives with nitrogen atom(s) incorporated into the carbon skeleton have been considered. These N-bearing molecules are interesting candidates for astronomical research since they possess a larger permanent dipole moment than purely carbon-based PAHs. Most of the data reported in the literature deal with rotationally unresolved data. During the last decade, high-resolution microwave spectroscopy initiated high resolution studies of this broad family of molecules.[c] Recent advances in laboratory techniques permitted to provide interesting new results to rotationally resolve the IR/Far-IR vibrational bands of these relatively large C-bearing molecules[d],in particular, making use of synchrotron radiation as the IR continuum source of high resolution Fourier transform (FT) spectrometers. We will present an overview of the synchrotron-based high resolution FTIR spectroscopy of 5 aza-derivatives of naphthalene (isoquinoline, quinoline, quinoxaline, quinazoline, [1,5] naphthyridine) using a room temperature long path absorption cell at the French facility SOLEIL. In support to the rovibrational analysis of these FIR spectra, very accurate anharmonic DFT calculations were performed.[e]

---

[a] Also at: Institut des Sciences Moléculaires d'Orsay, UMR 8214 CNRS-Université Paris Sud 91405 Orsay Cedex

[b] A. Leger, J. L. Puget, *Astron. Astrophys.* **137**, L5-L8 (1984); L. J. Allamandola et al. *Astrophys. J.* **290**, L25-L28 (1985).

[c] Z. Kisiel et al. *J. Mol. Spectrosc.* **217**, 115 (2003); S. Thorwirth et al. *Astrophys. J.* **662**, 1309 (2007); D. McNaughton et al. *J. Chem. Phys.* **124**, 154305 (2011).

[d] S. Albert et al. *Faraday Discuss.* **150**, 71-99 (2011); B. E. Brumfield et al. *Phys. Chem. Lett.* **3**, 1985-1988 (2012); O. Pirali et al. *Phys. Chem. Chem. Phys.* **15**, 10141 (2013).

[e] M. Goubet, O. Pirali, J. *Chem. Phys.*, **140**, 044322 (2014).

MF16   *Post-Deadline Abstract*   **5:50 – 6:05**

THE HYPERFINE STRUCTURE OF ALUMINUM MONOXIDE, AlO

A. BREIER, THOMAS BÜCHLING, THOMAS GIESEN, *Institute of Physics, University Kassel, Kassel, Germany*; JÜRGEN GAUSS, *Institut für Physikalische Chemie, University of Mainz, Mainz, Germany.*

Small metal-containing molecules were produced in a laser ablation supersonic jet apparatus. The products were investigated by means of millimeter/submillimeter wave spectroscopy and optical spectra were recorded with a high-resolution grating spectrometer (HR2000+, OceanOptics).

This method has been applied to study AlO produced from laser ablation of solid aluminum seeded in helium-buffer gas enriched with 2% of nitrogenous oxide. The adiabatically expanding dilute gas mixture is probed by monochromatic radiation of frequencies up to 400 GHz (WR2.8x3, Virginia Diodes Inc.). The measurements reveal the hyperfine structure of a linear molecule in Hund's case $b_{\beta S}$ due to the nuclear spin of aluminum. With the present measurements, new high accurate line positions for future astronomical observations and more accurate molecular parameters are available. The new data were compared to high level ab initio calculations performed by the group of J. Gauss.

MG. Small molecules

Monday, June 16, 2014 – 1:30 PM

Room: 100 Noyes Laboratory

Chair: Gary Eden, University of Illinois, Urbana, IL, USA

MG01 1:30 – 1:45

ROUNDING AND UNCERTAINTIES IN PARAMETERS DETERMINED FROM FITS TO EXPERIMENTAL DATA,

or

A FAILURE TO ROUND DATA-ANALYSIS FIT PARAMETERS PROPERLY MAY MAKE THEM USELESS

ROBERT LE ROY, *Department of Chemistry, University of Waterloo, Waterloo, ON, Canada.*

Almost no physically interesting physico/chemical parameter is determined directly from a measurement. Rather, they are determined by performing a least-squares fit of some model to a set of data. Unfortunately, there seems to be no commonly accepted set of 'best practices' for determining how to round off such fitted parameter values to a minimum number of significant digits while ensuring that they retain the ability to reproduce the experimental data within their uncertainties. This sometimes results in lists of fitted parameters with no quoted uncertainties that have 2–3 times as many significant digits as the data being fitted, or to the results of an analysis being defined by parameters that are (unnecessarily?) quoted to more digits than normal computer double precision, which makes those results difficult or impossible to apply. Alternatively, it may also lead to fitted parameters being 'over-rounded' so that the model no longer accurately represents the experimental data. This presentation describes a 'best practice' to address these problems, offers a general-purpose least-squares fitting program that applies it, and provides an illustrative application of this approach in a study of the $A\,^1\Sigma_u^+ - X\,^1\Sigma_g^+$ system of Mg_2.

MG02 1:47 – 2:02

FULL EMPIRICAL POTENTIAL CURVES AND IMPROVED DISSOCIATION ENERGIES FOR THE $X\,^1\Sigma^+$ AND $A\,^1\Pi$ STATES OF CH^+

YOUNG-SANG CHO, ROBERT LE ROY, *Department of Chemistry, University of Waterloo, Waterloo, ON, Canada.*

CH^+ has been a species of interest since the dawn of molecular astrophysics,[a] and it is an important intermediate in combustion processes. In the domain of 'conventional' spectroscopy there have been a number of studies of low v' and v'' portions of the $A\,^1\Pi - X\,^1\Sigma^+$ band system of various isotopologues, and Amano recently reported microwave measurements of the ground-state $R(0)$ lines of $^{12}CH^+$, $^{13}CH^+$ and $^{12}CD^+$.[b] In addition, Helm *et al.*[c] used photodissociation spectroscopy to observe transitions to very high-J' tunneling-predissociation levels (shape resonances) involving $v(A) = 0 - 10$, for many of which they also measured the photo-fragment kinetic energy release. More recently Hechtfischer *et al.* used photodissociation spectroscopy of 'Feschbach resonance' levels at very high $v'(A)$ and low J' to obtain the first direct measurement of the $^{12}CH^+$ dissociation energy with near-spectroscopic accuracy (± 1.1 cm^{-1}).[d] However, to date, all analyses of the data for this system had been performed using traditional band-constant or Dunham-expansion fits to data for the lowest vibrational levels,[e] and there have been no attempts to combine the 'conventional' low-v data with the high-J' and high-v' photodissociation data in a single treatment. The present work has addressed this problem by performing a Direct-Potential-Fit (DPF) analysis that obtains full analytic potential energy functions for the $X\,^1\Sigma^+$ and $A\,^1\Pi$ states of CH^+ that are able to account for all of the available data (on average) within their uncertainties.

[a] A.E. Douglas and G. Herzberg, Astrophys. J. **94**, 381 (1941).

[b] T. Amano, Astrophys. J. Lett. 716, L1 (2010)

[c] H. Helm, P.C. Crosby, M.M. Graff and J.T. Mosley, Phys. Rev. **A 25**, 304 (1982)

[d] U. Hechtfischer and C. J. Williams, M. Lange, J. Linkemann, D. Schwalm, R. Wester, A. Wolf and D. Zajfman, J.Chem.Phys. **117**, 8754 (2002).

[e] H.S.P. Müller, Astron. Astrophys. **514**, L7 (2010)

MG03 2:04 – 2:19

ACCURATE ANALYTIC POTENTIAL FUNCTIONS FOR THE $A\,^3\Pi_1$ and $X\,^1\Sigma^+$ STATES OF IBr

<u>TOKIO YUKIYA</u>, NOBUO NISHIMIYA, MASAO SUZUKI, *Faculty of Engineering, Tokyo Polytechnic University, Atsugi, Japan*; ROBERT LE ROY, *Department of Chemistry, University of Waterloo, Waterloo, ON, Canada.*

Spectra of IBr in various wavelength regions have been measured by a number of researchers using traditional diffraction grating and microwave methods, as well as using high-resolution laser techniques combined with a Fourier transform spectrometer.[a,b,c,d,e] In a previous paper at this meeting, we reported a preliminary determination of analytic potential energy functions for the $A\,^3\Pi_1$ and $X\,^1\Sigma^+$ states of IBr from a direct-potential-fit (DPF) analysis of all of the data available at that time.[f] That study also confirmed the presence of anomalous fluctuations in the v–dependence of the first differences of the inertial rotational constant, $\Delta B_v = B_{v+1} - B_v$ in the $A\,^3\Pi_1$ state for vibrational levels with $v'(A)$ in the mid 20's. However, our previous experience in a recent study of the analogous $A\,^3\Pi_1 - X\,^1\Sigma_g^+$ system of Br_2 suggested that the effect of such fluctuations may be overcome if sufficient data are available.[g] The present work therefore reports new measurements of transitions to levels in the $v'(A) = 23 - 26$ region, together with a new global DPF analysis that uses "robust" least-squares fits[h] to average properly over the effect of such fluctuations in order to provide an optimum delineation of the underlying potential energy curve(s).

[a] L.E.Selin,Ark. Fys. **21**,479(1962).

[b] E. Tiemann and Th. Moeller, Z. Naturforsch. A **30**,986 (1975).

[c] E.M. Weinstock and A. Preston, J. Mol. Spectrosc. **70**, 188 (1978).

[d] D.R.T. Appadoo, P.F. Bernath, and R.J. Le Roy, Can. J. Phys. **72**, 1265 (1994).

[e] N. Nishimiya, T. Yukiya and M. Suzuki, J. Mol. Spectrosc. **173**, 8 (1995).

[f] T. Yukiya, N. Nishimiya, and R.J. Le Roy, Paper MF12 at the 65th Ohio State University International Symposium on Molecular Spectroscopy, Columbus, Ohio, June 20-24, 2011.

[g] T. Yukiya, N. Nishimiya, Y. Samajima, K. Yamaguchi, M. Suzuki, C.D. Boone, I. Ozier and R.J. Le Roy, J. Mol. Spectrosc. **283**, 32 (2013)

[h] J.K.G. Watson, J. Mol. Spectrosc. **219**, 326 (2003).

MG04 2:21 – 2:36

OBSERVATION OF THE FORBIDDEN TRANSITIONS BETWEEN THE $A^1\Pi_u$ AND $b^3\Sigma_g^-$ STATES OF C_2

<u>WANG CHEN</u>, JIAN TANG, KENTAROU KAWAGUCHI, *Graduate School of Natural Science and Technology, Okayama University, Okayama, Japan.*

In the last symposium,[a] we reported that a global fit simultaneously for the Phillips band system ($A^1\Pi_u - X^1\Sigma_g^+$) and the Ballik-Ramsay band system ($b^3\Sigma_g^- - a^3\Pi_u$) was carried out to deperturb the spin-orbit interaction between the $X^1\Sigma_g^+$ state and the $b^3\Sigma_g^-$ state of C_2. As the result, the energy gap between the $a^3\Pi_u$ state and the $X^1\Sigma_g^+$ state was obtained as 720.0 cm^{-1}, which is quite larger than the previous value[b] of about 716.7 cm^{-1} (converted from 718.3 cm^{-1} after the definition of the Hamiltonian for the $^3\Pi$ state is corrected by adding one B value on all the diagonal elements as the one[c] we use widely today). This newly determined singlet-triplet energy gap showed that the $X^1\Sigma_g^+(v = 6)$ level and the $b^3\Sigma_g^-(v = 3)$ level cross at $J = 2$ with the energy difference of only 0.07 cm^{-1} before the spin-orbit interaction is considered, which makes the singlet-triplet mixing nearly 50-50%. Therefore, we thought that the forbidden transitions related to this mixing should be observable. When rechecking the previously observed FTIR emission spectrum in the study of the CH radical,[d] where the emission spectrum of C_2 appeared to be very strong, we found that the allowed $A^1\Pi_u(v = 4) - X^1\Sigma_g^+(v = 6)$ transitions of C_2 around 3950 cm^{-1} were accompanied by the forbidden $A^1\Pi_u(v = 4) - b^3\Sigma_g^-(v = 3)$ transitions with $J'' = 2$, and two such forbidden transitions were identified clearly with the similar intensities as the corresponding allowed transitions. The observation of the forbidden transitions exactly at the predicted positions means that our deperturbation analysis was successful.

[a] W. Chen, J. Tang, and K. Kawaguchi, 68th OSU International Symposium on Molecular Spectroscopy, WJ11 (2013).

[b] C. Amiot, J. Chauville, and J. -P. Maillard, J. Mol. Spectrosc. <u>75</u>, 19 (1979).

[c] J. M. Brown and A. J. Merer, J. Mol. Spectrosc. <u>74</u>, 488 (1979).

[d] P. N. Ghosh, M. N. Deo, and K. Kawaguchi, Astrophys. J. <u>525</u>, 539 (1999).

MG05 2:38 – 2:53

ROTATIONAL ANALYSIS OF HIGH RESOLUTION F. T. SPECTRUM OF a' $^3\Sigma$- a $^3\Pi$ TRANSITION OF CS MOLECULE

MADHAV DAS SAKSENA, *Deonar, A-10 Basera, Off Din-Quarry Road, Mumbai, Maharashtra, India*; K SUNANDA, *Atomic and Molecular Physics , Bhabha Atomic Research center, Mumbai, Maharastra, India*; M N DEO, *High Pressure and Synchrotron Radiation Physics Division, Bhabha Atomic Research Centre, Mumbai, Maharashtra, India*; KENTAROU KAWAGUCHI, *Graduate School of Natural Science and Technology , Okayama University, Okayama, Japan.*

The F. T. Spectrum of CS molecule was recorded with Bruker IFS 120 HR spectrometer at a spectral resolution of 0.03 cm^{-1} using liquid nitrogen cooled InSb detector in the region 10500 - 1800 cm^{-1}. Intense spectrum of CS radical was excited by DC discharge of mixture of CS_2 (120 mTorr) and He (2 Torr) in flowing condition. Two hours integration time was used for obtaining a good S/N ratio. For the first time seven bands of a' $^3\Sigma$ - a $^3\Pi$ transition of CS molecule are observed lying between 8000 - 4800 cm^{-1} region. Rotational analysis of these bands, viz. 7-1, 6-1, 5-0, 4-0, 3-0, 2-0, and 3-1 will be presented.

MG06 2:55 – 3:10

DEPERTURBATION STUDIES OF d $^3\Delta$ - a $^3\Pi$ TRANSITION OF CS MOLECULE

K SUNANDA, *Atomic and Molecular Physics , Bhabha Atomic Research center, Mumbai, Maharastra, India*; M N DEO, *High Pressure and Synchrotron Radiation Physics Division, Bhabha Atomic Research Centre, Mumbai, Maharashtra, India*; MADHAV DAS SAKSENA, *Deonar, A-10 Basera, Off Din-Quarry Road, Mumbai, Maharashtra, India*; KENTAROU KAWAGUCHI, *Graduate School of Natural Science and Technology , Okayama University, Okayama, Japan.*

The F. T. Spectrum of CS molecule was recorded with Bruker IFS 120 HR spectrometer at a spectral resolution of 0.03 cm^{-1} using liquid nitrogen cooled InSb detector in the region 10500 - 1800 cm^{-1}. Intense spectrum of CS radical was excited by DC discharge of mixture of CS_2 (120 mTorr) and He (2 Torr) in flowing condition. Two hours integration time was used for obtaining a good S/N ratio. The recorded spectrum is more intense compared to previous studies, therefore, it has been possible to excite lower values of v' and v" for the d $^3\Delta_i$ - a $^3\Pi_r$ transition. The bands of three sub-systems occur with varying intensity. The following bands have been rotationally analysed, viz. 1-0, 2-1, 3-2 (d $^3\Delta_3$ - a $^3\Pi_2$); 2-0, 3-1, 4-2 (d $^3\Delta_2$ - a $^3\Pi_1$); and 1-1, 3-3 (d $^3\Delta_1$ - a $^3\Pi_0$). The d $^3\Delta_i$ is state is highly perturbed. Using a deperturbation program PGOPHER (C. M. Western, Univ. of Bristol) the molecular constants of the two states have been derived.

MG07 3:12 – 3:27

CO$^+$ AND C_2 SPECTRA GENERATED BY CO_2 ATMOSPHERIC PRESSURE GLOW DISCHARGES IN MICROCHANNELS

CHUL SHIN, *Department of Electrical and Computer Engineering, University of Illinois at Urbana-Champaign, Urbana, IL, USA*; ZHEN DAI, *Department of Materials Science and Engineering, University of Illinois at Urbana-Champaign, Urbana, IL, USA*; THOMAS J. HOULAHAN, JR., SUNG-JIN PARK, GARY EDEN, *Department of Electrical and Computer Engineering, University of Illinois at Urbana-Champaign, Urbana, IL, USA.*

Intense emission in the near-ultraviolet and visible from the Comet Tail and Swan bands of CO$^+$ and C_2, respectively, has been observed from glow discharges produced in CO_2 at atmospheric pressure. Generated within 200–500 μm microchannels fabricated in nanoporous alumina, the microchannel plasmas are spatially homogeneous, diffuse glows. As the CO_2 flow rate through the microchannels is varied, the visible/UV spectra change dramatically and the chemical kinetics of this fascinating spectrum will be discussed.

MG08   **3:29 – 3:44**

STARK AND ZEEMAN EFFECT STUDY OF THE [18.6]3.5 - X(1)4.5 BAND OF URANIUM MONOFLUORIDE, UF.

COLAN LINTON, *Department of Physics, University of New Brunswick, Fredericton, NB, Canada*; ALLAN G. ADAM, *Department of Chemistry, University of New Brunswick, Fredericton, NB, Canada*; TIMOTHY STEIMLE, *Department of Chemistry and Biochemistry, Arizona State University, Tempe, AZ, USA*.

A high resolution spectrum of the [18.6]3.5 - X(1)4.5 band of UF was obtained using the laser ablation specrometer at Arizona State University (ASU). The rotational structure showed significant perturbations in the upper state. Examination of the Stark and Zeeman splittings of the lowest J lines in electric fields up to 3.4 kV/cm and a magnetic field of 1650 Gauss yielded permanent electric dipole moments, μ_e, of 1.99 and 1.87 Debye and magnetic g-factors, g_e of 3.28 and 3.26 for the ground and excited states respectively.

The above experimental results will be discussed in terms of the configurational composition of the ground and excited states and compared with recent theoretical calculations[a].

[a]Ivan O. Antonov and Michael C. Heaven, J. Phys. Chem. A 2013, 117, 9684-9694

Intermission

MG09   **4:01 – 4:16**

ANALYSIS OF $2\nu_3$ BAND OF HTO

KAORI KOBAYASHI, HIROKI MAKI, TAKUYA YAMAMOTO, *Department of Physics, University of Toyama, Toyama, Japan*; MASANORI HARA, YUJI HATANO, *Hydrogen Isotope Research Center, University of Toyama, Toyama, Japan*; HIROYUKI OZEKI, *Department of Environmental Science, Toho University, Funabashi, Japan*.

Tritium is a radioactive isotope of hydrogen. Tritium released into natural enviroment is said to be converted into mostly HTO. The detection of HTO is important from the viewpoint of basic science as well as its radioactivity. Spectroscopy is a good tool for detection, however, high-resolution spectroscopy studies are still limited. The microwave study were carried out and the molecular constants of the ground state were determined. [a] All fundamental ν_1, ν_2 and the ν_3 bands of HTO were reported. [b],[c],[d] At 1.38 micron region, overtone and combination bands are expected. In this study, we prepared a new double wall cell for safe handling of highly concentrated tritiated water and carried out the near-infrared measurement. More than 100 transitions were observed and most of them were assigned to belong to the $2\nu_3$ band based on the previous quantum chemical calculations. [e] We will report the current status of the analysis.

[a]P. Helminger, F. C. De Lucia, W. Gordy, P. A. Staats and H. W. Morgan, *Phys. Rev. A* **10**, 1072 (1974).
[b]S. D. Cope, D. K. Russell, H. A. Fry, L. H. Jones, and J. E. Barefield, *J. Mol. Spectrosc.* **127**, 464 (1988).
[c]P. P. Cherrier, P. H. Beckwith, and J. Reid, *J. Mol. Spectrosc.* **121**, 69 (1987).
[d]M. Tine, D. Kobor, I. Sakho, and L. H. Coudert, *J. Mod. Phys.* **3**, 1945 (2012).
[e]M. J. Down, J. Tennyson, M. Hara, Y. Hatano, and K. Kobayashi, *J. Mol. Spectrosc.* **289**, 35 (2013).

54

MG10 4:18 – 4:33

LIF SPECTROSCOPY OF JET COOLED MgOH

MASARU FUKUSHIMA, TAKASHI ISHIWATA, *Information Sciences, Hiroshima City University, Hiroshima, Japan.*

We have generated MgOH in supersonic free jet expansions, and observed the laser induced fluorescence (LIF) of the $\tilde{A}\,^2\Pi - \tilde{X}\,^2\Sigma^+$ transition. We have reported rotational analyses of the bending vibronic bands, 2_0^n, $n = 0$, 2, and 4,[a] and it has been found that the spin-orbit constant of the bending vibronic levels, $\tilde{A}\,^2\Pi\,(0n^00)\,\mu\,^2\Pi$, increases with increasing the bending vibrational quantum number, n. This observation is interpreted as the lower level possessing primarily bent character and the higher level possessing linear character. The bending potential surface of the $\tilde{A}\,^2\Pi$ state is thought to be the reason for the present observation and interpretation. This potential surface is thought to be fairly flat and anharmonic with bent and linear geometries at the r_e and r_0 structures, respectively. We also observed the other two vibronic bands at almost the middle between the 2_0^2 and 2_0^4 bands. One of the two has a similar rotational structure with that of the origin band, 0_0^0, and is assigned to be the Mg–OH stretching vibronic band, $\tilde{A}\,^2\Pi\,(00^01) - \tilde{X}\,^2\Sigma^+\,(00^00)$. The rotational structure of another vibronic band is very complex, and it seems to overlap, at least, two vibronic bands, $^2\Pi -$ and $^2\Sigma^{(+)} - {}^2\Sigma^+$. More precise analysis is underway.

[a]M. Fukushima and T. Ishiwata, The 21st Colloquium on High-Resolution Molecular Spectroscopy, F8 (2009).

MG11 4:35 – 4:50

ANALYSIS OF BOUND-FREE AND BOUND-BOUND EMISSION SPECTRA OF SCANDIUM MONOIODIDE PRODUCED BY THE PHOTODISSOCIATION OF ScI_3

WENTING WENDY CHEN, THOMAS C. GALVIN, THOMAS J. HOULAHAN, JR., GARY EDEN, *Department of Electrical and Computer Engineering, University of Illinois at Urbana-Champaign, Urbana, IL, USA.*

We report emission spectra of scandium monoiodide (resolution of 0.7 Å) generated by the photodissociation of of ScI_3 at 248 nm. Emission from a bound-free transition of ScI near 900 nm was recorded, as were bound-bound transitions in the range of 400 nm to 800 nm. Comparison of experimental investigation and simulations of both bound-bound vibrational progressions and bound-free emission will be presented as well.

MG12 4:52 – 5:07

MILLIMETER-WAVE SPECTROSCOPY OF OSSO

MARIE-ALINE MARTIN-DRUMEL[a], *I. Physikalisches Institut, Universität zu Köln, Köln, Germany*; JENNIFER VAN WIJNGAARDEN, *Department of Chemistry, University of Manitoba, Winnipeg, MB, Canada*; OLIVER ZINGSHEIM, SVEN THORWIRTH, FRANK LEWEN, STEPHAN SCHLEMMER, *I. Physikalisches Institut, Universität zu Köln, Köln, Germany.*

Sulfur is the element with the largest number of known binary oxides[b] and as such has attracted the curiosity of chemists and physicists for decades. In particular, the simpler ones are of great interest for diverse scientific disciplines like structural and theoretical chemistry, astrochemistry, atmospheric chemistry and molecular physics. Out of those, the simplest sulfur rich oxides and dioxides S_2O and S_2O_2 have been studied spectroscopically to some extent in the past[c] but still pose challenging problems to future gas-phase investigations.

In the present study, the pure rotational spectrum of S_2O_2 has been investigated in the ground and ν_3 states. In addition, ground state transitions of $OS^{34}SO$ were observed. OSSO was produced in a radio frequency discharge through SO_2. Experimental measurements have been supported by high-level CCSD(T) calculations. An extensive set of molecular parameters has been derived.

[a]Present address: Harvard-Smithsonian Center for Astrophysics, Cambridge, MA 02138, USA
[b]Steudel, *Top. Curr. Chem.* 231, 203–230 (2003)
[c]Thorwirth et al., *J. Mol. Struct.* 795, 1-3 (2006), and refs. therein

MG13   **5:09 – 5:19**

CALCULATION OF ANHARMONICITIES IN OVERTONE MODES AND SMALL-CLUSTER SHIFTS OF SPHERICAL-TOP MOLECULES

JAVIER D. FUHR, <u>JUAN FIOL</u>, EDUARDO CORTIZO, PABLO D. FAINSTEIN, DANIEL E FREGENAL, TOMÁS GUOZDEN, ENRIQUE KAÚL, PABLO KNOBLAUCH, ALBERTO LAMAGNA, PABLO MACEIRA, GUILLERMO ROZAS, MARTÍN ZARCO, *Subgerencia Aplicaciones de Tecn. Láser, Centro Atómico Bariloche, San Carlos de Bariloche, Río Negro, Argentina.*

Harmonic and anharmonic vibrational self-consistent field (VSCF) calculations were employed to investigate the fundamental and overtone modes of SF_6 molecules. Determination of the Potential Energy Surface (PES) on a multidimensional grid of more than 65000 nodes was performed, and a system of 1D coupled-equations was solved. Corrections to the harmonic approximation for the frequencies of the fundamental modes and their overtones were obtained. Ab-initio calculations to the interaction potential between two molecules as a function of their position and orientation, and the corresponding energies for dimer formation, have been computed. Finally, the effect of dimerization on the molecular frequencies is investigated.

MG14   **5:21 – 5:36**

MM-WAVE ROTATIONAL SPECTRUM OF METHYL NITRATE

<u>JESSICA THOMAS</u>, IVAN MEDVEDEV, *Department of Physics, Wright State University, Dayton, OH, USA*; DAVID DOLSON, *Department of Chemistry, Wright State University, Dayton, OH, USA.*

Methyl nitrate (CH_3NO_3), is a toxic liquid known for it's explosive properties. It is metabolically expressed in trace amounts in exhaled human breath and is a potential candidate for interstellar detection. Previous microwave studies of methyl nitrate have yielded a handful line transitions in its vibrational ground state in the 8-34 GHz range. This paper discusses the high-resolution spectrum of methyl nitrate in 210-270 GHz range, and extends the spectroscopic assignment of its rotational transitions in the ground and first excited vibrational states.

MG15   **5:38 – 5:53**

PURE ROTATIONAL SPECTROSCOPY OF VINYL MERCAPTAN

<u>MARIE-ALINE MARTIN-DRUMEL</u>[a], OLIVER ZINGSHEIM, SVEN THORWIRTH, HOLGER S. P. MÜLLER, FRANK LEWEN, STEPHAN SCHLEMMER, *I. Physikalisches Institut, Universität zu Köln, Köln, Germany.*

Vinyl mercaptan (ethenethiol, CH_2=CHSH) exists in the gas phase in two distinct rotameric forms, *syn* (planar) and *anti* (quasi-planar in the ground vibrational state). The microwave spectra of these two isomers were investigated previously[b] however not exceeding frequencies of about 65 GHz.

In the present investigation, the pure rotational spectra of both species have been investigated at millimeter wavelengths. Vinyl mercaptan was produced in a radiofrequency discharge through a constant flow of ethanedithiol at low pressure. Both *syn* and *anti* rotamers were observed and new extensive sets of molecular parameters were obtained.

Owing to its close structural relationship to vinyl alcohol and the astronomical abundance of complex sulfur-bearing molecules, vinyl mercaptan is a plausible candidate for future radio astronomical searches.

---

[a]Present address: Harvard-Smithsonian Center for Astrophysics, Cambridge, MA 02138, USA
[b]M. Tanimoto et al. *J. Mol. Spectrosc.* 78, 95–105 & 106–119 (1979)

MH. Mini-symposium: Spectroscopy in Kinetics and Dynamics

Monday, June 16, 2014 – 1:30 PM

Room: B102 Chemical and Life Sciences

Chair: Mitchio Okumura, California Institute of Technology, Pasadena, CA, USA

MH01   *Journal of Molecular Spectroscopy Review Lecture*   1:30 – 2:00

ROAMING AND SPECTROSCOPY

JOEL BOWMAN, *Department of Chemistry, Emory University, Atlanta, GA, USA.*

Accurate ab initio theoretical/computational work on dynamics and spectroscopy begins with a potential energy surface (PES). My talk therefore begins with a brief review of progress we have made in developing accurate ab initio global PESs for reaction dynamics. "Roaming" is an unusual alternate pathway to reaction products that was found in the unimolecular dissociation of H_2CO by running roughly 100 000 trajectories on such a PES. The signatures of roaming were seen in the spectroscopic detection of the rotational states of CO correlated with translational energy distribution of the H_2.

I will discuss roaming in NO_3 photodissociation to $NO+O_2$ and give a short history of the topic. In particular I will recount how poor Franck-Condon factors in pioneering LIF detection experiments in 1997 of the low-lying vibrational states of O_2 plus the assumption of a "prior" vibrational distribution led to the wrong conclusions about the O_2 vibrational-state distribution. Later more sophisticated experiments obtained the correct vibrational distribution, which led to the (correct) speculation about roaming in this system.

I conclude with some comments about roaming wavefunctions and will wonder aloud about ways to detect wavefunctions spectroscopically. (Roaming wavefunctions have been reported by Hua Guo and co-workers for the MgH_2.)

MH02   2:05 – 2:20

THE SIGNATURE OF A ROAMING MECHANISM IN THE THERMAL DECOMPOSITION OF ETHYL NITRITE: CHIRPED-PULSE MILLIMETER-WAVE EXPERIMENT AND KINETIC MODELING

KIRILL PROZUMENT, *Department of Chemistry, Wayne State University, Detroit, MI, USA*; YURY V. SULEIMANOV, BEAT BUESSER, WILLIAM H. GREEN, *Department of Chemical Engineering, Massachusetts Institute of Technology, Cambridge, MA, USA*; ARTHUR SUITS, *Department of Chemistry, Wayne State University, Detroit, MI, USA*; ROBERT W FIELD, *Department of Chemistry, MIT, Cambridge, MA, USA.*

Roaming dynamics is a ubiquitous mechanism for unimolecular decomposition as well as for more complex chemical reactions. Chirped Pulse Fourier-Transform spectroscopy, developed recently by Pate and coworkers, is capable of nearly instant acquisition of rotational spectra spanning broad, $\sim$ 10 GHz, regions with unprecedented sensitivity and meaningful relative intensities. We report on the use of a Chirped-Pulse millimeter-Wave (CPmmW) spectrometer, operating in the 60 – 100 GHz region, to simultaneously detect and quantify multiple reaction products. Ethyl nitrite, CH_3CH_2ONO, molecules diluted in argon at 0.1% undergo thermal decomposition in the hot (1000 – 1800 K) flash pyrolysis reactor and cooled in the supersonic expansion for CPmmW detection. The branching ratios of the CH_2O, CH_3CHO, and HNO products are deduced from the CPmmW spectra. Kinetic modeling is the key tool used in this work to gain an understanding of the complex chemistry that occurs in the pyrolysis reactor. We use the Reaction Mechanism Generator that includes thousands of possible reactions and quantifies the relative significance of various paths. We find that, whereas the calculated branching of CH_2O and CH_3CHO is in a reasonable agreement with experiment, an addition of a relatively fast unimolecular reaction $CH_3CH_2ONO \rightarrow CH_3CHO + HNO$ is required in order for the model to describe the measured yield of HNO. Because the corresponding unimolecular decomposition via a tight transition state was calculated to be too slow to make a significant contribution, we believe that this reaction is proceeding via the roaming mechanism.

MH03

ROLES OF LARGE AMPLITUDE MOTIONS IN THE DYNAMICS OF THE PROTON TRANSFER REACTION $H_3^+ + H_2 \rightarrow H_5^+ \rightarrow H_3^+ + H_2$

ZHOU LIN, ANNE B McCOY, *Department of Chemistry and Biochemistry, The Ohio State University, Columbus, OH, USA.*

$H_3^+ + H_2 \rightarrow H_5^+ \rightarrow H_3^+ + H_2$ is a prototypical proton transfer reaction of astrochemical significance. Developing a deeper understanding of its reaction mechanism will provide important physical insights into similar and more complicated proton transfer reactions. The large amplitude vibrational motions in the H_5^+ collision complex lead to the scrambling of its five protons, which complicates theoretical descriptions of this molecular ion. Using minimized energy path diffusion Monte Carlo, the one-dimensional zero-point corrected potential energy surface is mapped out as a function of the chosen reaction coordinate.[a] The evolution of the wave function along this reaction coordinate is characterized by the energy and the probability amplitude associated to it. This methodology is also extended to allow for the investigation of excited states in selected vibrational modes that are important in the proton scrambling dynamics. In particular, the vibrations of interest are ones that evolve between large amplitude motions in H_5^+ and rotations of the H_3^+ and H_2 fragments as the reaction proceeds. In addition, as H_5^+ has five protons we must account for the anti-symmetry of the total nuclear wave function, which places restrictions to the possible pathways involved in the proton scrambling dynamics.[b] The implications of these symmetry restrictions are also discussed.

[a]C. E. Hinkle and A. B. McCoy, J. Phys. Chem. Lett., 1, 562 (2010)
[b]M. Quack, Mol. Phys., 34, 477 (1977)

MH04

FULL-DIMENSIONAL FRANCK-CONDON FACTORS IN THE HARMONIC NORMAL MODE BASIS FOR THE $\tilde{A}\,^1A_u$—$\tilde{X}\,^1\Sigma_g^+$ TRANSITION OF ACETYLENE

BARRATT PARK, ROBERT W FIELD, *Department of Chemistry, MIT, Cambridge, MA, USA.*

Methods developed by J. K. G. Watson for calculation of Franck-Condon Factors in systems undergoing linear $\leftrightarrow$ bent electronic transitions in the harmonic normal mode basis have been extended to the acetylene $\tilde{A}\,^1A_u$—$\tilde{X}\,^1\Sigma_g^+$ transition in full dimension. Because the intensity of the overlap accumulates away from linear geometry, the Hamiltonian of the linear $\tilde{X}$ state may be approximately separated into 3 rotations and $3N-6$ vibrations, resulting in a one-to-one correspondence between the normal modes in the linear and bent geometries. The calculated results reproduce experimental intensities quantitatively only at low quanta of vibrational excitation due to the exclusion of anharmonic effects. However, the qualitative results explain a number of observations that were previously not understood.

A change of basis to local bending modes of the $\tilde{X}$ state has been performed to investigate Franck-Condon access to zero order bright states with extreme local bend excitation. These states are known to emerge above 12 quanta of bend excitation and are of interest because the local bending mode lies along the reaction coordinate in the acetylene $\rightleftharpoons$ vinylidene isomerization. The results indicate that the best strategy for reaching extreme local benders involves Stimulated Emission Pumping from $\tilde{A}$-state levels with high excitation in ν_3' and ν_4', contrary to existing semi-classical arguments that ν_6' grants the best access to local bend states.

MH05   2:56–3:11

CORRELATION BETWEEN THE SHAPE AND THE EXCITATION OF THE ANGULAR MOMENTUM FOR THE HCN/HNC MOLECULE

GEORG MELLAU, *Physikalisch Chemisches Institut, Justus Liebig Universitat Giessen, Giessen, Germany*; ROBERT W FIELD, *Department of Chemistry, MIT, Cambridge, MA, USA.*

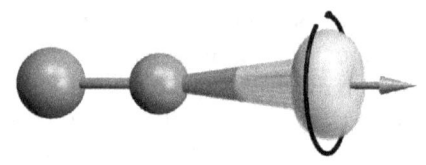

The [H,C,N] molecular system with the two linear isomers, HCN and HNC, is one of the most important model systems where it is possible to conduct eigenstate-resolved studies that encode the internal dynamics of strongly bound molecules at excitation energies relevant for chemical reactions.

The geometric shape of a molecule in its vibrationless state is an intuitive and useful chemical concept. In highly excited vibrational states we expect a significant change of this molecular shape. For triatomic linear HAB molecules this change depends on the two stretching and the bending vibrational excitations.

In highly excited bending states a fourth dynamical parameter becomes important: the vibrational angular momentum, l. The corresponding classical picture is the rotation of the hydrogen atom around the A-B core figure axis. The highest possible excitation for a given (v_2, l) bending state is the state with $l = v_2$. These are the states for which we expect the most significant dynamical effects due to the vibrational angular momentum.

In this work we study the correlation between the molecular shape and the rovibrational eigenenergies of the [H,C,N] molecule in the $v_1 v_2^{v_2} v_3$ states with the highest possible excitation of the vibrational angular momentum. The eigenergies we use have been measured using infrared HOTGAME (HOT GAs Molecular Emission) spectroscopy or come from the vibrational assignment of *ab initio* energy lists.

MH06   3:13–3:28

DIRECT OBSERVATION OF b_2 VIBRATIONAL LEVELS IN THE $^1B_2$ $\tilde{C}$ STATE OF SO_2: PRECISE MEASUREMENT OF ν_3 LEVEL STAGGERINGS

BARRATT PARK, JUN JIANG, CARRIE WOMACK, PETER RICHTER, ROBERT W FIELD, *Department of Chemistry, MIT, Cambridge, MA, USA*; ANDREW RICHARD WHITEHILL, SHUHEI ONO, *Earth, Atmospheric, and Planetary Sciences, MIT, Cambridge, MA, USA.*

The $^1B_2$ $\tilde{C}$ STATE OF SO_2 has been the subject of extensive investigation because it is important in the atmospheric photodissociation of SO_2. The state has a double-minimum potential in the dissociation coordinate, ν_3, arising from vibronic interactions, leading to a staggering of vibrational levels with v_3 odd vs. even. We report the first direct observations of the v_3 fundamental and of other levels with b_2 vibrational symmetry (odd v_3). Our work has made use of LIF, IR-UV double resonance, and coherent MODR techniques. Implications of the precision measurement of v_3 staggerings to the determination of double-minimum potential barrier and to vibronic coupling will be discussed.

Intermission

MH07   3:45–4:00

VIBRATIONAL STATES AT THE DISSOCIATION LIMIT PROTECTED FROM VIBRATIONAL ENERGY FLOW

MARTIN GRUEBELE, *Department of Chemistry, University of Illinois at Urbana-Champaign, Urbana, IL, USA.*

I will discuss the detection of sharp vibrational transitions with up to 40 quanta of total excitation above the dissociation limit of thiophosgene ($SCCl_2$). A 6-D CASSCF calculation and filtered diagonalization of eigenstates to compute dilution factors for bright states explains the 'survival' of these states and their expected number quantitatively.

MH08   4:02 – 4:17

TIME-RESOLVED FREQUENCY COMB SPECTROSCOPY OF TRANSIENT FREE RADICALS IN THE MID-INFRARED SPECTRAL REGION

BRYCE J BJORK, *Department of Physics, JILA - University of Colorado, Boulder, CO, USA*; ADAM J. FLEISHER, *Materials Measurement Laboratory - Chemical Sciences Division, National Institute of Standards and Technology, Gaithersburg, MD, USA*; BRYAN CHANGALA, *Department of Physics, JILA - University of Colorado, Boulder, CO, USA*; THINH QUOC BUI, *Division of Chemistry and Chemical Engineering, California Institute of Technology, Pasadena, CA, USA*; KEVIN COSSEL, *JILA, National Institute of Standards and Technology and Univ. of Colorado Department of Physics, University of Colorado, Boulder, CO, USA*; MITCHIO OKUMURA, *Division of Chemistry and Chemical Engineering, California Institute of Technology, Pasadena, CA, USA*; JUN YE, *JILA, National Institute of Standards and Technology and Univ. of Colorado Department of Physics, University of Colorado, Boulder, CO, USA*.

The chemical kinetics of transient free radicals, such as HOCO and Criegee intermediates, play important roles in combustion and atmospheric processes. Establishing accurate kinetics models for these complex systems require knowledge of the reaction rates and lifetimes of all molecules along a particular reaction pathway. However, standard spectroscopic techniques lack a combination of sensitivity, frequency resolution, and adequate temporal resolution to survey these reactions on the μs timescale. To answer this challenge, we have developed time-resolved frequency comb spectroscopy (TRFCS). This novel technique allows for the detection of transient intermediates with high time-resolution and sensitivity while also permitting the direct determination of rotational state distributions of all relevant molecules. We demonstrate this technique in the mid-infrared spectral region, at 3.7 μm, by studying the photolysis of deuterated acrylic acid. We simultaneously observe the time-dependent concentrations of photoproducts trans-DOCO, HOD, and D_2O, identified through their unique rovibrational structure, with 5×10^{10} molecules cm^{-3} sensitivity, and with a time resolution of 25 μs. We aim to apply this technique to detect directly the formation of the DOCO intermediate in the OD + CO chemical reaction at atmospherically relevant pressures, in order to validate statistical rate models of this reaction.

MH09   4:19 – 4:34

CHIRPED PULSE MICROWAVE SPECTROSCOPY IN PULSED UNIFORM SUPERSONIC FLOWS

CHAMARA ABEYSEKERA, JAMES OLDHAM, KIRILL PROZUMENT, BAPTISTE JOALLAND, *Department of Chemistry, Wayne State University, Detroit, MI, USA*; BARRATT PARK, ROBERT W FIELD, *Department of Chemistry, MIT, Cambridge, MA, USA*; IAN SIMS, *Institut de Physique de Rennes, Université de Rennes 1, Rennes, France*; ARTHUR SUITS, *Department of Chemistry, Wayne State University, Detroit, MI, USA*.

We present preliminary results describing the development of a new instrument that combines two powerful techniques: Chirped Pulse-Fourier Transform MicroWave (CP-FTMW) spectroscopy and pulsed uniform supersonic flows. It promises a nearly universal detection method that can deliver quantitative isomer, conformer, and vibrational level specific detection, characterization of unstable reaction products and intermediates and perform unique spectroscopic, kinetics and dynamics measurements.

We have constructed a new high-power K_a-band, 26–40 GHz, chirped pulse spectrometer with sub-MHz resolution, analogous to the revolutionary CP-FTMW spectroscopic technique developed in the Pate group at University of Virginia. In order to study smaller molecules, the E-band, 60–90 GHz, CP capability was added to our spectrometer. A novel strategy for generating uniform supersonic flow through a Laval nozzle is introduced. High throughput pulsed piezo-valve is used to produce cold (30 K) uniform flow with large volumes of 150 cm^3 and densities of 10^{14} molecules/cm^3 with modest pumping facilities. The uniform flow conditions for a variety of noble gases extend as far as 20 cm from the Laval nozzle and a single compound turbo-molecular pump maintains the operating pressure.

Two competing design considerations are critical to the performance of the system: a low temperature flow is needed to maximize the population difference between rotational levels, and high gas number densities are needed to ensure rapid cooling to achieve the uniform flow conditions. At the same time, collision times shorter than the chirp duration will give inaccurate intensities and reduced signal levels due to collisional dephasing of free induction decay. Details of the instrument and future directions and challenges will be discussed.

MH10 4:36 – 4:46

GAS-PHASE STRUCTURE DETERMINATION OF DIHYDROXYCARBENE, ONE OF THE SMALLEST STABLE SINGLET CARBENES

CARRIE WOMACK, *Department of Chemistry, MIT, Cambridge, MA, USA*; KYLE N CRABTREE, *Atomic and Molecular Physics, Harvard-Smithsonian Center for Astrophysics, Cambridge, MA, USA*; LAURA McCASLIN, *Department of Chemistry and Biochemistry, The University of Texas, Austin, TX, USA*; OSCAR MARTINEZ JR., *Atomic and Molecular Physics, Harvard-Smithsonian Center for Astrophysics, Cambridge, MA, USA*; ROBERT W FIELD, *Department of Chemistry, MIT, Cambridge, MA, USA*; JOHN F. STANTON, *Department of Chemistry, The University of Texas, Austin, TX, USA*; MICHAEL C McCARTHY, *Atomic and Molecular Physics, Harvard-Smithsonian Center for Astrophysics, Cambridge, MA, USA*.

Carbenes (R_1-C-R_2) are a reactive class of compounds, usually characterized by an electron-deficient divalent carbon atom, found in applications ranging from organic synthesis to gas phase oxidation chemistry. Carbenes with 2- or 3-atom substituents often undergo rapid unimolecular isomerization, but may be stabilized if these substituents are electron-donating. Dihydroxycarbene (HO-C̈-OH) is one of the smallest singlet carbenes to be afforded this stability, due to its two electron-donating hydroxyl groups. We report the first gas-phase detection and structural characterization of this reactive species, using a combination of Fourier transform microwave spectroscopy and high level electronic structure calculations. Detection in the gas phase indicates that it is fairly stable relative to its isomers, formic acid (HCOOH) and the simplest Criegee intermediate (CH_2OO), the latter of which has recently received a great deal of attention for its role in the atmospheric ozonolysis of alkenes. Our experimental results yield a precise structure of HO-C̈-OH, and we comment on upcoming experiments investigating its stability and reactivity with other common atmospheric species.

MH11 4:48 – 5:03

IR-DRIVEN DYNAMICS OF THE 3-AMINOPHENOL-AMMONIA COMPLEX

CORNELIA G HEID, W G MERRILL, AMANDA CASE, FLEMING CRIM, *Department of Chemistry, The Univeristy of Wisconsin, Madison, WI, USA*.

We report on gas-phase experiments investigating the predissociation and possible IR-driven isomerization of the 3-aminophenol-ammonia complex (3-AP-NH_3). A molecular beam of 3-AP-NH_3 is vibrationally excited with pulsed IR light, initiating an intramolecular vibrational redistribution and subsequent dissociation. The 3-AP fragment is then probed state-selectively via multiphoton ionization (REMPI) and time-of-flight mass spectrometry. Of particular interest is an IR-driven feature which we associate tentatively with a trans-cis isomerization process. We see clear correlation between the excitation of specific vibrational modes (namely the NH_3 symmetric and OH stretches) and the presence of this feature, as evidenced by IR-action and IR-depletion spectra. The feature persists atop a broader signal which we assign to the predissociation of the complex and whose cutoff in REMPI-action experiments provides an upper bound on the dissociation energy for 3-AP-NH_3.

MH12   5:05 – 5:20

VIBRATIONAL LEVELS AND RESONANCES ON A NEW POTENTIAL ENERGY SURFACE FOR THE GROUND ELECTRONIC STATE OF OZONE

STEVE ALEXANDRE NDENGUE, RICHARD DAWES, *Department of Chemistry, Missouri University of Science and Technology, Rolla, MO, USA*; XIAO-GANG WANG, TUCKER CARRINGTON, *Department of Chemistry, Queen's University, Kingston, ON, Canada.*

The isotopic ratios for ozone observed in laboratory and atmospheric measurements, known as the ozone isotopic anomaly,[1,2] have been an open question in physical and atmospheric chemistry for the past 30 years. The biggest limitation in achieving agreement between theory and experiment has been the availability of a satisfactory[3-5] ground state potential energy surface (PES). The presence of a spurious reef feature in the asymptotic region of most PESs has been associated with large discrepancies between calculated and observed rates of formation especially at low temperature. We recently proposed a new global potential energy surface for ozone[6,7] possessing 4 features that make it suitable for kinetics and dynamics studies: excellent equilibrium parameters, good agreement with experimental vibrational levels, accurate dissociation energy and a transition region with accurate topography (without the reef artifact). This PES has been used recently to simulate the temperature dependent exchange reaction (16O+16O2) with a quantum statistical model[6,7], and, for the first time, a negative temperature dependence which agrees with experiments was obtained, indicating the good quality of this global surface. A quantum description of the ozone exchange and recombination reaction requires knowledge of the resonances but also the rovibrational levels just below the dissociation. We present results of global 3-well vibrational-state calculations up to the dissociation threshold and (J = 0) resonances up to $1000 \, \text{cm}^{-1}$ beyond. The calculations were done using a large DVR basis (24 million functions) with a symmetry-adapted Lanczos algorithm as well as MCTDH. Results indicate the presence of localized bound states at energies close to the dissociation threshold beyond which some long-lived resonances follow, contrasted with a few delocalized bound states with density at large values of the stretching coordinates. References: 1- K. Mauersberger et al., Adv. At. Mol. Opt. Phys. 50, 1 (2005) 2- R. Schinke et al., Ann. Rev. Phys. Chem. 57, 625 (2006) 3- R. Siebert et al., J. Chem. Phys. 116, 9749 (2002) 4- M. Ayouz and D. Babikov, J. Chem. Phys. 138, 164311 (2013) 5- V.G. Tyuterev et al., J. Chem. Phys. 139, 134307 (2013) 6- R. Dawes et al., J. Chem. Phys. 135, 081102 (2011) 7- R. Dawes et al., J. Chem. Phys. 139, 201103 (2013)

MH13   5:22 – 5:37

SHORT-LIVED ELECTRONICALLY-EXCITED DIATOMIC MOLECULES COOLED VIA SUPERSONIC EXPANSION FROM A PLASMA MICROJET

THOMAS J. HOULAHAN, JR., RUI SU, GARY EDEN, *Department of Electrical and Computer Engineering, University of Illinois at Urbana-Champaign, Urbana, IL, USA.*

Using a pulsed plasma microjet to generate short-lived, electronically-excited diatomic molecules, and subsequently ejecting them into vacuum to cool via supersonic expansion, we are able to monitor the cooling of molecules having radiative lifetimes as low as 16 ns. Specifically, we report on the rotational cooling of He_2 molecules in the $d^3\Sigma_u^+$, $e^3\Pi_g$, and $f^3\Sigma_u^+$ states, which have lifetimes of 25 ns, 67 ns, and 16 ns, respectively. The plasma microjet is driven with a 2.6 kV, 140 ns high-voltage pulse (risetime of 20 ns) which, when combined with a high-speed optical imaging system, allows the nonequilibrium rotational distribution for these molecular states to be monitored as they cool from 1200 K to below 250 K with spatial and temporal resolutions of below 10 μm and 10 ns, respectively. The spatial and temporal resolution afforded by this system also allows the observation of excitation transfer between the $f^3\Sigma_u^+$ state and the lower lying $d^3\Sigma_u^+$ and $e^3\Pi_g$ states. The extension of this method to other electronically excited diatomics with excitation energies >5 eV will also be discussed.

MH14

CAVITY-ENHANCED ULTRAFAST TRANSIENT ABSORPTION SPECTROSCOPY

YUNING CHEN, *Department of Chemistry, Stony Brook University, Stony Brook, NY, USA*; MELANIE ROBERTS REBER, KEVIN KELEHER, *Department of Physics and Astronomy, State University of New York, Stony Brook, NY, USA*; THOMAS K ALLISON, *Department of Chemistry, Stony Brook University, Stony Brook, NY, USA*.

We introduce cavity enhanced ultrafast transient absorption spectroscopy, which employs frequency combs and high-finesse optical cavities. Sub-100 fs pulses with a repetition rate of 90 MHz are generated by a home-built Ytterbium fiber laser. The amplified light has a power up to 10 W, which is used to pump an optical parametric oscillator, followed by second-harmonic generation(SHG) that converts the wavelength from near-IR to visible. A pump comb at 530 nm is separately generated by SHG. Both pump and probe combs are coupled into high-finesse cavities. Compared to the conventional transient absorption spectroscopy method, the detection sensitivity can be improved by a factor of $\left(\frac{\mathcal{F}}{\pi}\right)^2 \sim 10^5$, where $\mathcal{F}$ is the finesse of cavity. This ultrasensitive technology enables the direct all-optical dynamics study in molecular beams. We will apply the cavity enhanced ultrafast transient absorption spectroscopy to investigate the dynamics of visible chromophores and then extend the wavelength to mid-IR to study vibrational dynamics of small hydrogen-bonded clusters.

MI. Radicals

Monday, June 16, 2014 – 1:30 PM

Room: 112 Chemistry Annex

Chair: Dmitry G. Melnik, The Ohio State University, Columbus, OH, USA

MI01   1:30 – 1:45

VIBRONIC EMISSION SPECTROSCOPY OF BENZYL-TYPE RADICALS GENERATED BY CORONA DISCHARGE

EUN HYE YI, YOUNG YOON, SANG LEE, *Department of Chemistry, Pusan National University, Pusan, Korea.*

Benzyl radical is a prototypical aromatic free radical and has been the subject of numerous spectroscopic studies. On the other hand, ring-substituted benzyl radicals, benzyl-type radicals, have received less attention due to the difficulties associated with production in corona discharge and analysis of spectra. We report vibronic emission spectra of hetero halogen multi-substituted benzyl radicals generated by corona discharge of corresponding toluene derivatives using a pinhole-type glass nozzle, from which visible vibronic emission spectra were recorded using a long-path monochromator. The spectra show nice features of strongest origin band and a series of vibronic bands in the lower energies originating from the vibrationless D_1 state. From the analysis of the spectra observed, we determined the energies of the $D_1 \rightarrow D_0$ electronic transition and vibrational mode frequencies in the ground electronic state.[a] On the other hand, all substituted benzyl radicals show the origin bands shifted to red region with respect to the parental benzyl radical at 22002 cm^{-1}. The shifts of multi-substituted benzyl radicals can be well estimated using the method developed from mono-substituted benzyl radicals as well as the positions of nodal point and mutual orientation of substituents, which could be useful for scientists to set a proper scanning range of their spectrometers for the spectroscopic observation of transient molecules. In this presentation, we will discuss the substituent effect[b][c] on electronic transition energy and the experimental technique developed in this laboratory.

---

[a] Y. W. Yoon and S. K. Lee, *J. Phys. Chem. A* **117**, 2485 (2013).

[b] Y. W. Yoon, S. Y. Chae, and S. K. Lee, *Chem. Phys. Lett.* **584**, 37 (2013).

[c] Y. W. Yoon and S. K. Lee, *Chem. Phys. Lett.* **570**, 29 (2013).

MI02   *Post-Deadline Abstract*   1:47 – 2:02

HIGHER ELECTRONIC EXCITED STATES OF JET-COOLED AROMATIC HYDROCARBON RADICALS: 1-PHENYLPROPARGYL, 1-NAPHTHYLMETHYL, 2-NAPHTHYLMETHYL AND 9-ANTHRACENYLMETHYL

GERARD O'CONNOR, GABRIELLE VICTORIA GRACE WOODHOUSE, TYLER TROY, *School of Chemistry, The University of Sydney, Sydney, NSW, Australia*; KLAAS NAUTA, TIMOTHY SCHMIDT, *School of Chemistry, UNSW, Sydney, NSW, Australia.*

The $D_0 \leftarrow D_1$ transitions of many aromatic resonance stabilised radicals (RSRs) have been observed in the gas-phase in recent years. This work has been primarily motivated by the suggestion that such molecules may be carriers of the diffuse interstellar bands (DIBs). Most gas- phase studies of these molecules have focused on the $D_0 \leftarrow D_1$ electronic transitions, primarily due to experimental limitations. These transitions are generally weak, a feature of odd- alternate hydrocarbon radicals, with intensity instead going to an electronically similar higher energy transition. This presentation will focus on higher electronic transitions with calculated intensity $f > 10^{-2}$. Experimental data will be presented for observed strong transitions of three benzilic polycyclic aromatic hydrocarbons (PAHs) radicals' 1-naphthylmethyl, 2-naphthylmethyl and 9-anthracenylmethyl. Experimental data will also be presented of a strong state of the aromatic/aliphatic RSR 1-phenylpropargyl. Trends in this experimental and theoretical data will be used to predict the spectroscopic properties of larger RSR molecules, and the relevance of these higher electronic states to astronomical observations will be discusses.

MI03   2:04 – 2:19

INFRARED LASER SPECTROSCOPY OF THE HELIUM-SOLVATED ALLYL AND ALLYL PEROXY RADICALS

CHRIS MORADI, CHRISTOPHER LEAVITT, BRAD ACREY, GARY DOUBERLY, *Department of Chemistry, University of Georgia, Athens, GA, USA.*

Infrared spectra in the C-H stretch region are reported for the allyl (CH_2CHCH_2) and allyl peroxy ($CH_2=CH-CH_2OO$) radicals solvated in superfluid helium nanodroplets. Nine bands in the spectrum of the allyl radical have resolved rotational substructure. We have assigned three of these to the ν_1 (a_1), ν_3 (a_1), and ν_{13} (b_2) C-H stretch bands and four others to the $\nu_{14}/(\nu_{15}+2\nu_{11})$ (b_2) and $\nu_2/(\nu_4+2\nu_{11})$ (a_1) Fermi dyads, and an unassigned resonant polyad is observed in the vicinity of the ν_1 band. Experimental coupling constants associated with Fermi dyads are consistent with quartic force constants obtained from density functional theory computations. The peroxy radical was formed within the He droplet via the reaction between allyl and O_2 following the sequential pick-up of the reactants. Five stable conformers are predicted for the allyl peroxy radical, and a computed two-dimensional potential surface for rotation about the CC-OO and CC-CO bonds reveals multiple isomerization barriers greater than 300 cm^{-1}. Nevertheless, the C-H stretch infrared spectrum is consistent with the presence of a single conformation following the allyl + O_2 reaction within helium droplets.

MI04   2:21 – 2:36

STUDY OF THE CH_2I + O_2 REACTION WITH A STEP-SCAN FOURIER-TRANSFORM INFRARED ABSORPTION SPECTROMETER: SPECTRA OF THE CRIEGEE INTERMEDIATE CH_2OO AND DIOXIRANE(?)

YU-HSUAN HUANG, YUAN-PERN LEE, *Applied Chemistry, National Chiao Tung University, Hsinchu, Taiwan.*

The Criegee intermediates are carbonyl oxides that play key roles in ozonolysis of unsaturated organic compounds. This mechanism was first proposed by Criegee in 1949, but the first direct observation of the simplest Criegee intermediate CH_2OO in the gaseous phase has been reported only recently using photoionization mass spectrometry.[a] Our group has reported the low-resolution IR spectra of CH_2OO, produced from the reaction of CH_2I + O_2, with a second-generation step-scan Fourier-transfom IR absorption spectrometer.[b] The spectral assignments were based on comparison of observed vibrational wavenumbers and rotational contours with theoretical predictions. Here, we report the IR absorption spectra of CH_2OO at a resolution of 0.32 cm^{-1}, showing partially rotationally-resolved structures. The origins of the ν_3, ν_4, ν_6, and ν_8 vibrational modes of CH_2OO are determined to be 1434.1, 1285.7, 909.2, and 847.3 cm^{-1}, respectively. With the analysis of the vibration-rotational spectra, we provide a definitive assignment of these bands to CH_2OO. The observed vibrational wavenumbers indicate a zwitterionic contribution to this singlet biradical showing a strengthened C-O bond and a weakened O-O bond. This zwitterionic character results to an extremely rapid self reaction via a cyclic dimer to form $2H_2CO + O_2$ ($^1\Delta_g$). Another group of weak transient IR bands centered at 1231.5, 1213.3, and 899.8 cm^{-1} are also observed. These bands might be contributed from dioxirane, which was postulated to be another important intermediate that might be isomerized from the Criegee intermediate in the reaction of O_3 with 1-alkenes.

[a]O. Welz, J. D. Savee, D. L. Osborn, S. S.Vasu, C. J. Percival, D. E. Shallcross, and C. A. Taatjes, Science **335**, 204 (2012).
[b]Y.-T. Su, Y.-H. Huang, H. A.Witek, and Y.-P. Lee, Science **340**, 174 (2013).

MI05   2:38 – 2:53

THE ELECTRONIC SPECTRUM OF BH_2 REVISITED

MOHAMMED GHARAIBEH, *Department of Chemistry, University of Kentucky, Lexington, KY, USA;* FUMIE SUNAHORI, *Department of Chemistry and Physics, Franklin College, Franklin, IN, USA;* DENNIS CLOUTHIER, *Department of Chemistry, University of Kentucky, Lexington, KY, USA;* RICCARDO TARRONI, *Dipartimento di Chimica Fisica ed Inorganica, Università di Bologna, Bologna, Italy.*

The $\tilde{A}^2B_1(\Pi_u) - \tilde{X}^2A_1$ linear-bent electronic transition of the BH_2 free radical from 11700 - 14600 cm^{-1} was previously studied by Herzberg and Johns in 1967.[a] In the current work, we have performed extensive LIF and emission studies of the electronic spectrum of jet-cooled $^{11}BH_2$, $^{10}BH_2$, $^{11}BD_2$, and $^{10}BD_2$ up to 21100 cm^{-1}. High resolution studies of $^{11}BD_2$ have allowed us to refine the ground state geometry. We have also used *ab initio* calculations and variational methods to predict the ro-vibronic energy levels of this Renner-Teller system. The results agree very well with the experimental data, confirming and extending our assignments.

[a]G. Herzberg and J. W. C. Johns *Proc. R. Soc. Lond. A* **298**(142), 1967.

MI06 2:55 – 3:10

THE X_2BO and X_2BS (X = HYDROGEN OR HALOGEN) FREE RADICALS

DENNIS CLOUTHIER, ROBERT GRIMMINGER, BING JIN, *Department of Chemistry, University of Kentucky, Lexington, KY, USA*; PHILLIP SHERIDAN, *Department of Chemistry and Biochemistry, Canisius College, Buffalo, NY, USA*.

The electronic spectra of the X_2BO and X_2BS free radicals have been studied by a combination of experimental and theoretical techniques. Experimentally, we have succeeded in preparing some of these species in a pulsed discharge jet and detecting them by laser-induced fluorescence and emission spectroscopy through the $\tilde{B}^2A_1 - \tilde{X}^2B_2$ transition. The radicals exhibit emission transitions down to the ground state and the low-lying $\tilde{A}^2B_1$ electronic state. We have also used high level *ab initio* theory [CCSD(T)/aug-cc-pV5Z] to calculate the properties of the ground and excited states and simulate the observed spectra. Experiment and theory agree that the radicals are planar, C_{2v} symmetry species in all the three combining states, with only small changes in geometry on electronic excitation.

MI07 3:12 – 3:27

OBSERVATION OF PURE ROTATIONAL SPECTRA OF SiCCN BY FOURIER-TRANSFORM MICROWAVE SPECTROSCOPY

HIROYA UMEKI, MASAKAZU NAKAJIMA, YASUKI ENDO, *Department of Basic Science, The University of Tokyo, Tokyo, Japan*.

Pure rotational spectra of SiCCN ($\tilde{X}\,^2\Pi_{3/2}$) have been observed using Fourier-transform microwave (FTMW) spectroscopy in the frequency region 13 to 35 GHz. The SiCCN radical was produced in a supersonic jet by discharging a mixture gas, 0.2% $SiCl_4$ and 0.2% CH_3CN diluted in Ar. The effective rotational constant $B_{eff,3/2}$, the centrifugal distortion constant D, and the hyperfine coupling constants, $a + (b + c)/2$ and eQq_0, were determined with a standard deviation of the fit to be 6 kHz. Determined B and eQq_0 are consistent with those derived from ab initio calculations. Λ-type doublings were not resolved for the observed spectra.

MI08 3:29 – 3:44

PERTURBATION FACILITATED OPTICAL OPTICAL DOUBLE RESONANCE INVESTIGATION OF THE QUINTET MANIFOLD OF C_2 BY APPLYING TWO-COLOR FOUR-WAVE MIXING

PETER BORNHAUSER, *General Energy, Paul Scherrer Institute, Villigen, Switzerland*; ROBERTO MARQUARDT, *Laboratoire de Chimie Quantique, Institut de Chimie, Université de Strasbourg, 67008 Strasbourg, France*; PETER RADI, *General Energy, Paul Scherrer Institute, Villigen, Switzerland*.

The potential of four-wave mixing spectroscopy for deperturbation studies has been demonstrated by an analysis of the spin-orbit and L-uncoupling interaction between the $d\,^3\Pi_g, v = 4$ and the $b\,^3\Sigma_g^-, v = 16$ states of C_2.[a] The double-resonance method provides unambiguous assignments of perturbed transitions by intermediate level labeling. Furthermore, the sensitivity of the method unveiled extra transitions that originate from the perturbing $b\,^3\Sigma_g^-, v = 16$ state. A following study[b] has successfully applied the method to deperturb the $d\,^3\Pi_g, v = 6$ state of the dicarbon and lead to the discovery of the first high-spin state of C_2. The energetically lowest quintet ($^5\Pi_g$) has been characterized by applying a conventional Hamiltonian. The detailed study unraveled major issues of the so-called high-pressure band of C_2 which were initially observed back in 1910[c] and later observed in numerous experimental environments.

In this work we take into account our recent studies on tri-carbon[d] where we used perturbation-facilitated two-color resonant four-wave mixing spectroscopy to access the (dark) triplet manifold of C_3 from the singlet $\tilde{X}\,^1\Sigma_g^+$ ground state *via* "gate-way" levels (i.e. singlet-triplet mixed levels). In a similar way, we performed for this study perturbation-facilitated optical-optical double-resonance experiments to access the first excited quintet state of C_2 *via* "gate-way states" in the perturbed $d\,^3\Pi_g, v = 6$. The newly found $^5\Pi_u$ state is characterized at rotational resolution by performing a least-squares fit of the observed transitions to a $^5\Pi_u$ - $^5\Pi_g$ Hamiltionian. The work represents a rare case of a successful analysis of a quintet manifold of a molecule exhibiting a singlet ground state ($^1\Sigma_g^+$).

---

[a] P. Bornhauser, G. Knopp, T. Gerber, and P.P. Radi, Journal of Molecular Spectroscopy 262, 69 (2010).

[b] P. Bornhauser, Y. Sych, G. Knopp, T. Gerber, and P.P. Radi, J. Chem. Phys. 134, 044302 (2011).

[c] A. Fowler, Monthly Notices of the Royal Astronomical Society 70, 484 (1910).

[d] Y. Sych, P. Bornhauser, G. Knopp, Y. Liu, T. Gerber, R. Marquardt, and P.P. Radi, J. Chem. Phys. 139, 154203 (2013).

66

MI09   **3:46 – 4:01**

ANALYSIS OF BANDS OF THE 405 nm ELECTRONIC TRANSITION OF C_3Ar

YEN-CHU HSU, YI-JEN WANG, ANTHONY MERER[a], *Institute of Atomic and Molecular Sciences, Academia Sinica, Taipei, Taiwan.*

Bands of the C_3Ar complex can be observed near almost all the bands of the $\tilde{A}^1\Pi_u$ - $\tilde{X}^1\Sigma_g^+$ transition of C_3. The strongest bands of C_3Ar form close-lying pairs. Rotational analyses have been carried out for the bands at 25025 and 25029 cm^{-1}(near the 02^-0-000 band of C_3) and 25426 and 25430 cm^{-1}(near the 04^-0-000 band). Each pair consists of a type A and a type C band of an asymmetric top, where the upper states interact by b-axis Coriolis coupling; this represents the lifting of the degeneracy of the Π state in the lower symmetry of the complex. Only K = even lower state levels are found, showing that C_3Ar has the shape of a distorted letter T. The Ar atom lies 3.82 Å from the centre of mass of the C_3 part. Emission spectra have been recorded and lifetimes measured for several C_3Ar upper state levels. The assignment of the emission bands is complicated by significant intramolecular relaxation in the upper states, which populates mainly the lowest level of each local potential minimum of the upper state; however the variation of the upper state well depth (binding energy) with vibrational quantum number can then be determined.

---

[a]Also at Department of Chemistry, University of British Columbia, Vancouver, B.C., Canada

Intermission

MI10   **4:18 – 4:33**

VIBRONIC INTERACTION AND VIBRATIONAL ASSIGNMENT FOR NO_3 IN THE GROUND ELECTRONIC STATE

EIZI HIROTA, *The Central Office, The Graduate University for Advanced Studies, Hayama, Kanagawa, Japan.*

The strongest IR band of NO_3 appears at 1492 cm^{-1}and has been assigned traditionally to the N-O degenerate stretching ν_3 mode. In 2007 Stanton proposed the ν_3 to be about 500 cm^{-1}lower, i.e. it is located around 1000 cm^{-1}, based on theoretical calculations. Jacox and collaborators supported this proposal, on the basis of their IR spectra observed in Ne matrix, and reassigned the 1492 cm^{-1}band to $\nu_3 + \nu_4$. The traditional vibrational assignment is referred to as Assignment I and the Stanton-Jacox one to as Assignment II, and thus the upper state of the 1492 cm^{-1}band Z is ν_3 and $\nu_3 + \nu_4$ for Assignment I and II, respectively.

Kawaguchi, Ishiwata, and Hirota (KIH) have been making much effort to settle which assignment is correct, by observing and analyzing FTIR spectra. They thought in 2009 that the observation of hot bands from the in-plane ONO degenerate bending ν_4 state to the Z state will make it possible for KIH to select the correct assignment among the two. Namely for Assignment I only one hot band of E - E type (i.e. $\nu_3 - \nu_4$) will appear, whereas three bands for Assignment II: E - E, A_1 - E, and A_2 - E. It was straightforward to detect and assign the E - E type hot band, because the upper state is Z in common with that of the 1492 cm^{-1}band. After careful searching for the spectra, KIH arrived at a conclusion that there is only one A - E type hot band present, which is difficult to reconcile with Assignment II, and the observed A - E hot band is reasonably ascribed to $2\nu_2 - \nu_4$ in Assignment I.

The NO_3 radical has been thought to be subjected to strong vibronic interaction. This view originated from an anomalous ν_4 progression appearing in the NO_3^- photoelectron spectra by Neumark et al.; they explained this observation in terms of Herzberg-Teller (H-T) effect with a sizable interaction parameter. However, KIH did not observe any anomalous features in the ν_4 vibration-rotation structure, which would be caused by the huge H-T perturbation as presumed by Neumark. Hirota found that the ν_4 progression of Neumark can be explained by the coupling of the unpaired electron orbital angular momentum with that of the ν_4 mode in the ground electronic state.

MI11 4:35 – 4:50

VIBRONIC ANALYSIS OF THE $\tilde{A}^2 E''$ STATE OF NO_3 RADICAL

TERRANCE JOSEPH CODD, *Department of Chemistry and Biochemistry, The Ohio State University, Columbus, OH, USA*; JOHN F. STANTON, *Department of Chemistry, The University of Texas, Austin, TX, USA*; TERRY A. MILLER, *Department of Chemistry and Biochemistry, The Ohio State University, Columbus, OH, USA*.

A moderate resolution spectrum of the $\tilde{A}^2 E''$ state of the NO_3 radical has been obtained using jet-cooled cavity ringdown spectroscopy. The analysis of the vibronic structure of this spectrum has been undertaken using a quadratic Jahn-Teller Hamiltonian. All observed transitions have been assigned and unperturbed frequencies and Jahn-Teller coupling constants were determined by fitting the experimental spectrum using nonlinear least squares regression. We find evidence of strong Jahn-Teller coupling in this electronic state, particularly in the ν_3 mode. These results are compared to electronic structure calculations including terms up to fourth order. Calculated and experimental spectra are presented and critically compared.

MI12 4:52 – 5:07

ROVIBRONIC ANALYSIS OF THE e' BANDS IN THE $\tilde{A}^2 E''$ STATE OF NO_3 RADICAL

HENRY TRAN, TERRANCE JOSEPH CODD, DMITRY G. MELNIK, MOURAD ROUDJANE, TERRY A. MILLER, *Department of Chemistry and Biochemistry, The Ohio State University, Columbus, OH, USA*.

The vibronic structure of the NO_3 radical has been the subject of much recent research in our group.[a] We have also collected several high resolution spectra of transitions to the $\tilde{A}^2 E''$ state. Parallel bands, with a_1'' symmetry, have been satisfactorily fit using an oblate symmetric top Hamiltonian with spin rotation. Some lines were seen to be perturbed and it is likely that this is the result of random perturbations from levels originating from the ground electronic state. The perpendicular bands, which have e' symmetry, are not satisfactorily described using this Hamiltonian. In particular the rotational structure of the e' levels has many more transitions than in the oblate top model predicts. For this reason we have developed two different rovibronic Hamiltonians for the analysis of the vibronically degenerate levels. Both include spin-orbit, coriolis, spin-rotation, and Jahn-Teller distortion terms. However, they are derived starting from two different limiting cases. In Case 1 the Hamiltonian is built by assuming first a D_{3h} configuration and then perturbations are added. Case 2 starts at the statically distorted, low symmetry geometry and introduces interactions among the vibronic levels. In the case of Jahn-Teller coupling that is neither very weak nor very strong these models should both adequately describe the observed spectra. These models and preliminary analysis of several e' bands are presented.

---

[a]Codd, T. et al. 68th Int. Symp. Molec. Spec. (2012)

MI13

ROTATIONALLY-RESOLVED HIGH-RESOLUTION LASER SPECTROSCOPY OF THE $B\ ^2E' \leftarrow X\ ^2A'_2$ TRANSITION OF $^{14}NO_3$ RADICAL

SHUNJI KASAHARA, *Molecular Photoscience Research Center, Kobe University, Kobe, Japan*; KOHEI TADA, *Graduate School of Science, Kobe University, Kobe, Japan*; TAKASHI ISHIWATA, *Information Sciences, Hiroshima City University, Hiroshima, Japan*; EIZI HIROTA, *The Central Office, The Graduate University for Advanced Studies, Hayama, Kanagawa, Japan.*

Rotationally-resolved high-resolution fluorescence excitation spectra and the Zeeman effects of the 662 nm band, which is called as the 0-0 band of the $B\ ^2E' \leftarrow X\ ^2A'_2$ electronic transition, of $^{14}NO_3$ have been observed. [a] Sub-Doppler excitation spectra were measured by crossing a single-mode laser beam perpendicular to a collimated radical beam, which was formed by the heat decomposition of $^{14}N_2O_5$; $^{14}N_2O_5 \rightarrow\ ^{14}NO_3 +\ ^{14}NO_2$. The typical linewidth was 30 MHz and the absolute wavenumber was calibrated with accuracy 0.0001 cm^{-1} by measurement of the Doppler-free saturation spectrum of iodine molecule and fringe pattern of the stabilized etalon. In the observed spectra, only the rotational line pairs from the $X\ ^2A'_2(v'' = 0, K'' = 0, N'' = 1, F_1$ and $F_2)$ levels are assigned, but the other rotational lines were not found yet. In this work, we expanded the measurement of the Zeeman splittings for the other rotational lines, which are predicted their position by using the combination differences calculated from the reported molecular constants. [b] From the observed Zeeman patterns, we have assigned unambiguously several transition lines for the 0-0 band. Additionally, we have measured the rotationally-resolved high-resolution spectra of vibrational excited levels of the $B\ ^2E'$ state, which lies 770 cm^{-1} and 948 cm^{-1} above the 0-0 band, and found many tiny rotational lines in these vibronic bands.

[a]K. Tada, W. Kashihara, S. Kasahara, M. Baba, T. Ishiwata, and E. Hirota, *The 68th OSU Symposium*, WJ04 (2013).
[b]R. Fujimori, N. Shimiza, J. Tang, T. Ishiwata, and K. Kawaguchi, *J. Mol. Spectrosc.*, **283**, 10 (2013).

MI14

ROTATIONALLY-RESOLVED HIGH-RESOLUTION LASER SPECTROSCOPY OF THE $B\ ^2E' \leftarrow X\ ^2A'_2$ TRANSITION OF $^{15}NO_3$ RADICAL

KOHEI TADA, *Graduate School of Science, Kobe University, Kobe, Japan*; SHUNJI KASAHARA, *Molecular Photoscience Research Center, Kobe University, Kobe, Japan*; TAKASHI ISHIWATA, *Information Sciences, Hiroshima City University, Hiroshima, Japan*; EIZI HIROTA, *The Central Office, The Graduate University for Advanced Studies, Hayama, Kanagawa, Japan.*

Nitrate radical (NO_3) has two electronic excited states: $A\ ^2E''$ and $B\ ^2E'$ near the electronic ground state: $X\ ^2A'_2$. These three electronic states can vibronically interact each other. Therefore, NO_3 is one of the great subjects for understanding intramolecular interactions in polyatomic radicals. High-resolution fluorescence excitation spectrum and its magnetic effect of the 662 nm band, which is assigned as the 0 - 0 band of the $B\ ^2E' \leftarrow X\ ^2A'_2$ transition, of $^{14}NO_3$ have been observed.[a] The observed $^{14}NO_3$ spectrum was too complicated to be analyzed rotationally because of less rotational regularity. In this work, we observed high-resolution fluorescence excitation spectrum of the 662 nm band of $^{15}NO_3$. The observed region was 15080 - 15103 cm^{-1}. We also observed the Zeeman splitting of intense rotational lines for unambiguous rotational assignment. A part of the observed rotational lines was successfully assigned by using ground state combination differences calculated from the reported molecular constants[b] and the observed Zeeman patterns. The effective molecular constants of the excited state were determined under the oblate symmetric-top model.

[a]K. Tada, W. Kashihara, S. Kasahara, M. Baba, T. Ishiwata, and E. Hirota, The 68th OSU Symposium, WJ04 (2013).
[b]R. Fujimori, N. Shimizu, J. Tang, T. Ishiwata, and K. Kawaguchi, J. Mol. Spectrosc., 283, 10 (2013).

MI15

MULTISTATE VIBRONIC HAMILTONIAN FOR THE NITRATE RADICAL

JOHN F. STANTON, *Department of Chemistry, The University of Texas, Austin, TX, USA.*

The quasidiabatic model of Köppel, Domcke and Cederabaum has proven to be a powerful tool for computing intensities and level positions in electronic spectra involving strong vibronic coupling effects. The model has been used, as parametrized by very high-level equation-of-motion coupled cluster (EOM-CC) calcaultions, is demonstrated by applications to the NO_3 radical.

MI16   6:00 – 6:15

AB INITIO CALCULATION FOR THE SPIN-ORBIT SPLITTINGS OF THE NITRATE RADICAL (NO_3)

LAN CHENG, <u>JOHN F. STANTON</u>, *Department of Chemistry, The University of Texas, Austin, TX, USA.*

In this work we present a quantum-chemical calculation of the electronic spin-orbit splittings for the spatially degenerate low-lying E' and E'' states of the nitrate radical (NO_3). The calculation is based on a degenerate perturbation theory using scalar-relativistic equation of motion ionization potential (EOMIP) coupled-cluster singles and doubles (CCSD) wave functions together with spin-orbit matrix elements constructed in the framework of exact two-component (X2C) theory. The computed SO splittings are discussed in view of recent experiments by Babu et al. for the $B\,^2E'$ state and Miller et al. for the X^2A_2' state. Calculations of splittings based on vibronic, rather than simply electronic, wave functions are also discussed.

MI17   *Post-Deadline Abstract*   6:17 – 6:27

FOUR WAVE MIXING SPECTROSCOPY OF THE NO_3 $\tilde{B}\,^2E' - \tilde{X}\,^2A_2'$ transition

<u>MASARU FUKUSHIMA</u>, TAKASHI ISHIWATA, *Information Sciences, Hiroshima City University, Hiroshima, Japan.*

The $\tilde{B}\,^2E' - \tilde{X}\,^2A_2'$ electronic transition of NO_3 generated in a supersonic free jet expansion was investigated by four wave mixing (4WM) spectroscopy. The degenerated 4WM and laser induced fluorescence (LIF) spectra around the 0_0^0 band region were measured simultaneously. The D4WM spectrum shows broad band features for the 0_0^0 band similar to that of the LIF spectrum. The broad 0_0^0 band does not consist of one sub-band, but of several bands. The intensity distribution of the sub-bands of the D4WM spectrum is similar, but not identical to that of the LIF spectrum.

MJ. Analytical, Combustion, Plasma
Monday, June 16, 2014 – 1:30 PM
Room: 274 Medical Sciences Building

Chair: Albert Ratner, The University of Iowa, Iowa City, IA, USA

MJ01   1:30 – 1:45

HEADSPACE ANALYSIS OF VOLATILE COMPOUNDS USING SEGEMENTED CHIRPED-PULSE FOURIER TRANSFORM MM-WAVE SPECTROSCOPY

BRENT HARRIS, AMANDA STEBER, <u>BROOKS PATE</u>, *Department of Chemistry, The University of Virginia, Charlottesville, VA, USA.*

A chirped-pulse Fourier transform mm-wave spectrometer has been tested in analytical chemistry applications of headspace analysis of volatile species. A solid-state mm-wave light source (260-290 GHz) provides 30-50 mW of power. This power is sufficient to achieve optimal excitation of individual transitions of molecules with dipole moments larger than about 0.1 D. The chirped-pulse spectrometer has near 100% measurement duty cycle using a high-speed digitizer (4 GS/s) with signal accumulation in an FPGA. The combination of the ability to perform optimal pulse excitation and near 100% measurement duty cycle gives a spectrometer that is fully optimized for trace detection. The performance of the instrument is tested using an EPA sample (EPA VOC Mix 6 – Supelco) that contains a set of molecules that are fast eluting on gas chromatographs and, as a result, present analysis challenges to mass spectrometry. The ability to directly analyze the VOC mixture is tested by acquiring the full bandwidth (260-290 GHz) spectrum in a "high dynamic range" measurement mode that minimizes spurious spectrometer responses. The high-resolution of molecular rotational spectroscopy makes it easy to analyze this mixture without the need for chemical separation. The sensitivity of the instrument for individual molecule detection, where a single transition is polarized by the excitation pulse, is also tested. Detection limits in water will be reported. In the case of chloromethane, the detection limit (0.1 microgram/L), matches the sensitivity reported in the EPA measurement protocol (EPA Method 524) for GC/MS.

MJ02   1:47 – 2:02

HIGH SPEED, ULTRASENSITIVE TRACE GAS SENSING

<u>DAVID A. LONG</u>, ADAM J. FLEISHER, *Material Measurement Laboratory, National Institute of Standards and Technology, Gaithersburg, MD, USA*; DAVID F. PLUSQUELLIC, *Physical Measurement Laboratory, National Institute of Standards and Technology, Boulder, CO, USA*; JOSEPH HODGES, *Material Measurement Laboratory, National Institute of Standards and Technology, Gaithersburg, MD, USA.*

I will describe a variety of cavity-enhanced spectroscopic techniques, including frequency-agile rapid scanning spectroscopy (FARS) and heterodyne-detected cavity ring-down spectroscopy (HD-CRDS), which we have recently developed for rapid, ultrasensitive absorption measurements. Scanning rates that are limited only by the cavity response time itself as well as noise-equivalent detection limits as low as 6×10^{-14} cm^{-1} Hz$^{-1/2}$ have been achieved. I will discuss the application of these techniques to current problems in atmospheric science including recent infrared measurements of analytes which are present at ultra-trace concentrations.

MJ03   2:04 – 2:19

CHEMICAL ANALYSIS OF EXHALED HUMAN BREATH USING HIGH RESOLUTION MM-WAVE ROTATIONAL SPECTRA

<u>TIANLE GUO</u>, DANIELA BRANCO, JESSICA THOMAS, IVAN MEDVEDEV, *Department of Physics, Wright State University, Dayton, OH, USA*; DAVID DOLSON, *Department of Chemistry, Wright State University, Dayton, OH, USA*; HYUN-JOO NAM, *Department of Bioengineering, University of Texas at Dallas, Dallas, TX, USA*; KENNETH O, *Electrical Engineering, University of Texas at Dallas, Dallas, TX, USA.*

High resolution rotational spectroscopy enables chemical sensors that are both sensitive and highly specific, which is well suited for analysis of expired human breath. We have previously reported on detection of breath ethanol, methanol, acetone, and acetaldehyde using THz sensors[a]. This paper will outline our present efforts in this area, with specific focus on our ongoing quest to correlate levels of blood glucose with concentrations of a few breath chemicals known to be affected by elevated blood sugar levels. Prospects, challenges and future plans will be outlined and discussed.

---

[a]Fosnight, A.M., B.L. Moran, and I.R. Medvedev, Chemical analysis of exhaled human breath using a terahertz spectroscopic approach. Applied Physics Letters, 2013. 103(13): p. 133703-5.

MJ04

THZ/MM-WAVE SPECTROSCOPIC SENSORS, CATALOGS, AND UNCATALOGUED LINES

IVAN MEDVEDEV, *Department of Physics, Wright State University, Dayton, OH, USA*; CHRISTOPHER F. NEESE, FRANK C. DE LUCIA, *Department of Physics, The Ohio State University, Columbus, OH, USA.*

Analytical chemical sensing based on high resolution rotational molecular spectra has been recognized as a viable technique for decades. We recently demonstrated a compact implementation of such a sensor [a]. Future generations of these sensors will rely on automated algorithms for quantification of chemical dilutions based on their spectral libraries, as well as identification of spectral features not present in spectral catalogs. Here we present an algorithm aimed at detection of unidentified lines in complex molecular species based on spectroscopic libraries developed in our previous projects. We will discuss the approaches suitable for data mining in feature-rich rotational molecular spectra.

[a]Neese, C.F., I.R. Medvedev, G.M. Plummer, A.J. Frank, C.D. Ball, and F.C. De Lucia, "A Compact Submillimeter/Terahertz Gas Sensor with Efficient Gas Collection, Preconcentration, and ppt Sensitivity." Sensors Journal, IEEE, 2012. 12(8): p. 2565-2574

MJ05

IDENTIFICATION OF FORGED BANK OF ENGLAND 20 GBP BANKNOTES USING IR SPECTROSCOPY

EMILY SONNEX, *Department of Chemistry, University of Reading, Reading, United Kingdom.*

Bank of England notes of 20 GBP denomination have been studied using infrared spectroscopy in order to generate a method to identify forged notes. A principal aim of this work was to develop a method so that a small, compact ATR FTIR instrument could be used by bank workers, police departments or others such as shop assistants to identify forged notes in a non-lab setting. The ease of use of the instrument is the key to this method, as well as the relatively low cost. The presence of a peak at 1400 cm^{-1} from the blank paper section of a forged note proved to be a successful indicator of the note's illegality for the notes that we studied. Moreover, differences between the spectra of forged and genuine 20 GBP notes were observed in the ν(OH) (ca. 3500 cm^{-1}), ν(C-H) (ca. 2900 cm^{-1}) and ν(C=O) (ca. 1750 cm^{-1}) regions of the IR spectrum recorded for the polymer film covering the holographic strip. In cases where these simple tests fail, we have shown how an infrared microscope can be used to further differentiate genuine and forged banknotes by producing infrared maps of selected areas of the note contrasting inks with background paper. Further to this, with an announcement by the Bank of England to produce polymer banknotes in the future, the work has been extended using Australian polymer banknotes to show that the method would be transferable.

72

MJ06   2:55 – 3:10

INTERFACIAL PROCESSES IN MODEL LITHIUM ION SYSTEMS PROBED WITH VIBRATIONAL SUM FREQUENCY GENERATION SPECTROSCOPY

BRUNO G NICOLAU, NATALIA GARCIA REY, DANA DLOTT, *Department of Chemistry, University of Illinois at Urbana-Champaign, Urbana, IL, USA.*

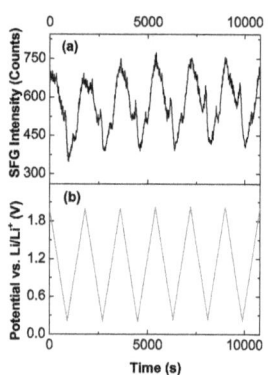

Vibrational sum frequency generation (SFG) spectroscopy was used to probe electrochemical processes taking place at the interface between metal anodes and the liquid phase in model lithium ion systems.

Lithium ion batteries have been extensively studied and characterized by numerous techniques. However, the mechanisms behind many properties are still unclear due to the lack of techniques that can directly probe them in situ. The formation of the electrode passivating layer known as solid-electrolyte interphase (SEI) is one such example. During the first charging cycle of a battery, some of the electrolyte undergoes reduction at the electrode surface forming an electrically isolating barrier that prevents the subsequent reduction of more electrolyte molecules.

The SFG selection rules suppress signals from molecules in centrosymmetric environments such as electrolyte layers, so SFG is a selective probe of interfacial environments such as the SEI.

In this study, ethylene carbonate's (EC) response to potential cycling was observed. EC is commonly used as a high permittivity solvent in batteries and is widely believed to be the main component of the SEI in its reduced form, lithium ethylene dicarbonyl. EC's carbonyl stretch (1850 cm^{-1}) was measured in conjunction with cyclic voltammetry experiments. The SFG intensity showed remarkable agreement with the changing potential, as seen in the figure below. The shoulders on each side of the peaks in (a) are especially interesting, as they correspond to the potentials where lithium metal is oxidized and reduced. Vibrational modes found at 1300-1400 cm^{-1}, usually assigned to the reduced form of EC, are also being studied in order to provide more information on the nature of the SEI.

MJ07   3:12 – 3:27

DISTINCTIONS IN THE RAMAN SPECTROSCOPY FEATURES OF WO_3 MATERIALS WITH INCREASING TEMPERATURE

RAUL F GARCIA-SANCHEZ, PRABHAKAR MISRA, *Department of Physics and Astronomy, Howard University, Washington, DC, USA.*

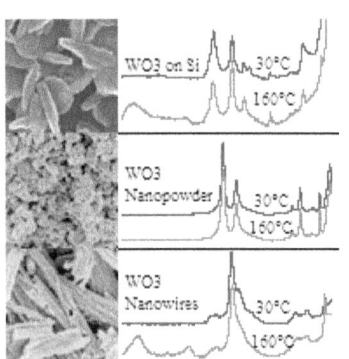

Metal oxides are widely used in gas sensor applications due to their low cost, easy production and selectivity. Tungsten Oxide (WO_3) is one of the most used metal oxides in the detection of Nitrogen gases (NO_x). The purpose of this research is to determine if the Raman features of a metal oxide gas sensor can serve as tools to make estimates regarding the sensor capabilities related to the target gases. This research will be used for gas sensing of oxidizing/reducing toxic gases (i.e. H_2S, NO_x, SO_2, etc.) and finding the effect that temperature, gas concentration, type of gas, exposure time and other variables have on the Raman spectra of metal oxides. In this experiment, the temperature was increased from $30 - 160\,^\circ\text{C}$ and the Raman data was taken using a 780 nm infrared laser. In two of the samples, WO_3 on Silicon substrate and WO_3 nanopowder, we found vibrational modes at 807, 716 and 271 cm^{-1}, which are indicators of a monoclinic WO_3 structure. The WO_3 nanowires samples exhibit the O-W-O bond stretching feature is present and asymmetric stretching of the W-O bonds occurs, resulting in a $750\,\text{cm}^{-1}$ band. The intensity of Raman features such as $750\,\text{cm}^{-1}$ for nanowires and 492 and 670 cm^{-1} for WO_3 on Silicon substrate begins to decay as temperature increases. Additionally, the vibrational modes related to O-H and W-OH become more pronounced as temperature increases due to those bonds reacting more strongly to the temperature change than the normal W-O bonds related to the original lattice structure. Finally, all samples have low-frequency phonon mode markers associated with temperature change, and in most cases these change as temperature increases. The understanding of the thermal effects will help develop theoretical models for the identification of specific metal oxide-gas relationships and provide a supplemental way of observing gas adsorption in addition to current conductivity measurements.

Intermission

MJ08 3:44 – 3:59

PROBING THIN FILMS AND MONOLAYERS ON GOLD WITH LARGE AMPLITUDE TEMPERATURE JUMPS

YUXIAO SUN, CHRISTOPHER M BERG, DANA DLOTT, *Department of Chemistry, University of Illinois at Urbana-Champaign, Urbana, IL, USA.*

A methodology to probe localized vibrational transitions of self-assembled monolayers (SAMs) adsorbed on gold films using vibrational sum-frequency generation (SFG) is described. The gold film is subjected to heating from a 400nm pump laser, exposing the adsorbed molecules to a temperature jump in the 30-175 ° K range, calibrated using ultrafast reflectance measurements of the gold compared to steady state oven heating . SAMs of alkyl thiols as well as nitro functionalized aryl thiols were deposited and temperature jumped while be observed with SFG, monitoring the symmetric and asymmetric methyl vibrations as well as nitro vibrations.

The amplitude, center, and width of the transitions were measured and provide information about delay and orientation of the molecules, as well as providing an indicator of the overall monolayer state. All transitions probed exhibited overshoot decay plateau patterns, attributed to a fast hot electron process directly exciting the probed transitions, followed by a slower bulk heating process causing monolayer disordering. This leads to a shift in the average angle of the terminal methyl, manifesting itself as a change in the amplitude of the vibration.

These techniques will be applied to thin films of energetic materials to study reactions to temperature jumps. HMX is known to have a peak in sensitivity as δ-HMX transitions to β-HMX at high temperatures, but fairly little information about the reason for this is known. This technique should be able to probe that process and provide data that can be used with computational models to gain some understanding of the process.

MJ09 4:01 – 4:16

USING LASER-DRIVEN FLYER-PLATES TO STUDY MECHANOCHEMICAL SHOCK WAVE ENERGY DISSIPATION

WILLIAM L. SHAW, DANA DLOTT, *Department of Chemistry, University of Illinois at Urbana-Champaign, Urbana, IL, USA.*

Mechanically induced chemical reactions that attenuate the energy of a shock wave have been investigated. To accomplish this, we have built a tabletop system for launching and monitoring laser-driven flyer-plates. Using a 2.5 J pump laser and a diffractive beam homogenizer we drive planar aluminum plates at speeds of up to 4 km/s. Flyer launch through impact is monitored using a photonic Doppler velocimiter (PDV) which probes the center 60 μm of the 700 μm diameter flyer-plate. Sustained 1D shock duration ranges from 4-20 ns. The aluminum thickness can be chosen as 25, 50, 75, or 100 μm thick. This results in a trade off between the maximum velocity we can achieve and the shock wave duration produced in the target material.

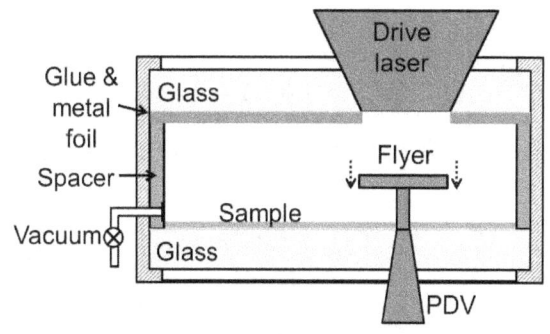

The laser-driven flyer-plates are used to drive mechanochemical reactions in materials specifically designed to undergo endothermic, negative volume reactions with the goal of attenuating the shock wave energy in a new paradigm for shock wave dissipation. Material reactions are probed *in situ* using high-speed diagnositics including time-resolved fluorescence spectroscopy captured by a streak camera. Recovered samples are analysed with UV/VIS, FT-IR, and fluorescence spectroscopy. Shock wave dissipation is quantified using our PDV by monitoring an embedded Au mirror behind the target compound.

MJ10 4:18–4:33

SIMULTANEOUS QUANTIFICATION OF OH AND HO$_2$ IN DIMETHYL ETHER OXIDATION USING FARADAY ROTATION SPECTROSCOPY

BRIAN BRUMFIELD, *Department of Electrical Engineering, Princeton University, Princeton, NJ, USA*; XUELIANG YANG, JOSEPH LEFKOWITZ, YIGUANG JU, *Department of Mechanical and Aerospace Engineering, Princeton University , Princeton , NJ, USA*; GERARD WYSOCKI, *Department of Electrical Engineering, Princeton University, Princeton, NJ, USA.*

OH and HO$_2$ are key radical species that control the autoignition and flame chemistry of fuels. Quantification of these radicals in the low-temperature oxidation of fuels is challenging due to their low concentrations. Strong spectral interference from more abundant non-radical species can further complicate accurate quantification of OH and HO$_2$. Faraday Rotation Spectroscopy (FRS), a laser-based diagnostic that exploits magneto-optical properties of paramagnetic radical species, can overcome these technical challenges to provide sensitive and selective *in situ* quantification of radicals.

Previously we have been able to illustrate the strengths of FRS in quantification of HO$_2$ radicals in the low-temperature oxidation of dimethyl ether[a,b,c]. Recently we have constructed a dual-wavelength FRS system capable of simultaneous *in situ* measurement of OH and HO$_2$. A DFB diode laser operating at 2.8 μm is used to target the Q(1.5e) and Q(1.5f) transitions in the fundamental vibrational band of the $^2\Pi_{3/2}$ ground electronic state of OH. An EC-QCL operating at 7.1 μm is used to target a Q-branch spectral feature in the ν_2 vibrational band of HO$_2$. Concentrations of the target species are extracted from the measured spectra through fitting of an FRS spectral model. Based on preliminary retrievals, 3σ detection limits of <1 ppmv for OH and HO$_2$ have been estimated from the non-linear least-squares fitting results. In this talk I will discuss the application of dual-wavelength FRS for sensitive measurement of OH and HO$_2$ radicals generated by oxidation of dimethyl ether in a flow reactor over a 520 K - 1050 K temperature range.

[a]B. Brumfield et al., *J. Phys. Chem. Lett.*, **4**, 872 (2013).
[b]B. Brumfield et al., "Dual Modulation Faraday Rotation Spectroscopy of HO$_2$ in a Flow Reactor" *Accepted in Optics Letters* (2014).
[c]N. Kurimoto et al., "Quantitative Measurements of HO$_2$ / H$_2$O$_2$ and Intermediate Species in Low and Intermediate Temperature Oxidation of Dimethyl Ether", *Submitted to The 35th International Symposium on Combustion.*

MJ11 4:35–4:50

CHIRPED PROBE PULSE FEMTOSECOND COHERENT ANTI-STOKES RAMAN SCATTERING FOR TURBULENT COMBUSTION DIAGNOSTICS

CLARESTA N. FINEMAN, ROBERT P. LUCHT, *Mechanical Engineering, Purdue University, West Lafayette, IN, USA.*

Chirped probe pulse (CPP) femtosecond (fs) coherent anti-Stokes Raman scattering (CARS) thermometry at 5 kHz has been successfully applied for single-laser-shot flame temperature measurements in a mildly turbulent hydrogen-air jet diffusion flame, sooting methane-air jet diffusion flame, and most recently a turbulent combustor of practical interest. Measurements were performed at various heights and radial locations within each flame and resulted in temperatures ranging from 300 K to 2400 K. In the turbulent combustor every laser shot produced some resonant CARS signal; no loss of signal due to beam steering, pressure fluctuations, or shear layer density gradients was noticeable. Furthermore, the measurement volume spatial resolution is better than has previously been reported for other CARS experiments. Flame temperature measurements compare well with those previously reported in similar flames. These results indicate high repetition rate CPP fs-CARS is an excellent technique for the study of turbulent combustion.

MJ12

FTIR ANALYSIS OF FLOWING AFTERGLOW FROM A HIGH-FREQUENCY SPARK DISCHARGE

ALLEN WHITE, *Department of Mechanical Engineering, Rose-Hulman Institute of Technology, Terre Haute, IN, USA*; GARY M HIEFTJE, STEVE RAY, KEVIN PFEUFFER, *Department of Chemistry, Indiana University, Bloomington, IN, USA.*

Plasmas are often used as ionization sources for ambient mass spectrometry (AMS). Here, the flowing afterglow of a novel high-energy spark discharge system, operated in nitrogen at high repetition rates, is investigated as a source for AMS. The spark discharge here is the same as that of an automobile ignition circuit.Combustion in automobile engines is initiated by a spark ignition system that is designed to deliver short-duration,high-voltage sparks to multiple engine cylinders. The arrangement utilized in this study is a modified discharge configuration designed to produce similarly short-duration, high-voltage discharges. It consists of an automotive ignition coil that is activated by a spark initiation circuit that discharges in turn into a cell with neutral gas input flow and ultimately into the collection orifice of a mass spectrometer. The discharge voltage is approximately 40kV at 800 Hz. High-frequency spark discharges in a nitrogen flow produce reagent ions such as NO+. In order to better evaluate the effectiveness of the discharge in producing reagent ions, an FTIR is utilized to measure IR active species such as nitric oxide, hydroxide, ozone, and water in the afterglow of the spark discharge during variation of discharge parameters. Time-resolved IR emission spectra provide additional insight into the reagent ion production mechanisms.

MJ13

EXPERIMENTAL EXAMINATION OF THE THERMOACOUSTIC INSTABILITY OF A LOW SWIRL FLAME WITH PLANAR LASER INDUCED FLUORESCENCE OF OH

JIANAN ZHANG, KELSEY KAUFMAN, ALBERT RATNER, *Mechanical and Industrial Engineering, The University of Iowa, Iowa city, IA, USA.*

Thermoacoustic combustion instability results from the coupling between oscillating heat release and fluctuating pressure inside of a combustion chamber. In the current work, thermoacoustic instability in a low swirl burner is investigated for lean premixed conditions. Measurement of the heat release is a very important aspect of thermoacoustic instability, and in the current experiment the local heat release information is captured with a method based on Planar Laser Induced Fluorescence of the OH radical (OH-PLIF). This is then combined with the pressure signal to quantify the level of thermal-acoustic coupling. The specific goal is to examine the global and local flame response to velocity (5 – 10 m/s) and driving pressure amplitude (up to 1.12% of atmospheric pressure) changes. The root mean square of a non-dimensional Rayleigh index (R_{RMS}) was analyzed as the indicator of the global response of flame to acoustic perturbation with different amplitudes. The result shows that the coupling level increases with the forcing amplitude in the beginning. However, when the forcing level is high enough, the coupling saturates. Local response is also examined using a locally-weighted R_{RMS}, focusing on the contribution of the positive and negative coupling regions to the global response. At low velocities, the positive and negative structures play similar roles. However, as velocity is increased, the positive structures become more dominant.

MJ14   5:26 – 5:41

ABSORPTION SPECTROSCOPY IN THE 4.4-4.6 μm INFRARED WAVELENGTH RANGE FOR THE 10 KHZ HIGH-SPEED MEASUREMENT OF CO AND CO_2 CONCENTRATIONS IN COMBUSTING ENVIRONMENTS.

MATTHEW L. FOTIA, *Advanced Concepts Group - Combustion Branch, Air Force Research Laboratory, Wright-Patterson AFB, OH, USA*; BRIAN C. SELL, JOHN HOKE, *Innovative Scientific Solutions Inc., Dayton, OH, USA*; FRED SCHAUER, *Advanced Concepts Group - Combustion Branch, Air Force Research Laboratory, Wright-Patterson AFB, OH, USA*.

An instrument has been developed to make 10 kHz *in situ* combustion gas measurements of carbon monoxide (CO) and carbon dioxide (CO_2) concentrations. Operating in both the 4.40 and 4.58 μm wavelength ranges allows for the fundamental molecular absorption bands of both molecules to be utilized.

Such concentration measurements allow for the determination of total combustion efficiency of a particular process, which has engineering implications when considering the energy available from a combustion process to be utilized for propulsion purposes.

A brief discussion of the initial calibration of the sensor with a calibrated diffusion flame, Hencken burner, and pressure-concentration cell is made with the main focus of the current work being the application of the instrument to examine the structure of propagating detonation waves.

MJ15   5:43 – 5:58

SPECTROSCOPIC STUDIES OF A LOW-TEMPERATURE ATMOSPHERIC PLASMOID ANALOGOUS TO BALL LIGHTNING

SCOTT E. DUBOWSKY, DAVID M. FRIDAY, KEVIN C. PETERS, RICHARD H. PERRY, *Department of Chemistry, University of Illinois at Urbana-Champaign, Urbana, IL, USA*; ZHANGJI ZHAO, *Department of Physics, University of Illinois at Urbana-Champaign, Urbana, IL, USA*; BRADLEY DEUTSCH, ROHIT BHARGAVA, *Department of Chemical and Biomolecular Engineering, University of Illinois at Urbana-Champaign, Urbana, IL, USA*; JUI-NUNG LIU, *Department of Electrical and Computer Engineering, University of Illinois at Urbana-Champaign, Urbana, IL, USA*; BENJAMIN J. McCALL, *Departments of Chemistry and Astronomy, University of Illinois at Urbana-Champaign, Urbana, IL, USA*.

Atmospheric-pressure, low-temperature plasmas exist in nature in the form of ball lightning, and last year a natural ball lightning event was finally observed with scientific equipment.[a] Production of ball lightning in the laboratory dates back to Tesla's work at Colorado Springs.[b] Today, Tesla's "fireballs" are easily produced in the laboratory by discharging kiloJoules of energy slightly above an electrolyte solution via a metal electrode.[c] For the sake of clarity, those plasmas produced using this technique are referred to as "plasmoids." Valuable information is obtained from previous experiments, such as the identification of water clusters and the temperature of the interior of plasmoids.[c] We perform mass spectrometry and Fourier-transform infrared emission spectroscopy in an effort to characterize these plasmoids. We present, to our knowledge, the first mass spectrometric data and infrared emission spectra of plasmoid discharges. Mass spectrometry reveals the presence of small protonated water clusters [$H^+(H_2O)_2$, $H^+(H_2O)_3$] and nitrogen-containing molecules [NO^+, NO^+-H_2O]. IR spectra exhibit signals observed in the water emission region (1300-2000 cm^{-1}, 3000-4000 cm^{-1}), and signals in several other regions of interest. Fundamental properties of these plasmoids including the electron energy distribution function, component densities, and collisional cross sections will be discussed.

[a]Cen, J.; Yuan, P.; Xue, S. *Phys. Rev. Lett.* **2014**, *112*, 035001
[b]Tesla, N. *Colorado Springs Notes 1899-1900*; Marinčić, A., Ed.; Nolit: Beograd, Yugoslavia, **1978**; pp 368-370
[c]Friday, D.M.; Broughton, P.B.; Lee, T.A.; Schutz, G.A.; Betz, J.N.; Lindsay, C.M. *J. Phys. Chem. A* **2013**, *117* (**39**), 9931-9940

MK. Linelists, Lineshapes, Collisions
Monday, June 16, 2014 – 1:30 PM
Room: 217 Noyes Laboratory

Chair: Keeyoon Sung, Jet Propulsion Laboratory/Caltech, Pasadena, CA, USA

MK01   1:30–1:45

ARE YOUR SPECTROSCOPIC DATA BEING USED?

IOULI E GORDON, LAURENCE S ROTHMAN, *Atomic and Molecular Physics, Harvard-Smithsonian Center for Astrophysics, Cambridge, MA, USA*; JONAS WILZEWSKI, *Physics, Technische Universität Dresden, Dresden, Germany.*

Spectroscopy is an established and indispensable tool in science, industry, agriculture, medicine, surveillance, etc.. The potential user of spectral data, which is not available in HITRAN[a] or other databases, searches the spectroscopy publications. After finding the desired publication, the user very often encounters the following problems: 1) They cannot find the data described in the paper. There can be many reasons for this: nothing is provided in the paper itself or supplementary material; the authors are not responding to any requests; the web links provided in the paper have long been broken; etc. 2) The data is presented in a reduced form, for instance through the fitted spectroscopic constants. While this is a long-standing practice among spectroscopists, there are numerous serious problems with this practice, such as users getting different energy and intensity values because of different representations of the solution to the Hamiltonian, or even just despairing of trying to generate usable line lists from the published constants.

Properly providing the data benefits not only users but also the authors of the spectroscopic research. We will show that this increases citations to the spectroscopy papers and visibility of the research groups. We will also address the quite common issue when researchers obtain the data, but do not feel that they have time, interest or resources to write an article describing it. There are modern tools that would allow one to make these data available to potential users and still get credit for it. However, this is a worst case scenario recommendation, i.e., publishing the data in a peer-reviewed journal is still the preferred way.

[a]L. S. Rothman, I. E. Gordon, et al. "The HITRAN 2012 molecular spectroscopic database," JQSRT 113, 4-50 (2013).

MK02   1:47–2:02

ROVIBRATIONAL LINE LISTS FOR NINE ISOTOPOLOGUES OF CO SUITABLE FOR MODELLING AND INTERPRETING SPECTRA AT VERY HIGH TEMPERATURES AND DIVERSE ENVIRONMENT

GANG LI, IOULI E GORDON, LAURENCE S ROTHMAN, *Atomic and Molecular Physics, Harvard-Smithsonian Center for Astrophysics, Cambridge, MA, USA*; YAN TAN, SHUI-MING HU, *Hefei National Laboratory for Physical Science at Microscale, University of Science and Technology of China, Hefei, China*; SAMIR KASSI, ALAIN CAMPARGUE, *UMR5588 LIPhy, Université Grenoble 1/CNRS, Saint Martin D'heres, France.*

In order to improve and extend existing HITRAN[a] and HITEMP[b] data for carbon monoxide, the ro-vibrational line lists were computed for all transitions of nine isotopologues of the CO molecule, namely $^{12}C^{16}O$, $^{12}C^{17}O$, $^{12}C^{18}O$, $^{13}C^{16}O$, $^{13}C^{17}O$, $^{13}C^{18}O$, $^{14}C^{16}O$, $^{14}C^{17}O$, and $^{14}C^{18}O$ in the electronic ground state up to $v = 41$ and $J = 150$. Line positions and intensity calculations were carried out using a newly-determined piece-wise dipole moment function (DMF) in conjunction with the wavefunctions calculated from a previous experimentally-determined potential energy function of Coxon and Hajigeorgiou[c]. Ab initio calculations and a direct-fit method which simultaneously fits all the reliable experimental ro-vibrational matrix elements has been used to construct the piecewise dipole moment function.
To provide additional input parameters into the fit, new Cavity Ring Down Spectroscopy experiments were carried out to enable measurements of the lines in the 4-0 band with low uncertainty (Grenoble) as well as the first measurements of lines in the 6-0 band (Hefei).
Accurate partition sums have been derived through direct summation for a temperature range from 1 to 9000 Kelvin. A complete set of broadening and shift parameters is also provided and now include parameters induced by CO_2 and H_2 to aid planetary applications.

[a]L. S. Rothman, I. E. Gordon, et al. "The HITRAN 2012 molecular spectroscopic database," JQSRT 113, 4-50 (2013).
[b]L. S. Rothman, I. E. Gordon, et al. "HITEMP, the high-temperature molecular spectroscopic database," JQSRT 111, 2139-2150 (2010).
[c]J. Coxon and P. Hajigeorgiou. "Direct potential fit analysis of the $X^1\Sigma^+$ ground state of CO," J. Chem. Phys. 121, 2992-3008 (2004).

78

MK03 2:04 – 2:19

CH$_4$, C$_2$H$_4$, SF$_6$ AND CF$_4$ CALCULATED SPECTROSCOPIC DATABASES FOR THE VIRTUAL ATOMIC AND MOLECULAR DATA CENTRE

<u>VINCENT BOUDON</u>, CHRISTIAN WENGER, ROMAIN SURLEAU, MAUD LOUVIOT, MBAYE FAYE, *Laboratoire ICB, CNRS/Université de Bourgogne, DIJON, France*; MAUD ROTGER, LUDOVIC DAU-MONT, DAVID A. BONHOMMEAU, VLADIMIR TYUTEREV, *Laboratoire GSMA, CNRS / Université de Reims Champagne-Ardenne, REIMS, France*; YAYE AWA BA, MARIE-LISE DUBERNET, *LERMA2, CNRS UMR8812, Observatoire de Paris, MEUDON, France.*

Two spectroscopic relational databases, denoted MeCaSDa and ECaSDa, have been implemented for methane and ethene, and included in VAMDC (Virtual Atomic and Molecular Data Centre, http://portal.vamdc.eu/vamdc_portal/home.seam) [1]. These databases collect calculated spectroscopic data from the accurate analyses previously performed for the electronic ground state of methane, ethene, and some of their isotopologues: ^{12}CH$_4$, ^{13}CH$_4$, and ^{12}C$_2$H$_4$ [2]. Both infrared absorption and Raman scattering lines are included. The polyad structures are reported and the transitions are precisely described by their energy, their intensity and the full description of the lower and upper states involved in the transitions.

Very recently, we also built on the same model two new databases, named SHeCaSDa and TFMeCaSDa for the SF$_6$ and CF$_4$ greenhouse gas molecules, respectively.

The relational schemas of these four databases are equivalent and optimized to enable the better compromise between data retrieval and compatibility with the XSAMS (XML Schema for Atoms, Molecules, and Solids) format adopted within the VAMDC European project.

[1] Y. A. Ba, Ch. Wenger *et al.*, *J. Quant. Spectrosc. Radiat. Transfer*, **130**, 62 (2013).

[2] M.-L. Dubernet, V. Boudon *et al.*, *J. Quant. Spectrosc. Radiat. Transfer*, **111**, 2151 (2010).

MK04 2:21 – 2:36

HOT EXPERIMENTAL ABSORPTION SPECTRA OF CH$_4$ IN THE PENTAD AND OCTAD REGION

<u>ROBERT J. HARGREAVES</u>, MICHAEL DULICK, PETER F. BERNATH, *Department of Chemistry and Biochemistry, Old Dominion University, Norfolk, VA, USA.*

We present comprehensive line lists of CH$_4$ at high temperatures for the pentad and octad region (2400–5000 cm^{-1}). These spectra improve on our previous emission measurements for this region[a] by using a new quartz sample cell in conjunction with a tube furnace (pictured). Ten temperatures have been recorded from room temperature up to 1000°C and our technique involves the acquisition of four separate Fourier transform infrared spectra at each temperature, thus accounting for both the emission and absorption of the molecule and the cell. By combining these four spectra we obtain true transmission spectra of hot CH$_4$ in this region. Analysis of this set of spectra enables the production of a line list that includes the position, intensity and empirical lower state energy. Our spectra and line lists can be used directly to model planetary atmospheres and brown dwarfs.

[a]Hargreaves, R.J., Beale, C.A., Michaux, L., Irfan, M., & Bernath, P.F. 2012, *ApJ*, 757, 46

MK05 2:38 – 2:53

PREDICTING ROTATION-VIBRATION LEVELS OF ISOTOPICALLY SUBSTITUTED MOLECULES: WATER AS AN EXAMPLE

OLEG POLYANSKY, *Department of Physics and Astronomy, University College London, London, IX, United Kingdom*; ALEKSANDRA KYUBERIS, *Microwave Spectroscopy, Institute of Applied Physics, Nizhny Novgorod, Russia*; LORENZO LODI, JONATHAN TENNYSON, *Department of Physics and Astronomy, University College London, London, IX, United Kingdom*; NIKOLAY FEDOROVICH ZOBOV, *Microwave Spectroscopy, Institute of Applied Physics, Nizhny Novgorod, Russia.*

We present the results of prediction of energy levels of various water isotopologues using the fit of mass independent part of potential energy surface (PES) of $H_2^{16}O$ only. Fit was done in such a way that mass dependent part of PES - adiabatic correction of $H_2^{16}O$ was used during the fit and also non-adiabatic mass-dependent correction to kinetic energy. When the predicted levels of other isotopologues have been calculated, the same mass independent surface as for $H_2^{16}O$ was used and both adiabatic and non-adiabatic corrections of the corresponding isotopologues have been employed. As a result the discrepancy between predicted and observed energy levels of $H_2^{17}O$ and $H_2^{18}O$ were almost of the same value as for $H_2^{16}O$ for the levels of $H_2^{17}O$ and $H_2^{18}O$ known experimentally. It is natural to suppose, that at least for the levels of $H_2^{16}O$ included in the fit, the same accuracy of prediction of $H_2^{17}O$ and $H_2^{18}O$ levels should be expected. Thus, the procedure transferring experimental knowledge of the major isotopologues to the minor isotopologues without the fitting of the levels of these minor isotopologues , has been developed. For other isotopologues such as $D_2^{16}O$ and HDO the discrepancy are less perfect, but still within about 0.1 cm^{-1}. These results provide us with a very accurate tool for the prediction of energy levels of minor isotopologues.

MK06 2:55 – 3:10

NEW HIGH PRECISION LINELIST OF H_3^+

JAMES N. HODGES, ADAM J. PERRY, CHARLES MARKUS, PAUL A JENKINS II, G. STEPHEN KOCHERIL, *Department of Chemistry, University of Illinois at Urbana-Champaign, Urbana, IL, USA*; BENJAMIN J. McCALL, *Departments of Chemistry and Astronomy, University of Illinois at Urbana-Champaign, Urbana, IL, USA.*

As the simplest polyatomic molecule, H_3^+ serves as an ideal benchmark for theoretical predictions of rovibrational energy levels. By strictly *ab initio* methods, the current accuracy of theoretical predictions is limited to an impressive one hundredth of a wavenumber,[a] which has been accomplished by consideration of relativistic, adiabatic, and non-adiabatic corrections to the Born-Oppenheimer PES. More accurate predictions rely on a treatment of quantum electrodynamic effects, which have improved the accuracies of vibrational transitions in molecular hydrogen to a few MHz.[b] High precision spectroscopy is of the utmost importance for extending the frontiers of *ab initio* calculations, as improved precision and accuracy enable more rigorous testing of calculations. Additionally, measuring rovibrational transitions of H_3^+ can be used to predict its forbidden rotational spectrum. Though the existing data can be used to determine rotational transition frequencies, the uncertainties are prohibitively large.[c] Acquisition of rovibrational spectra with smaller experimental uncertainty would enable a spectroscopic search for the rotational transitions.

The technique Noise Immune Cavity Enhanced Optical Heterodyne Velocity Modulation Spectroscopy, or NICE-OHVMS has been previously used to precisely and accurately measure transitions of H_3^+, CH_5^+, and HCO^+ to sub-MHz uncertainty.[d] A second module for our optical parametric oscillator has extended our instrument's frequency coverage from 3.2-3.9 μm to 2.5-3.9 μm. With extended coverage, we have improved our previous linelist by measuring additional transitions.

[a]O. L. Polyansky, *et al. Phil. Trans. R. Soc. A* (2012), **370**, 5014–5027.
[b]J. Komasa, *et al. J. Chem. Theor. Comp.* (2011), **7**, 3105–3115.
[c]C. M. Lindsay, B. J. McCall, *J. Mol. Spectrosc.* (2001), **210**, 66–83.
[d]J. N. Hodges, *et al. J. Chem. Phys.* (2013), **139**, 164201.

Intermission

MK07   3:27 – 3:37

PRECISE AND STABLE FREQUENCY SOURCE, AND MEASUREMENT OF ^{130}Te$_2$ REFERENCE LINES FROM 443 TO 451 NM

JAMES COKER, DAVID LA MANTIA, JOHN FURNEAUX, JEFFREY GILLEAN, *Homer L Dodge Department of Physics and Astronomy, University of Oklahoma, Norman, OK, USA.*

A precise, repeatable and stable optical frequency source is required for many modern spectroscopy experiments. Frequency combs have proven invaluable to many, but are not obtainable for others due to their high cost. Using a GPS disciplined oscillator, a stabilized Fabry-Pérot cavity, a relatively low-cost wavemeter and standard RF equipment, we have achieved a reliable laser system with a 10^{-9} or better frequency uncertainty at a fraction of the cost. With this system we have measured approximately 3000 transitions in ^{130}Te$_2$ continuously between 664 and 676 THz to $\sim$ 0.0001 cm^{-1} precision. The system is described in detail, and the possibility of improving our knowledge of the excited states of ^{130}Te$_2$ is considered.

MK08   3:39 – 3:49

ELECTRONIC TRANSITIONS ($BO_u^+ \leftarrow XO_g^+$) AND BANDHEAD FITTING FOR ^{130}Te$_2$ IN THE INFRARED

DAVID LA MANTIA, JAMES COKER, JOHN FURNEAUX, JEFFREY GILLEAN, *Homer L Dodge Department of Physics and Astronomy, University of Oklahoma, Norman, OK, USA.*

The electronic spectrum of ^{130}Te$_2$ serves as a frequency standard for many spectroscopic investigations. We scanned the molecule in the region 664 to 676 THz using a tunable diode laser to create an atlas of transition lines, in line with the previous investigations of *Cariou, et al.*[a] The $BO_u^+ \leftarrow XO_g^+$ transition was studied in great detail using the precise data for the X band from *Verges, et al.*[b] Using this data, the number of vibrational bandheads was identified. This allowed the rotational parameters B, D and H to be precisely obtained for each bandhead. These results are combined to obtain the appropriate spectroscopic parameters for the B_0 electronic band. The results of this investigation will be presented.

[a]Cariou, J. and Luc, P. "Atlas Du Spectre D'Absorption De La Molecule De Tellure." Laboratoire Aime, Cotton CNRS II 91405 Orsay, France. 1980.

[b]Verges, J. "The Laser Induced Fluorescence Spectrum of Te$_2$ Studied by Fourier Transformation Spectrometry. *Physica Scripta. Vol. 25, 338-350, 1982.*

MK09   3:51 – 4:06

NEW LINE LISTS FOR ROVIBRATIONAL AND ROTATIONAL TRANSITIONS WITHIN THE NH X$^3\Sigma^-$ AND OH X$^2\Pi$ GROUND STATES

JAMES S.A. BROOKE, *Department of Chemistry, University of York, York, United Kingdom*; PETER F. BERNATH, *Department of Chemistry and Biochemistry, Old Dominion University, Norfolk, VA, USA*; COLIN WESTERN, *School of Chemistry, University of Bristol, Bristol, United Kingdom*; GANG LI, *Atomic and Molecular Physics, Harvard-Smithsonian Center for Astrophysics, Cambridge, MA, USA*; GERRIT GROENENBOOM, *Institute for Molecules and Materials (IMM), Radboud University Nijmegen, Nijmegen, Netherlands.*

A new line list for rovibrational and rotational transitions within the NH X$^3\Sigma^-$ ground state has been created, including line intensities in the form of Einstein A and f-values, for all possible bands up to v'=6. The intensities are based on a new dipole moment function (DMF), which has been calculated using the internally contracted MRCI method with an aug-cc-pV6Z basis set. The programs RKR1, LEVEL and PGOPHER were used to calculate line positions and intensities using the most recent spectroscopic observations and the new DMF, and including the rotational dependence on the matrix elements. The Hund's case (b) matrix elements from the LEVEL output have been transformed to the case (a) form required by PGOPHER.

Equivalent calculations have been performed for the OH X$^2\Pi$ ground state. This includes a new DMF calculation using the internally contracted MR-ACPF method with an aug-cc-pV6Z basis set. A similar line list has been produced for rovibrational and rotational transitions for all possible transitions up to v'=13.

MK10 4:08 – 4:23

DETERMINATION OF PRESSURE BROADENING AND SHIFTS FOR THE FIRST OVERTONE $2 \leftarrow 0$ OF HCL

BRIAN DROUIN, TIMOTHY J CRAWFORD, *Jet Propulsion Laboratory, California Institute of Technology, Pasadena, CA, USA*; BRENNAN M. COFFEY, *College of Engineering, University of Colorado Boulder, Boulder, CO, USA.*

Precise and accurate positions of the line centers of HCl (v = 2 $\leftarrow$ 0) have not been reported for quite some time[1], however, using methane as a transferrable standard, we aim to report line positions to an order of magnitude better than previously characterized and enable the use of HCl as a secondary frequency standard. The line centers for the first overtone of (v =2 $\leftarrow$ 0) in the spectral region 5500 - 5900 cm^{-1} have been measured simultaneously with the $2\nu_3$ band of methane using the JPL Bruker IFS 125HR. The positions are determined through multispectrum analysis to 1×10^{-5} cm^{-1}. Further analysis using the multispectrum fitting software Labfit allows the determination of the self-broadening, self-shift and Dicke narrowing parameters to high precision as well. This potential new standard is compared to the HITRAN 2012 line parameters.

[1] Guelachvilli, G. Opt. Comm. 19, 150-154. (1976).

MK11 4:25 – 4:40

CONCENTRATION DEPENDENCE OF LINE SHAPES IN THE $\nu_1 + \nu_3$ BAND OF ACETYLENE

MATTHEW CICH, *Department of Chemistry, Stony Brook University, Stony Brook, NY, USA*; DAMIEN FORTHOMME, GREGORY HALL, C. McRAVEN, TREVOR SEARS[a], *Department of Chemistry, Brookhaven National Laboratory, Upton, NY, USA.*

Using an extended cavity diode laser locked to a frequency comb, the line shape of the P(11) line in the $\nu_1 + \nu_3$ combination band of acetylene has been studied as a function of varying concentration of the absorber in nitrogen. Mixture concentrations of 1, 5 and 10% at 296 K and pressures between a few Torr and one atmosphere were made and the measurements analyzed using two different speed-dependent broadening models. These experiments are designed to test the additivity of contributions to pressure broadening and shift in speed-dependent line shape modeling, *i.e.* whether the lineshape parameters follow partial pressure weighting in the binary mixtures. P(11) is relatively isolated with respect to underlying hot band transitions and neighboring transitions of the same band, but it was found that the accurate positions of underlying hot band transitions were crucial to the successful modeling of the observed line shapes, even though these lines are typically 100-1000 times weaker than P(11) itself and are many Doppler line widths removed from the line center. Positions of the hot band lines quoted in the HITRAN database, which are derived from the analysis of high resolution FTIR spectra, are of the order of 10's of MHz in error. In parallel work, we have measured the positions of many of these lines by saturation dip spectroscopy. Progress in the analysis of the data and the new saturation dip line center measurements will be reported.

Acknowledgments: Work at Brookhaven National Laboratory was carried out under Contract No. DE-AC02-98CH10886 with the U.S. Department of Energy and supported by its Office of Basic Energy Sciences, Division of Chemical Sciences, Geosciences and Biosciences.

---

[a] Also, Department of Chemistry, Stony Brook University, Stony Brook, NY 11794
[a] C. P. McRaven, *et al.* Paper RI05, 68th International Symposium on Molecular Spectroscopy, 2013

MK12 4:42 – 4:57

OBSERVATIONS OF DICKE NARROWING AND SPEED-DEPENDENCE IN CO2 LINESHAPES NEAR 2060NM

THINH QUOC BUI, *Division of Chemistry and Chemical Engineering, California Institute of Technology, Pasadena, CA, USA*; DAVID A. LONG, *Material Measurement Laboratory, National Institute of Standards and Technology, Gaithersburg, MD, USA*; CYGAN AGATA, *Institute of Physics, Faculty of Physics, Astronomy and Informatics, Nicolaus Copernicus University, Torun, Poland*; VINCENT SIRONNEAU, *Material Measurement Laboratory, National Institute of Standards and Technology, Gaithersburg, MD, USA*; DANIEL HOGAN, *Department of Applied Physics, Stanford University, Stanford, CA, USA*; PRIYANKA MILINDA RAPUSINGHE, *Physical Sciences, Cameron University, Lawton, OK, USA*; MITCHIO OKUMURA, *Division of Chemistry and Chemical Engineering, California Institute of Technology, Pasadena, CA, USA*.

To achieve NASA's strategic scientific mission of monitoring global atmospheric CO_2 at a target precision of 0.25%, we perform laser based measurements and lineshape studies of the reference atmospheric target CO_2 band (20013)←(00001) centered at 2060nm to meet the precision requirements of current (ACOS/GOSAT/TCCON) and future (OCO-2/OCO-3/ASCENDS) remote sensing applications. We utilize a frequency-stabilized cavity ringdown spectrometer, (FS-CRDS), to provide high precision and high spectral resolution for probing non-Voigt lineshape features (Dicke narrowing, speed dependence) of CO_2 currently unaccounted for in databases like HITRAN. We discuss our results for two R-branch lines R(24) and R(30) near 2055nm, which display concurrent collisional narrowing and speed dependence effects.

MK13 4:59 – 5:14

RAPID AND ACCURATE CALCULATION OF A SPEED DEPENDENT SPECTRAL LINE SHAPE

D. REED BEVERSTOCK, *Department of Physics, College of William and Mary, Williamsburg, VA, USA*; KENDRA LETCHWORTH WEAVER, *Physics, Cornell University, Ithaca, NY, USA*; D. CHRIS BENNER, *Department of Physics, College of William and Mary, Williamsburg, VA, USA*.

Use of the Voigt profile with the Lorentz width allowed to vary with the speed of collision has been hampered by the lack of fast accurate algorithms. Such an algorithm has been written assuming a quadratic dependence of the Lorentz width upon the speed of collision that is accurate to one part in 10 000 and is generally only a factor of four or so slower than the equivalent Voigt calculation with the Letchworth and Benner algorithm.[a] The only exception to the accuracy is far from line center near the Doppler limit when the speed dependent parameter is quite large. At this point the spectral line has fallen by at least 17 orders of magnitude from the line center and is generally insignificant. Gauss-Hermite quadrature of third to seventeenth order, Taylor series expansion about precomputed points and spline interpolation are used in the computation of both the real and imaginary parts for various regions.[b]

[a]Kendra L. Letchworth and D. Chris Benner, JQSRT 107 (2007) 173-192.
[b]This work was funded by the Jet Propulsion Laboratory and National Science Foundation.

MK14 5:16–5:31

A THEORETICAL MODEL FOR WIDE-BAND INFRARED-ABSORPTION MOLECULAR SPECTRA AT ANY PRESSURE: FICTION OR REALITY?

JEANNA BULDYREVA, *Institute UTINAM, UMR CNRS 6213, University of Franche-Comte, Besancon, France*; JEAN VANDER AUWERA, *Service de Chimie Quantique et Photophysique, Université Libre de Bruxelles, Brussels, Belgium.*

Various atmospheric applications require modeling of infrared absorption by the main atmospheric species in wide ranges of frequencies, pressures and temperatures. For different pressure regimes, different mechanisms are responsible for the observed intensities of vibration-rotation line manifolds, and the structure of the bands changes drastically when going from low to high densities. Therefore, no universal theoretical model exists presently to interpret simultaneously collapsed band-shapes observed at very high pressures and isolated-line shapes recorded in sub-atmospheric regimes. Using CO_2 absorption spectra as an example, we introduce some improvements in the non-Markovian Energy-Corrected Sudden model, developed for high-density spectra of arbitrary tensorial rank[a] and generalized recently to parallel and perpendicular infrared absorption bands[b], and test the applicability of this approach for the case of nearly Doppler pressure regime via comparisons with recently recorded experimental intensities[c].

[a] J.V. Buldyreva and L. Bonamy, Phys. Rev. A 60(1), 370-376 (1999).

[b] J. Buldyreva and L. Daneshvar, J. Chem. Phys. 139, 164107 (2013).

[c] L. Daneshvar, T. Földes, J. Buldyreva, J. Vander Auwera, J. Quant. Spectrosc. Radiat. Transfer 2014 (to be submitted).

MK15 5:33–5:48

COLLISION-INDUCED SPECTRA: AN AVENUE TO INVESTIGATE MICROSCOPIC-SCALE DIFFUSION IN FLUIDS

WOUTER A. HERREBOUT, BENJAMIN J. VAN DER VEKEN, *Department of Chemistry, University of Antwerpen, Antwerpen, Belgium*; ALEXANDER KOUZOV, *Department of Physics, Saint-Petersburg State University, St. Petersburg, Russia.*

New data on the IR spectra induced by intermolecular interactions in liquid cryogenic mixtures at T=89 K (O_2 in LAr and LN_2 and binary O_2-Ar solutions in LN_2) are reported. The induced fundamental bands appear as diffuse pedestals (with FWHH$\approx$100 cm^{-1}) on which weak, paradoxically sharp lines (FWHH$\approx$2 cm^{-1}) develop at the 2326 and 1552 cm^{-1} frequencies of the free-molecule vibrational transitions in N_2 and O_2, respectively. In LAr and LN_2 these lines were carefully separated and studied at varied O_2 concentrations up to c =0.23 mole fractions (mf). While the 1552 cm^{-1} line scales as $c[O_2]^2$ and thus is induced by the O_2-O_2 interactions in a bulk of cryosolvent (Ar, N_2), the 2326 cm^{-1} feature varies linearly with $c[O_2]$ and hence is caused by interaction of a guest (O_2) with a vibrating host (N_2). The impurity induction mechanism was further supported by our data on the binary O_2-Ar solutions in LN_2 recorded at the fixed $c[O_2]$ (0.03 and 0.06 mf) and the varied $c[Ar] \leq$ 0.2 mf. Both series revealed the same (linear) enhancement of the sharp N_2 line by argon, in an accord with our previous studies of the Ar-LN_2 system[a]. The results suggest that the resonance 2326 cm^{-1} feature is primarily due to the local distortion of the first coordination sphere around a vibrating N_2 by a guest molecule. We also notice that the resonance lines should be due to the dispersion- and overlap-induced dipole moments independent on the rotational degrees of freedom[b]. As our previous studies of the H_2-LNe system showed[c], the unusual line sharpness is a conspicuous manifestation of the relative solvent-solute and solute-solute translations dramatically retarded in a liquid by a fast velocity relaxation, an effect directly related to the microscopic-scale diffusion. The collision-induced spectra thus open up new vistas for studies of microscopic liquid dynamics.

[a] W.A. Herrebout, A.A. Stolov, E.J. Sluyts, and B.J. van der Veken, Chem. Phys. Lett. **295**, 223 (1998).

[b] J.E. Bohr and K.L.C. Hunt , J. Chem. Phys. **86**, 5441 (1987).

[c] W. A. Herrebout, B. J. van der Veken, and A. P. Kouzov, J. Chem. Phys. **137**, 084509 (2012).

TA. Astronomy

Tuesday, June 17, 2014 – 8:30 AM

Room: 116 Roger Adams Lab

Chair: David E. Woon, University of Illinois, Urbana, IL, USA

TA01   8:30 – 8:45

TIME-DOMAIN TERAHERTZ SPECTROSCOPY (0.3 - 7.5 THz) OF MOLECULAR ICES OF SIMPLE ALCOHOLS

BRETT A. McGUIRE, *Division of Chemistry and Chemical Engineering, California Institute of Technology, Pasadena, CA, USA*; SERGIO IOPPOLO, *Geological and Planetary Sciences , California Institute of Techonolgy, Pasadena, CA, USA*; MARCO A. ALLODI, *Division of Chemistry and Chemical Engineering, California Institute of Technology, Pasadena, CA, USA*; XANDER DE VRIES, *Theoretical Chemistry, University of Nijmegen, Nijmegen, Netherlands*; IAN FINNERAN, BRANDON CARROLL, GEOFFREY BLAKE, *Division of Chemistry and Chemical Engineering, California Institute of Technology, Pasadena, CA, USA.*

We have recently constructed a time-domain TeraHertz (THz) spectrometer for the study of molecular ices in the far-infrared. Here, we present the results of a study of amorphous and crystalline ices of simple alcohols from methanol (CH_3OH) through butanol ($CH_3(CH_2)_3OH$) in the region of 0.3 - 7.5 THz. We examine the effects of the length and degree of branching of the carbon chain on the observed spectra arising from the bulk, large-amplitude motions which are prominent in this spectral region. We also discuss these results in an astrochemical context: the application of these spectra to astronomical observations of interstellar ices with Herschel PACS/SPIRE and SOFIA.

TA02   8:47 – 9:02

THz TIME-DOMAIN SPECTROSCOPY OF COMPLEX INTERSTELLAR ICE ANALOGS

SERGIO IOPPOLO, *Geological and Planetary Sciences , California Institute of Techonolgy, Pasadena, CA, USA*; BRETT A. McGUIRE, MARCO A. ALLODI, *Division of Chemistry and Chemical Engineering, California Institute of Technology, Pasadena, CA, USA*; XANDER DE VRIES, *Theoretical Chemistry, University of Nijmegen, Nijmegen, Netherlands*; IAN FINNERAN, BRANDON CARROLL, GEOFFREY BLAKE, *Division of Chemistry and Chemical Engineering, California Institute of Technology, Pasadena, CA, USA.*

It is generally accepted that complex organic molecules (COMs) form on the icy surface of interstellar grains. Our ability to identify interstellar complex species in the ices is affected by the limited number of laboratory analogs that can be compared to the huge amount of observational data currently coming from international astronomical facilities, such as the Herschel Space Observatory, SOFIA, and ALMA. We have recently constructed a new THz time-domain spectroscopy system to investigate the spectra of interstellar ice analogs in a range that fully covers the spectral bandwidth of the aforementioned facilities (0.3 – 7.5 THz). The system is coupled to a FT-IR spectrometer to monitor the ices in the mid-IR (4000 – 500 cm^{-1}). This talk focuses on the laboratory investigation of the composition and structure of the bulk phases of interstellar ice analogs (i.e., H_2O, CO_2, CO, CH_3OH, NH_3, and CH_4) compared to more complex molecules (e.g., HCOOH, CH_3COOH, CH_3CHO, $(CH_3)_2CO$, $HCOOCH_3$, and $HCOOC_2H_5$). The ultimate goal of this research project is to provide the scientific community with an extensive THz ice database, which will allow quantitative studies of the ISM, and potentially guide future astronomical observations of species in the solid phase.

TA03 9:04 – 9:19

TIME-DOMAIN TERAHERTZ SPECTRSOCOPY OF POLYCYCLIC AROMATIC HYDROCARBONS

BRANDON CARROLL, MARCO A. ALLODI, *Division of Chemistry and Chemical Engineering, California Institute of Technology, Pasadena, CA, USA*; SERGIO IOPPOLO, *Division of Geological and Planetary Sciences, California Institute of Technology, Pasadena, CA, USA*; IAN FINNERAN, BRETT A. McGUIRE, GEOFFREY BLAKE, *Division of Chemistry and Chemical Engineering, California Institute of Technology, Pasadena, CA, USA*.

Polycyclic aromatic hydrocarbons (PAHs) present themselves as a strong candidate as carriers of the unidentified infrared features (UIRs). As UIR carriers, PAHs may account for up to 20% of the interstellar carbon budget and may play key roles in many chemical and physical processes in the ISM, and yet our inability to definitively detect PAHs hinders our ability to evaluate the role they may play. A possible solution is observations in the TeraHertz (THz) regime, where observed transitions are specific to each molecule. Recent advances in THz technology have enabled both laboratory spectroscopy and astronomical observations in this region. A first step in both laboratory and astronomical studies of PAHs is the acquisition of spectra of pure PAH samples. Here, we present the THz time-domain spectra (0.3 - 7 THz) of several PAHs, including naphthalene, anthracene, and pyrene, and discuss the utility of these spectra for future laboratory and astronomical studies.

TA04 9:21 – 9:36

PROBING GAS PHASE CHEMISTRY ABOVE ICE SURFACES WITH MILLIMETER/SUBMILLIMETER SPECTROSOCPY

AJ MESKO, IAN C WAGNER, *Department of Chemistry, Emory University, Atlanta, GA, USA*; STEFANIE N MILAM, *Astrochemistry, NASA Goddard Space Flight Center, Greenbelt, MD, USA*; SUSANNA L. WIDICUS WEAVER, *Department of Chemistry, Emory University, Atlanta, GA, USA*.

Chemical reactions involving the icy mantles of interstellar dust grains have been invoked in astrochemical models to explain the formation of complex organic molecules in interstellar clouds. Interstellar ices can act as a substrate to encourage reactions in three ways: reactions within the bulk ice, reactions between mobile species on the ice surface, or gas-phase reactions that are initiated by thermal desorption or photodesorption of the ice. We are building a new experiment that uses millimeter/submillimeter absorption spectroscopy to probe the gas-phase chemistry directly above the ice surface during thermal- or photo-processing. We will present the experimental design and preliminary results for pure water ices and water+ methanol ice mixtures.

TA05 9:38 – 9:53

MILLIMETER/SUBMILLIMETER SPECTROSCOPY OF PREBIOTIC MOLECULES FORMED FROM THE O(^{1}D) INSERTION INTO METHYLAMINE

BRIAN HAYS, ALTHEA A. M. ROY, SUSANNA L. WIDICUS WEAVER, *Department of Chemistry, Emory University, Atlanta, GA, USA*.

Astrochemical models of interstellar chemistry predict the formation of many complex molecules of prebiotic interest, including aminomethanol ($HOCH_2NH_2$). Aminomethanol has been proposed as the gas-phase interstellar precursor to glycine, the simplest amino acid, in star-forming regions. Aminomethanol is therefore a potential tracer of prebiotic interstellar chemistry. However, the laboratory spectrum of aminomethanol remains elusive because it is unstable under typical laboratory conditions. A new (sub)millimeter spectrometer is being used to study the reaction between O(^{1}D) and methylamine to form aminomethanol. O(^{1}D) is produced via laser photolysis of ozone in a fused silica tube, where it reacts with methylamine before a supersonic expansion. The insertion reaction of O(^{1}D) with methylamine to form aminomethanol is highly exothermic, leading to a mixture of additional reaction products that have been identified through their rotational spectroscopic signatures. Here we will present the experimental setup, observed reaction products, and initial results towards the characterization of aminomethanol. Comparisons will also be made with observational spectra from several star-forming regions.

TA06   9:55 – 10:10

ASTRONOMICAL APPLICATIONS OF NEW LINE LISTS FOR CN, C_2 AND THEIR ISOTOPOLOGUES

PETER F. BERNATH, *Department of Chemistry and Biochemistry, Old Dominion University, Norfolk, VA, USA*; CHRIS SNEDEN, *Department of Astronomy, The University of Texas at Austin, Austin, TX, USA*; JAMES S.A. BROOKE, RAM RAM, *Department of Chemistry, University of York, York, United Kingdom.*

For cool stellar and substellar objects, atomic lines weaken, and detailed elemental and isotopic abundances are often derived from molecular absorption features. We have embarked on a project to provide molecular line lists by combining experimental observations for line positions with ab initio calculations for line strengths. So far we have results for MgH (A-X and B-X transitions)[a][b], C_2 (Swan system)[c][d], CN (red and violet systems)[e], CP (A-X transition)[f], NH (vibration-rotation bands) and OH (Meinel system)[g]. This talk will briefly describe the new line lists for the Swan system ($d^3\Pi$-$a^3\Pi$) of C_2 and $^{12}C^{13}C$, and the red ($A^2\Pi$-$X^2\Sigma^+$) and violet ($B^2\Sigma^+$-$X^2\Sigma^+$) systems of CN, ^{13}CN and C^{15}N. Applications to the spectra of carbon-enhanced metal-poor stars, the K-giant Arcturus, the metal-rich open cluster NGC 6791, the Sun and comets will be presented.

[a]E. GharibNezhad, A. Shayesteh and P. F. Bernath, Mon. Notices R. Astro. Soc. 432, 2043-2047 (2013)

[b]K. H. Hinkle, L. Wallace, R. S. Ram, P. F. Bernath, C. Sneden and S. Lucatello, Astrophys. J. Suppl. 207, 26 (7pp) (2013)

[c]J. S. A. Brooke, P. F. Bernath, T. W. Schmidt and G. B. Bacskay, J. Quant. Spectrosc. Rad. Trans. 124, 11-20 (2013)

[d]R. S. Ram, J. S. A. Brooke, P. F. Bernath, C. Sneden and S. Lucatello, Astrophys. J. Suppl. 211, 5 (7pp) (2014)

[e]J. S. A. Brooke, R. S. Ram, C. M. Western, G. Li, D. W. Schwenke and P. F. Bernath, Astrophys. J. Suppl. 210, 23 (15pp) (2014)

[f]R. S. Ram, J. S. A. Brooke, C.M. Western and P. F. Bernath, J. Quant. Spectrosc. Rad. Transfer (in press)

[g]J. S. A. Brooke et al., this meeting, P301

TA07   10:12 – 10:27

MILLIMETER/SUBMILLIMETER STUDIES OF IONS AND RADICALS OF ASTROPHYSICAL INTEREST USING A HOLLOW CATHODE SPECTROMETER

TREVOR CROSS, NADINE WEHRES, MARY RADHUBER, ANNE CARROLL, SUSANNA L. WIDICUS WEAVER, *Department of Chemistry, Emory University, Atlanta, GA, USA.*

Ions and radicals are important in astrochemical models because they act as key reaction intermediates in the interstellar medium. However, much laboratory work remains to determine the rotational spectra of most ions and radicals of astrophysical interest. This is especially true in the millimeter/submillimeter range, where small sample quantities limit spectral signal intensities. Hollow–cathode discharges have previously been used to create and study ions and radicals of astrophysical interest, but most of these instruments have been coupled with infrared spectrometers. We have developed a hollow–cathode spectrometer to investigate ions and radicals using (sub)millimeter spectroscopy. Spectrometer performance has been benchmarked using the N_2H^+ molecular ion, which has a known rotational spectrum. Initial results from these benchmarking studies, as well as new spectral results for other molecular targets, will be presented.

Intermission

TA08

THE MM-WAVE ROTATIONAL SPECTRUM OF GLYCOLIC ACID

ZBIGNIEW KISIEL, LECH PSZCZÓŁKOWSKI, EWA BIAŁKOWSKA-JAWORSKA, *ON2, Institute of Physics, Polish Academy of Sciences, Warszawa, Poland*; STEVEN B CHARNLEY, *Astrochemistry, NASA Goddard Space Flight Center, Greenbelt, MD, USA.*

Glycolic acid, $HOCH_2COOH$ is the simplest α-hydroxy acid. It is as yet undetected in the interstellar medium, but is known to be present in carbonaceous meteorites and in residues from UV-photolysed interstellar ice analogue mixtures. Prior rotational spectroscopy has been carried out up to 40 GHz for the main, SSC conformer, [a,b] and up to 72 GHz for the weaker, AAT, conformer.[c]

Presently we report the analysis of the rotational spectrum of glycolic acid on the basis of broadband measurements performed up to 318 GHz, and updated spectroscopic constants for the ground state and the first two excited states of the low-frequency ν_{21} torsional mode. We have used the AABS package to assign multiple further excited vibrational states of the SSC conformer. In particular, we have been able to assign the highly perturbed triad of ν_{14}, ν_{20} and $3\nu_{21}$ states. The triad has been fitted down to experimental accuracy with a coupled fit, which allowed us to pin down the hitherto elusive frequency of the ν_{21} mode. The experimental results make an interesting comparison with those of anharmonic force field calculations. We have also been able to extend the measurements for the AAT conformer.

[a]C.E.Blom, A.Bauder, *Chem. Phys. Lett.*, **82**, 492 (1981), *J. Am. Chem. Soc.*, **104**, 2993 (1982).
[b]H.Hasegawa, O.Ohashi, I.Yamaguchi, *J. Mol. Spectrosc.*, **82**, 205 (1982).
[c]P.D.Godfrey, F.M.Rodgers, R.D.Brown, *J. Am. Chem. Soc.*, **119**, 2232 (1997).

TA09

ROTATIONAL SPECTRA OF UREA IN ITS GROUND AND FIRST EXCITED VIBRATIONAL STATES

JESSICA THOMAS, IVAN MEDVEDEV, *Department of Physics, Wright State University, Dayton, OH, USA*; ZBIGNIEW KISIEL, *ON2, Institute of Physics, Polish Academy of Sciences, Warszawa, Poland.*

Urea is an important terrestrial bio-molecule, which has been tentatively detected in the interstellar medium [a]. To match the much improved range and sensitivities of modern sub-millimeter telescopes a broad laboratory assay of rotational transitions needs to be recorded in order to aid in the definitive identification of this molecule. This paper focuses on the spectroscopic assignment of the rotational transitions of urea in the 207-500 GHz range which belong to its ground and first excited vibrational states.

[a]Remijan, A.J., L.E. Snyder, B.A. McGuire, H.-L. Kuo, L.W. Looney, D.N. Friedel, G.Y. Golubiatnikov, F.J. Lovas, V.V. Ilyushin, E.A. Alekseev, S.F. Dyubko, B.J. McCall, and J.M. Hollis, Observational Results of a Multi-Telescope Campaign in Search of Interstellar Urea [NH22CO]. The Astrophysical Journal, 2014. 783(2): p. 77

TA10

THE LOWEST VIBRATIONAL STATES OF UREA FROM THE ROTATIONAL SPECTRUM

ZBIGNIEW KISIEL, *ON2, Institute of Physics, Polish Academy of Sciences, Warszawa, Poland*; JESSICA THOMAS, IVAN MEDVEDEV, *Department of Physics, Wright State University, Dayton, OH, USA.*

The urea molecule, $(NH_2)_2CO$, has a complex potential energy surface resulting from a combination of the NH_2 torsion and NH_2 inversion motions. This leads to a distribution of lowest vibrational states that is expected to be significantly different from the more familiar picture from simple inversion or normal mode models.[a,b]

The broadband 207-500 GHz spectrum of urea recorded in Dayton has signal to noise sufficient for assignment of rotational transitions in excited vibrational states up to at least 500 cm^{-1}. In addition to the previously reported analysis of the ground and the lowest excited state we have been able to assign transitions in at least five other excited vibrational states. Strongly perturbed transitions in a close doublet of such states have been fitted to within experimental accuracy with a coupled fit and a splitting in the region of 1 cm^{-1}. These assignments combined with vibrational energy estimates from relative intensity measurements allow for empirical discrimination between different models for the energy level manifestation of the large amplitude motions in urea.[b]

[a]P.D.Godfrey, R.D.Brown, A.N.Hunter *J. Mol. Struct.*, **413-414**, 405-414 (1997).
[b]N.Inostroza, M.L.Senent, *Chem. Phys. Lett.*, **524**, 25 (2012).

TA11   11:35 – 11:50

HIGH RESOLUTION MEASUREMENTS AND ELECTRONIC STRUCTURE CALCULATIONS OF A DIAZANAPH-THALENE

SÉBASTIEN GRUET[a], *AILES beamline, Synchrotron SOLEIL, Saint Aubin, France*; MANUEL GOUBET, *Laboratoire PhLAM, Université de Lille 1, Villeneuve de Ascq, France*; OLIVIER PIRALI, *Institut des Sciences Moléculaires d'Orsay, Université Paris-Sud, Orsay, France.*

Polycyclic Aromatic Hydrocarbons (PAHs) have long been suspected to be the carriers of so called Unidentified Infrared Bands (UIBs).[b] Most of the results published in the literature report rotationally unresolved spectra of pure carbon as well as heteroatom-containing PAHs species. To date for this class of molecules, the principal source of rotational informations is ruled by microwave (MW) spectroscopy [c] while high resolution measurements reporting rotational structure of the infrared (IR) vibrational bands are very scarce. Recently, some high resolution techniques provided interesting new results to rotationally resolve the IR and far-IR bands of these large carbonated molecules of astrophysical interest.[d] One of them is to use the bright synchrotron radiation as IR continuum source of a high resolution Fourier transform (FTIR) spectrometer. We report the very complementary analysis of the [1,6] naphthyridine (a N-bearing PAH) for which we recorded the microwave spectrum at the PhLAM laboratory (Lille) and the high resolution far-infrared spectrum on the AILES beamline at synchrotron facility SOLEIL. MW spectroscopy provided highly accurate rotational constants in the ground state to perform Ground State Combinations Differences (GSCD) allowing the analysis of the two most intense FT-FIR bands in the 50-900 cm^{-1} range. Moreover, during this presentation the negative value of the inertial defect in the GS of the molecule will be discussed.

[a] Also at: Institut des Sciences Moléculaires d'Orsay, UMR 8214 CNRS-Université Paris Sud 91405 Orsay Cedex
[b] A. Leger, J. L. Puget, *Astron. Astrophys.* **137**, L5-L8 (1984); L. J. Allamandola et al. *Astrophys. J.* **290**, L25-L28 (1985).
[c] Z. Kisiel et al. *J. Mol. Spectrosc.* **217**, 115 (2003); S. Thorwirth et al. *Astrophys. J.* **662**, 1309 (2007); D. McNaughton et al. *J. Chem. Phys.* **124**, 154305 (2011).
[d] S. Albert et al. *Faraday Discuss.* **150**, 71-99 (2011); B. E. Brumfield et al. *Phys. Chem. Lett.* **3**, 1985-1988 (2012); O. Pirali et al. *Phys. Chem. Chem. Phys.* **15**, 10141 (2013).

TA12   11:52 – 12:07

Cis-METHYL VINYL ETHER: THE ROTATIONAL SPECTRUM UP TO 600 GHz

LUCIE KOLESNIKOVÁ, ADAM M DALY, JOSÉ L. ALONSO, *Grupo de Espectroscopia Molecular, Lab. de Espectroscopia y Bioespectroscopia, Unidad Asociada CSIC, Universidad de Valladolid, Valladolid, Spain.*

Astronomical observation of dimethyl ether,[a] methyl ethyl ether[b] and vinyl alcohol[c] places the methyl vinyl ether among the species of potential interstellar relevance. The millimeter and submillimeter-wave transitions pertaining to the vibrational ground state and the first excited states of the methoxy, ν_{24}, and methyl, ν_{23}, torsional modes and the in-plane bending mode, ν_{16}, of the *cis*-methyl vinyl ether have been measured and analyzed in the frequency region from 50 to 600 GHz. A significant Fermi-type and Coriolis interactions between the $v_{24} = 1$ and $v_{23} = 1$ states have been observed and the rotational spectra were analyzed using an effective two-state Hamiltonian explicitly involving corresponding coupling operators. A sets of spectroscopic constants for the ground state as well as for all three excited states reproducing the observed spectrum within the experimental uncertainty provide sufficiently precise information for the astronomical search for methyl vinyl ether.

[a] Z. Peeters, S. D. Rodgers, S. B. Charnley, L. Schriver-Mazzuoli, A. Schriver, J. V. Keane, and P. Ehrenfreund, *Astron. & Astrophys.* **2006**, *445*, 197.
[b] G. W. Fuchs, U. Fuchs, T. F. Giesen, F. Wyrowski, *Astron. & Astrophys.* **2005**, *444*, 521.
[c] B. E. Turner, A. J. Apponi, *Astrophys. J. Lett.* **2001**, *561*, 207.

TB. Instrument/Technique Demonstration
Tuesday, June 17, 2014 – 8:30 AM
Room: 100 Noyes Laboratory

Chair: Kevin Lehmann, The University of Virginia, Charlottesville, VA, USA

TB01 8:30 – 8:45

CHARACTERISATION AND CONTROL OF COLD CHIRAL COMPOUNDS

CHRIS MEDCRAFT, THOMAS BETZ, V. ALVIN SHUBERT, DAVID SCHMITZ, SIMON MERZ, MELANIE SCHNELL, *CoCoMol, Max-Planck-Institut für Struktur und Dynamik der Materie, Hamburg, Germany.*

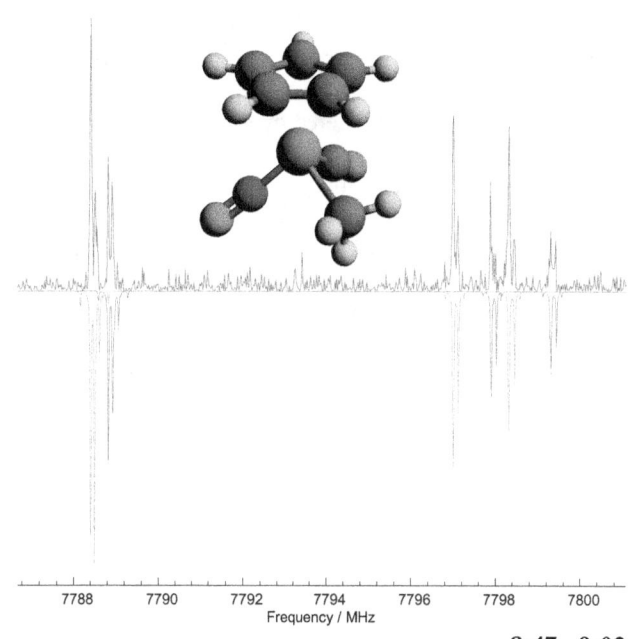

The rotational spectrum of the chiral molecule $CpReCH_3(CO)(NO)$ (Cp=cyclopentadienyl) has been measured between 2 and 8.5 GHz using a chirped-pulse FTMW spectrometer. Rotational and nuclear quadrupole constants for the nitrogen and both rhenium isotopes (^{185}Re & ^{187}Re) have determined. Initial assignment was hampered by the large coupling constants for the rhenium nuclei (χ_{xx}=720.21 and 680.55 MHz for ^{185}Re and ^{187}Re, respectively) and the barely resolved splitting due to the nitrogen nucleus. Large (100-300 MHz) off-diagonal quadrupole coupling terms (χ_{ab} and χ_{bc}) for the rhenium nuclei also complicated the fitting. Relativistic corrections were required to reproduce experimental results in *ab initio* calculations.

This molecule is of particular interest due to a high energy difference between enantiomers due to parity violation. For the related molecule $CpRe(CO)I(NO)$, a frequency difference in rotational transitions of a few hertz is anticipated. A high-resolution, cavity-based microwave spectrometer is being commissioned with the aim of approaching this level of precision.

TB02 8:47 – 9:02

MOLECULAR STRUCTURE AND CHIRALITY DETERMINATION FROM PULSED-JET FOURIER TRANSFORM MICROWAVE SPECTROSCOPY

SIMON LOBSIGER, CRISTOBAL PEREZ, LUCA EVANGELISTI, NATHAN A SEIFERT, BROOKS PATE, KEVIN LEHMANN, *Department of Chemistry, The University of Virginia, Charlottesville, VA, USA.*

Fourier transform microwave (FTMW) spectroscopy has been used for many years as one of the most accurate methods to determine gas-phase structures of molecules and small molecular clusters. In the last years two pioneering works ushered in a new era applications. First, by exploiting the reduced measurement time and the high sensitivity, the development of chirped-pulse CP-FTMW spectrometers[a] enabled the full structural determination of molecules of increasing size as well as molecular clusters. Second, and more recently, Patterson et al.[b] showed that rotational spectroscopy can also be used for enantiomer-specific detection. Here we present an experimental approach that combines both in a single spectrometer. This set-up is capable to rapidly obtain the full heavy-atom substitution structure using the CP-FTMW features. The inclusion of an extra set of broadband horns allows for a chirality-sensitive measurement of the sample.[c,d] The measurement we implement is a three-wave mixing experiment that uses time-separated pulses to optimally create the chiral coherence – an approach that was proposed recently.[e] Using samples of R-, S- and racemic Solketal, the physical properties of the three-wave mixing experiment were studied. This involved the measurement of the corresponding nutation curves (molecular signal intensity vs excitation pulse duration) to demonstrate the optimal pulse sequence. The phase stability of the chiral signal, required to assign the absolute stereochemistry, has been studied as a function of the measurement signal-to-noise ratio using a "phasogram" method.

[a]G. G. Brown, B. C. Dian, K. O. Douglass, S. M. Geyer, S. T. Shipman, B. H. Pate, Rev. Sci. Instrum. 2008, 79, 053103.
[b]D. Patterson, M. Schnell, J. M. Doyle, Nature 2013, 497, 475–477.
[c]D. Patterson, J. M. Doyle, Phys. Rev. Lett. 2013, 111, 023008.
[d]V. A. Shubert, D. Schmitz, D. Patterson, J. M. Doyle, M. Schnell, Angew. Chem. Int. Ed. 2014, 53, 1152–1155.
[e]J.-U. Grabow, Angew. Chem. 2013, 125, 11914 – 11916; Angew. Chem. Int. Ed. 2013, 52, 11698 –11700.

TB03   9:04 – 9:19

LIICG - A NEW METHOD FOR ROTATIONAL AND RO-VIBRATIONAL SPECTROSCOPY AT 4K

LARS KLUGE, ALEXANDER STOFFELS[a], SANDRA BRUENKEN, OSKAR ASVANY, STEPHAN SCHLEMMER, *I. Physikalisches Institut, Universität zu Köln, Köln, Germany.*

Since many years low temperature ion trapping techniques are successfully used in our laboratories in combination with sensitive action spectroscopy schemes (Laser Induced Reactions) to measure high resolution ro-vibrational and rotational spectra of gas-phase molecular ions. Here we present a further development of a LIR method first introduced for recording rotationally resolved electronic spectra of N_2^+ [b]. This new method, called LIICG (Light Induced Inhibition of Complex Growth), makes use of state specific He-attachment rates to stored cold molecular ions. We have recently demonstrated its applicability to rotational and ro-vibrational spectroscopy of C_3H^+ and CH_5^+ [c]. The measurements were performed in recently completed 4K 22-pole ion trap instruments. Ionic species are produced in a storage ion source and are mass selected before they enter the trap. For spectroscopy normally a few thousand ions are stored at 4K together with He at high number densities (around 10^{14} cm^{-3}). Under these conditions He attaches to the ions via ternary collision processes. As we will show, this attachement process is hindered by exciting a rotational or ro-vibrational transition, likely because the attachment rates for He are slower for higher rotational or ro-vibrational levels. So by exciting the bare ion the number of ion- He complexes at equilibrium is reduced. In this way the spectrum of the bare ion can be recorded by counting the number of ion-He complexes as a function of frequency. To test the new method we chose well known rotational ground state transitions of CO^+, HCO^+ and CD^+. In particular CD^+ appeared to be a good candidate for understanding the new method in detail, due to its strong LIICG signal and its simple rotational spectrum. In this contribution we will explain the LIICG scheme and its underlying kinetics using the example of CD^+. We will show effects of different experimental conditions on the signal (e.g. He number density, temperature, radiation power...) to explain our kinetic model. Beside these tests we will present measurements of new rotational transitions of C_3H^+, CD_2H^+ and CH_2D^+, demonstrating LIICG as a general spectroscopic method.

[a] also: Radboud University Nijmegen, Institute for Molecules and Materials (IMM), Nijmegen, Netherlands
[b] Chakrabarty et al. 2013, J. Phys. Chem. Lett., 4, 4051
[c] S. Brünken et al. 2014, ApJL, 783, L4; O. Asvany et al. 2014, ApPhB, 114, 203-211

TB04   9:21 – 9:31

METHODS DEVELOPMENT FOR SPECTRAL SIMPLIFICATION OF ROOM-TEMPERATURE ROTATIONAL SPECTRA

ERIN B KENT, STEVEN SHIPMAN, *Department of Chemistry, New College of Florida, Sarasota, FL, USA.*

Room-temperature rotational spectra are dense and difficult to assign, and so we have been working to develop methods to accelerate this process. We have tested two different methods with our waveguide-based spectrometer, which operates from 8.7 to 26.5 GHz. The first method, based on previous work by Medvedev and De Lucia[a], was used to estimate lower state energies of transitions by performing relative intensity measurements at a range of temperatures between -20 and +50 °C. The second method employed hundreds of microwave-microwave double resonance measurements to determine level connectivity between rotational transitions. The relative intensity measurements were not particularly successful in this frequency range (the reasons for this will be discussed), but the information gleaned from the double-resonance measurements can be incorporated into other spectral search algorithms (such as autofit or genetic algorithm approaches) via scoring or penalty functions to help with the spectral assignment process.

[a] I.R. Medvedev, F.C. De Lucia, Astrophys. J. 656, 621-628 (2007).

TB05   9:33 – 9:48

DELIVERING MICROWAVE SPECTROSCOPY TO THE MASSES: A DESIGN OF A LOW-COST MICROWAVE SPECTROMETER OPERATING IN THE 18-26 GHZ FREQUENCY RANGE

AMANDA STEBER, BROOKS PATE, *Department of Chemistry, The University of Virginia, Charlottesville, VA, USA.*

Advances in chip-level microwave technology in the communications field have led to the possibilities of low cost alternatives for current Fourier transform microwave (FTMW) spectrometers. Many of the large, expensive microwave components in a traditional design can now be replaced by robust, mass market monolithic microwave integrated circuits (MMICs). "Spectrometer on a board" designs are now feasible that offer dramatic cost reduction for microwave spectroscopy. These chip-level components can be paired with miniature computers to produce compact instruments that are operable through USB. A FTMW spectrometer design using the key MMIC components that drive cost reduction will be presented. Two dual channel synthesizers (Valon Technology Model 5008), a digital pattern generator (Byte Paradigm Wav Gen Xpress), and a high-speed digitizer \ arbitrary waveform generator combination unit (Tie Pie HS-5 530 XM) form the key components of the spectrometer for operation in the 18-26.5 GHz range. The design performance is illustrated using a spectrometer that is being incorporated into a museum display for astrochemistry. For this instrument a user interface, developed in Python, has been developed and will be shown.

Intermission

TB06   10:05 – 10:20

PRECISION FREQUENCY MEASUREMENT OF N_2O TRANSITIONS NEAR 4.5 μm AND ABOVE 150 μm

WEI-JO TING, CHUN-HUNG CHANG, SHIH-EN CHEN, *Department of Physics, National Tsing Hua University, Hsinchu, Taiwan*; HSUAN CHEN CHEN, *Institute of Photonics Technologies, National Tsing Hua University, Hsinchu, Taiwan*; JOW-TSONG SHY, *Frontier Research Center on Fundamental and Applied Sciences of Matters, National TsingHua University, Hsinchu, Taiwan*; BRIAN DROUIN, ADAM M DALY, *Jet Propulsion Laboratory, California Institute of Technology, Pasadena, CA, USA.*

Frequency measurements are given for the 10^00-00^00 band of N_2O near 4.5 μm and for pure rotational transitions beyond 151.5 μm. The infrared measurements utilize a periodically-poled lithium niobate (PPLN) based difference frequency generation (DFG) source locked to the saturated absorption center of an N_2O absorption line. The DFG frequency is calibrated by an optical frequency comb and an iodine hyperfine transition. We report 44 transitions ranging from $J = 1 - 100$ for both the $P-$ and $R-$ branches and the accuracy is better than 10 kHz for most transitions. In addition, 175 pure rotational transitions have been measured including 33 measurement with sub-Doppler precision (< 3 kHz), and 142 Doppler limited measurements. These are combined with other precision rotational and vibrational measurements to provide improved quantum mechanical parameters as well as frequency calibration tables for the N_2O bands near 4.5 μm.

TB07   10:22 – 10:37

SUBMILLIMETER SPECTROSCOPIC DIAGNOSTICS IN SEMICONDUCTOR PROCESSING PLASMAS

YASER H. HELAL, CHRISTOPHER F. NEESE, FRANK C. DE LUCIA, *Department of Physics, The Ohio State University, Columbus, OH, USA*; PAUL R. EWING, *Applied Materials, Austin, TX, USA*; PHILLIP J. STOUT, QUENTIN WALKER, MICHAEL D. ARMACOST, *Applied Materials, Sunnyvale, CA, USA.*

Submillimeter absorption spectroscopy was used to study semiconductor processing plasmas. Abundances and temperatures of molecules, radicals, and ions can be determined without altering any of the properties of the plasma. The behavior of these measurements provides useful applications in monitoring process steps. A summary of such applications will be presented, including etching and cleaning endpoint detection.

TB08   10:39 – 10:54

INTERFERENCE EFFECTS IN NONLINEAR VIBRATIONAL SPECTROSCOPY FROM MULTILAYERED MATERIAL INTERFACES

DANIEL B. O'BRIEN, AARON M. MASSARI, *Chemistry Department, University of Minnesota, Minneapolis, MN, USA.*

Vibrational sum frequency generation spectroscopy (VSFG) is a popular approach to obtaining molecular information from the interfaces of liquids and submonolayer adsorbates. A new challenge becomes apparent when applying this ultrafast technique to thin films in which two interfaces are present in the focal volume of the experiment. Although the signal levels from these two interfaces can be adjusted to some extent by beam geometries, their overall interferences can remain significant. Rather than viewing this as a two-interface problem, we present experimental work from our group in which we utilize this interference as a tool to determine molecular information from both interfaces. In particular, we present data showing the influence of charge accumulation at a buried interface and the possibilities of using mixed beam polarizations to improve global data fitting.

TB09   10:56 – 11:11

MIXED POLARIZATION VIBRATIONAL SUM FREQUENCY GENERATION SPECTRA OF ORGANIC SEMICONDUCTING THIN FILMS

PATRICK KEARNS, ZAHARA SOHRABPOUR, AARON M. MASSARI, *Chemistry Department, University of Minnesota, Minneapolis, MN, USA.*

The buried interface of an organic semiconductor at the dielectric has a large on influence on the function of organic field effect transistors (OFETs). The use of vibrational sum frequency generation (VSFG) to obtain structural and orientational information on the buried interfaces of organic thin films has historically been complicated by the signals from other interfaces in the system. A thin film of N,N'-Dioctyl-3,4,9,10-perylenedicarboximide (PTCDI-C8) was deposited on a SiO_2 dielectric to simulate the interfaces found in OFETs. We will show how probing the sample with a varying mixture of linear polarizations in the experimental setup can deconvolute contributions to the overall signal from multiple interfaces.

TB10   11:13 – 11:28

SHOCK COMPRESSION INDUCED HOT SPOTS IN ENERGETIC MATERIAL DETECTED BY THERMAL IMAGING MICROSCOPY

MING-WEI CHEN, DANA DLOTT, *Department of Chemistry, University of Illinois at Urbana-Champaign, Urbana, IL, USA.*

The chemical reaction of powder energetic material is of great interest in energy and pyrotechnic applications since the high reaction temperature. Under the shock compression, the chemical reaction appears in the sub-microsecond to microsecond time scale, and releases a large amount of energy. Experimental and theoretical research progresses have been made in the past decade, in order to characterize the process under the shock compression. However, the knowledge of energy release and temperature change of this procedure is still limited, due to the difficulties of detecting technologies. We have constructed a thermal imaging microscopy apparatus, and studied the temperature change in energetic materials under the long-wavelength infrared (LWIR) and ultrasound exposure.[a][b] Additionally, the real-time detection of the localized heating and energy concentration in composite material is capable with our thermal imaging microscopy apparatus. Recently, this apparatus is combined with our laser driven flyer plate system[c] to provide a lab-scale source of shock compression to energetic material. A fast temperature increase of thermite particulars induced by the shock compression is directly observed by thermal imaging with 15-20 μm spatial resolution. Temperature change during the shock loading is evaluated to be at the order of $10^9 K/s$, through the direct measurement of mid-wavelength infrared (MWIR) emission intensity change. We observe preliminary results to confirm the hot spots appear with shock compression on energetic crystals, and will discuss the data and analysis in further detail.

[a]M.-W. Chen, S. You, K. S. Suslick, and D. D. Dlott, *Rev. Sci. Instr.*, **85**, 023705 (2014),
[b]M.-W. Chen, S. You, K. S. Suslick, and D. D. Dlott, *Appl. Phys. Lett.*, **104**, 061907 (2014)
[c]K. E. Brown, W. L. Shaw, X. Zheng, and D. D. Dlott, *Rev. Sci. Instr.*, **83**, 103901 (2012)

TB11   *Post-Deadline Abstract*   **11:30 – 11:45**

INFRARED SPECTROMETRIC DIAGNOSIS OF MALIGNANCY: TRENDS AND CONCERNS

<u>MOHAMMADREZA KHANMOHAMMADI</u>, AMIR BAGHERI GARMARUDI, *Department of Chemistry, Imam Khomeini International University, Qazvin, Iran.*

Biostructure disorders such as uncontrolled cell division, invasive cell growth into adjacent tissue and metastaic implantation to other body sites is called "Cancer". Diagnosis of cancer is one of the most important aspects of medical researches and has become as the main aim of biomedical detective investigations. "Misdiagnosis" is the main drawback in most of the introduced diagnostic routes, while some of these methods are invasive and expensive. Medical analysis of biochemicals relies upon the ability of the analytical techniques to identify qualitatively and quantitatively the ingredients of bio-related samples. The new trend in medical diagnostics is to avoid the chemical analyses which may depend on a reaction between given biochemical and any introduced agent which can be destructive. On the other hand, the invasive sampling approaches such as tissue biopsy by surgery is the other concern. Current efforts in diagnosis of illness patterns are focused on studies which may be non-destructive and deal with describing a biochemical based on its properties e.g. spectral characteristics. The other important aspect is the capability for extraction of the most informative data from the huge amount of raw data from a biochemical sample, is of high importance. Biodiagnostics is a very tough medical area, because the main signs of illness at an early stage may be very subtle chemical variations , while the composition of disease related biochemical depends on many factors, not just whether the person is affected by a disease but also age, sex, nutrition, genetics, lifestyle etc. Recently infrared spectrometry has been proposed as a rapid, accurate and sensitive technique in biomedical researches especially for detection of illness patterns. Research activities for cancer detection by infrared spectroscopy, role of data processing on quality assurance of obtained results and also the role of novel proposals such as bio-fluid analysis for diagnostic aims are needed to be discussed, while there are some concerns in this regard. However, it can be concluded that infrared spectrometry as a novel green approach in clinical oncology and cancer research. Considering these complexities, the role of pattern recognition seems more critical for diagnosis of an illness case. It is very hard to obtain a sufficiently large and representative sample with suitable controls.

TC. Mini-symposium: Spectroscopy in Kinetics and Dynamics
Tuesday, June 17, 2014 – 8:30 AM
Room: B102 Chemical and Life Sciences

Chair: Ian Sims, Université de Rennes 1, Rennes, France

TC01   *INVITED TALK*   8:30 – 9:00

O(^{1}D) REACTION WITH METHANE STUDIED BY STATE RESOLVED SCATTERING DISTRIBUTION MEASUREMENTS OF METHYL RADICALS

TOSHINORI SUZUKI, *Graduate School of Science, Kyoto University, Kyoto, Japan.*

The scattering distributions of state-selected methyl radicals are measured for the O(^{1}D$_2$) reaction with methane using a crossed molecular beam ion imaging method at collision energies of 0.9 – 6.8 kcal/mol. The results are compared with the reaction with deuterated methane to examine the isotope effects. The scattering distributions exhibit contributions from both the insertion and abstraction pathways respectively on the ground and excited-state potential energy surfaces. Insertion is the main pathway, and it provides a strongly forward-enhanced angular distribution of methyl radicals. Abstraction is a minor pathway, causing backward scattering of methyl radicals with a discrete speed distribution. From the collision energy dependence of the abstraction/insertion ratio, the barrier height for the abstraction pathway is estimated for O(^{1}D$_2$) with CH$_4$ and CD$_4$, respectively. The insertion pathway of the O(^{1}D$_2$) reaction with CH$_4$ has a narrower angular width in the forward scattering and a larger insertion/abstraction ratio than the reaction with CD$_4$, which indicate that the insertion reaction with CH$_4$ has a larger cross section and a shorter reaction time than the reaction with CD$_4$. Additionally, while the insertion reaction with CD$_4$ exhibits strong angular dependence of the CD$_3$ speed distribution, CH$_3$ exhibits considerably smaller dependence. The result suggests that, although intramolecular vibrational redistribution (IVR) within the lifetime of the methanol intermediate is restrictive in both isotopomers, relatively more extensive IVR occurs in CD$_3$OD than CH$_3$OH, presumably due to the higher vibrational state density.

TC02   9:05 – 9:20

COLLISION DYNAMICS OF EXCITED SODIUM MOLECULES

BURCIN S BAYRAM, PHILLIP ARNDT, *Physics, Miami University, Oxford, OH, USA*; CEYLAN GUNEY, *Physics, Istanbul University, Istanbul, Turkey*; JACOB McFARLAND, *Physics, Miami University, Oxford, OH, USA.*

Collision cross section for transfer of anisotropy arising from collisions between electronically excited sodium dimer and ground level argon atoms has been examined. The experimental method is based on a polarization spectroscopy using a sophisticated resonant cw-pump-stimulated emission probe technique. Measurement of polarization from analysis of the emitted light is a very powerful method gaining information about the inelastic collision process between the electronically excited molecules and other collision partners. From the measurement, anisotropy-dependent polarization spectra of the Na$_2^*$ with Ar has been investigated.[a]

[a]The author gratefully acknowledges financial support from the National Science Foundation (Grant No. NSF-PHY-1309571).

TC03   9:22 – 9:37

SPECTROSCOPY AND DISSOCIATION DYNAMICS OF THE NO$_3^+$: A T-PEPICO STUDY

KANA TAKEMATSU, *Division of Chemistry and Chemical Engineering, California Institute of Technology, Pasadena, CA, USA*; GUSTAVO A. GARCIA, *DESIRS beamline, Synchrotron SOLEIL, Gif-sur-Yvette, France*; JOHN F. STANTON, *Department of Chemistry, The University of Texas, Austin, TX, USA*; LAURENT NAHON, *DESIRS Beamline, Synchrotron SOLEIL, Saint Aubin, France*; MITCHIO OKUMURA, *Division of Chemistry and Chemical Engineering, California Institute of Technology, Pasadena, CA, USA.*

The spectroscopy of the nitrate cation NO$_3^+$ remains poorly understood. It has a symmetric D$_{3h}$ ground state with an IP = 12.55 eV and is predicted to have four low lying electronic states, three of E symmetry exhibiting the Jahn-Teller Effect. There have been two low resolution experiments - by photoionization spectroscopy and photoelectron spectroscopy, but evidence for the excited states is controversial. We report preliminary studies on the Threshold PhotoElectron-PhotoIon Coincidence (T-PEPICO) spectrum of the radical cation NO$_3^+$ using the DELICIOUS3 coincidence spectrometer on the DESIRS beamline at the Soleil Synchrotron. Photoelectron and photoion imaging allow us to identify the transitions to the excited states, extract the NO$_3^+$ signal from the ionization of the dominant background species NO$_2$ and N$_2$O$_5$, and observe vibronic spectra and dissociation dynamics of the electronic excited states.

TC04   **9:39 – 9:54**

ULTRAVIOLET PHOTODISSOCIATION DYNAMICS OF THE CYCLOHEXYL RADICAL

MICHAEL LUCAS, YANLIN LIU, JINGSONG ZHANG, *Department of Chemistry, University of California, Riverside, Riverside, Ca, USA.*

The ultraviolet (UV) photodissociation dynamics of the cyclohexyl (c-C_6H_11) radical was studied for the first time using the high-n Rydberg atom time-of-flight (HRTOF) technique. The cyclohexyl radical was produced by the 193 nm photodissociation of chlorocyclohexane and bromocyclohexane and was examined in the photolysis region of 232-262 nm. The H-atom photofragment yield (PFY) spectrum contains a broad peak centering around 250 nm, in good agreement with the UV absorption spectra of cyclohexyl. The translational energy distributions of the H-atom loss product channel, $P(E_T)$'s, show a large translational energy release peaking at 45 kcal/mol. The fraction of average translational energy in the total excess energy, $<f_T>$, is in the range of 0.45-0.57 from 232-262 nm. The H-atom product angular distribution is anisotropic with a positive β parameter in the range of 0.3-1.0, indicating a dissociation time scale faster than one rotation period of the radical. The translational energy release and anisotropy of the H-atom loss product channel are significantly larger than those expected for a statistical unimolecular dissociation of a hot radical, thus showing a non-statistical dissociation mechanism of this large radical. The dissociation mechanism is consistent with direct dissociation on a repulsive excited state surface or on the repulsive part of the ground state surface to produce cyclohexene + H, possibly mediated by conical intersection.

TC05   **9:56 – 10:11**

ULTRAVIOLET PHOTODISSOCIATION DYNAMICS OF THE 3-CYCLOHEXENYL RADICAL

MICHAEL LUCAS, YANLIN LIU, RAQUEL BRYANT, JASMINE MINOR, JINGSONG ZHANG, *Department of Chemistry, University of California, Riverside, Riverside, Ca, USA.*

The ultraviolet (UV) photodissociation dynamics of the cyclohexenyl radical (c-C_6H_9) was studied for the first time in the photolysis region of 232-262 nm using the high-n Rydberg atom time-of-flight (HRTOF) technique. The cyclohexenyl radical was produced by the 193 nm photodissociation of 3-chlorocyclohexene and 3-bromocyclohexene. The H-atom photofragment yield (PFY) spectrum contains a broad peak centering around 250 nm, in good agreement with the UV absorption spectra of the $^2B_1 \leftarrow {}^2A_2$ transition in cyclohexenyl. The translational energy distributions of the H-atom loss product channel, $P(E_T)$'s, for cyclohexenyl show a modest translational energy release peak at $\sim$ 10 kcal/mol. The fraction of average translational energy in the total excess energy, $<f_T>$, is $\sim$ 0.16 from 232-262 nm. The H-atom product angular distribution is isotropic with a β parameter $\sim$ 0. The dissociation mechanism is a statistical unimolecular dissociation of a hot radical following internal conversion from the excited electronic state to produce the lowest energy product, H + cyclohexadiene. The dissociation mechanisms of the cyclohexenyl radical and cyclohexyl radical will be compared.

Intermission

TC06

THERMAL DECOMPOSITION OF BENZYL RADICAL VIA MULTIPLE ACTIVE PATHWAYS

<u>GRANT BUCKINGHAM</u>, *Department of Chemistry and Biochemistry, University of Colorado, Boulder, CO, USA*; DAVID ROBICHAUD, *Biomass Molecular Science , National Renewable Energy Laboratory , Golden, CO, USA*; THOMAS ORMOND, *Department of Chemistry and Biochemistry, University of Colorado, Boulder, CO, USA*; MARK R NIMLOS, *Biomass Molecular Science , National Renewable Energy Laboratory , Golden, CO, USA*; JOHN W DAILY, *Department of Mechanical Engineering, University of Colorado Boulder, Boulder, CO, USA*; BARNEY ELLISON, *Department of Chemistry and Biochemistry, University of Colorado, Boulder, CO, USA*.

The thermal decomposition of benzyl radical ($C_6H_5CH_2$) has been investigated using a combination infrared absorption spectroscopy in a neon matrix and 118.2 (10.487 eV) photoionization mass spectrometry. Both techniques are coupled with a heated tubular reactor to allow temperature control over the decomposition to indicate relative barrier heights of fragmentation pathways. Three possible chemical mechanisms have been considered. 1) Ring expansion to cycloheptatrienyl radical (C_7H_7) with subsequent breakdown to HCCH and C_5H_5, 2) isomerization to the substituted five-membered ring fulvenallene ($C_5H_4=C=CH_2$), which is of interest to kinetic theorists and finally 3) hydrogen shift to form methyl-substituted phenyl radical, which can then form ortho-benzyne, diacetylene and other fragments. Benzyl radical is generated from two precursors, $C_6H_5CH_2CH_3$ and $C_6H_5CH_2Br$, and both lead to the appearance of HCCH and C_5H_5. At slightly hotter temperatures peaks are observed at m/z 90, presumed to be $C_5H_4=C=CH_2$, and 89, potentially the substituted propargyl $C_5H_4=C=CH$. Additionally, decomposition of isotopically substituted parent molecules $C_6H_5CD_2CD_3$ and $C_6D_5CH_2CH_3$ indicates C_7H_7 as an intermediate due to H/D ratios in fragment molecules.

TC07

THE NICKEL ASSISTED DECOMPOSITION OF PENTANAL IN THE GAS PHASE AT VARIOUS INTERNAL ENERGIES

<u>ADAM MANSELL</u>, *Chemistry, Baylor University, Waco, TX, USA*; DARRIN BELLERT, *Chemistry Department, Baylor University, Waco, TX, USA*.

The rate constants for the dissociation of Ni^+Pentanal at various internal energies (15000 cm^{-1}-18800 cm^{-1}) were measured using a custom time of flight instrument. Clusters are generated in a large source chamber by ablating the surface of a rotating nickel rod with an excimer and entraining the ablated metal atoms in a helium gas plume slightly doped with pentanal vapor. The molecular beam enters a Wiley-Mclaren type acceleration grid, and cations are accelerated along a 1.8 meter long time-of-flight mass spectrometer (TOFMS). At the other end of the TOF is a sector and a detector. The sector allows ions of a particular kinetic energy through to the detector. If an ion breaks apart in the time it takes to reach the sector, the mass (and therefore kinetic energy) is reduced, and the sector can be set to allow these fragment ions to reach the detector (fig 2). In our experiment, the energy required to initiate the breakdown is provided by a laser pulse between 15000 cm^{-1} and 18800 cm^{-1}.

TC08

STRONG-FIELD INDUCED DISSOCIATIVE IONIZATION OF VINYL BROMIDE PROBED BY FEMTOSECOND EXTREME ULTRAVIOLET (XUV) TRANSIENT ABSORPTION SPECTROSCOPY

MING-FU LIN[a], *Department of Chemistry, University of California at Berkeley, Berkeley, CA, USA*; DANIEL NEUMARK[b], STEPHEN R. LEONE[c], *Department of Chemistry, The University of California, Berkeley, CA, USA*; OLIVER GESSNER, *UXSL, Chemical Sciences Division, Lawrence Berkeley National Laboratory, Berkeley, CA, USA.*

A table-top high harmonic XUV light source (50 eV to 70 eV) has been successfully utilized to explore the ultrafast dynamics of vinyl bromide ($CH_2=CHBr$) with electronic state specificity and elemental sensitivity. Strong-field ionization (SFI) provides a method to produce ions in different ionic states. The production and dissociation dynamics of these ionic states are investigated by femtosecond XUV transient absorption spectroscopy. The XUV photons probe the time-dependent spectroscopic features associated with transitions of the Br (3d) inner-shell electrons to vacancies in molecular and atomic valence orbitals. The experimental observation shows that two ionic states are produced by SFI. The first ionic excited state is dissociative, leading to C-Br bond dissociation which is observed in real time as a shift in the absorption energy. The results offer powerful new insights about orbital-specific electronic processes in high field ionization, coupled vibrational relaxation and dissociation dynamics, and the correlation of valence hole-state location and dissociation in polyatomic molecules, all probed simultaneously by ultrafast table-top XUV spectroscopy.

[a] 1. UXSL, Chemical Sciences Laboratory, Lawrence Berkeley National Laboratory; 2. CLSL, Department of Chemistry, UIUC
[b] UXSL, Chemical Sciences Division, Lawrence Berkeley National Laboratory
[c] 1. UXSL, Chemical Sciences Division, Lawrence Berkeley National Laboratory; 2. Department of Physics, University of California at Berkeley

TC09

VELOCITY MAP IMAGING STUDIES OF NON-CONVENTIONAL METHANETHIOL PHOTOCHEMISTRY

BENJAMIN W. TOULSON, JONATHAN ALANIZ, CRAIG MURRAY, *Department of Chemistry, University of California, Irvine, Irvine, CA, USA.*

Velocity map imaging (VMI) in combination with state-selective resonance enhanced multiphoton ionization (REMPI) has been used to study the photodissociation dynamics of methanethiol following excitation to the first and second singlet electronically excited states. Formation of sulfur atoms, in both the singlet and triplet manifolds, is observed and can be attributed to primary dissociation of the parent molecule. We will report the nascent photofragment velocity distributions, and hence the internal energy of the methane co-fragment. Sulfur atom quantum yields are benchmarked against a known standard to evaluate the significance of this pathway. The role of non-conventional photochemical mechanisms such as roaming-mediated intersystem crossing, previously observed in methylamine photochemistry,[a] will be discussed.

[a] James O. Thomas, Katherine E. Lower, and Craig Murray, The Journal of Physical Chemistry Letters, 2012, 3 (10), 1341-1345.

TD. Biology, natural substances
Tuesday, June 17, 2014 – 8:30 AM
Room: 112 Chemistry Annex

Chair: Isabelle Kleiner, CNRS et Universités Paris Est et Paris Diderot, Créteil, France

---

TD01   8:30 – 8:45

CONFORMATION-SPECIFIC IR AND UV SPECTROSCOPY OF THE AMINO ACID GLUTAMINE: AMIDE-STACKING AND HYDROGEN BONDING IN AN IMPORTANT RESIDUE IN NEURODEGENERATIVE DISEASES

PATRICK S. WALSH, JACOB C. DEAN, TIMOTHY S. ZWIER, *Department of Chemistry, Purdue University, West Lafayette, IN, USA.*

Glutamine plays an important role in several neurodegenerative diseases including Huntington's disease (HD) and Alzheimer's disease (AD). An intriguing aspect of the structure of glutamine is its incorporation of an amide group in its side chain, thereby opening up the possibility of forming amide-amide H-bonds between the peptide backbone and side chain. In this study the conformational preferences of two capped gluatamines Z(carboxybenzyl)-Glutamine-X (X=OH, NHMe) are studied under jet-cooled conditions in the gas phase in order to unlock the intrinsic structural motifs that are favored by this flexible sidechain. Conformational assignments are made by comparing the hydride stretch (3100-3700 cm^{-1}) and amide I and II (1400-1800 cm^{-1}) resonant ion-dip infrared spectra with predictions from harmonic frequency calculations. Assigned structures will be compared to previously published results on both natural and unnatural residues. Particular emphasis will be placed on the comparison between glutamine and unconstrained γ-peptides due to the similar three-carbon spacing between backbone and side chain in glutamine to the backbone spacing in γ-peptides. The ability of the glutamine side-chain to form amide stacked conformations will be a main focus, along with the prevalence of extended backbone type structures.[a]

---

[a]W. H. James, III, C W. Müller, E. G. Buchanan, M. G. D. Nix, L. Guo, L. Roskop, M. S. Gordon, L. V. Slipchenko, S. H. Gellman, and T. S. Zwier, *J. Am. Chem. Soc.*, **2009**, *131*(40), 14243-14245.

TD02   8:47 – 9:02

SURVEYING THE HYDROGEN BONDING LANDSCAPE OF AN ACHIRAL, α-AMINO ACID: CONFORMATION SPECIFIC IR AND UV SPECTROSCOPY OF 2-AMINOISOBUTYRIC ACID

JOSEPH R GORD, DANIEL M. HEWETT, *Department of Chemistry, Purdue University, West Lafayette, IN, USA*; MATTHEW A. KUBASIK, *Department of Chemistry and Biochemistry, Fairfield University, Fairfield, CT, USA*; TIMOTHY S. ZWIER, *Department of Chemistry, Purdue University, West Lafayette, IN, USA.*

2-Aminoisobutyric acid (Aib) is an achiral, α-amino acid having two equivalent methyl groups attached to C$_\alpha$. Extended Aib oligomers are known to preferentially adopt a 3$_{10}$-helical structure in the condensed phase.[a] Here, we take a simplifying step and focus on the intrinsic folding propensities of Aib by looking at a single, capped Aib structure and then extending to longer oligomers in the gas phase, free from the influence of solvent molecules and cooled in a supersonic expansion. Resonant two-photon ionization and IR-UV holeburning will be used to record single-conformation UV spectra using the Z-cap as UV chromophore. Resonant ion-dip infrared (RIDIR) spectroscopy provides single-conformation IR spectra in the OH stretch, NH stretch, amide I and amide II regions. Two conformational isomers have been identified for the smallest unit in the study, Z-Aib-OH, and four conformational isomers were seen for Z-Aib-Aib-OH, with widely-varying IR spectral patterns. In addition to investigating the conformational dependence on oligomer length, this work also studies the steric and electrostatic impact of different capping groups, R-X where X = -OH, -OMethyl, and -OtButyl. These caps are considered here for the case of Z-Aib-Aib-X. Extension to larger Z-(Aib)$_n$-X oligomers will shed light on the extent to which the solution phase preference for 3$_{10}$-helix formation is retained in the gas phase, and when its onset first appears. When possible ^{13}C isotopomers will be used to assist with the assignments and modulate the coupling between amide I fundamentals.

---

[a]Toniolo, C.; Bonora, G. M.; Barone, V.; Bavoso, A.; Benedetti, E.; Di Blasio, B.; Grimaldi, P.; Lelj, F.; Pavone, V.; Padone, C., Conformation of Pleionomers of α-Aminoisobutyric Acid. *Macromolecules* **1985**, *18*, 895-902.

TD03   **9:04 – 9:14**

ROTATIONAL STUDY OF NATURAL AMINO ACID GLUTAMINE

MARCELINO VARELA, <u>CARLOS CABEZAS</u>, JOSÉ L. ALONSO, *Grupo de Espectroscopia Molecular, Lab. de Espectroscopia y Bioespectroscopia, Unidad Asociada CSIC, Universidad de Valladolid, Valladolid, Spain.*

Recent improvements in laser ablation molecular beam Fourier transform microwave spectroscopy (LA-MB-FTMW) have allowed the investigation of glutamine ($COOH$-$CH(NH_2)$-CH_2-CH_2-$CONH_2$), a natural amino acid with a long polar side chain. One dominant structure has been detected in the rotational spectrum. The nuclear quadrupole hyperfine structure of two ^{14}N nuclei has been totally resolved allowing the conclusive identification of the observed species.

TD04   **9:16 – 9:26**

ROTATIONAL SPECTRUM OF TRYPTOPHAN

<u>M. EUGENIA SANZ</u>[a], CARLOS CABEZAS, SANTIAGO MATA, JOSÉ L. ALONSO, *Grupo de Espectroscopia Molecular, Lab. de Espectroscopia y Bioespectroscopia, Unidad Asociada CSIC, Universidad de Valladolid, Valladolid, Spain.*

The rotational spectrum of the natural amino acid tryptophan has been observed using a recently constructed LA-MB-FTMW spectrometer, specifically designed to optimize the detection of heavier molecules at a lower frequency range. Independent analyses of the rotational spectra of individual conformers have conducted to a definitive identification of two different conformers of tryptophan, with one of the observed conformers never reported before. The experimental values of the ^{14}N nuclear quadrupole coupling constants have been found capital in the discrimination of the conformers. Both observed conformers are stabilized by a O-H$\cdots$N hydrogen bond in the side chain and a N–H$\cdots\pi$ interaction forming a chain that reinforces the strength of hydrogen bonds through cooperative effects.

[a]Department of Chemistry, King's College London, London, UK

TD05   **9:28 – 9:43**

REMPI AND DOUBLE RESONANCE SPECTROSCOPY OF L-β-HOMOTRYPTOPHAN IN GAS PHASE

<u>HYUK KANG</u>, *Department of Chemistry, Ajou University, Suwon, Korea.*

Resonance enhance multiphoton ionization (REMPI) and UV-UV double resonance spectra of L-β-homotryptophan (HTrp) was obtained. Hydrochloride salt of HTrp was applied on a lateral surface of a graphite disk, desorbed by 1064 nm nanosecond laser pulses, and subsequently cooled down by expanding gas from a pulsed valve. Formation of neutral HTrp by laser-desorption was confirmed by its mass spectrum. The REMPI spectrum shows several peaks in a UV region similar to that of tryptophan. UV-UV hole-burning spectrum was obtained and compared with quantum mechanical calculation. Based on the experimental and computational results, conformation preference of HTrp is discussed.

Intermission

TD06   **10:00 – 10:15**

THE CONFORMATIONAL LANDSCAPE OF SERINOL

M. EUGENIA SANZ, DONATELLA LORU, *Department of Chemistry, King's College London, London, United Kingdom*; ISABEL PEÑA, JOSÉ L. ALONSO, *Grupo de Espectroscopia Molecular, Lab. de Espectroscopia y Bioespectroscopia, Unidad Asociada CSIC, Universidad de Valladolid, Valladolid, Spain.*

The rotational spectrum of the amino alcohol serinol $CH_2OH–CH(NH_2)–CH_2OH$, which constitutes the hydrophilic head of the lipid sphingosine, has been investigated using chirped-pulsed Fourier transform microwave spectroscopy in combination with laser ablation[a]. Five different forms of serinol have been observed and conclusively identified by the comparison between the experimental values of their rotational and ^{14}N quadrupole coupling constants and those predicted by ab initio calculations. In all observed conformers several hydrogen bonds are established between the two hydroxyl groups and the amino groups in a chain or circular arrangement. The most abundant conformer is stabilised by O–H···N and N–H···O hydrogen bonds forming a chain rather than a cycle. One of the detected conformers presents a tunnelling motion of the hydrogen atoms of the functional groups similar to that observed in glycerol[b].

[a]S. Mata, I. Peña, C. Cabezas, J. C. López, J. L. Alonso, J. Mol. Spectrosc. 2012, 280, 91
[b]V. V. Ilyushin, R. A. Motiyenko, F. J. Lovas, D. F. Plusquellic, J. Mol. Spectrosc. 2008, 251, 129.

TD07   **10:17 – 10:27**

THE STRUCTURE OF PHENYLGLYCINOL

ALCIDES SIMAO, ISABEL PEÑA, CARLOS CABEZAS, JOSÉ L. ALONSO, *Grupo de Espectroscopia Molecular, Lab. de Espectroscopia y Bioespectroscopia, Unidad Asociada CSIC, Universidad de Valladolid, Valladolid, Spain.*

The most abundant conformer of the amino alcohol D-phenylglycinol has been observed in gas phase using broadband chirped pulse Fourier transform microwave spectroscopy (CP-FTMW) and laser ablation molecular beam Fourier transform microwave spectroscopy (LA-MB-FTMW). The rotational spectra corresponding to seven monosubstituted ^{13}C, one monosubstituted ^{15}N and one monosubstituted ^{18}O species have been observed in their natural abundance, and the r_s structure has been derived. The observed conformer is stabilized by O-H···N, N-H···π intramolecular hydrogen bond network.

TD08   **10:29 – 10:44**

A NUCLEOSIDE UNDER OBSERVATION IN THE GAS PHASE: A ROTATIONAL STUDY OF URIDINE

ISABEL PEÑA, JOSÉ L. ALONSO, *Grupo de Espectroscopia Molecular, Lab. de Espectroscopia y Bioespectroscopia, Unidad Asociada CSIC, Universidad de Valladolid, Valladolid, Spain.*

The nucleoside of uridine has been placed in the gas phase by laser ablation and the most stable C2'-anti conformation characterized by broadband chirped pulse (CP-FTMW) and narrowband molecular beam Fourier transform microwave (LA-MB-FTMW) spectroscopies. The quadrupole hyperfine structure, originated by two ^{14}N nuclei, has been completely resolved. Intramolecular hydrogen bonds involving uracil and ribose moieties have been found to play an important role in the stabilization of the nucleoside.

TD09   **10:46 – 11:01**

THE CONFORMATIONAL BEHAVIOUR OF GLUCOSAMINE

ISABEL PEÑA, LUCIE KOLESNIKOVÁ, CARLOS CABEZAS, CELINA BERMÚDEZ, MATÍAS BERDAKIN, ALCIDES SIMAO, JOSÉ L. ALONSO, *Grupo de Espectroscopia Molecular, Lab. de Espectroscopia y Bioespectroscopia, Unidad Asociada CSIC, Universidad de Valladolid, Valladolid, Spain.*

A laser ablation method has been successfully used to vaporize the bioactive amino monosaccharide D-glucosamine. Three cyclic α-$^4C_1$ pyranose forms have been identified using a combination of CP-FTMW and LA-MB-FTMW spectroscopy. Stereoelectronic hyperconjugative factors, like those associated with anomeric or gauche effects, as well as the cooperative OH···O, OH···N and NH···O chains, extended along the entire molecule, are the main factors driving the conformational behavior. All observed conformers exhibit a counter-clockwise arrangement (*cc*) of the network of intramolecular hydrogen bonds. The results are compared with those recently obtained for D-glucose.[a]

[a]J. L. Alonso, M. A. Lozoya, I. Peña, J. C. López, C. Cabezas, S. Mata, S. Blanco, *Chem. Sci.* **2014**, *5*, 515.

TD10   **11:03 – 11:13**

THE DIPEPTIDE ALA-GLY IN THE GAS PHASE

CELINA BERMÚDEZ, MARCELINO VARELA, CARLOS CABEZAS, ISABEL PEÑA, JOSÉ L. ALONSO, *Grupo de Espectroscopia Molecular, Lab. de Espectroscopia y Bioespectroscopia, Unidad Asociada CSIC, Universidad de Valladolid, Valladolid, Spain.*

The dipeptide Ala-Gly has been examined in gas phase by laser ablation molecular beam Fourier transform microwave (LA-MB-FTMW) spectroscopy in the frequency region 3-12 GHz. Three rotamers have been detected in the supersonic expansion. The quadrupole hyperfine structure of two ^{14}N (I=1) nuclei has been totally resolved allowing the conclusive identification of one conformer.

TD11   **11:15 – 11:30**

STRUCTURES AND ELECTRONIC SPECTRA OF THE CAFFEINE AND ITS HYDRATED CLUSTERS: A COMPUTATIONAL STUDY

VIPIN BAHADUR SINGH[a], *Department of Physics, Udai Pratap Autonomous College, Varanasi, India.*

Extensive ab initio and Density Functional Theory (DFT) calculations were employed to characterize the ground state structures of caffeine and its hydrated clusters. The geometry optimizations of possible rotamers of bare caffeine (differs only in torsional angles of the methyl groups) using MP2 and DFT (M06, M06-2X, X3LYP, B3LYP and B3PW91) methods employing the basis set 6-311++G(d,p), yield four stable structures. Maximum relative energy between the conformers was found to be 4.44 KJ/Mol at MP2/6-311++G(d,p) level which amended the previously reported theoretical value 1.84 KJ/Mol. We have also determined the five lowest energy optimized structures of the 1:1 complex of caffeine with water $caff_1(water)_1$, four O-bonded and one N-bonded, amongst which the two most stable O-bonded caffeine monohydrates were reported for the first time. Interestingly, vertical excitation energy to the lowest excited state S_1 ($1\pi\pi^*$) for the O-bonded $caff_1(water)_1$ clusters involving isolated carbonyl were blue shifted whereas that involving conjugated carbonyl were red shifted. The absence of the red shifted bands in the R2PI spectra is expected due to the fast radiationless decay of the complexes involving conjugated carbonyl to the electronic ground state.

[a]single authorship

TD12   **11:32 – 11:37**

FLUORESCENCE SWITCH FOR SELECTIVELY SENSING COPPER AND HISTIDINE IN BOTH VITRO AND LIVING CELLS

XIAOJING WANG, *Department of Chemistry, University of Science and Technology of China, Hefei, China.*

One new synthetic probes for the detection of copper and Histidine in both vitro and living cells. In the absence of metal ions, the new established probes exhibits comparable fluorescence to that of free FITC. In the presence of metal ions, probes selectively coordinates with Cu^{2+}, causing its fluorescence emission quenched via photoinduced electron transfer. Interestingly, as-formed complex selectively responds to $_L$-His among the 20 natural AAs by turning its fluorescence on. Using this dualfunctional probe, we also sequentially imaged Cu^{2+} and $_L$-His in living cells. Our new probe could be applied for not only environment monitoring or biomolecule detections, but also disease diagnoses in the near future.

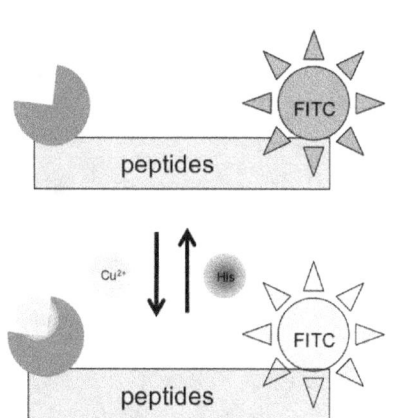

102

TD13

BASED ON THE NATIONAL SCALE SOIL SPECTRAL DATABASE NITROGEN CONTENT INVERSION

QIANLONG WANG, *College of Environmental and Resource Sciences, Zhejiang University, Hangzhou, Zhejiang, China.*

Fully mining the valid information in soil spectral library, establishing a strong universal inversion model to predict soil total nitrogen (TN) content is one of the important applications of high spectral direction. Studies using partial least squares regression (PLSR) global model, locally weighted regression (LWR) and fuzzy K-means clustering methods combined with PLSR (FKMC-PLSR). 1661 soil samples were collected from 13 provinces in China, which include Tibet, Xinjiang, Heilongjiang, and Hainan. The samples represent 17 soil groups of the Chinese Soil (Genetic) Classification System, and Zhejiang province 104 paddy soil samples to predict. The results show that, under the national scale PLSR global model for high TN values underestimated the prevalence of samples to be predicted, resulting in low overall predictive accuracy; LWR (R^2 = 0.76, RPD_{P2} = 2.1), especially FKMC-PLSR (R^2 = 0.82, RPD_{P3} = 2.4) than the local model PLSR (R^2 = 0.64, RPD_{P1} = 1.4) global model can more accurately inversion TN content. The results can take advantage of the national scale to establish stability and universal spectral database higher content of soil TN forecasting model to provide the necessary information.

TD14

GREY INCIDENCE ANALYSIS (GIA): A NEW LOCAL METHOD FOR MODELLING CHINESE SOIL VIS-NIR SPECTRAL LIBRARY TO PREDICT SOIL TOTAL NITROGEN CONTENT

QIANLONG WANG, *College of Environmental and Resource Sciences, Zhejiang University, Hangzhou, Zhejiang, China.*

This paper introduces a new approach called grey incidence analysis(GIA), by which high accuracy prediction model can be established combined with partial least squares regression(PLSR) to deal with Chinese soil vis-NIR spectral library and estimate soil total nitrogen content in local area. Using spectral matching algorithm such as Mahalanobis distance, spectral angle model (SAM) and spectral correlation fitting(SCF), fuzzy k-means clustering method only based on the spectral data without considering soil total nitrogen content in the library. Soil total nitrogen content have grey uncertainty relationship with each vis-NIR spectral band(400-2500nm). This method combine the soil total nitrogen value with spectrum data when performing spectral matching. In this study, 1661 soil samples in the library were collected from 13 provinces in China, which include Tibet, Xinjiang, Heilongjiang, and Hainan. The samples represent 17 soil groups of the Chinese Soil (Genetic) Classification System. After air-drying and sieving, the diffuse reflectance spectra of the samples were measured under laboratory conditions in the range between 400 and 2500 nm using a portable vis-NIR spectrometer. Hyperspectral inversion model was built based on 104 paddy soil samples in Zhejiang province to predict the soil total nitrogen content. The results show that the GIA-PLSR method presents great potential for predicting soil total nitrogen content in large soil vis-NIR library. The prediction accuracy: R^2 = 0.897, $RMSE_P$ = 0.028, RPD_P = 3.151. This study also show that soil vis-NIR spectroscopy combined with TN value can be used to further improve the prediction performance of spectral models.

TE. Structure determination

Tuesday, June 17, 2014 – 8:30 AM

Room: 274 Medical Sciences Building

Chair: Stewart E. Novick, Wesleyan University, Middletown, CT, USA

TE01   **8:30 – 8:45**

HIGH-RESOLUTION INFRARED SPECTROSCOPY SLIT-JET COOLED HYDROXYMETHYL RADICAL (CH_2OH): CH SYMMETRIC STRETCHING MODE

FANG WANG, CHIH-HSUAN CHANG, *JILA, UCB-NIST, Boulder, CO, USA*; DAVID NESBITT, *Department of Chemistry, JILA CU-NIST, Boulder, CO, USA.*

Hydroxymethyl radical (CH_2OH) plays an important role in combustion and environmental chemistry as a reactive intermediate. Reisler's group published[a] the first rotationally resolved spectroscopy of CH_2OH with determined band origins for fundamental CH symmetric stretch state, CH asymmetric stretch state and OH stretch state, respectively. Here CH_2OH was first studied via sub-Doppler infrared spectroscopy in a slit-jet supersonic discharge expansion source. Rotationally resolved direct absorption spectra in the CH symmetric stretching mode were recorded. As a result of the low rotational temperature and sub-Doppler linewidths, the tunneling splittings due to the large amplitude of COH torsion slightly complicate the spectra. Each of the ground vibration state and the CH symmetric stretch state includes two levels. One level, with a 3:1 nuclear spin statistic ratio for $K_a=0^+/K_a=1^+$, is labeled as "+". The other tunneling level, labeled as "-", has $K_a=0^-/K_a=1^-$ states with 1:3 nuclear spin statistics. Except for the $K_a=0^+ \leftarrow 0^+$ band published before [b], more bands ($K_a=1^+ \leftarrow 1^+$, $K_a=0^- \leftarrow 0^-$ and $K_a=1^- \leftarrow 1^-$) were identified. The assigned transitions were fit to a Watson A-reduced symmetric top Hamiltonian to improve the accuracy of the band origin of CH symmetric state. The rotational parameters for both ground and CH symmetric stretch state were well determined.

[a] L. Feng, J. Wei and H. Reisler, J. Phys. Chem. A, Vol. 108.

[b] M. A. Roberts, E. N. Sharp-Williams and D. J. Nesbitt, J. Phys. Chem. A 2013, 117, 7042-7049

TE02   **8:47 – 8:57**

FOURIER-TRANSFORM MICROWAVE AND MILLIMETERWAVE SPECTROSCOPY OF CH_2IBr IN ITS GROUND VIBRATIONAL STATE

KOTOMI TANIGUCHI, SHOHEI SAKAI, HIROYUKI OZEKI, *Department of Environmental Science, Toho University, Funabashi, Japan*; TOSHIAKI OKABAYASHI, *Graduate School of Science and Technology, Shizuoka University, Shizuoka, Japan*; WILLIAM C. BAILEY, *Department of Chemistry-Physics, Kean University (Retired), New Jersey, USA*; DENIS DUFLOT, STEPHANE BAILLEUX, *Laboratoire PhLAM, Université de Lille 1, Villeneuve de Ascq, France.*

Halo-substituted methanes constitute a class of molecules that are important in various fields, from spectroscopy to quantum-chemical calculations. They are also gaining interest due to their potential adverse impact on the atmospheric chemistry.[a]

In the series of the CH_2IX iodomethanes where $X = \{F, Cl, Br\}$, only the rotational spectra of CH_2IF [b] and CH_2ICl [c] have been published. We present our investigations on the high-resolution rotational spectroscopy of the two bromine isotopologues of bromoiodomethane, $CH_2I^{79}Br$ and $CH_2I^{81}Br$.

Due to the lack of spectroscopic information available for this compound, high-level quantum-chemical calculations were essential to guide the microwave and millimeterwave spectral assignments of both μ_a- and μ_b-type transitions. They provided rotational and centrifugal distortion constants (quartic and sextic), as well as the quadrupole-coupling tensor of the iodine ($I_I = 5/2$) and bromine ($I_{Br} = 3/2$) nuclei.

More than 1900 lines have been analyzed, leading to an accurate determination of molecular constants for both isotopologues. The experimental structure (r_0) of the title species has been derived from the two sets of rotational constants.

[a] S.B. acknowledges support from the Laboratoire d'Excellence CaPPA (**C**hemical **a**nd **P**hysical **P**roperties of the **A**tmosphere) through contract ANR-10-LABX-005 of the Programme d'Investissement d'Avenir.

[b] C. Puzzarini, G. Cazzoli, J. C. López, J. L. Alonso, A. Baldacci, A. Baldan, S. Stopkowicz, L. Cheng and J. Gauss, J. Chem. Phy. 62, 174312 (2011).

[c] S. Bailleux, H. Ozeki, S. Sakai, T. Okabayashi, P. Kania and D. Duflot, J. Mol. Spectrosc. 270, 51 (2011).

104

GAS PHASE MICROWAVE MEASUREMENTS OF MONO-FLUOROBENZOIC ACIDS.

ADAM M DALY, *Jet Propulsion Laboratory, California Institute of Technology, Pasadena, CA, USA*; STEPHEN G. KUKOLICH, *Department of Chemistry and Biochemistry, University of Arizona, Tucson, AZ, USA.*

We report the rotational and distortion constants of 3 conformers of 2-fluorobenzoic acid, 2 conformers of 3-fluorobenzoic acid and a single conformer of 4-fluorobenzoic acid fitted from the assignment of pure rotational transitions measured in the microwave region from 4-12 GHz. We also recorded the microwave spectrum and assigned the b-dipole transitions of a very large dimer of 3-fluorobenzoic acid which has very small rotational constants. The b-dipole transitions were split and assigned to the two states 0+ and 0- vibrational states with rotational constants: $A^+ =$ 1157.01939(75) MHz, $B^+ = 95.45061(199)$ MHz and $C^+ = 88.21514(124)$ and $A^- = 1157.02249(46)$, $B^- = 95.45110(74)$ and $C^- 88.24425(70)$. This large dimer is a milestone in our groups efforts at "climbing" the ladder to very large dimers that contain dynamics information.

THE COMBINED ORTHO / PARA HYDROGEN ASSIGNMENTS IN H_2 METAL CHLORIDES

DANIEL A. OBENCHAIN, *Department of Chemistry, Wesleyan University, Middletown, CT, USA*; G. S. GRUBBS II, *Department of Chemistry, Missouri University of Science and Technology, Rolla, MO, USA*; DEREK S. FRANK, HERBERT M. PICKETT, STEWART E. NOVICK, *Department of Chemistry, Wesleyan University, Middletown, CT, USA.*

The rotational spectra of H_2-AgCl and H_2-AuCl have been measured using a cavity FTMW spectrometer equipped with a laser ablation source. A combination of isotopic substitution, including HD and D_2 substitutions, and the spin-spin interaction of *ortho* hydrogen were used to determine the structures of these species. Trends in these structures and the strengths of the H_2 interaction will be discussed.

Previous work with hydrogen containing complexes have shown that separate spectra are observed for the both the *ortho* and *para* hydrogen species.[a;b;c] In this work, *ortho* and *para* hydrogen are assigned together. The a-axis in the present species is coincident with internal rotation axis of hydrogen. This symmetry, along with a covalent interaction of the H_2 with the metal chlorides, allows for a straightforward global assignment of the *ortho* and *para* species. The differences in the present study from the previous works will be discussed, as well as the assignment of the combined *ortho* and *para* fits.

[a] Y. Zhenhong, K. J. Higgins, W. Klemperer, M. C. McCarthy and P. Thaddeus, *J. Chem. Phys.*, **123**(2005) 221106.

[b] J. M. Michaug, W. C. Topic, W. Jäger, *J. Phys. Chem. A.*, **115**(2011) 9456.

[c] M. Ishiguro, K Harada, K. Tanaka, Y. Sumiyoshi, Y. Endo, *Chem. Phys. Lett.*, **554**(2012) 33.

TE05 9:33 – 9:48

CHIRPED PULSE AND CAVITY FOURIER TRANSFORM MICROWAVE (CP-FTMW AND FTMW) INVESTIGATIONS INTO 3-BROMO-1,1,1,2,2-PENTAFLUOROPROPANE; A MOLECULE OF ATMOSPHERIC INTEREST

NICHOLAS FORCE, DAVID JOSEPH GILLCRIST, CASSANDRA C. HURLEY, FRANK E MARSHALL, NICHOLAS A. PAYTON, THOMAS D. PERSINGER, N. E. SHREVE, G. S. GRUBBS II, *Department of Chemistry, Missouri University of Science and Technology, Rolla, MO, USA.*

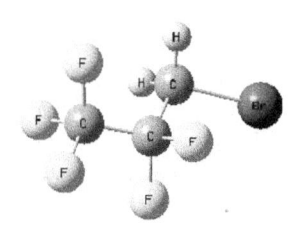

The microwave spectrum of the molecule 3-bromo-1,1,1,2,2-pentafluoropropane has been measured on a newly constructed CP-FTMW spectrometer along with a FTMW spectrometer relocated from Oxford University to Missouri S&T. 3-bromo-1,1,1,2,2-pentafluoropropane has been cited as a possible safer alternative for replacing CFCs as refrigerants and this is the first of a series of studies to understand the chemistry of 3-bromo-1,1,1,2,2-pentafluoropropane with other atmospheric cleaning agents. Rotational constants, centrifugal distortion parameters, and nuclear quadrupole coupling constants will be discussed.

The CP-FTMW utilized in this experiment will be discussed in great detail. The new machine has been assembled to directly create and digitize up to 18 GHz signals without the need of mixing on the broadcast or detection side of the experiment allowing for the elimination of many microwave components typically needed in both CP-FTMW and FTMW experiments.

Intermission

TE06 10:05 – 10:20

MICROWAVE SPECTRUM AND MOLECULAR STRUCTURE OF THE ARGON-(E)-1-CHLORO-1,2-DIFLUOROETHYLENE COMPLEX

MARK D. MARSHALL, HELEN O. LEUNG, HANNAH TANDON, JOSEPH P. MESSINGER, ELI MLAVER, *Chemistry Department, Amherst College, Amherst, MA, USA.*

Previous studies of argon complexes with fluoroethylenes have revealed a preference for a geometry that maximizes the contact of the argon atom with heavy atoms on the fluoroethylene.[a] We have observed a continuation of this trend when one of the fluorine atoms is replaced by chlorine. The argon-(E)-1-chloro-1,2-difluoroethylene complex provides two competing heavy atom cavities, FCCF and FCCl, and the opportunity to examine whether the number of heavy atoms or the associated increase in polarizability is determinative of structure. The 5.6 – 18.1 GHz chirped-pulse Fourier transform microwave spectrum of this species provides initial assignments and predictions for spectra obtained in a more sensitive and higher precision Balle-Flygare instrument. Transitions for both the ^{35}Cl and ^{37}Cl isotopologues are observed and analyzed to provide geometric parameters for this non-planar complex. The spectrum is consistent with the argon atom located in the FCCl cavity, and the structure agrees well with *ab initio* predictions. Comparisons are made with Ar-1-chloro-1-fluoroethylene, (Z)-1-chloro-2-fluoroethylene, and Ar-vinyl chloride.

[a]Z. Kisiel, P.W. Fowler, and A.C. Legon, *J. Chem. Phys.* **95**, 2283 (1991).

TE07 10:22 – 10:37

THE EFFECT OF PROTIC ACID IDENTITY ON THE STRUCTURES OF COMPLEXES WITH VINYL CHLO-
RIDE: FOURIER TRANSFORM MICROWAVE SPECTROSCOPY AND MOLECULAR STRUCTURE OF THE VINYL
CHLORIDE-HYDROGEN CHLORIDE COMPLEX

JOSEPH P. MESSINGER, HELEN O. LEUNG, MARK D. MARSHALL, *Chemistry Department, Amherst College, Amherst, MA, USA.*

In all previous examples of complexes formed between protic acids and haloethylenes, we have observed similar modes of binding regardless of the specific identity of the acid, HF, HCl, or HCCH. Although details of the structures, such as hydrogen bond length and amount of deviation from linearity, do reflect the strength of the interaction and show clear correlations with the gas-phase acidity, the complexes of a given haloethylene with any of the acids have identical structural motifs. Vinyl chloride, on the other hand, has been observed to adopt different modes of binding in its interactions with HF and HCCH. The HF complex, reported two years ago, has a geometry with HF interacting across the double bond of vinyl chloride and forming a secondary interaction with the hydrogen *cis* to the chlorine atom, but in the complex with acetylene, reported last year, HCCH locates at one end of the vinyl chloride with the secondary interaction occurring with the geminal hydrogen atom. This variety continues and is expanded in the vinyl chloride-HCl complex. *Ab initio* theory predicts a complex that has the HCl molecule interacting across the double bond, but located out of the vinyl chloride plane. The microwave spectrum of the most abundant isotopologue of this complex is consistent with theoretical predictions and additionally shows the presence of large amplitude motion connecting two equivalent structures.

TE08 10:39 – 10:54

MICROWAVE SPECTRUM OF THE HNO_3-HCOOH COMPLEX

BECCA MACKENZIE, CHRIS DEWBERRY, KEN LEOPOLD, *Chemistry Department, University of Minnesota, Minneapolis, MN, USA.*

The interconversion between two equivalent configurations of a system via exchange of protons is an important process with implications in areas ranging from chemical dynamics to molecular biology. In this work, we present microwave spectroscopic results and *ab initio* calculations on the doubly hydrogen bonded complex HNO_3-HCOOH. Spectra of seven isotopologues of the system confirm a doubly hydrogen bonded geometry, analogous to that of the well studied carboxylic dimers. Unlike most carboxylic acid dimers, however, the two hydrogen bond lengths (and their associated heavy atom distances) are substantially different, causing the double proton transfer process to be accompanied by significant heavy atom motion. Splittings in the a-type rotational spectrum are observed but disappear for HNO_3-HCOOD, indicating a tunneling motion of protons across the hydrogen bonds. *Ab initio* calculations of the binding energy and intermolecular potential surface are compared with those on the more thoroughly studied carboxylic dimers and the role of heavy atom motion is discussed. Spectroscopic constants provide accurate structural information and quantify the degree to which the electronic structure of the HNO_3 is altered upon complexation.

TE09 10:56 – 11:11

IMPACT FT-MW SPECTROSCOPY OF ORGANIC RINGS: INVESTIGATION OF THE CONFORMATIONAL
LANDSCAPE

DENNIS WACHSMUTH, JENS-UWE GRABOW, *Institut für Physikalische Chemie und Elektrochemie, Gottfried-Wilhelm-Leibniz-Universität, Hannover, Germany*; ALBERTO LESARRI, *Department Quimica Fisica y Quimica Inorganica, Universidad de Valladolid, Valladolid, Spain.*

Organic rings are common building blocks in biologically active molecules. In contrast to six-membered rings, cycloalkanes with seven or more ring atoms have a less rigid structure which allows for conformational diversity and complex internal dynamics. The broadband *in-phase/quadrature-phase-modulation passage-acquired-coherence technique* (IMPACT) FT-microwave spectroscopy is the ideal experimental method to gain insight into the conformational freedom and structure of these cyclic motifs.

The combination of high spectral resolution (< 10 kHz) and frequency accuracy with a 1 GHz wideband acquisition range per experimental pulse (over the entire 2 to 26.5 GHz range) makes the IMPACT spectrometer a time-saving tool when it comes to the investigation of fine or hyperfine effects and/or wide tunelling splittings. In this communication we will present the rotational spectra of several organic seven-membered rings, reporting on their conformational preferences and molecular structures.

TE10   **11:13 – 11:28**

MOLECULAR CHIRALITY: ENANTIOMER DIFFERENTIATION BY HIGH-RESOLUTION SPECTROSCOPY

EIZI HIROTA, *The Central Office, The Graduate University for Advanced Studies, Hayama, Kanagawa, Japan.*

I have demonstrated that triple resonance performed on a three-rotational-level system of a chiral molecule of C_1 symmetry exhibits signals opposite in phase for different enantiomers, thereby making enantiomer differentiation possible by microwave spectroscopy[a]. This prediction was realized by Patterson et al. on 1,2-propanediol and 1,3-butanediol[b]. We thus now add a powerful method: microwave spectroscopy to the study of chiral molecules, for which hitherto only the measurement of optical rotation has been employed. Although microwave spectroscopy is applied to molecules in the gaseous phase, it is unprecedentedly superior to the traditional method: polarimeter in resolution, accuracy, sensitivity, and so on, and I anticipate a new fascinating research area to be opened in the field of molecular chirality. More versatile and efficient systems should be invented and developed for microwave spectroscopy, in order to cope well with new applications expected for this method

For C_2 and C_n ($n \geq 3$)chiral molecules, the three-rotational-level systems treated above for C_1 molecules are no more available within one vibronic state. It should, however, be pointed out that, if we take into account an excited vibronic state in addition to the ground state, for example, we may encounter many three-level systems. Namely, either one rotational transition in the ground state is combined with two vibronic transitions, or such a rotational transition in an excited state may be connected through two vibronic transitions to a rotational level in the ground state manifold.

The racemization obviously plays a crucial role in the study of molecular chirality. However, like many other terms employed in chemistry, this important process has been "defined" only in a vague way, in other words, it includes many kinds of processes, which are not well classified on a molecular basis. I shall mention an attempt to obviate these shortcomings in the definition of racemization and also to clarify the implicit assumptions made in Hund's paradox[c].

[a]E. Hirota, 3rd Molecular Science Symposium, Nagoya, September 2009, E. Hirota, Proc. Jpn. Acad. Ser. B, 88, 120 (2012).
[b]D. Patterson, M. Schnell and J. M. Doyle, Nature 497, 475 (2013), D. Patterson and J. M. Doyle, Phys. Rev. Lett. 111, 023008 (2013).
[c]F. Hund, Z. Phys. 43, 805 (1927).

TE11   **11:30 – 11:45**

A PEPTIDE CO SOLVENT IN A CHIRALITY INDUCTION MODEL SYSTEM: BROADBAND ROTATIONAL SPECTROSCOPY OF THE 2,2,2-TRIFLUOROETHANOL- -PROPYL ENE OXIDE ADDUCT

JAVIX THOMAS, YUNJIE XU, *Department of Chemistry, University of Alberta, Edmonton, AB, Canada.*

Chirality induction in a model system, i.e. the 2,2,2-trifluoroethanol (TFE)- -propylene oxide (PO) adduct, was investigated in detail using chirped pulse and cavity based Fourier transform microwave spectroscopy, complemented with high level ab initio calculations. Hydrogen-bonding interaction of TFE with the permanently chiral PO molecule results in eight binary TFE- -PO diastereomers. Rotational spectra of four of them were observed experimentally and unambiguously assigned and identified. Unlike the TFE dimer where an extreme case of chirality synchronization was previously reported, diastereomers due to both the g+ and g- forms of TFE were observed, indicating that the tunneling between the two isoenergetic gauche forms of TFE was quenched. Comparison to the previous studies reveals that perfluorination increases the hydrogen-bonding energy by about 70% over its ethanol counterpart. TFE- -PO serves as a prototype system for chirality induction which leads to chirality amplification rather than a system with chirality synchronization.

TE12   **11:47 – 11:57**

CONORMATONAL AND VIBRATIONAL ANALYSIS OF DIBENZO-18-CROWN-6.[a]

A A EL-AZHARY, N. A. AL-JALLAL, *Chemistry, King Saud University, Riyadh, Riyadh, Saudi Arabia.*

We performed conformational analysis of dibenzo-18-crown-6 (db18c6) using the CONFLEX conformational search method. Ab initio computations were performed for the predicted conformations in the gas phase and solution using the PCM model. The computations were performed at the B3LYP, PBE1PBE and MP2 levels and for some selected low energy conformations at the G3MP2 level. The DFT functionals were used with and without the dispersion correction known as the DFT-D3. The vibrational, IR and Raman, spectra of db18c6 were measured. Comparison between the measured and calculated vibrational spectra using scaled-quantum-mechanical force field of some of the low energy conformations aided in the prediction in what conformation db18c6 exits. Use of the dispersion correction seemed to have almost no effect on the calculated vibrational spectra.

[a]Supported by NPST program by King Saud University, Project Number: 12-ENV2680-02.

TF. Mini-symposium: Astronomical Molecular Spectroscopy in the Age of ALMA

Tuesday, June 17, 2014 – 1:30 PM

Room: 116 Roger Adams Lab

Chair: Geoffrey Blake, California Institute of Technology, Pasadena, CA, USA

TF01	*INVITED TALK*	1:30 – 2:00

PHYSICS AND CHEMISTRY OF STAR AND PLANET FORMATION IN THE ALMA ERA

EDWIN BERGIN, *Department of Astronomy, University of Michigan, Ann Arbor, MI, USA.*

ALMA will open up new avenues of exploration encompassing the wide range of star formation in our galaxy and peering into the central heart of planet-forming circumstellar disks. As we seek to explore the origins of stars and planets molecular emission will be at the front and center of many studies probing gas physics and chemistry. In this talk I will discus some of the areas where we can expect significant advances due to the increased sensitivity and superb spatial resolution of ALMA. In star-forming cores, a rich chemistry is revealed that may be the simpler molecular precursors to more complex organics, such as amino acids, seen within primitive rocks in our own solar system. ALMA will provide new information regarding the relative spatial distribution within a given source for a host of organics, sampling tens to hundreds of transitions of a variety of molecules, including presumably new ones. In this area there is a rich synergy with existing ground and space-based data, including Herschel/Spitzer. Here the increased sampling of sources to be enabled by ALMA should bring greater clarity toward the key products of interstellar chemistry and further constrain processes. On smaller Solar System scales, for over a decade most observations of planet-forming disks focused on the dust thermal continuum emission as a probe of the gas content and structure. ALMA will enable reliable and direct studies of gas to explore the evolving physics of planet-formation, the gas dissipation timescales (i.e. the upper limit to the timescale for giant planet birth), and also the chemistry. It is this chemistry that sets the composition of gas giants and also influences the ultimate composition of water and organic materials that are delivered to terrestrial worlds. Here I will show how we can use molecular emission to determine the gas thermal structure of a disk system and the total gas content - key astrophysical quantities. This will also enable more constrained chemical studies that will seek to determine whether the chemistry of planetary birth is universal and similar to our own.

TF02

UBIQUITOUS ARGONIUM, ArH$^+$, IN THE DIFFUSE INTERSTELLAR MEDIUM

P. SCHILKE, HOLGER S. P. MÜLLER, C. COMITO, A. SANCHEZ-MONGE, *I. Physikalisches Institut, Universität zu Köln, Köln, Germany*; D. A. NEUFELD, NICK INDRIOLO, *Department of Physics and Astronomy, Johns Hopkins University, Baltimore, MD, USA*; EDWIN BERGIN, *Department of Astronomy, University of Michigan, Ann Arbor, MI, USA*; D. C. LIS, *Cahill Center for Astronomy and Astrophysics, California Institute of Technology, Pasadena, CA, USA*; MARYVONNE GERIN, *LERMA, Observatoire de Paris, Paris, France*; J. H. BLACK, *Onsala Space Observatory, Chalmers University of Technology, Onsala, Sweden*; M. G. WOLFIRE, *Department of Astronomy, University of Maryland, College Park, MD, USA*; JOHN PEARSON, *Jet Propulsion Laboratory, California Institute of Technology, Pasadena, CA, USA*; KARL MENTEN, B. WINKEL, *Millimeter- und Submillimeter-Astronomie, Max-Planck-Institut für Radioastronomie, Bonn, NRW, Germany.*

ArH$^+$ is isoelectronic with HCl. The $J = 1 - 0$ and $2 - 1$ transitions of ^{36}ArH$^+$ near 617.5 and 1234.6 GHz, respectively, have been identified very recently as emission lines in spectra obtained with *Herschel* toward the Crab Nebula supernova remnant.[a] On Earth, ^{40}Ar is by far the most abundant isotope, being almost exclusively formed by the radioactive decay of ^{40}K. However, ^{36}Ar is the dominant isotope in the Universe.

In the course of unbiased line surveys of the massive and very luminous Galactic Center star-forming regions Sagittarius B2(M) and (N) with the high-resolution instrument HIFI on board of *Herschel*, we detected the $J = 1 - 0$ transition of ^{36}ArH$^+$ as a moderately strong absorption line initially associated with an unidentified carrier.[b] In both cases, the absorption feature is unique in its appearance at all velocity components associated with diffuse foreground molecular clouds, together with its conspicuous absence at velocities related to the denser sources themselves. Model calculations are able to reproduce the derived ArH$^+$ column densities and suggest that argonium resides in the largely atomic, diffuse interstellar medium with a molecular fraction of no more than $\sim 10^{-4}$. The ^{38}ArH$^+$ isotopolog was also detected.

Subsequent observations toward the continuum sources W51, W49, W31C, and G34.3+0.1 resulted in unequivocal detections of ^{36}ArH$^+$ absorption. Hence, argonium is a good probe of the transition zone between atomic and molecular gas, in particular in combination with OH$^+$ and H$_2$O$^+$, whose abundances peak at a molecular fraction of ~ 0.1. Moreover, argonium is a good indicator of an enhanced cosmic ray ionization rate. Therefore, it may be prominent toward, e.g., active galactic nuclei (AGNs) in addition to supernova remnants.

---

[a]M. J. Barlow et al., *Science* **342** (2013) 1343.

[b]H. S. P. Müller et al., Proceedings of the IAU Symposium 297, 2013, "The Diffuse Interstellar Bands", Eds. J. Cami & N. Cox.

TF03

ACCURATE LABORATORY MEASUREMENTS OF VIBRATION-ROTATION TRANSITIONS OF ^{36}ArH$^+$ and ^{38}ArH$^+$

MAITE CUETO, *Molecular Physics, Instituto de Estructura de la Materia (IEM-CSIC), Madrid, Spain*; JOSE CERNICHARO, *Departamento de Astrofísica, Centro de Astrobiología CAB, CSIC-INTA, Madrid, Spain*; VICTOR JOSE HERRERO, ISABEL TANARRO, JOSE LUIS DOMENECH, *Molecular Physics, Instituto de Estructura de la Materia (IEM-CSIC), Madrid, Spain.*

The protonated Ar ion ^{36}ArH$^+$ has recently been identified in space,[a] in the Crab Nebula, from Herschel spectra. Its R(0) and R(1) transitions lie at 617.5 and 1234.6 GHz, respectively, where atmospheric transmission is rather poor, even for a site as good as that of ALMA. As an alternative, especially after the end of the Herschel mission, rovibrational transitions of ArH$^+$ could be observed in absorption against bright background sources such as the galactic center, or other objects.

We report on accurate laboratory wavenumber measurements of 19 lines of the $v = 1 - 0$ band of ^{36}ArH$^+$ and ^{38}ArH$^+$, using a hollow cathode discharge cell, a difference frequency laser spectrometer and Ar with natural isotopic composition. Of those lines, only eight had been reported before and with much less accuracy.[b] The data have also been used in a Dunham-type global fit of all published laboratory data (IR and sub-mm) of all isotopologues.[c]

---

[a]Barlow et al., Science, 342, 1343 (2013)

[b]R.R. Filgueira and C.E. Blom, J. Mol. Spectrosc., 127, 279 (1988)

[c]M. Cueto et al, Astrophys. J. Lett, 783, L5 (2014)

TF04   2:39 – 2:54

THE ALMA ORION BAND 6 SCIENCE VERIFICATION DATA SPECTRAL LINE SURVEY - CONTENT AND DIS-COVERY AVAILABLE FOR ALL SCIENCE

ANTHONY REMIJAN, *ALMA, National Radio Astronomy Observatory, Charlottesville, VA, USA*; SUZANNA K. RANDALL, *EU ALMA Regional Center, European Space Agency, Garching, Germany*; CATHERINE VLA-HAKIS, *Commissioning and Science Verification, ALMA, Santiago, Chile.*

This presentation will provide an overview of the ALMA Science Verification observations taken toward the Orion KL region - the most massive star forming region to the Solar System at a distance of $\sim$450 pc - in a 30 GHz total bandwidth spectral survey in ALMA Band 6 from 214 - 246 GHz frequency range (lower 2/3 of the total frequency coverage available in ALMA Band 6). In total, over 70000 spectral channels were observed over the 30 GHz of bandwidth. To make a fully cleaned image of these $\sim$70000 channels would require (at the very least) several thousand unique clean boxes for the molecular emission. This is because the molecular emission is not uniformly distributed but rather concentrated toward several distinct regions and these regions all contain distinct chemistry which lead to unique molecular emission features toward each site. A rough approximation of the spectrum taken toward the Orion "hot core" pointing position gives $\sim$20 lines per 200 MHz of bandwidth above a signal-to-noise ratio of 5 (the noise level in the maps is $\sim$50 mJy/beam) for a total of $\sim$4000 spectral features. To date these data have resulted in 9 referred publications since 2012. In just over an hour, ALMA has collected more interferometric data toward the Orion KL region at 1mm wavelengths than all other arrays

TF05   2:56 – 3:11

THE ALMA SPECTRUM OF IRC+10216

JOSE CERNICHARO, MARCELINO AGÚNDEZ, FABIEN DANIEL, *Departamento de Astrofísica, Centro de Astrobiología CAB, CSIC-INTA, Madrid, Spain*; ARANCHA CASTRO-CARRIZO, *IRAM, IRAM, Grenoble, France*; NURIA MARCELINO, *ALMA, National Radio Astronomy Observatory, Charlottesville, VA, USA*; CHRISTINE JOBLIN, *IRAP, Université de Toulouse 3 - CNRS, Toulouse, France*; MICHEL GUÉLIN, *Institut de Radioastronomie Millimétrique, Observatoire de Paris, Paris, France*; JAVIER GOICOECHEA, *Departamento de Astrofísica, Centro de Astrobiología CAB, CSIC-INTA, Madrid, Spain.*

We report the detection in IRC+10216 of lines of HNC J = 3-2 pertaining to nine excited vibrational states with energies up to $\sim$5300 K. The spectrum, observed with ALMA, also shows a surprising large number of narrow, unidentified lines that arise in the vicinity of the star. The HNC data are interpreted through a 1D-spherical non-local radiative transfer model, coupled to a chemical model that includes chemistry at thermochemical equilibrium for the innermost regions and reaction kinetics for the external envelope. Although unresolved by the current early ALMA data, the radius inferred for the emitting region is $\sim$0.06 (i.e., 3 stellar radii), similar to the size of the dusty clumps reported by IR studies of the innermost region (r < 0.3). The derived abundance of HNC relative to H_2 is $10^{-8} <$ X(HNC) $< 10^{-6}$, and drops quickly where the gas density decreases and the gas chemistry is dominated by reaction kinetics. Merging HNC data with that of molecular species present throughout the inner envelope, such as vibrationally excited HCN, SiS, CS, or SiO, should allow us to characterize the physical and chemical conditions in the dust formation zone.

The interpretation of ALMA observations of C-rich evolved stars will require spectroscopic studies of highly excited states of HCN and HNC including accurate determination of l-doubling frequencies for all combination vibrational levels involving the bending mode. Several l-doubling transitions for which laboratory data allow the determination of their frequencies have been identified in the spectrum of IRC+10216.

TF06

^{13}C-METHYL FORMATE IN ORION-KL: ALMA OBSERVATIONS AND SPECTROSCOPIC CHARACTERIZATION

<u>CÉCILE FAVRE</u>, *Department of Astronomy, University of Michigan, Ann Arbor, MI, USA*; MIGUEL CARVA-JAL, *Dpto. Fisica Aplicada, Unidad Asociada CSIC, Facultad de Ciencias Experimentales, Universidad de Huelva, Huelva, Spain*; DAVID FIELD, *Department of Physics and Astronomy, University of Aarhus, Aarhus, Denmark*; EDWIN BERGIN, JUSTIN NEILL, *Department of Astronomy, University of Michigan, Ann Arbor, MI, USA*; NATHAN CROCKETT, *Geological and Planetary Sciences , California Institute of Techonolgy, Pasadena, CA, USA*; JES JØRGENSEN, SUZANNE BISSCHOP, *Centre for Star and Planet Formation, Niels Bohr Institute and Natural History Museum of Denmark, University of Copenhagen, Copenhagen, Denmark*; NATHALIE BROUILLET, DIDIER DESPOIS, ALAIN BAUDRY, *Laboratoire d'Astrophysique de Bordeaux, Université de Bordeaux, Floirac, France*; ISABELLE KLEINER, *Universités Paris Est et Paris Diderot , Laboratoire Interuniversitaire des systèmes atmosphériques (LISA), CNRS, Creteil, France*; L. MARGULÈS, T. R. HUET, JEAN DEMAISON, *UMR8523 CNRS - Université Lille 1, Laboratoire PhLAM, Villeneuve d'Ascq, France.*

Determination of elemental isotopic ratios is valuable for understanding the chemical evolution of interstellar material. Until now the ^{12}C/^{13}C ratio has predominantly been measured in simple species such as CO, CN and H_2CO and, becomes larger with increasing distance from the Galactic Center. We have investigated the carbon isotopic ratio for methyl formate $HCOOCH_3$, and its isotopologues $H^{13}COOCH_3$ and $HCOO^{13}CH_3$ addressing the issue whether the ^{12}C/^{13}C ratio is the same for both simple and large molecules. Using ALMA science verification observations of Orion-KL and the spectroscopic characterization of the complex $H^{13}COOCH_3$ and $HCOO^{13}CH_3$ species that we have performed, we have 1) confirmed the detection of the ^{13}C-methyl formate species in Orion-KL and, 2) image for the first time their spatial distribution. I will present some of these results. In particular, our analysis shows that the ^{12}C/^{13}C isotope ratio in methyl formate toward the Compact Ridge and Hot Core-SW components that are associated with Orion-KL are, for both the ^{13}C-methyl formate isotopologues, commensurate with the well-known ^{12}C/^{13}C ratio of the simple species CO. Our findings suggest that grain surface chemistry very likely prevails in the formation of methyl formate main and ^{13}C isotopologues.

TF09

THE MOLECULAR COMPLEXITY OF G34.3+0.2

<u>DOUGLAS FRIEDEL</u>, *Department of Astronomy, University of Illinois at Urbana-Champaign, Urbana, IL, USA.*

Recent observations of the Orion-KL region[1,2,3] have shown that the chemical distribution in the region is much more complex than originally thought. There are not just one nitrogen rich core and one diffuse oxygen rich region. But rather, at higher resolution, each of these regions breaks up into smaller more compact components associated with individual heating/energy sources. Additionally, one molecular species, acetone [$(CH_3)_2CO$], has a distinctly different distribution from any other large molecular species. These results cannot be explained by current chemical models. In order to expand our understanding of the chemistry in complex regions like Orion-KL, we have observed four additional high mass star forming regions: W3, G34.3+0.2, W75N, and W51 e1/e2 at several spatial resolutions (1" - 5"). The results of these multi-resolution observations (with an emphasis on G34.3), a comparison to the results from Orion-KL, and their implications for astrochemical models, will be presented.

[1] Friedel, D. N. & Snyder, L. E. "High-Resolution λ=1 mm CARMA Observations of Large Molecules in Orion-KL." Astrophysical Journal, 2008, 673, 962

[2] Widicus Weaver, S. L. & Friedel, D. N. "Complex Organic Molecules at High Spatial Resolution toward ORION-KL. I. Spatial Scales." Astrophysical Journal Supplements, 2012, 201, 16

[3] Friedel, D. N. & Widicus Weaver, S. L. "Complex Organic Molecules at High Spatial Resolution toward ORION-KL. II. Kinematics." Astrophysical Journal Supplements, 2012, 201, 17

Intermission

112

TF10   4:02 – 4:12

STAR FORMATION NEAR SGR A* AND THE ROLE OF COSMIC RAYS IN GALACTIC CENTER MOLECULAR CLOUDS

FARHAD YUSEF-ZADEH, *Physics and Astronomy, Northwestern University, Evanston, IL, USA.*

I will present recent study of molecular line observations of the Galactic center. On a scale of 2pc from Sgr A*, I show signatures of protostellar outflows within 2pc of Sgr A* based on SiO (5-4), SiO (2-1) line data taken with ALMA, SMA and CARMA. On a larger scale, I discuss the origin of widespread SiO and methanol emission detected from Galactic center molecular clouds. A time-dependent chemical model in which cosmic rays drive the gas chemistry is discussed in order to account for high abundance of SiO and methanol and the high molecular gas temperature.

TF11   4:14 – 4:24

MOLECULAR SPECTRAL LINES IN FILAMENTARY INFRARED DARK CLOUDS

XING LU, QIZHOU ZHANG, *Radio and Geoastronomy Division, Harvard-Smithsonian Center for Astrophysics, Cambridge, MA, USA*; HAUYU BAOBAB LIU, *Academia Sinica Institute of Astronomy and Astrophysics, Academia Sinica, Taipei, Taiwan.*

Many infrared dark clouds (IRDCs) in our Galaxy have filamentary structures, and some of them present converging filaments to a central hub, known as hub-filament systems (HFSs). These filaments could play a crucial role in feeding gas to the star forming regions at the hub. We analyzed NH_3 (J,K)=(1,1) and (2,2) spectral lines data obtained with the Very Large Array (VLA) towards five filamentary IRDCs, and derived the gas temperature based on the line ratios. Furthermore, with the Submillimeter Array (SMA) in the compact and sub-compact configurations, we obtained dust emission and spectra lines at 1.3 mm towards these sources. We found filamentary structures in both dust continuum and spectral line emission, with a characteristic width of 0.1 pc and length of 1 pc. The dust emission is consistent with the infrared extinction features, indicating the existence of dense and cold gas, while massive dust cores are usually associated with the hubs. Complex organic molecules including CH_3OH are found towards the dust cores. In particular, optically-thin intermediate density gas tracers, such as $C^{18}O$, reveal a possible trend of gas infall along filaments towards hubs. This is consistent with the scenario that dense gas is accreted onto dense cores through filaments and form high-mass star clusters.

TF12   4:26 – 4:41

THE CARMA LARGE-AREA STAR-FORMATION SURVEY: CLASSY

LESLIE LOONEY, MANUEL FERNANDEZ-LOPEZ, DOMINIQUE M. SEGURA-COX, *Department of Astronomy, University of Illinois at Urbana-Champaign, Urbana, IL, USA*; LEE MUNDY, SHAYE STORM, KATHERINE LEE, *Department of Astronomy, University of Maryland, College Park, MD, USA*; HÉCTOR ARCE, *Department of Physics, Yale University, New Haven, CT, USA.*

The spectroscopy of molecular clouds probes their structure and kinematics from large to small spatial scales and covering a range of environments are of fundamental importance to understanding how clouds evolve to form stars. CARMA has the unique ability to survey the gas participating in star formation in nearby clouds on scales from parsecs to $\sim$1,000 AU. We will present the results from the CARMA Large Area Star-formation Survey (CLASSy) Key Project. CLASSy has mapped 3 fields in Perseus (NGC 1333, Barnard 1, and L1451) and 2 fields in Serpens (Serpens Main and Serpens South) totaling 700 square-arcminutes in HCN, HCO^+, and N_2H^+ J=1-0 emission lines (dense gas tracers) to: 1) test the predictions of turbulence-driven star formation, 2) test if magnetic fields are dynamically important in turbulent clouds, 3) clarify the relationship between dense cores, their surrounding cloud, and the local YSOs, and 4) study core evolution. The data products from CLASSy will be available to the community.

TF13   **4:43 – 4:58**

TURBULENCE AND HEATING OF MOLECULAR CLOUDS IN THE GALACTIC CENTER

NATALIE O BUTTERFIELD, CORNELIA LANG, *Department of Physics and Astronomy, University of Iowa, Iowa City, IA, USA*; BETSY MILLS, *NRAO, NRAO, Socorro, NM, USA*; DOMINIC A. LUDOVICI, *Department of Physics and Astronomy, University of Iowa, Iowa City, IA, USA.*

Molecular gas temperatures in the Galactic Center have been shown to much higher than the gas temperatures of molecular clouds in the disk. These Galactic Center clouds also show large line widths characteristic of turbulence. However, the origin of this heating and turbulence is not well known. In order to investigate this question we analyzed two Galctic Center molecular clouds that showed these characteristic: the G0.10-0.08 cloud and the M0.25+0.01 cloud. We observed these clouds using the VLA at K (25 GHz) and Ka (36 GHz) bands, both of which contain multiple molecular transitions including NH_3, CH_3OH and HC_3N. Using multiple transitions of NH_3, we determined that the rotational gas temperature in the clouds was $\sim$90-100 K. We also discovered multiple 36 GHz CH_3OH class I masers in both the G0.10-0.08 and M0.25+0.01 clouds, $\sim$50 and $\sim$80 respectively. Since these masers trace shocked gas, this indicates that some of this heating and turbulence is caused by these strong shocks. We also present images of the HC_3N line which is a high density tracer and shows dense cores at the center of both clouds.

TF14   **5:00 – 5:15**

GROUND AND AIRBORNE OBSERVATIONS OF INTERSTELLAR HYDRIDES: NEW RESULTS FROM APEX AND SOFIA

FRIEDRICH WYROWSKI, *Millimeter- und Submillimeter-Astronomie, Max-Planck-Institut für Radioastronomie, Bonn, NRW, Germany.*

Hydrides are key ingredients of interstellar chemistry since they are the initial products of chemical networks that lead to the formation of more complex molecules. The fundamental rotational transitions of light hydrides fall into the submillimeter wavelength range and are therefore difficult to observe from the ground. Here we report on new observation of hydrides using the Atacama Pathfinder Experiment telescope (APEX) at the excellent Chajnantor site at 5100m altitude and the Stratospheric Observatory for Infrared Astronomy (SOFIA). Detections of SH,SH^+, H_2O, OH, OH^+, NH_2, NH_3 in diffuse clouds and the envelopes of star forming regions will be discussed.

TF15   **5:17 – 5:27**

MAGNETIC FIELDS AND STAR FORMATION - CN ZEEMAN MAPPING

RICHARD CRUTCHER, *Department of Astronomy, University of Illinois at Urbana-Champaign, Urbana, IL, USA.*

Zeeman observations are essential since they provide the only direct measurement of magnetic field strengths in molecular clouds. Earlier single-dish CN Zeeman N=1-0 measurements of field strengths toward high-mass star formation regions such as W3OH and DR21OH were with 23 arcsec resolution; needed now are higher angular resolution Zeeman maps of magnetic field strengths. Our CARMA Zeeman-effect commissioning observations reported here were carried out for the N=2-1 CN transitions with $\tilde{3}$ arcsec resolution toward W3OH and DR21OH. These are the first interferometric CN Zeeman observations of dense molecular cores. Results will be presented with discussion of the astrophysical implications and the implications for future ALMA observations.

TF16   **5:29 – 5:44**

SURFACE CHEMISTRY UPDATE OF THE KOSMA-τ PDR CODE

MARKUS RÖLLIG, SILKE ANDREE-LABSCH, VOLKER OSSENKOPF, *I. Physikalisches Institut, Universität zu Köln, Köln, Germany.*

Numerical PDR codes are commonly limited to a pure gas-phase chemistry, with the exception of the Formation of H2. For most species this is reasonably well assumption given the usual physical conditions in photondominated regions. However, for some species, such as H2O, O2, or CH3OH, the pure gas-phase computation is insufficient.

We present a recent update of the chemistry included in the Cologne PDR code KOSMA-τ, where we included a network of grain surface reactions into our chemical computation scheme.

TF07   5:46 – 6:01

MOLECULES IN DUSTY IR-LUMINOUS GALAXIES WITH ALMA

FRANCESCO COSTAGLIOLA, *Radioastronomy and galactic structure, Instituto de Astrofísica de Andalucía, Granada, Spain.*

Evidence is now mounting that most of the activity in some luminous infrared galaxies takes place in their compact obscured nuclei (CON), regions of less than 100 pc in diameter, which harbor large amounts of warm (T>100 K) molecular material (N(H$_2$)> 10^{24} cm^{-2}). The combined effect of warm, shielded gas and intense infrared radiation produce rich molecular spectra, which make these objects unique laboratories to study molecular excitation in extreme environments. Also, recent studies have shown that such compact nuclei may drive extremely young (1-2 Myr) molecular outflows, and are thus ideal targets to study AGN/starburst feedback processes. Here I will present some first results from Cycle 0 ALMA observations of two obscured nuclei. In the prototypical CON of NGC 4418 we proposed a 170 GHz-wide spectral scan in bands 3, 6 and 7, aimed at obtaining a template for the molecular chemistry and excitation in compact galaxy cores. In NGC 1377, the galaxy with the highest far-IR/radio ratio observed to date, we successfully mapped the molecular outflow in CO 1-0.

TF08   6:03 – 6:18

HOT AMMONIA IN LUMINOUS HIGH-MASS STAR FORMING REGIONS IN THE GALAXY

CIRIACO GODDI, *Joint Institute for VLBI in Europe, Dwingeloo, Netherlands.*

The formation process of high-mass (O-B type) stars is still a matter of debate. While recent numerical simulations have demonstrated that the radiation feedback can be overcome if accretion occurs through a circumstellar disk, observational evidences of accreting O-type young stars is lacking. In this context, imaging of highly-excited molecular lines from high-density tracers are key, as they probe the hottest and densest gas at small radii from O-type forming stars, i.e. in centrifugally-supported disks and/or infalling envelopes. Traditionally, this has been done using molecular transitions in the (sub)mm regime. Despite the leap in sensitivity and angular resolution enabled by ALMA, submm transitions can only probe outer parts of disks/envelopes where dust optical depth is low, but cannot penetrate the innermost regions, where dust emission is optically thick and, hence, dominates over molecular emission at any given frequency. High-angular resolution observations of optically-thin, highly-excited molecular lines at cm-wavelengths provide the best tool for probing the inner part of accretion flows around massive protostars, and hence the best complement to submm studies with ALMA. In this context, ammonia is an excellent "thermometer" of dense molecular gas and it can trace excitation up to temperatures of 2000 K by observing its inversion transitions within a relatively narrow frequency range, 20-40 GHz. The new broadband JVLA Ka-band receivers provide the unique opportunity to observe simultaneously a range of highly-excited (metastable) ammonia transitions, which are sensitive to gas of different temperatures and densities. I will report initial results from an imaging survey of hot-cores in the Galaxy in the ammonia lines from (6,6) to (13,13) with the JVLA.

TG. Mini-symposium: Beyond the Mass-to-Charge Ratio: Spectroscopic Probes of the Structures of Ions
Tuesday, June 17, 2014 – 1:30 PM
Room: 100 Noyes Laboratory

Chair: Michael A Duncan, University of Georgia, Athens, GA, USA

TG01   *INVITED TALK*   1:30 – 2:00

STATE-TO-STATE SPECTROSCOPY AND DYNAMICS OF IONS AND NEUTRALS BY PHOTOIONIZATION AND PHOTOELECTRON METHODS

CHEUK-YIU NG, *Department of Chemistry, The University of California, Davis, CA, USA.*

Recent advances in high-resolution photoionization, photoelectron, and photodissociation studies based on single-photon vacuum ultraviolet (VUV) and two-color infrared (IR)-VUV, visible (VIS)-ultraviolet (UV), and VUV-VUV laser excitations are illustrated with selected examples. We show that VUV laser photoionization coupled with velocity-map-imaging (VMI)-threshold photoelectron (VMI-TPE) detection can achieve comparable energy resolutions, but higher detection sensitivities than those observed in VUV laser pulsed field ionization-photoelectron (PFI-PE) measurements. For molecules with known intermediate states, IR-VUV and VIS-UV excitation schemes are highly sensitive for rovibronically selected and resolved PFI-PE studies. The successful applications of the VUV-PFI-PE, VUV-VMI-TPE and VIS-UV-PFI-PE methods to state-resolved and state-to-state photoelectron studies of transient radicals and transitional metal-containing molecules are highlighted. The most recently established VUV-VUV pump-probe time-slice VMI-photoion method is shown to be promising for state-to-state photodissociation studies of small molecules relevant to planetary atmospheres and for the fundamental understanding of photodissociation dynamics.

TG02   2:05 – 2:20

ROVIBRONICALLY SELECTED AND RESOLVED LASER PHOTOIONIZATION AND PHOTOELECTRON STUDIES OF TRANSITION METAL CARBIDES, NITRIDES, AND OXIDES.

ZHIHONG LUO, YIH-CHUNG CHANG, HUANG HUANG, CHEUK-YIU NG, *Department of Chemistry, The University of California, Davis, CA, USA.*

Transition metal (M) carbides, nitrides, and oxides (MX, X = C, N, and O) are important molecules in astrophysics, catalysis, and organometallic chemistry. The measurements of the ionization energies (IEs), bond energies, and spectroscopic constants for MX/MX$^+$ in the gas phase by high-resolution photoelectron methods represent challenging but profitable approaches to gain fundamental understandings of the electronic structures and bonding properties of these compounds and their cations. We have developed a two-color laser excitation scheme for high-resolution pulse field ionization photoelectron (PFI-PE) measurements of MX species. By exciting the neutral MX species to a single rovibronic state using a visible laser prior to photoionization by a UV laser, we have obtained fully rotational resolved PFI-PE spectra for TiC$^+$, TiO$^+$, VCH$^+$, VN$^+$, CoC$^+$, ZrO$^+$, and NbC$^+$. The unambiguous rotational assignments of these spectra have provided highly accurate IE values for TiC, TiO, VCH, VN, CoC, ZrO, and NbC, and spectroscopic constants for their cations.

TG03   2:22 – 2:37

MULTIPLE ISOMERS OF La(C$_4$H$_6$) FORMED IN REACTIONS OF La ATOM WITH SMALL HYDROCARBONS

WENJIN CAO, DILRUKSHI HEWAGE, DONG-SHENG YANG, *Department of Chemistry, University of Kentucky, Lexington, KY, USA.*

La(C$_4$H$_6$) was observed from the reactions of laser-vaporized La atom with propene (CH$_2$CHCH$_3$),isobutene [CH$_2$C(CH$_3$)2],1-butyne (CHCC$_2$H$_5$), and 2-butyne (CH$_3$CCCH$_3$) in a metal cluster beam source and investigated by mass-analyzed threshold ionization (MATI) spectroscopy in combination with quantum chemical calculations and spectral simulations. La(cyclobutene) [La(CH$_2$CHCHCH$_2$)] and La(trimethylenemethane) [LaC(CH$_2$)$_3$] were identified from the La + propene reaction. The formation of the two isomers involves two steps: the first step is the reaction of La+ CH$_2$CHCH$_3$ with the products of La(CH$_2$) + C$_2$H$_4$, and the second step is the reaction of LaCH$_2$ + CH$_2$CHCH$_3$ which preduces La(C$_4$H$_6$) + H$_2$. For the La + isobutene reaction, La(trimethylenemethane) was formed by dehydrogenation. For the reaction of La + 1- and 2-butynes, preliminary data analysis suggested that La(C$_4$H$_6$) was formed by simple association. All these structural isomers have similar geometries in the neutral and singly charged ion states, as evidenced by the very strong origin bands and short Franck-Condon profiles in the MATI spectra.

TG05   2:39 – 2:54

LANTHANUM ATOM-MEDIATED BOND ACTIVATION, COUPLING, AND CYCLIZATION OF 1,3-BUTADIENE PROBED BY MASS-ANALYZED THRESHOLD IONIZATION SPECTROSCOPY

<u>DILRUKSHI HEWAGE</u>, DONG-SHENG YANG, *Department of Chemistry, University of Kentucky, Lexington, KY, USA.*

Activation of small hydrocarbons by transition metal centers has gained much attention because of its applications in chemical catalysis. A series of $La(C_nH_m)$ complexes (n= 2, 4, 5 or 6 and m= 0, 2, 6, 5 or 8) were produced by La atom-mediated activation of 1,3-butadiene ($CH_2 = CH - CH = CH_2$) in a supersonic molecular beam. Molecular structures and electronic states of the reaction products were investigated by mass-analyzed threshold ionization spectroscopy and theoretical calculations. In this work we will discuss the formation, binding, and structures of $La(C_2H_2)$, $La(C_4H_6)$, and $La(C_6H_6)$. $La(C_2H_2)$ is formed by the C-C bond activation and H migration in a primary reaction; $La(C_4H_6)$, a metallacycle, is formed through an association reaction; and $La(C_6H_6)$ is produced by the C-C bond coupling and cyclization in a secondary reaction. The organic moiety in $La(C_6H_6)$ is a distorted benzene molecule. All three complexes have a two-fold metal binding mode and a doublet ground electronic state, with $La(C_2H_2)$ in C_{2v} point group and $La(C_4H_6)$ and $La(C_6H_6)$ in C_s.

TG06   2:56 – 3:11

CHARACTERIZATION OF THE RETINAL CHROMOPHORE IN THE GAS-PHASE VIA PHOTOISOMERIZATION ACTION SPECTROSCOPY

<u>KATHERINE JEAN CATANI</u>, *School of Chemistry, University of Melbourne, Melbourne, VIC, Australia;* NEVILLE J COUGHLAN, BRIAN D ADAMSON, EVAN BIESKE, *School of Chemistry, The University of Melbourne, Melbourne, Victoria, Australia.*

The wavelength dependence for the photoisomerization of the isolated retinal protonated n-butylamine Schiff base (RPSB) is explored in the gas phase using a new technique that combines laser spectroscopy and ion mobility spectrometry. The technique involves exposing electrosprayed ions to tunable laser radiation as they are introduced into the drift region of an ion mobility spectrometer. Ions that absorb laser radiation photoisomerize, resulting in a detectable change in their drift speed through N_2 buffer gas. Without laser irradiation, 4 peaks are observed in the arrival time distribution of RPSB. The most intense and slowest peak is assigned as the all-*trans* isomer by comparison with the calculated collision cross sections. With laser radiation, there is a clear depletion of the all-*trans* isomer peak between 440 to 660 nm, corresponding to the $S_1 \leftarrow S_0$ transition, with a maximum effect at 615 ± 5 nm. There is also evidence of photoisomerization below 450 nm associated with the onset of the $S_2 \leftarrow S_0$ transition. The photoisomerization action spectrum of RPSB is expected to mimic its absorption spectrum and should prove useful for calibrating theoretical descriptions of the isolated chromophore.

TG07   3:13 – 3:28

VIBRATIONAL SPECTROSCOPY OF TRANSIENT DIPOLAR RADICALS VIA AUTODETACHMENT OF DIPOLE-BOUND STATES OF COLD ANIONS

DAO-LING HUANG, HONG-TAO LIU, PHUONG DIEM DAU, LAI-SHENG WANG, *Department of Chemistry, Brown University, Providence, RI, USA.*

High-resolution vibrational spectroscopy of transient species is important for determining their molecular structures and understanding their chemical reactivity. However, the low abundance and high reactivity of molecular radicals pose major challenges to conventional absorption spectroscopic methods. The observation of dipole-bound states (DBS) in anions extend autodetachment spectroscopy to molecular anions whose corresponding neutral radicals possess a large enough dipole moment (>2.5 D).[1,2] However, due to the difficulty of assigning the congested spectra at room temperature, there have been only a limited number of autodetachment spectra via DBS reported. Recently, we have built an improved version of a cold trap[3] coupled with high-resolution photoelectron imaging.[4] The first observation of mode-specific auotodetachment of DBS of cold phenoxide have shown that not only vibrational hot bands were completely suppressed, but also rotational profile was observed.[5] The vibrational frequencies of the DBS were found to be the same as those of the neutral radical, suggesting that vibrational structures of dipolar radicals can be probed via DBS.[5] More significantly, the DBS resonances allowed a number of vibrational modes with very weak Frank-Condon factors to be "lightened" up via vibrational autodetachment.[5] Recently, our first high-resolution vibrational spectroscopy of the dehydrogenated uracil radical, with partial rotational resolution, via autodetachment from DBS of cold deprotonated uracil anions have been reported.[6] Rich vibrational information is obtained for this important radical species. The resolved rotational profiles also allow us to characterize the rotational temperature of the trapped anions for the first time.[6]

Reference
1 K. R. Lykke, D. M. Neumark, T. Andersen, V. J. Trapa, and W. C. Lineberger, J. Chem. Phys. **87**, 6842 (1987).
2 D. M. Wetzel, and J. I. Brauman, J. Chem. Phys. **90**, 68 (1989).
3 P. D. Dau, H. T. Liu, D. L. Huang, and L. S. Wang, J. Chem. Phys. **137**, 116101 (2012).
4 I. Leon, Z. Yang, and L. S. Wang, J. Chem. Phys. **138**, 184304 (2013).
5 H. T. Liu, C. G. Ning, D. L. Huang, P. D. Dau, and L. S. Wang, Angew. Chem. Int. Ed. **52**, 8976 (2013).
6 H. T. Liu, C. G. Ning, D. L. Huang, and L. S. Wang, Angew. Chem. Int. Ed. (accepted)

TG08   3:30 – 3:45

UV PHOTOFRAGMENTATION SPECTROSCOPY OF MODEL LIGNIN-ALKALI ION COMPLEXES: EXTENDING LIGNOMICS INTO THE SPECTROSCOPIC REGIME

JACOB C. DEAN, NICOLE L BURKE, JOHN R. HOPKINS, JAMES REDWINE, BIDYUT BISWAS, P. V. RAMACHANDRAN, SCOTT A McLUCKEY, TIMOTHY S. ZWIER, *Department of Chemistry, Purdue University, West Lafayette, IN, USA.*

Lignin is a heteroaromatic biopolymer that is an essential component in the cell wall of plants. The structural and chemical properties of lignin provide plants with macroscopic structural rigidity, and protection against microbial invasion leading to subsequent cell wall degradation. For this reason, lignin presents a major inhibition to the efficient harvesting of biomass. Given the variability of lignin composition and structure among species, environment, etc., the field of "lignomics" seeks to "sequence" lignin oligomers into constituent unit types (H, G, S) and linkages. This is predominantly done by means of tandem mass-spectrometry, by first generating a library of characteristic fragmentation pathways built from collision-induced dissociation of model dilignol ions, and applying them to the interpretation of fragmentation in larger ions. While these methods have proven powerful, UV photofragmentation spectroscopy of lignin ions cooled in a 22-pole cold ion trap provides an alternative approach to lignomics based on fragmentation following resonant UV excitation. This approach serves as a complimentary method to pure MS^n-based methods with the potential for unveiling dissociation pathways only accessed by UV excitation. Further, the multichromophoric nature of lignin enables site-selectivity for the energy imparted into the molecule/ion when differentiation of the site absorptions may be possible. IR spectroscopy of the cold ions can be used for detailed analysis of the preferred conformations and binding sites of metal cations. UV spectroscopy and photofragmentation mass spectrometry has been carried out on the model (G-type) β-O-4 and β-β dilignol linkages complexed with Li^+ and Na^+. The UV spectral signatures were found to vary between dilignols and metal complexes, and unique photofragmentation pathways were observed among the four complexes. IR spectroscopy in the OH stretch region was used as a probe of the conformation and binding preferences. In all cases, the dominating factor driving these structures was metal$^+$-oxygen electrostatic interactions with the alkali metal ion bound at the linkage between the lignol sub-units.

TG09   3:47 – 4:02

ISOMERIC EFFECTS ON FRAGMENTATIONS OF CROTONALDEHYDE AND METHACROLEIN IN LOW-ENERGY
ELECTRON–MOLECULE COLLISIONS

ARUP K. GHOSH, APARAJEO CHATTOPADHYAY, TAPAS CHAKRABORTY, *Physical Chemistry, Indian
Association for the Cultivation of Science, Kolkata, India.*

Low kinetic energy (KE) electrons can induce various chemical events when allowed to interact with molecules of different sizes and complexities. We would report in this talk the fragmentation behavior of two isomeric molecules, crotonaldehyde
(CA) and methacrolein (MA) upon impinging with low-energy electrons of kinetic energy in the range of 10-70 eV and the
fragmentations have been probed by quadrupole mass spectrometry. At 10 eV, the mass spectra of both CA and MA show
intense molecular ion peak at m/z 70, and also the peaks for H and CO loss channels at m/z 69 and m/z 42, respectively. A
distinct peak for MA at m/z 41 (HCO loss) is observed at 10 eV whose intensity increases more rapidly than the case of CA.
For MA, the CO loss channel is more preferred, while for CA it is the H loss channel. Electronic structure calculations reveal
that these preferences are associated with relative energetic positions of the transition states, and the details will be presented
in the talk.

Intermission

TG10   4:19 – 4:34

STRUCTURAL ISOMERIZATION OF THE GAS PHASE 2-NORBORNYL CATION REVEALED WITH INFRARED
SPECTROSCOPY AND COMPUTATIONAL CHEMISTRY

DANIEL MAUNEY, JONATHAN MOSLEY, MICHAEL A DUNCAN, *Department of Chemistry, University of
Georgia, Athens, GA, USA.*

The non-classical structure of the 2-norborny cation ($C_7H_{11}^+$) which was
at the center of "the most heated chemical controversy of our time"[a] has been
observed in the condensed phase and recently using X-ray crystallography.
However, no gas phase vibrational spectrum has been collected.

The $C_7H_{11}^+$ cation is produced via H_3^+ protonation of norbornene
by pulsed discharge in a supersonic expansion of H_2/Ar. Ions are mass-
selected and probed using infrared photodissociation spectroscopy. Due to
high exothermicity, protonation via H_3^+ leads to a structural isomerization to
the global minimum structure 1,3-dimethylcyclopentenyl (DMCP$^+$). Experiments are currently being conducted to find softer protonation techniques that
could lead to the authentic 2-norbornyl cation.

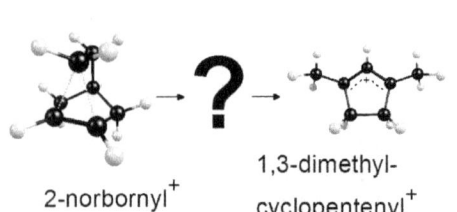

2-norbornyl$^+$ 1,3-dimethyl-
                              cyclopentenyl$^+$

[a]Schleyer,P.v.R. et. al.; Stable Carbocation Chemistry, John Wiley & Sons,Inc.; New York, 1997, Cahpter 2

TG11   **4:36 – 4:51**

COLD, GAS-PHASE UV AND IR SPECTROSCOPY OF PROTONATED LEUCINE ENKEPHALIN AND ITS ANA-
LOGUES

NICOLE L BURKE, JAMES REDWINE, JACOB C. DEAN, SCOTT A McLUCKEY, <u>TIMOTHY S. ZWIER</u>,
Department of Chemistry, Purdue University, West Lafayette, IN, USA.

The conformational preferences of peptide backbones and the resulting hydrogen bonding patterns provide critical biochemical information regarding the structure-function relationship of peptides and proteins. The spectroscopic study of
cryogenically-cooled peptide ions in a mass spectrometer probes these H-bonding arrangements and provides information
regarding the influence of a charge site. Leucine enkephalin, a biologically active endogenous opiod peptide, has been extensively studied as a model peptide in mass spectrometry. This talk will present a study of the UV and IR spectroscopy of
protonated leucine enkephalin $[YGGFL+H]^+$ and two of its analogues: the sodiated $[YGGFL+Na]^+$ and C-terminally methyl
esterified $[YGGFL\text{-}OMe+H]^+$ forms. All experiments were performed in a recently completed multi-stage mass spectrometer outfitted with a cryocooled ion trap. Ions are generated via nano-electrospray ionization and the analyte of interest is
isolated in a linear ion trap. The analyte ions are trapped in a 22-pole ion trap held at 5 K by a closed cycle helium cryostat
and interrogated via UV and IR lasers. Photofragments are trapped and isolated in a second LIT and mass analyzed. Double-
resonance UV and IR methods were used to assign the conformation of $[YGGFL+H]^+$, using the NH/OH stretch, Amide
I, and Amide II regions of the infrared spectrum. The assigned structure contains a single backbone conformation at vibra-
tional/rotational temperatures of 10 K held together with multiple H-bonds that self-solvate the NH_3^+ site. A "proton wire"
between the N and C termini reinforces the H-bonding activity of the COO-H group to the F-L peptide bond, whose cleavage
results in formation of the b_4 ion, which is a prevalent, low-energy fragmentation pathway for $[YGGFL+H]^+$. The reinforced
H-bonding network in conjunction with the mobile proton theory may help explain the prevalence of the b_4 pathway. In order
to elucidate structural changes caused by modifying this H-bonding activity, structural analogues were investigated. Determining the $[YGGFL+Na]^+$ structure will lend insight as to the impact of the ammonium group and methyl esterification of the
C-terminus eliminates the carboxy proton. The talk will also report on high resolution, cold UV spectra, non-conformation
specific IR gain spectra and conformation specific IR dip spectra for the analogues.

TG12   **4:53 – 5:03**

INTRAMOLECULAR HYDROGEN BONDING MOTIFS IN DEPROTONATED GLYCINE PEPTIDES REVEALED BY
CRYOGENIC ION VIBRATIONAL SPECTROSCOPY

BRETT MARSH, <u>ERIN M. DUFFY</u>, MICHAEL SOUKUP, JIA ZHOU, ETIENNE GARAND, *Department of
Chemistry, University of Wisconsin, Madison, WI, USA.*

We present the infrared spectra of deprotonated glycine chains, $(Gly_n - H)^-$, where n=1-4, obtained by cryogenic ion
vibrational spectroscopy. Comparisons between the experimental and calculated spectra reveal the dependence on chain
length of the hydrogen bonding motifs within these species. First, an interaction between the terminal carboxylate and the
neighboring amide $N - H$ is present in all the peptide structures. In the longer $(Gly_3 - H)^-$ and $(Gly_4 - H)^-$ structures,
the peptide chain folds such that the terminal NH_2 group also interacts with the carboxylate. Both of these also display
an interaction between the terminal NH_2 and the neighboring amide $N - H$. Finally, $(Gly_4 - H)^-$ is the first among these
species to contain an amide-amide interaction. Analysis of the $N - H$ peak positions elucidates the interplay among these
different hydrogen bonds, especially around the negatively charged carboxylate moiety.

TG13   5:05 – 5:20

A GAS PHASE INVESTIGATION OF $CuOH(H_2O)^+$ AND Cu(II) OLIGOGLYCINE WATER OXIDATION CATALYSTS

BRETT MARSH, JIA ZHOU, ETIENNE GARAND, *Department of Chemistry, University of Wisconsin, Madison, WI, USA.*

The abundance of copper and copper compounds within the crust of the Earth and the well characterized coordination chemistry of copper species has led to investigation of copper as a catalyst for several different chemical reactions. Among the most notable of these has been the application of copper containing compounds to water oxidation. The chemistry used in this reaction has thus far been dominated by expensive materials containing the rare metals ruthenium or iridium. In this presentation we study several copper species which have attracted interest in this field including $CuOH(H_2O)_n^+$ clusters and Cu(II) oligoglycine complexes. We find that in the case of $CuOH(H_2O)_n^+$ clusters that the undercoordinated $CuOH(H_2O)^+$ cluster has a strong affinity towards activating D_2 while the higher coordinated $CuOH(H_2O)_2^+$ and $CuOH(H_2O)_3^+$ clusters show no propensity for the activation of D_2. In the case of the Cu(II) oligoglycine complexes we find that the charge environment strongly affects the diagnostic CO stretch frequencies, creating a clear spectroscopic fingerprint for assessing the charge and interactions within these systems. Despite the small size (6-12 atoms) of the $CuOH(H_2O)_n^+$ clusters electronic structure methods including DFT and MP2 give poor agreement with the experimental results while DFT calculations on the relatively large Cu(II) oligoglycine species show excellent agreement with experiment.

TG14   5:22 – 5:32

GAS PHASE SPECTRA AND STRUCTURAL DETERMINATION OF GLUCOSE 6 PHOSPHATE USING CRYOGENIC ION VIBRATIONAL SPECTROSCOPY

STEVEN J. KREGEL, JONATHAN VOSS, *Department of Chemistry, University of Wisconsin–Madison, Madison, WI, USA*; BRETT MARSH, *Department of Chemistry, University of Wisconsin, Madison, WI, USA*; ETIENNE GARAND, *Department of Chemistry, The Univeristy of Wisconsin, Madison, WI, USA.*

Glucose-6-Phosphate (G6P) is one member of a class of simple phosphorylated sugars that are relevant in biological processes. We have acquired a gas phase infrared spectrum of $G6P^-$ using cryogenic ion vibrational spectroscopy (CIVS) in a home-built spectrometer. The experimental spectrum was compared with calculated vibrational spectra from a systematic conformer search. For both of the α and β anomers, results show that only the lowest energy conformers are present in the gas phase. If spectral signatures for similar sugars could be cataloged, it would allow for conformer-specific determination of mixture composition, for example, for glycolyzation processes.

TG15   5:34 – 5:49

EXTENSIVE FREQUENCY COMB VELOCITY MODULATION SPECTROSCOPY OF ThF^+ FOR USE IN THE JILA ELECTRON EDM EXPERIMENT

DAN GRESH, KEVIN COSSEL, JUN YE, ERIC CORNELL, *JILA, National Institute of Standards and Technology and Univ. of Colorado Department of Physics, University of Colorado, Boulder, CO, USA.*

The metastable $^3\Delta_1$ state in trapped HfF^+ is being used for an ongoing measurement of the electron electric dipole moment (eEDM)[a]. ThF^+, which has a larger effective electric field and a longer-lived $^3\Delta_1$ state, offers increased sensitivity for an eEDM measurement. Recently, the Heaven group has spectroscopically studied the low-lying states of ThF^+ [b]. However, to date there is no detailed information available about technically-accessible laser transitions in the near-infrared region of the spectrum, which are necessary for state preparation and detection in an eEDM experiment. By applying the technique of frequency comb velocity modulation spectroscopy (VMS)[c] to ThF^+ we can acquire 150 cm^{-1} of continuous, ion-sensitive spectra with 150 MHz resolution in 25 minutes. Here, we report on extensive broadband, high-resolution survey spectroscopy of ThF^+ in the near-IR where we have observed and accurately fit several rovibronic transitions. In addition, we have observed and characterized numerous rovibronic transitions from an unknown thoriated species of molecular ions.

[a] H. Loh, K. C. Cossel, M. C. Grau, K.-K. Ni, E. R. Meyer, J. L. Bohn, J. Ye, E. A. Cornell, Science 342, 1220 (2013).
[b] B. J. Barker, I. O. Antonov, M. C. Heaven, K. A. Peterson, J. Chem. Phys. 136, 104305 (2012).
[c] L. C. Sinclair, K. C. Cossel, T. Coffey, J. Ye, E. A. Cornell, PRL 107, 093002 (2011).

TG16   **5:51 – 6:06**

MONITORING THERMODYNAMIC EQUILIBRIUM PROCESSES AT 10 K: CONFORMATIONAL ISOMERIZATION AND PHOTOCHROMISM OF O_4^+ IN ARGON MATRICES.[a]

RYAN M LUDWIG, <u>DAVID T MOORE</u>, *Chemistry Dept., Lehigh University, Bethlehem, PA, USA.*

Bands corresponding to structural isomers of matrix-isolated O_4^+ are observed upon deposition of ions into argon matrices doped with moderate (0.1-1%) concentrations of O_2. These bands have been assigned based on previous matrix isolation spectroscopy,[b] as well as high-level computational studies.[c] In the current work, these bands are observed upon co-deposition of Cu^- and Ar^+ ions at low-energies. The Cu^- is present only as a non-interacting counter-ion, as is verified by studies using exclusively high-energy Ar^+ beams; in this case, the spectroscopy of the O_4^+ species is completely equivalent, however there is now also an intense peak corresponding to O_4^- counter-ion species. Following deposition at 20 K, the matrices are cooled to 10 K, where the FTIR spectra show a band at 1119 cm^{-1} for the trans-O_4^+ isomer, and a doublet at 1329/1331 cm^{-1}, corresponding to the cyclic-O_4^+ isomer, based on earlier work. There is also a band at 1186 cm^{-1} that was previously assigned to a larger O_6^+ complex. A temperature series taken in 1 K increments between 10 and 20 K reveal two reversible interconversion processes: the 1119 cm^{-1} band decreases between 10 and 14 K while a new band grows in at 1242 cm^{-1}, and the 1186 band shows a similar interconversion between 11 and 16 K with the 1331 cm^{-1} peak of the cyclic-O_4^+ doublet, while the 1329 cm^{-1} peak diminishes and broadens over the same temperature range. The interconverting peak pairs can be converted into equilibrium constants based on relative changes in integrated intensities, and the associated van't Hoff plots show linear trends with ΔH values in the range expected based on computational work. Finally, the 1186 cm^{-1} and 1331 cm^{-1} peak pair exhibit strong photochromism at 10 K: irradiation with red light converts 1186 to 1331, while irradiation with blue light shifts the equilibrium in the other direction. In both cases the phenomena is completely reversible and reproducible, with the original intensity ratio being restored after a few minutes once the light has been switched off. Irradiation has no effect on the other spectra bands, indicating that the 1329 and 1331 cm^{-1} bands cannot arise from the same species, as previously thought.

[a]Funding support from NSF CAREER Award CHE-0955637 is gratefully acknowledged
[b]Jacox, M.E.; Thompson, W.E.; J. Chem. Phys. 100, 750 (1994); Zhou, M.-F., et al.; J. Chem. Phys. 110, 9450 (1999).
[c]Lindh, R.; Barnes, L.; J. Chem. Phys. 100, 224 (1994).

TG04   **6:08 – 6:23**

PULSED-FIELD IONIZATION AND ELECTRONIC STATES OF $La(C_6H_5X)$ (X = H, CH_3, AND OH) COMPLEXES

<u>TAO JIN</u>, DONG-SHENG YANG, *Department of Chemistry, University of Kentucky, Lexington, KY, USA.*

$La(C_6H_5X)$ (X=H, CH_3, and OH) complexes were produced in a laser vaporization molecular beam source and studied using pulsed-field ionization electron or ion spectroscopy and theoretical calculations. Adiabatic ionization energies and vibrational frequencies were measured from the electron or ion spectra. Metal binding sites and electronic states of the complexes were determined by comparing the spectroscopic measurements with the quantum chemical calculations and spectral simulations. The binding and structures of these complexes will be compared with those of the corresponding scandium species reported previously by our group.

TH. Mini-symposium: Spectroscopy in Kinetics and Dynamics

Tuesday, June 17, 2014 – 1:30 PM

Room: B102 Chemical and Life Sciences

Chair: Ahmed Heikal, University of Minnesota Duluth, Duluth, MN, USA

TH01   **1:30 – 1:45**

FEMTOSECOND TIME AND ANGLE-RESOLVED PHOTOELECTRON SPECTROSCOPY OF AQUEOUS SOLUTIONS

TOSHINORI SUZUKI, *Graduate School of Science, Kyoto University, Kyoto, Japan.*

We present the femtosecond time- and angle-resolved photoemission spectroscopy of a liquid beam of aqueous solution. A liquid laminar flow 25 micron in diameter is introduced into a high-vacuum photoelectron spectrometer and interrogated using the pump-probe method. The photoelectron kinetic energy distribution is measured by rotating the probe laser polarization with respecto to an electron detection axis with a small detecion solid angle. The observed time- and angle-resolved photoelectron kinetic energies exhibit electron binding energies and characters of transient electronic states of solutes near the liquid surface. The method provides novel information for understanding solution chemistry.

TH02   **1:47 – 2:02**

EXCITED STATE STUDY OF $CuCl_4{}^{2-}$ COMPLEX IN ACETONITRILE

ANDREY S MERESHCHENKO[a], *Faculty of Chemistry, Saint-Petersburg State University, Saint-Petersburg, Russia*; ALEXANDER TARNOVSKY, *Chemistry and Center for Photochemical Sciences, Bowling Green State University, Bowling Green, OH, USA.*

In this work, we reported detailed description of the ultrafast dynamics of a $CuCl_4{}^{2-}$ complex in acetonitrile upon excitation into all possible Ligand Field (LF) states due to transitions between d-orbitals of the Cu^{2+} ion, and two most intense Ligand-to-Metal Charge Transfer (LMCT) transitions. The LF states are found to be nonreactive states, while excitation into the LMCT states leads to the ionic dissociation of $CuCl_4{}^{2-}$ complex. Upon 2000 nm excitation, $CuCl_4{}^{2-}$ complex was promoted into the lowest lying ^2E LF excited state. Vibrational relaxation of the $CuCl_4{}^{2-}$ complex in ^2E state finishes within the first 500 fs, and is followed by internal conversion to the $^2B_2$ ground state with a lifetime of about 1 ps. Excitation of the $CuCl_4{}^{2-}$ complex into the $^2B_1$ LF excited state upon 1500 nm excitation results in its fast relaxation into the ^2E state in less than 50 fs, which then relaxes to the electronic ground state. The $^2A_1$ LF state of the $CuCl_4{}^{2-}$ complex is formed upon 1100 and 1300 nm excitation directly relaxes to the $^2B_2$ ground state with a lifetime of about 5 ps, in contrast to the $^2B_1$ state, which nonradietively decays to the ground state cascading through the intermediate ^2E state. Vibrational relaxation of the $CuCl_4{}^{2-}$-complex in $^2A_1$ LF state finishes within first 800 fs. Internal conversion dynamics of the LF excited states was explained using the energy gap law approach in the strong coupling limit. In our experiments, $CuCl_4{}^{2-}$ complex was also excited into $^2E(Cl(n) - Cu(dxy))$ and $^2E(Cu-Cl(\pi,\sigma) - Cu(dxy))$ LMCT states using 420 and 310 nm excitation wavelengths, respectively. After formation, $^2E(Cl(n) -Cu(dxy))$ LMCT state undergoes internal conversion to the lower-lying vibrationally hot LF excited states, which then either nonradiatively decay to the ground state, or undergo Cu-Cl ionic bond dissociation, forming $CuCl_3{}^-$ and Cl^- ions. These species do not recombine until the longest measured time delay, 1 ns. The $^2E(Cu-Cl(\pi,\sigma) - Cu(dxy))$ LMCT state undergoes internal conversion to the lower-lying vibrationally hot LF excited states within 300 fs cascading through the lower-lying LMCT states, and Cu-Cl ionic bond dissociation. The vibrational relaxation of formed LF states takes about 5-7 ps, and more than 10 ps for excitation into $^2E(Cl(n) - Cu(dxy))$ and $^2E(Cu-Cl(\pi,\sigma) - Cu(dxy))$ LMCT states, respectively.

---

[a] Andrey S. Mereshchenko acknowledges Saint-Petersburg-State University for the postdoctoral fellowship (12.50.1562.2013)

TH03 $\hspace{10cm}$ **2:04 – 2:19**

DETECTION OF INTRAMOLECULAR CHARGE TRANSFER AND DYNAMIC SOLVATION IN EOSIN B BY FEMTOSECOND TWO-DIMENSIONAL ELECTRONIC SPECTROSCOPY

SOUMEN GHOSH, JEROME D. ROSCIOLI, WARREN F. BECK, *Department of Chemistry, Michigan State University, East Lansing, MI, USA.*

We have employed 2D electronic photon echo spectroscopy to study intramolecular charge-transfer dynamics in eosin B. After preparation of the first excited singlet state (S_1) with 40-fs excitation pulses at 520 nm, the nitro group ($-NO_2$) in eosin B undergoes excited state torsional motion towards a twisted intramolecular charge transfer (TICT) state. As the viscosity of the surrounding solvent increases, the charge-transfer rate decreases because the twisting of the $-NO_2$ group is hindered. These conclusions are supported by the time evolution of the 2D spectrum, which provides a direct measure of the the ground-to-excited-state energy gap time-correlation function, $M(t)$. In comparison to the inertial and diffusive solvation time scales exhibited by eosin Y, which lacks the nitro group, the $M(t)$ function for eosin B exhibits under the same conditions an additional component on the 150-fs timescale that arises from quenching of the S_1 state by crossing to the TICT state. These results indicate that 2D electronic spectroscopy can be used as a sensitive probe of the rate of charge transfer in a molecular system and of the coupling to the motions of the surrounding solvent. (Supported by grant DE-SC0010847 from the Department of Energy, Office of Basic Energy Sciences, Photosynthetic Systems program.)

TH04 $\hspace{10cm}$ **2:21 – 2:36**

COLLECTIVE VIBRATIONS OF WATER-SOLVATED HYDROXIDE IONS INVESTIGATED WITH BROADBAND 2DIR SPECTROSCOPY

ARITRA MANDAL, *Department of Chemistry, MIT, Cambridge, MA, USA*; KRUPA RAMASESHA, *Department of Chemistry, University of California at Berkeley, Berkeley, CA, USA*; LUIGI DE MARCO, *Department of Chemistry, MIT, Cambridge, MA, USA*; MARTIN THÄMER, ANDREI TOKMAKOFF, *Department of Chemistry, The University of Chicago, Chicago, IL, USA.*

The infrared spectra of aqueous solutions of NaOH and other strong bases exhibit a broad continuum absorption for frequencies between 800-3500 cm-1, which is attributed to the strong interactions of the hydroxide ion with its solvating water molecules. To provide molecular insight into the origin of the broad continuum absorption feature, we have performed ultrafast pump-probe and 2DIR experiments on aqueous NaOH by exciting the O—H stretch vibrations and probing the response from 1350-3800 cm-1 using a newly developed sub-70 fs broadband mid-infrared source. These experiments, in conjunction with harmonic vibrational analysis of OH–(H2O)n clusters, reveal that O—H stretch vibrations of aqueous hydroxides arise from coupled vibrations of multiple water molecules solvating the ion. We classify the vibrations of the hydroxide complex by symmetry defined by the relative phase of vibrations of the O—H bonds hydrogen bonded to the ion. Although spectral broadening does not allow us to distinguish 3- and 4-coordinate ion complexes, we find a resolvable splitting between asymmetric and symmetric stretch vibrations, and assign the 2850 cm-1 peak infrared spectra of aqueous hydroxides to asymmetric stretch vibrations.

TH05 $\hspace{10cm}$ **2:38 – 2:53**

DYNAMICS OF MODEL HYDRAULIC FRACTURING LIQUID STUDIED BY TWO-DIMENSIONAL INFRARED SPECTROSCOPY

KIM DALEY, KEVIN J KUBARYCH, *Department of Chemistry, University of Michigan, Ann Arbor, MI, USA.*

The technique of two-dimensional infrared (2DIR) spectroscopy is used to expose the chemical dynamics of various concentrations of polymers and their monomers in heterogeneous mixtures. An environmentally relevant heterogeneous mixture, which inspires this study, is hydraulic fracturing liquid (HFL). Hydraulic fracking is a technique used to extract natural gas from shale deposits. HFL consists of mostly water, proppant (sand), an emulsifier (guar), and other chemicals specific to the drilling site. Utilizing a metal carbonyl as a probe, we observe the spectral dynamics of the polymer, guar, and its monomer, mannose, and compare the results to see how hydration dynamics change with varying concentration. Another polymer, Ficoll, and its monomer, sucrose, are also compared to see how polymer size affects hydration dynamics. The two results are as follows: (1) Guar experiences collective hydration at high concentrations, where as mannose experiences independent hydration; (2) no collective hydration is observed for Ficoll in the same concentration range as guar, possibly due to polymer shape and size. HFL experiences extremely high pressure during natural gas removal, so future studies will focus on how increased pressure affects the hydration dynamics of polymers and monomers.

TH06   2:55 – 3:10

DYNAMIC STUDIES OF BOTH NON-EQUILIBRIA AND EQUILIBRIA PHENOMENA IN SILICA SOL-GEL MATE-
RIALS

CHRISTOPHER JERALD HUBER, AARON M. MASSARI, *Chemistry Department, University of Minnesota,
Minneapolis, MN, USA.*

Silica sol-gel syntheses are among the most widely used and studied synthetic methods in the materials field. However
there lacks a fundamental understanding of how these materials gel and what is happening dynamically inside of the pores
of these materials. In this study, we adapt a typical silica sol gel synthesis so as to introduce an intrinsic probe (a SiH
stretch) throughout the material. Using these probes and two-dimensional infrared spectroscopy (2DIR) we have monitored
the molecular motions (solution dynamics) of the solution as it approaches gel formation. After gel formation we can then
study the solution dynamics inside the pores of these materials.

TH07   3:12 – 3:22

ENERGY TRANSFER IN A SYNTHETIC DENDRON-BASED LIGHT HARVESTING SYSTEM

LEA NIENHAUS, MARTIN GRUEBELE, *Department of Chemistry, University of Illinois at Urbana-
Champaign, Urbana, IL, USA.*

Single molecule experiments based on Förster resonance energy transfer (FRET) are now capable of detecting energy funnel-
ing in branched molecules. Here we present the synthesis, as well
as the optical characterization of a dendron coupled to two donor
dyes (Cy3) and one acceptor dye (Cy5). Characterization of the dendron by ensemble absorption and emission spectroscopy shows that
the molecule is capable of light harvesting; yielding a FRET signal
from the acceptor that is greater than expected for a single donor.
Additionally, we investigated an energy transfer cascade upon UV
excitation of the conjugated backbone, resulting in several competing energy transfer pathways with the same total energy transfer as
direct FRET. The first pathway is FRET from the backbone to Cy3
and resulting FRET to Cy5, with the competing pathway that allows
direct energy transfer to Cy5 from the backbone via superexchange.
Structural simulations in solution, as well as direct imaging by scanning tunneling microscopy show that the dyes can fold over onto
the dendron, creating a heterogeneous distribution of conformations
suitable for imaging single molecule studies of light harvesting.

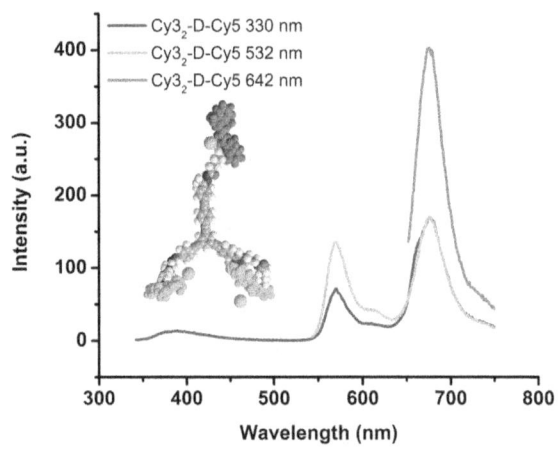

Intermission

TH08   3:39 – 3:44

EXCITED STATE DYNAMICS OF SUBSTITUTED TETRAPHENYLPORPHYRINS AND THEIR ADDUCTS WITH
FULLERENE

ANDREY S MERESHCHENKO[a], ALEXEY V POVOLOTSKIY, ALEXANDER S KONEV, YURIY S
TVER'YANOVICH, ALEXANDER F KHLEBNIKOV, *Faculty of Chemistry, Saint-Petersburg State University,
Saint-Petersburg, Russia.*

In this work, the photochemistry of tetraphenylporphyrin adducts with fullerene and corresponding porphyrins were
studied using nanosecond transient absorption spectroscopy upon excitation into Q-band of porphyrin. We found that pi-
system of the fullerene stabilizes the triplet excited state of porphyrin. Moreover, we found that charge transfer from porphyrin
to fullerene occurs from the triplet excited state.

[a]Author acknowledges Saint-Petersburg University for the postdoctoral fellowship. This study is carried out in the Center for optical and laser materials research of St.
Petersburg State University.

TH09

IDENTIFICATION OF A CRITICAL INTERMEDIATE IN GALVANIC EXCHANGE REACTIONS BY SINGLE-NANOPARTICLE RESOLVED KINETICS

JEREMY GEORGE SMITH, PRASHANT JAIN, *Department of Chemistry, University of Illinois at Urbana-Champaign, Urbana, IL, USA.*

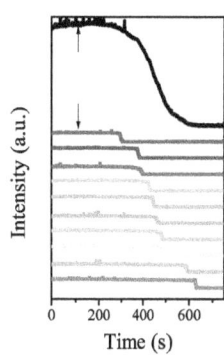

The realization of common materials transformations in nanocrystalline systems is fostering the development of novel nanostructures and allowing a deep look into the atomistic mechanisms involved. Galvanic corrosion is one such transformation. We studied galvanic replacement within individual metal nanoparticles by using plasmonic spectroscopy. This proved to be a powerful approach to studying materials transformations in the absence of ensemble averaging. Individual nanoscale units act as domains that can be interrogated optically in isolation, whereas the averaging of all such domains provides a bulk reaction trajectory. Single-nanoparticle reaction trajectories showed that a Ag nanoparticle exposed to Au^{3+} makes an abrupt transition into a nanocage structure. The transition is limited by a critical structural event, which we identified by electron microscopy to comprise the formation of a nanosized void, similar to the pitting process commonly observed in the corrosion of metals. Trajectories also revealed a surprisingly strong nonlinearity of the reaction kinetics, which we explain by a model involving the critical coalescence of vacancies into a growing void. The critical void size for galvanic exchange to spontaneously proceed was found to be 20 atomic vacancies. In the future we hope to extend this approach to examine a wide variety of materials transformations and chemical reactions.

TH10

ELECTRONIC GROUND AND EXCITED STATE SPECTRAL DIFFUSION OF A PHOTOCATALYST

LAURA M. KIEFER, JOHN T. KING, KEVIN J KUBARYCH, *Department of Chemistry, University of Michigan, Ann Arbor, MI, USA.*

$Re(bpy)(CO)_3Cl$ is a well studied CO_2 reduction catalyst, known for its ability as both a photosensitizer and a catalyst with a high quantum yield and product selectivity. The catalysis reaction is initiated by a 400 nm excitation, followed by an intersystem crossing (ISC) and re-equilibration in the lowest triplet state. We utilize the quasi-equilibrium nature of this long-lived triplet metal-to-ligand charge-transfer (^{3}MLCT) state to completely characterize the solvent dynamics using the technique of transient two-dimensional infrared (t-2DIR) spectroscopy to extract observables such as the frequency-frequency correlation function (FFCF), an equilibrium function. The electronic ground state solvent dynamics are characterized using equilibrium two-dimensional infrared spectroscopy (2D IR). Our technique allows us to independently observe the solvent dynamics of different electronic states and compare them.

In this study, three carbonyl stretching modes were utilized to probe both the intramolecular and solvent environments in each electronic state. In the electronic ground state, the totally symmetric mode exhibits pure homogeneous broadening and a lack of spectral dynamics, while the two other modes have similar FFCF decay times of $\sim$ 1.5 ps. In the ^{3}MLCT, however, all three modes experience similar spectral dynamics and have a FFCF decay time of $\sim$ 4.5 ps, three times slower than in the electronic ground state. Our technique allows us to directly observe the differences in spectral dynamics of the ground and excited electronic states and allows us to attribute the differences to specific origins such as solvent-solute coupling and molecular flexibility.

TH11   4:20 – 4:35

FEMTOSECOND NONLINEAR OPTICAL STUDIES OF RADIATIONLESS DECAY IN CAROTENOIDS AND IN THE PERIDININ–CHLOROPHYLL A PROTEIN

JEROME D. ROSCIOLI, SOUMEN GHOSH, MICHAEL M BISHOP, WARREN F. BECK, *Department of Chemistry, Michigan State University, East Lansing, MI, USA*; HARRY A. FRANK, *Department of Chemistry, University of Connecticut, Storrs, CT, USA.*

Femtosecond transient-grating spectroscopy with optical heterodyne detection was employed to observe separately the time evolution of the absorption and dispersion components of the third-order nonlinear optical signal following resonant excitation of the S_2 ($^1B_u{}^+$) states of β-carotene in benzonitrile and peridinin in methanol using 40-fs pulses centered at 520 nm. The absorption and dispersion components exhibit distinctively different time profiles owing to the population of intermediate states. An initial intermediate state is populated on an ultrashort (< 30 fs) time scale in both carotenoids. Owing to the fast red-shifting of the stimulated emission part of the S_2-state transient grating signal, we suggest that the intermediate state arises from vibrational displacements on the S_2-state potential surface that eventually yield twisted or bent conformations. Motions of the molecule of this type in the S_2-state would contribute to a mixing of the diabatic S_2 and S_1 electronic states and would promote the formation of intramolecular charge-transfer character. Both of these effects would enhance the efficiency of energy transfer from the S_1 state to the (B)Chl Q_y state in photosynthetic light-harvesting proteins. The time-resolved transient-grating spectra obtained for peridinin in the peridinin–chlorophyll a protein from *Amphidinium carterae* suggest a more rapid formation of the intermediate than for peridinin in methanol. This finding suggests that the conformation of the peridinin chromophore is controlled in the binding site to optimize the formation of the twisted intermediate upon excitation of the S_2 state. (Supported by grant DE-SC0010847 from the Department of Energy, Office of Basic Energy Sciences, Photosynthetic Systems program.)

TH12   4:37 – 4:52

THE EFFECT OF SULFUR SUBSTITUTION ON THE EXCITED-STATE DYNAMICS OF DNA AND RNA BASE DERIVATIVES

MARVIN POLLUM, *Department of Chemistry, Case Western Reserve University, Cleveland, OH, USA*; CARLOS E. CRESPO-HERNÁNDEZ, *Chemistry, Case Western Reserve, Cleveland, OH, USA.*

Substitution of oxygen by a sulfur atom in the natural DNA and RNA bases gives rise to a family of derivatives commonly known as the thiobases. Upon excitation with UV radiation, the natural bases are able to quickly and efficiently dissipate the imparted energy as heat to their surroundings. Thiobases, on the other hand, relax into a long-lived triplet excited state in quantum yields that approach unity. This finding has both fundamental and biological relevance because the triplet state plays a foremost role in the photochemistry of the thiobases, this is especially important in the current medicinal applications of thiobase derivatives. Using femtosecond transient absorption spectroscopy, we are able uncover the ultrafast dynamics leading to the population of this reactive triplet state. In particular, I will present our results on how the site of sulfur substitution and the degree of substitution impact these dynamics and I will compare these experimental results to some recent computational work. Pinning down the excited-state dynamics of the thiobases is important to furthering the understanding of dynamics in natural DNA/RNA bases, as well as to the discovery of thiobase derivatives with desirable therapeutic properties.

The authors acknowledge the CAREER program of the National Science Foundation (Grant No. CHE-1255084) for financial support.

TH13   **4:54 – 5:09**

PROTEIN DYNAMICS AND CONFORMATIONAL HETEROGENEITY CHARACTERIZED WITH TWO-DIMENSIONAL INFRARED SPECTROSCOPY

MEGAN THIELGES, EDWARD BASOM, JAMES SPEARMAN, *Department of Chemistry, Indiana University, Bloomington, IN, USA.*

Conformational heterogeneity and dynamics impact protein function, but their investigation is limited by the availability of methods for characterizing rapidly fluctuating protein states with both high spatial and temporal resolution. Multidimensional infrared spectroscopy is emerging as a powerful technique that directly probes the structural dynamics on fast timescales. To overcome the spectral congestion inherent to protein spectra that restricts application of infrared spectroscopy to protein systems, we incorporate into proteins vibrational probes with spectrally isolated frequencies and local-mode character that make possible rigorous analysis of protein environments and dynamics with infrared spectroscopy. In particular, both heme-bound carbon monoxide and nitriles selectively incorporated as unnatural amino acids are used as probes of cytochrome P450. 2D IR spectroscopy is used to investigate the impact of protein dynamics and conformational heterogeneity on the selectivity of its catalytic activity. Comparative studies of mutants of cytochrome P450 are used to unravel the contribution of specific residues to the protein's dynamics.

TH14   **5:11 – 5:26**

KINETIC TRAPPING OF METASTABLE AMINO ACID POLYMORPHS

GARTH SIMPSON, *Department of Chemistry, Purdue University, West Lafayette, IN, USA.*

Second harmonic generation (SHG) microscopy was integrated with synchrotron X-ray diffraction (XRD) to test the Ostwald Rule of Stages, in which is hypothesized that crystals dynamically transition through metastable polymorphs before settling on the most thermodynamically favored form. The presence or absence of metastable forms has historically been challenging to probe due to the stochastic randomness of crystal nucleation coupled with the relatively short time-frame over which the metastable forms may survive. In this work, inkjet printing of a racemic amino acid solutions results in rapid solvent evaporation, placing crystallization under kinetic rather than thermodynamic control. SHG microscopy is used to rapidly and selectively identify the positions of metastable crystal forms. Coupling this measurement with synchrotron XRD allows diffraction analysis to be performed on individual inkjet printed dots of only a few pg of total material, prepared from single 1 pL droplets. In studies of amino acids, we have shown that the homochiral crystals emerge when printed, while those same solutions exclusively generate the racemic co-crystals upon slow solvent evaporation.

TH15   **5:28 – 5:43**

BIOENERGETICS AND DIFFUSION IN THE CROWDED MILIEU OF LIVING CELLS

AHMED HEIKAL, *Chemistry and Biochemistry, University of Minnesota Duluth, Duluth, MN, USA.*

Intracellular nicotinamide adenine dinucleotide (NADH) is a key cofactor in energy metabolism pathways and a myriad of oxidation-reduction reactions in living cells. The crowded milieu of these cells with organelles and macromolecules influences many biological processes such as biomolecular diffusion, protein-protein and protein-substrate interactions, and protein folding. In this contribution, I will highlight our recent findings on the role of macromolecular crowding on biochemical reaction between NADH and selected dehydrogenases in both living cells and in controlled macromolecules-rich environment. In addition, multiscale diffusion (rotational and translational) of a small fluorophore will be used to understand the role of non-specific binding, heterogeneity in microenvironmental viscosity in crowded solutions. Our experimental approach is a combination of fluorescence lifetime imaging microscopy, time-resolved anisotropy and fluorescence correlation spectroscopy. The broader impacts of these results will be discussed within the context of energy metabolism and biophysics in the crowded milieu of living cells.

TI. Atmospheric science
Tuesday, June 17, 2014 – 1:30 PM
Room: 112 Chemistry Annex

Chair: Iouli E Gordon, Harvard-Smithsonian Center for Astrophysics, Cambridge, MA, USA

TI01   1:30–1:45

METHACROLEIN IN THE IR ATMOSPHERIC WINDOW: MM-WAVE AND FTIR SPECTROSCOPIES COMPLE-MENTED BY QUANTUM CALCULATIONS

OLENA ZAKHARENKO, JUAN-RAMON AVILES MORENO, HAYKAL IMANE, R. A. MOTIYENKO, T. R. HUET, *Laboratoire PhLAM, UMR8523 CNRS - Université Lille 1, Villeneuve d'Ascq, France*; OLIVIER PIRALI, *AILES beamline, Synchrotron SOLEIL, Saint Aubin, France.*

Methacrolein, $CH_2 = C(CH_3)CHO$ or MAC, is an important atmospheric molecule because it is a major product of the isoprene-OH reaction. Meanwhile the spectroscopic information on MAC is very scarse. [a] [b]

On the theoretical side, we have performed quantum calculations at different levels of theory (DFT and *ab initio*) to model the structure of the two conformers, the large amplitude motion associated with the methyl top, and the anharmonic vibrational structure.

On the experimental side, we have at first characterized the millimeter-wave spectrum of MAC in the 150-465 GHz range using the Lille frequency multiplication chain spectrometer. In particular the ground state has been analyzed up to J, $K_a = 37$, 17 and the first excited states are currently investigated.

Secondly, FTIR spectra have been recorded on the AILES beamline of SOLEIL using a long path cell, between 30 and 3500 cm^{-1} at medium resolution (0.5 cm^{-1}). A few bands of atmospheric interest have also been recorded at higher resolution (0.001 cm^{-1}). We will report the details of the vibrational analysis, as well as the molecular parameters derived from the analysis of the high resolution spectrum of the c-type band located around 930 cm^{-1}.

Support from the French Laboratoire d'Excellence CaPPA (Chemical and Physical Properties of the Atmosphere) through contract ANR-10-LABX-0005 of the Programme d'Investissements d'Avenir is acknowledged. The experiment on the AILES beam-line of the synchrotron SOLEIL was performed under project number 20130192.

[a] M. Suzuki and K. Kozima, *J. Molec. Spectrosc.* **38** (1971) 314

[b] J. R. Durig, J. Qiu, B. Dehoff and T. S. Little, *Spectrochimica Acta* **42A** (1986) 89

TI02   1:47–2:02

IR SPECTROSCOPY OF SELECTED ATMOSPHERIC MONOTERPENES AND OXYDATION PRODUCTS

JUAN-RAMON AVILES MORENO, T. R. HUET, MANUEL GOUBET, *Laboratoire PhLAM, UMR8523 CNRS - Université Lille 1, Villeneuve d'Ascq, France*; PASCALE SOULARD, PIERRE ASSELIN, *MONARIS UMR8233, CNRS - UNiversité Paris 6 UPMC, Paris, France*; ROBERT GEORGES, *IPR UMR6251, CNRS - Université Rennes 1, Rennes, France*; OLIVIER PIRALI, P. ROY, *AILES beamline, Synchrotron SOLEIL, Saint Aubin, France.*

Several monoterpenes are biogenic volatile organic compounds (BVOCS) present in the atmosphere. They can react with OH, O_3, NO_x, etc. to give rise to several oxydation and degradation products.

We have studied the gas phase spectroscopy of monocyclic (limonene, γ-terpinene) and bicyclic (α-pinene, β-pinene) atmospheric monoterpenes ($C_{10}H_{16}$), as well as two $C_{10}H_{14}O$ oxydation products (perillaldehyde, carvone). In the first step of this work, theoretical calculations and microwave spectroscopy were used in order to evidence the most stable conformers and their relative energies. [a] [b] [c]

In the present communication we will present the results of the IR study. Infrared spectra have been recorded on the FTIR spectrometer of the AILES beamline at synchrotron SOLEIL, using the Jet-AILES molecular beam and a long path cell. Special attention was given to the 700–1300 cm^{-1} atmospheric window, to the CH vibration region, and to the "finger print" FIR region. Quantum calculations have been performed at different levels of theory (DFT, ab initio). In particular anharmonic force fields were obtained in order to model the vibrational structures.

Support from the French Laboratoire d'Excellence CaPPA (Chemical and Physical Properties of the Atmosphere) through contract ANR-10-LABX-0005 of the Programme d'Investissements d'Avenir is acknowledged. The experiment on the AILES beam-line at synchrotron SOLEIL was performed under project number 20130192.

[a] J. R. Aviles Moreno, F. Partal Urena, J. J. Lopez Gonzalez and T. R. Huet, *C. Phys. Lett.* **473** (2009) 17

[b] J. R. Aviles Moreno, T. R. Huet, F. Partal Urena, J. J. Lopez Gonzalez, *Struc. Chem.* **24** (2013) 1163

[c] J. R. Aviles Moreno, E. Neeman, T. R. Huet, manuscript in preparation

TI03   **2:04 – 2:19**

HIGH-RESOLUTION INFRARED SPECTROSCOPY SLIT-JET COOLED HYDROXYMETHYL RADICAL (CH$_2$OH): OH STRETCHING MODE

FANG WANG, CHIH-HSUAN CHANG, *JILA, UCB-NIST, Boulder, CO, USA*; DAVID NESBITT, *Department of Chemistry, JILA CU-NIST, Boulder, CO, USA*.

Rotationally resolved direct absorption spectra in the OH stretching mode were recorded using difference frequency generation infrared spectrometer and slit-jet supersonic discharge expansion source. The spectra are really complicated because of the following reasons: a) CH$_2$OH was generated by discharging the expansion mixed gas CH$_3$OH and Cl$_2$ in carrier gas 70% Ne and 30% He. Large signals for CH$_3$OH are overlapped with that for CH$_2$OH in the OH stretching mode. b) The large amplitude of COH torsion for CH$_2$OH splits each vibration state into two states labeled as "+" and "-". The "+" state has energy levels with a 3:1 nuclear spin statistic ratio for $K_a=0^+/K_a=1^+$. The "-" state has $K_a=0^-/K_a=1^-$ levels with 1:3 nuclear spin statistics. c) The OH stretch fundamental is a hybrid *a/b*-type band. Both *a*-type and *b*-type bands were observed. As a result of the low rotational temperature and sub-Doppler linewiths, the spin-rotation structure was also resolved for *b*-type transitions. The spectra were assigned and fit to a Watson A-reduced symmetric top Hamiltonian to improve the accuracy of rotational parameters for the ground vibration state and determine the rotational parameters for the OH stretch state. The spin rotational constants and the tunneling splittings for both ground and OH stretch state were also determined.

TI04   **2:21 – 2:36**

VIBRONIC COUPLING OF $\tilde{B}^2A'$ ELECTRONIC STATE WITH THE $\tilde{X}^2A'$, $\tilde{A}^2A''$ TWOFOLD OF ISOPROPOXY RADICAL.

MOURAD ROUDJANE, *Department of Chemistry and Biochemistry, The Ohio State University, Columbus, OH, USA*; RABI CHHANTYAL-PUN, *Department of Chemistry, The Ohio State University, Columbus, OH, USA*; DMITRY G. MELNIK, TERRY A. MILLER, *Department of Chemistry and Biochemistry, The Ohio State University, Columbus, OH, USA*; JINJUN LIU, *Department of Chemistry, University of Louisville, Louisville, KY, USA*.

We performed rotational analyses of previously reported[a] vibronic bands belonging to $\tilde{B}^2A' \leftarrow \tilde{X}^2A'$ and $\tilde{B}^2A' \leftarrow \tilde{A}^2A''$ electronic transitions of isopropoxy radical. It is noted that certain vibronic bands belonging to both $\tilde{B}^2A' \leftarrow \tilde{X}^2A'$ and $\tilde{B}^2A' \leftarrow \tilde{A}^2A''$ electronic transitions exhibit unusual rotational contours inconsistent with the electronic symmetry designation of the connecting levels, and the orientation of the electronic transition dipole moment with respect to the principal axis system of the molecule is inconsistent with expectations from the molecule's electronic structure. A coupled, three-electronic-state, vibronic Hamiltonian has been used to account for vibronic interactions between the $\tilde{B}$ electronic state and the $\tilde{X}/\tilde{A}$ states, whereas an effective rotational Hamiltonian developed earlier[b] has been used to describe the rovibronic eigenstates within the $\tilde{X}^2A'$ and $\tilde{A}^2A''$ twofold. We show that inclusion of the vibronic coupling of the ground twofold to the upper electronic $\tilde{B}$ state is necessary to account for the observed rotational structure anomalies and present molecular parameters resulting from the rotational analysis of the vibronic spectra.

[a]R. Chhantyal-Pun and T. A. Miller, **TD03**, *68th Molecular Spectroscopy Symposium*, Columbus, 2013

[b]J. Liu, D. Melnik and T. A. Miller, *J. Chem. Phys.*, **139**, 094308, (2013)

TI05

DEVELOPMENT OF A NEAR-IR CAVITY ENHANCED ABSORPTION SPECTROMETER FOR THE DETECTION OF ATMOSPHERIC OXIDATION PRODUCTS AND ORGANOAMINES

NATHAN C EDDINGSAAS, BREANNA JEWELL, EMILY THURNHERR, *College of Science, Rochester Institute of Technology, Rochester, NY, USA.*

An estimated 10,000 to 100,000 different compounds have been measured in the atmosphere, each one undergoes many oxidation reactions that may or may not degrade air quality. To date, the fate of even some of the most abundant hydrocarbons in the atmosphere is poorly understood. One difficulty is the detection of atmospheric oxidation products that are very labile and decompose during analysis. To study labile species under atmospheric conditions, a highly sensitive, non-destructive technique is needed. Here we describe a near-IR incoherent broadband cavity enhanced absorption spectroscopy (IBBCEAS) setup that we are developing to meet this end. We have chosen to utilize the near-IR, where vibrational overtone absorptions are observed, due to the clean spectral windows and better spectral separation of absorption features. In one spectral window we can simultaneously and continuously monitor the composition of alcohols, hydroperoxides, and carboxylic acids in an air mass. In addition, we have used our CEAS setup to detect organoamines. The long effective path length of CEAS allows for low detection limits, even of the overtone absorption features, at ppb and ppt levels.

TI06

INFRARED SPECTROSCOPY OF HALOGENATED SPECIES FOR ATMOSPHERIC REMOTE SENSING

JEREMY J. HARRISON, *Department of Chemistry, University of York, York, United Kingdom.*

Fluorine- and chlorine-containing molecules in the atmosphere are very strong greenhouse gases, meaning that even small amounts of these gases contribute significantly to the radiative forcing of climate. Chlorofluorocarbons (CFCs) and hydrochlorofluorocarbons (HCFCs) are regulated by the 1987 Montreal Protocol because they deplete the ozone layer. Hydrofluorocarbons (HFCs), which do not deplete the ozone layer and are not regulated by the Montreal Protocol, have been introduced as replacements for CFCs and HCFCs. HFCs have global-warming potentials many times greater than carbon dioxide, and are increasing in the atmosphere at a very fast rate.

Various satellite instruments monitor many of these molecules by detecting infrared radiation that has passed through the Earth's atmosphere. However, the quantification of their atmospheric abundances crucially requires accurate quantitative infrared spectroscopy. This talk will focus on new and improved laboratory spectroscopic measurements for a number of important halogenated species.

TI07 3:12 – 3:27

PERFORMANCE OF A CRYOGENIC MULTIPATH HERRIOTT CELL VACUUM-COUPLED TO A BRUKER IFS-125HR SYSTEM

ARLAN MANTZ, *Department of Physics, Connecticut College, New London, CT, USA*; KEEYOON SUNG, *Jet Propulsion Laboratory, Science Division, California Institute of Technology, Pasadena, CA, USA*; TIMOTHY J CRAWFORD, LINDA BROWN, *Jet Propulsion Laboratory, California Institute of Technology, Pasadena, CA, USA*; MARY ANN H. SMITH, *Science Directorate, NASA Langley Research Center, Hampton, VA, USA.*

Accurate modeling of atmospheric trace gases requires detailed knowledge of spectroscopic line parameters at temperatures and pressures relevant to the atmospheric layers where the spectroscopic signatures form. Pressure-broadened line shapes, frequency shifts, and their temperature dependences, are critical spectroscopic parameters that limit the accuracy of state-of-the-art atmospheric remote sensing. In order to provide temperature dependent parameters from controlled laboratory experiments, a 20.946 ± 0.001 m long path Herriott cell and associated transfer optics were designed and fabricated at Connecticut College to operate in the near infrared using a Bruker 125 HR Fourier transform spectrometer. The cell body and gold coated mirrors are fabricated with Oxygen-Free High Conductivity (OFHC) copper. Transfer optics are through-put matched for entrance apertures smaller than 2 mm. A closed-cycle Helium refrigerator cools the cell and cryopumps the surrounding vacuum box. This new system and its transfer optics are fully evacuated to ˜ 10 mTorr (similar to the pressure inside the interferometer). Over a period of several months, this system has maintained extremely good stability in recording spectra at gas sample temperatures between 75 and 250 K. The absorption path length and cell temperatures are validated using CO spectra. The characterization of the Herriott cell is described along with its performance and future applications.[a] [b]

[a]We thank Drs. V. Malathy Devi and D. Chris Benner at The College of William and Mary for helpful discussion.

[b]Research described in this paper was performed at Connecticut College, the Jet Propulsion Laboratory, California Institute of Technology, and NASA Langley Research Center, under contracts and cooperative agreements with the National Aeronautics and Space Administration.

TI08 3:29 – 3:44

MEASUREMENTS AND MODELING OF $^{16}O^{12}C^{17}O$ SPECTROSCOPIC PARAMETERS AT 2 μm

DAVID JACQUEMART, *Chemistry/ MONARIS, CNRS, UMR 8233, Sorbonne Universités, UPMC Univ Paris 06, Paris, France*; KEEYOON SUNG, *Jet Propulsion Laboratory, Science Division, California Institute of Technology, Pasadena, CA, USA*; LINDA BROWN, *Jet Propulsion Laboratory, California Institute of Technology, Pasadena, CA, USA*; MAX COLEMAN, *Science Division, Jet Propulsion Laboratory/Caltech, Pasadena, CA, USA*; ARLAN MANTZ, *Department of Physics, Connecticut College, New London, CT, USA*; MARY ANN H. SMITH, *Science Directorate, NASA Langley Research Center, Hampton, VA, USA.*

Nearly 1000 line intensities of $^{16}O^{12}C^{17}O$ between 4604 and 5126 cm^{-1} were measured using an isotopically-enriched mixture sample having 40 % (determined by mass spectrometry). Spectra were recorded at 0.0056 cm^{-1} resolution with a Fourier transform spectrometer (Bruker IFS-125HR at JPL) configured to a Herriott cell with a 20.946 m absorption path. Since collisional narrowing effects were observed, the Rautian profile was systematically applied (instead of the Voigt profile) using a multispectrum retrieval procedure. Transition dipole moments and Herman-Wallis factors were derived for 15 bands, and a global comparison with theoretical calculations and predictions was obtained. Accuracies for the line intensities ranged between 2 – 3 % for strong bands and 6 – 30 % for weak bands. Retrieved line positions were calibrated using CO, HCl and some well-known $^{16}O^{12}C^{16}O$ transitions. For both measured and calculated line positions, the accuracies fell between $0.1 - 1 \times 10^{-3}$ cm^{-1}. Self-broadening was also obtained for a few bands. Complete line lists were generated to support atmospheric remote sensing of the Earth (*e.g.*, OCO-2 mission), Mars and Venus.[a]

[a]Research described in this paper was performed at Connecticut College, the Jet Propulsion Laboratory, and California Institute of Technology, and NASA Langley Research Center, under contracts and cooperative agreements with the National Aeronautics and Space Administration.

TI09 3:46–4:01

BROADBAND SPECTROSCOPY OF CO_2 BANDS NEAR 2 μm USING A FEMTOSECOND MODE-LOCKED LASER

ANDREW KLOSE, DANIEL L. MASER, GABRIEL YCAS, SCOTT DIDDAMS, *Time and Frequency Division, National Institute of Standards and Technology, Boulder, CO, USA*; NATHAN R. NEWBURY, IAN CODDINGTON, *Quantum Electronics and Photonics Division, National Institute of Standards and Technology, Boulder, CO, USA.*

The optical frequency comb provided in the output of a femtosecond, mode-locked laser has been employed for many applications, including broadband spectroscopic measurements of trace gases using a variety of detection techniques. One environmentally significant trace gas is CO_2, which has characteristic absorption bands near 1.6 μm and 2.0 μm. Continuous wave (cw) lasers have typically been used to measure CO_2 at atmospheric-level concentrations. However, a broadband frequency comb source can provide rapid, simultaneous and accurate measurements of multiple transitions without the need for mechanical scanning or frequency tuning. Previously, precision broadband spectroscopy was performed on CO_2 bands near 1.6 μm. However, the CO_2 absorption bands near 2 μm have nearly a ten-fold increase in line strength compared to the bands near 1.6 μm, making the 2 μm bands attractive candidates for precision measurements of CO_2 with improved signal-to-noise and reduced uncertainty.

Here, broadband quantitative spectroscopy of CO_2 bands near 2 μm is pursued. The source that was developed consists of an Er:fiber oscillator, Er:doped fiber amplifier, and highly nonlinear optical fiber, which generates a broadband spectrum spanning from 1 to 2.2 μm with an average power of 270 mW. Over 70 mW of the optical power is contained in the 1.8-2.2 μm region relevant to the CO_2 measurement. After generation, the laser light is passed through laboratory gas cells or open air where the absorption features from the sample gas are imprinted onto the laser light. Initial detection efforts involve a virtually imaged phased array- (VIPA-)based spectrometer whose output is subsequently imaged on a InSb array detector. The bandwidth of the measured spectrum is 50 nm, limited by the size of the detector array. The characteristics of the spectrometer, including the detection limits and temporal resolution, will be presented. In addition, the progress towards the use of the present spectrometer and related frequency comb technology for quantitative measurement of CO_2 on a 2 km open-air path on the NIST-Boulder campus will be discussed.

Intermission

TI10 4:18–4:33

FTS STUDIES OF THE [17]O ENRICHED ISOTOPOLOGUES OF CO_2 TOWARD CREATING A COMPLETE AND HIGHLY ACCURATE REFERENCE STANDARD

BEN ELLIOTT, *Jet Propulsion Laboratory, California Institute of Technology, Pasadena, CA, USA*; KEEYOON SUNG, *Jet Propulsion Laboratory, Science Division, California Institute of Technology, Pasadena, CA, USA*; LINDA BROWN, CHARLES MILLER, *Jet Propulsion Laboratory, California Institute of Technology, Pasadena, CA, USA.*

The proliferation and increased abilities of remote sensing missions for the monitoring of planetary atmospheric gas species has spurred the need for complete and accurate spectroscopic reference standards. As a part of our ongoing effort toward creating a global carbon dioxide (CO_2) frequency reference standard, we report new FTS measurements of the [17]O enriched isotopologues of CO_2. The first measurements were taken in the ν_3 region (2200 - 2450 cm^{-1}, 65 - 75 THz), have absolute calibration accuracies of 100 kHz (3E-6 cm^{-1}), comparable to the uncertainties for typical sub-millimeter/THz spectroscopy. Such high absolute calibration accuracy has become regular procedure for the cases of linear molecules such as CO_2 and CO for FTS measurements at JPL, and enables us to produce measured transition frequencies for entire bands with accuracies that rival those of early heterodyne measurements for individual beat notes. Additionally, by acquiring spectra of multiple carbon dioxide isotopologues simultaneously, we have begun to construct a self-consistent frequency grid based on CO_2 that extends from 20 - 200 THz. These new spectroscopic reference standards are a significant step towards minimizing CO_2 retrieval errors from remote sensing applications, especially those involving targets with predominantly CO_2 atmospheres such as Mars, Venus and candidate terrestrial exoplanets where minor isotopologues will make significant contributions to the radiance signals.

TI11 **4:35–4:50**

WATER VAPOR SELF-CONTINUUM BY CAVITY RING DOWN SPECTROSCOPY IN THE 1.6 MICRON TRANSPARENCY WINDOW

ALAIN CAMPARGUE, SAMIR KASSI, DIDIER MONDELAIN, *UMR5588 LIPhy, Université Grenoble 1/CNRS, Saint Martin D'heres, France.*

Since its discovery one century ago, a deep and unresolved controversy remains on the nature of the water vapor continuum. Several interpretations are proposed: accumulated effect of the distant wings of many individual spectral lines, metastable or true bound water dimers, collision-induced absorption. The atmospheric science community has largely sidestepped this controversy, and has adopted a pragmatic approach: most radiative transfer codes used in climate modelling, numerical weather prediction and remote sensing use the MT_CKD model which is a semi-empirical formulation of the continuum[a]. The MT_CKD cross-sections were tuned to available observations in the mid-infrared but in the absence of experimental constraints, the extrapolated near infrared (NIR) values are much more hazardous. Due to the weakness of the broadband absorption signal to be measured, very few measurements of the water vapor continuum are available in the NIR windows especially for temperature conditions relevant for our atmosphere. This is in particular the case for the 1.6 μm window where the very few available measurements show a large disagreement. Here we present the first measurements of the water vapor self-continuum cross-sections in the 1.6 μm window[b] by cavity ring down spectroscopy (CRDS). The pressure dependence of the absorption continuum was investigated during pressure cycles up to 12 Torr for selected wavenumber values. The continuum level is observed to deviate from the expected quadratic dependence with pressure. This deviation is interpreted as due to a significant contribution of water adsorbed on the super mirrors to the cavity loss rate. The pressure dependence is well reproduced by a second order polynomial. We interpret the linear and quadratic terms as the adsorbed water and vapour water contribution, respectively. The derived self-continuum cross sections, measured between 5875 and 6450 cm^{-1}, shows a minimum value around 6300 cm^{-1}. These cross sections will be compared to the existing experimental data and models, especially to recent FTS measurements and to the last version of the MT_CKD 2.5 model.

[a]Mlawer, E.J., V.H. Payne, J.L. Moncet, et al. (2012), Phil. Trans. R. Soc. A, 370, 2520–2556.
[b]Mondelain, D., A. Aradj, S. Kassi, et al. (2013), JQSRT, 130, 381–391.

TI12 **4:52–5:07**

AN ACCURATE AND COMPLETE EMPIRICAL LINE LIST FOR WATER VAPOR BETWEEN 5850 AND 7920 CM^{-1}

SEMEN MIKHAILENKO, *Atmospheric Spectroscopy Div., Institute of Atmospheric Optics, Tomsk, Russia*; DIDIER MONDELAIN, SAMIR KASSI, ALAIN CAMPARGUE, *UMR5588 LIPhy, Université Grenoble 1/CNRS, Saint Martin D'heres, France.*

An empirical line list has been constructed for "natural" water vapor at 296 K in the 5850 – 7920 cm^{-1}region. It was obtained by gathering separate line lists recently published on the basis of spectra recorded by high sensitivity Continuous Wave Cavity Ring Down Spectroscopy (CW-CRDS) of natural water, complemented with literature data for the strongest lines. The list includes 38318 transitions of four major water isotopologues ($H_2^{16}O$, $H_2^{18}O$, $H_2^{17}O$ and $HD^{16}O$) with an intensity cut-off of 1×10^{-29} cm/molecule at 296 K. The list is made mostly complete over the whole spectral region by including a large number of weak lines with positions calculated using experimentally determined energy levels and intensities obtained from variational calculations. In addition, we provide $HD^{18}O$ and $HD^{17}O$ lists in the same region for transitions with intensities larger than 1×10^{-29} cm/molecule. The $HD^{18}O$ and $HD^{17}O$ lists (1972 lines in total) were obtained using empirical energy levels available in the literature and variational intensities. The global list (40290 transitions) for water including the contribution of the six major isotopologues will be adopted for the next edition of the GEISA database in the region. The advantages and drawbacks of our list are discussed in comparison with the list provided for the same region in the 2012 edition of the HITRAN database. The direct comparison of the CRDS spectra to simulations based on the HITRAN list has revealed some insufficiencies which could easily be corrected: missing HDO lines, duplicated lines, inaccurate line positions or line intensities from variational calculations.

134

TI13   5:09 – 5:24

THE ROTATIONAL SPECTRUM OF HDO: ACCURATE SPECTROSCOPIC AND HYPERFINE PARAMETERS

GABRIELE CAZZOLI, *Dep. Chemistry 'Giacomo Ciamician', University of Bologna, Bologna, Italy*; VALERIO LATTANZI, *Chemistry G. Ciamician, University of Bologna, Bologna, Italy*; CRISTINA PUZZARINI, *Dep. Chemistry 'Giacomo Ciamician', University of Bologna, Bologna, Italy*; JÜRGEN GAUSS, *Institut für Physikalische Chemie, University of Mainz, Mainz, Germany.*

Due to its importance for atmospheric and astrophysical science, a spectroscopic investigation of the rotational spectrum of the singly deuterated water molecule (HDO) was performed, spanning a large frequency range: from the millimeter-wave region up to the THz frequency domain. Sub-Doppler resolution was obtained by exploiting the Lamb-dip technique, thus allowing us to resolve the hyperfine structure of the rotational lines due to deuterium and hydrogen. The experimental determination of the hyperfine parameters involved was supported by highly accurate quantum-chemical calculations. On the whole, the present measurements allowed us to carry out an accurate spectroscopic characterization of HDO.

TI14   5:26 – 5:41

O_2 ENERGY LEVELS, BAND CONSTANTS, POTENTIALS, FRANCK-CONDON FACTORS AND LINELISTS INVOLVING THE $X^3\Sigma_g^-$, $a^1\Delta_g$ AND $b^1\Sigma_g^+$ STATES

SHANSHAN YU, BRIAN DROUIN, CHARLES MILLER, *Jet Propulsion Laboratory, California Institute of Technology, Pasadena, CA, USA*; IOULI E GORDON, *Atomic and Molecular Physics, Harvard-Smithsonian Center for Astrophysics, Cambridge, MA, USA.*

The isotopically invariant Dunham fit of O_2 was updated with newly reported literature transitions to derive (1) the energy levels, band-by-band molecular constants and RKR potentials for the $X^3\Sigma_g^-$, $a^1\Delta_g$ and $b^1\Sigma_g^+$ states of the six O_2 isotopologues, $^{16}O^{16}O$, $^{16}O^{17}O$, $^{16}O^{18}O$, $^{17}O^{17}O$, $^{17}O^{18}O$ and $^{18}O^{18}O$; (2) the line positions and Franck-Condon factors for their $a^1\Delta_g - X^3\Sigma_g^-$, $b^1\Sigma_g^+ - X^3\Sigma_g^-$ and $a^1\Delta_g - b^1\Sigma_g^+$ band systems. The best available experimental and theoretical data were used as input for calculating intensities of lines involving the $X^3\Sigma_g^-$, $a^1\Delta_g$ and $b^1\Sigma_g^+$ states. The newly calculated positions and intensities are combined to provide HITRAN-format linelists.

TI15   5:43 – 5:58

THE OXYGEN A BAND

D. CHRIS BENNER, V. MALATHY DEVI, JIAJUN HOO, *Department of Physics, College of William and Mary, Williamsburg, VA, USA*; JOSEPH HODGES, DAVID A. LONG, *Material Measurement Laboratory, National Institute of Standards and Technology, Gaithersburg, MD, USA*; KEEYOON SUNG, *Jet Propulsion Laboratory, Science Division, California Institute of Technology, Pasadena, CA, USA*; BRIAN DROUIN, *Jet Propulsion Laboratory, California Institute of Technology, Pasadena, CA, USA*; MITCHIO OKUMURA, THINH QUOC BUI, *Division of Chemistry and Chemical Engineering, California Institute of Technology, Pasadena, CA, USA*; PRIYANKA RUPASINGHE, *Physical Sciences, Cameron University, Lawton, OK, USA.*

The oxygen A band is used for numerous atmospheric experiments, but spectral line parameters that sufficiently describe the spectrum to the level required by OCO2 and other high precision/accuracy experiments are lacking. Fourier transform spectra from the Jet Propulsion Laboratory and cavity ring down spectra from the National Institute of Standards and Technology were fitted simultaneously using the William and Mary multispectrum nonlinear least squares fitting technique[a] into a single solution including the entire band. In addition, photoacoustic spectra already available from the California Institute of Technology will be added to the solution. The three types of spectrometers are complementary allowing the strengths of each to fill in the weaknesses of the others. With this technique line positions, intensities, widths, shifts, line mixing, Dicke narrowing, temperature dependences and collision induced absorption have been obtained in a single physically consistent fit.[b]

[a]D. Chris Benner, C. P. Rinsland, V. M. Devi, M. A. H. Smith, and D. Atkins, JQSRT 1995;53:705-21.

[b]Part of the research described in this paper was performed at The College of William and Mary, the, Jet Propulsion Laboratory, California Institute of Technology, under contracts and cooperative agreements with the National Aeronautics and Space Administration and the Jet Propulsion Laboratory. Support for the National Institute of Standards and Technology was provided by the NIST Greenhouse Gas Measurements and Climate Research Program and a NIST Innovations in Measurement Science (IMS) award.

TI16 **6:00 – 6:15**

SELF- AND AIR-BROADENED LINE SHAPE PARAMETERS OF $^{12}CH_4$: 4500-4620 CM^{-1}

<u>V. MALATHY DEVI</u>, D. CHRIS BENNER, *Department of Physics, College of William and Mary, Williamsburg, VA, USA*; KEEYOON SUNG, *Jet Propulsion Laboratory, Science Division, California Institute of Technology, Pasadena, CA, USA*; LINDA BROWN, TIMOTHY J CRAWFORD, *Jet Propulsion Laboratory, California Institute of Technology, Pasadena, CA, USA*; MARY ANN H. SMITH, *Science Directorate, NASA Langley Research Center, Hampton, VA, USA*; ARLAN MANTZ, *Department of Physics, Connecticut College, New London, CT, USA*; ADRIANA PREDOI-CROSS, *Department of Physics and Astronomy, University of Lethbridge, Lethbridge, Canada.*

Accurate knowledge of spectral line shape parameters is important for infrared transmission and radiance calculations in the terrestrial atmosphere. We report the self- and air-broadened Lorentz widths, shifts and line mixing coefficients along with their temperature dependences for methane absorption lines in the 2.2 μm spectral region. For this, we obtained a series of high-resolution, high S/N spectra of 99.99% ^{12}C-enriched samples of pure methane and its dilute mixtures in dry air at cold temperatures down to 150 K using the Bruker IFS 125HR Fourier transform spectrometer at JPL. The coolable absorption cell had an optical path of 20.38 cm and was specially built to reside inside the sample compartment of the Bruker FTS[a]. The 13 spectra used in the analysis consisted of seven pure $^{12}CH_4$ spectra at pressures from 4.5 to 169 Torr and six air-broadened spectra with total sample pressures of 113-300 Torr and methane volume mixing ratios between 4 and 9.7%. These 13 spectra were fit simultaneously using the multispectrum least-squares fitting technique[b]. The results will be compared to existing values reported in the literature.[c]

[a]K. Sung, A. W. Mantz, L. R. Brown, *et al., J. Mol. Spectrosc.* **162** (2010) 124-134.

[b]D. C. Benner, C. P. Rinsland, V. Malathy Devi, M. A. H. Smith and D. Atkins, *JQSRT* **53** (1995) 705-721.

[c]Research described in this paper was performed at Connecticut College, the College of William and Mary, NASA Langley Research Center and the Jet Propulsion Laboratory, California Institute of Technology, under contracts and cooperative agreements with the National Aeronautics and Space Administration.

T.J. Instrument/Technique Demonstration

Tuesday, June 17, 2014 – 1:30 PM

Room: 274 Medical Sciences Building

Chair: Terry A. Miller, The Ohio State University, Columbus, OH, USA

TJ01   1:30 – 1:45

FAR-INFRARED BEAMLINE AT THE CANADIAN LIGHT SOURCE

BRANT E BILLINGHURST, TIM E MAY, *EFD, Canadian Light Source Inc., Saskatoon, Saskatchewan, Canada.*

The far-infrared beamline at the Canadian Light Source is a state of the art user facility, which offers significantly more far-infrared brightness than conventional globar sources. The infrared radiation is collected from a bending magnet through a 55 X 37 mrad^2 port to a Bruker IFS 125 HR spectrometer, which is equipped with a nine compartment scanning arm, allowing it to achieve spectral resolution better than 0.001 cm^{-1}. Currently the beamline can achieve signal to noise ratios up to 8 times that which can be achieved using a traditional thermal source. This talk will provide an overview of the the beamline, and the capabilities available to users, recent and planned improvements including the addition of a Glow Discharge cell and advances in Coherent Synchrotron Radiation. Furthermore, the process of acquiring access to the facility will be covered.

TJ02   1:47 – 2:02

COHERENT GENERATION OF BROADBAND PULSED LIGHT IN THE SWIR AND MWIR USING AN ALL POLARIZATION-MAINTAINING FIBER FREQUENCY COMB SOURCE

H. HOOGLAND, M. ENGELBRECHT, C. McRAVEN, R. HOLZWARTH, *Menlo Systems, GmbH, Martinsried, Germany*; A. THAI, D. SÁNCHEZ, S. L. COUSIN, M. HEMMER, M. BAUDISCH, *ICFO - Institut de Ciencies Fotoniques, Barcelona, Spain*; K. ZAWILSKI, P. G. SCHUNEMANN, *BAE Systems, Nashua, NH, USA*; J. BIEGERT[a], *ICREA - Instituciò Catalana de Recerca i Estudis Avancats, Barcelona, Spain.*

We report on an all polarization-maintaining, modelocked, fiber laser system which generates coherent broadband pulses centered at 2.03 μm with a spectral FWHM bandwidth of 60 nm and 360 mW. Using this frequency comb source, we generate phase-coherent, ultra-broadband pulses centered at 6.5 μm and spanning 5.5 μm to 8 μm with DFG in CdSiP$_2$.

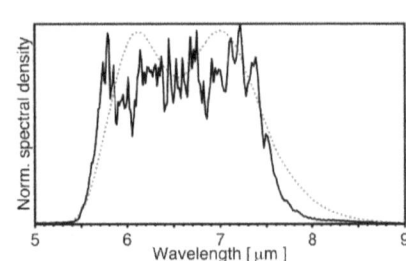

---

[a]Also, ICFO - Institut de Ciencies Fotoniques, Barcelona, Spain

TJ03 2:04 – 2:19

DUAL-COMB SPECTROSCOPY OF C_2H_2, CH_4 AND H_2O OVER 1.0 - 1.7 μm

KANA IWAKUNI, *Department of Physics, Faculty of Science and Technology, Keio University, Yokohama, Japan*; SHO OKUBO, HAJIME INABA, KAZUMOTO HOSAKA, ATSUSHI ONAE, *National Metrology Institute of Japan (NMIJ), Ntional Institute of Advanced Industrial Science and Technology (AIST), Tsukuba, Japan*; HIROYUKI SASADA, *Department of Physics, Faculty of Science and Technology, Keio University, Yokohama, Japan*; FENG-LEI HONG, *National Metrology Institute of Japan (NMIJ), Ntional Institute of Advanced Industrial Science and Technology (AIST), Tsukuba, Japan.*

A dual-comb spectrometer (DCS) has great advantages over the conventional FTIR in respect with the resolution and the measurement time. We reduce the relative linewidth of two optical frequency combs in our DCS to less than 1 Hz and extended the observable spectral bandwidth to be compatible with the FTIR. The figure shows the recorded spectrum of entire vibrational bands of $^{12}C_2H_2$ at 1.03 μm and 1.53 μm, CH_4 at 1.67 μm and H_2O at 1.46 μm. It takes 140 ms to record a time domain interferogram, from which the spectrum across over 1.0-1.7 μm is obtained by Fourier transformation. The interferogram is averaged more than 400,000 times successively to improve the signal to noise ratio. The horizontal axis is scaled by the absolute frequency and the transition frequencies are determined by fitting the absorption lines with the Voigt functions. The discrepancy from the previous sub-Doppler resolution measurements is typically a few MHz.

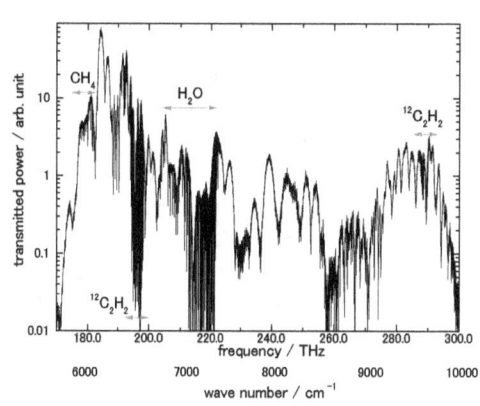

TJ04 2:21 – 2:36

MULTIPLEXED DETECTION OF CO_2 USING A NOVEL DUAL COMB SPECTROMETER

ADAM J. FLEISHER, DAVID A. LONG, JOSEPH HODGES, *Material Measurement Laboratory, National Institute of Standards and Technology, Gaithersburg, MD, USA*; DAVID F. PLUSQUELLIC, *Physical Measurement Laboratory, National Institute of Standards and Technology, Boulder, CO, USA.*

We report multiheterodyne-detected cavity-enhanced absorption spectroscopy (MH-CEAS) of CO_2 in the near-infrared using a novel frequency comb light source generated from the electro-optical modulation of a single continuous-wave laser. Using a pair of dual-drive Mach Zehnder modulators we create two frequency combs of slightly different comb mode spacing. The probe comb is coupled to a high-finesse optical cavity containing CO_2 and then combined with the local oscillator comb on a photodetector. This results in down-conversion of the optical comb to radio frequencies, therefore facilitating fast, multiplexed analysis of the CO_2 direct absorption signal. The noise-equivalent absorption coefficient for MH-CEAS as initially demonstrated here is 3×10^{-10} cm^{-1} $Hz^{-1/2}$ per spectral element for $\sim$50 simultaneously resolved channels of unique optical frequency.

TJ05   **2:38 – 2:53**

DOPPLER-FREE TWO-PHOTON ABSORPTION SPECTROSCOPY OF NAPHTHALENE ASSISTED BY AN OPTICAL FREQUENCY COMB

AKIKO NISHIYAMA, AYUMI MATSUBA, MASATOSHI MISONO, *Applied Physics, Fukuoka University, Fukuoka, Japan.*

Optical frequency combs are powerful tools for precise frequency measurements in various wavelength regions. The combs have been applied not only to metrology, but also to molecular spectroscopy. Recently, we studied high resolution spectroscopy of iodine molecule assisted by an optical frequency comb. [a] In the study, the comb was used for frequency calibration of a scanning dye laser.

In this study, we developed a frequency calibration scheme with a comb and an acousto-optic modulator to realize more precise frequency measurement in a wide frequency range. And the frequency calibration scheme was applied to Doppler-free two-photon absorption (DFTPA) spectroscopy of naphthalene. Naphthalene is one of the prototypical aromatic molecules, and its detailed structure and dynamics in excited states have been reported. We measured DFTPA spectra of $A^1B_{1u}(v_4 = 1) \leftarrow X^1A_g(v = 0)$ transition around 298 nm.

A part of obtained spectra is shown in the figure. The spectral lines are rotationally resolved and the resolution is about 100 kHz. The horizontal axis was calibrated by the developed frequency calibration system employing the comb. The uncertainties of the calibrated frequencies were determined by the fluctuations of the comb modes which were stabilized to a GPS-disciplined clock.

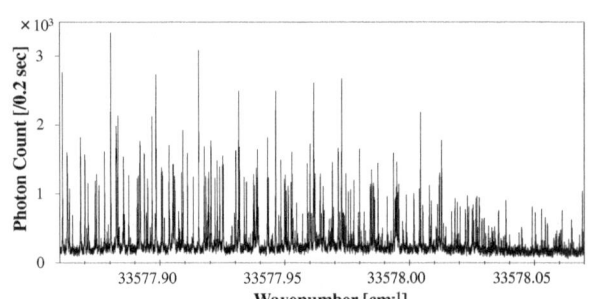

[a] A. Nishiyama, D. Ishikawa, and M. Misono, J. Opt. Soc. Am. B 30, 2107 (2013).

TJ06   **2:55 – 3:10**

MULTIPLEXED CHIRPED PULSE QUANTUM CASCADE LASER MEASUREMENTS OF AMMONIA AND OTHER SMALL MOLECULES

CRAIG PICKEN, NIGEL LANGFORD, GEOFFREY DUXBURY, *Department of Physics, University of Strathclyde, Glasgow, United Kingdom.*

Spectrometers based on Quantum Cascade (QC) lasers can be run in either continuous or pulsed operation. Although the instrumentation based upon the most recent versions of continuously operating QC lasers can have higher resolution than chirped lasers, using chirped pulse QC lasers can give an advantage when rapid changes in gas composition occur. For example, when jet engines are being tested, a variety of temperature dependent effects on the trace gas concentrations of the plume may be observed. Most pulsed QC lasers are operated in the down chirped mode, in which the chirp rate slows during the pulse. In our spectrometer the changes in frequency are recorded using two Ge etalons, one with a free spectral range of 0.0495 cm^{-1}, and the other with a fringe spacing of 0.0195 cm^{-1}.They can also be deployed in multiplex schemes in which two or more down-chirped lasers are used.

In this paper we wish to show examples of the use of multiplexed chirped pulse lasers to allow overlapping spectra to be recorded. The examples of multiplex methods used are taken partly from measurements of $^{14}NH_3$ and $^{15}NH_3$ in the region from 1630 to 1622 cm^{-1}, and partly from the use of other chirped pulse lasers operating in the 8 μm region. Among the effects seen are rapid passage effects caused by the rapid down-chirp, and the use of gases such as nitrogen to cause variation in the shape of the collisional broadened absorption lines.

Intermission

TJ07 3:27 – 3:42

BROADBAND HIGH-RESOLUTION SPECTROSCOPY WITH FABRY-PEROT QUANTUM CASCADE LASERS

YIN WANG, GERARD WYSOCKI, *Department of Electrical Engineering, Princeton University, Princeton, NJ, USA.*

Simultaneous spectroscopic detection of large molecules with broad ro-vibrational spectra, and small molecules with well-resolved narrow spectral lines requires both broadband optical frequency coverage (>50 cm^{-1}) and high resolution (<0.01 cm^{-1}) to perform accurate spectral measurements. With the advent of room temperature, high power, continuous wave quantum cascade lasers (QCLs), high resolution mid-IR spectrometers for field applications became feasible. So far to address the broadband spectral coverage, external cavity (EC) QCLs with >100 cm^{-1} tuning ranges have been spectroscopic sources of choice in the mid-IR; however EC-QCLs are rather complex opto-mechanical systems, which are vibration-sensitive, and construction of robust transportable systems is difficult. In this work we present a new method of performing broadband mid-IR spectroscopy using two free-running Fabry-Perot (FP) QCLs to perform multi-heterodyne down-conversion of optical signals to RF domain. The sample transmission spectrum probed by one multi-mode FP-QCL is down-converted to the RF domain through an optical multi-heterodyne process using a second FP-QCL as the local oscillator[a]. Both a broadband multi-mode spectral measurement as well as high-resolution (15 MHz or 0.0005 cm^{-1}) absorption spectroscopy of NH$_3$ and N$_2$O are demonstrated and show potential for all-solid-state FP-laser-based spectrometers for chemical sensing.

[a]Y. Wang, M. G. Soskind, W. Wang, and G. Wysocki, "High-resolution multi-heterodyne spectroscopy based on Fabry-Perot quantum cascade lasers," Appl Phys Lett 104, 0311141-0311145 (2014)

TJ08 3:44 – 3:59

DEVELOPMENT OF A FREQUENCY-STABILIZED MID-INFRARED EXTERNAL CAVITY-QCL CAVITY RING-DOWN SPECTROMETER

BRADLEY M. GIBSON, *Department of Chemistry, University of Illinois at Urbana-Champaign, Urbana, IL, USA*; BENJAMIN J. McCALL, *Departments of Chemistry and Astronomy, University of Illinois at Urbana-Champaign, Urbana, IL, USA.*

External cavity quantum cascade lasers (EC-QCLs) provide significantly enhanced wavelength tunability while maintaining the high output powers of traditional QCLs throughout the mid-infrared portion of the spectrum. However, the inclusion of wavelength-selective moving parts allows vibrational and acoustic noise to be coupled into the laser as frequency and power noise. This can be particularly troublesome for cavity-enhanced spectroscopy, as attempts to increase vibrational isolation may interfere with stable cavity alignment.

Here, we discuss our efforts to improve the frequency stability of our EC-QCL while maintaining tunability and consistent cavity alignment. A mid-infrared hollow silica waveguide is used to vibrationally isolate the laser from the cavity while maintaining alignment. To further increase frequency stability, the laser is side-of-fringe locked to an uncoated solid germanium etalon. Tunability is maintained by incrementing the angle of incidence upon the etalon using a piezo-driven mirror. Angle of incidence tuning and automated periodic re-locking are managed using a BeagleBone Black development board. This allows us to maintain essentially continuous frequency tuning between 1140-1220 cm^{-1} while stabilizing the laser's frequency to within 30 MHz. Other mid-infrared frequencies could easily be accessed by swapping QCL chips.

TJ09   **4:01 – 4:16**

HIGH-RESOLUTION SPECTROSCOPY OF THE ν_{16} BAND OF 1,3,5-TRIOXANE

BRADLEY M. GIBSON, NICOLE KOEPPEN, *Department of Chemistry, University of Illinois at Urbana-Champaign, Urbana, IL, USA*; BENJAMIN J. McCALL, *Departments of Chemistry and Astronomy, University of Illinois at Urbana-Champaign, Urbana, IL, USA.*

1,3,5-trioxane, often used as a solid fuel or source of formaldehyde, is a symmetric top of the C_{3v} group. Although the microwave[a] and low-resolution vibrational spectra[b] have been studied extensively, only the ν_{17} band near 1072 cm^{-1} has been observed with rotational resolution[c]. Here, we will present our studies of trioxane vapor from 1140-1220 cm^{-1}, covering the ν_{16} band at a resolution of approximately 30 MHz. Solid trioxane was heated, and the resulting vapor was entrained in a continuous supersonic expansion of argon. Continuous-wave cavity ringdown spectroscopy was then performed using a frequency-stabilized external cavity quantum cascade laser (EC-QCL) as the light source. In addition to providing new ro-vibrational transition frequencies of trioxane, the present work serves to validate our newly-developed EC-QCL spectrometer and will be used to evaluate the cooling performance of the sheath-flow supercritical fluid expansion source currently under development[d].

[a]Oka, T., Tsuchiya, K., Iwata, S., and Morino, Y. Microwave Spectrum of s-Trioxane. *Bull. Chem. Soc. Jpn.* **37** (1964), 4-7.

[b]Stair, A.T. Jr. and Nielsen, J. Rud. Vibrational Spectra of sym-Trioxane. *J. Chem. Phys.* **27** (1957), 402-407.

[c]Henninot, J-F., Bolvin, H., Demaison, J., and Lemoine, B. The Infrared Spectrum of Trioxane in a Supersonic Slit Jet. *J. Mol. Spect.* **152** (1992), 62-68.

[d]Gibson, B.M., Stewart, J.T., and McCall, B.J., contribution TJ14, presented at the 68th International Symposium on Molecular Spectroscopy, Columbus, OH, USA, 2013.

TJ10   **4:18 – 4:33**

DUAL WAVELENGTH CAVITY RINGDOWN SPECTROSCOPY FOR HIGH PRECISION METHANE ISOTOPE RATIO MEASUREMENTS

THINH QUOC BUI, LINHAN SHEN, *Division of Chemistry and Chemical Engineering, California Institute of Technology, Pasadena, CA, USA*; DANIEL HOGAN, *Department of Applied Physics, Stanford University, Stanford, CA, USA*; PIN CHEN, *Jet Propulsion Laboratory, Science Division, California Institute of Technology, Pasadena, CA, USA*; MITCHIO OKUMURA, *Division of Chemistry and Chemical Engineering, California Institute of Technology, Pasadena, CA, USA.*

We demonstrate a frequency stabilized cavity ringdown spectrometer capable of measuring simultaneous isotopes of methane ($^{12}CH_4$, $^{13}CH_4$, $^{12}CH_3D$) of enriched samples to high precision ($\delta D < 0.03\%$ and $\delta 13C < 0.01\%$). The spectrometer employs coupling of two orthogonally polarized CW lasers into a ringdown cavity for simultaneous spectral measurements over the full wavelength range of 1.45-1.65μm. In addition, we discuss the necessity of modeling methane lineshapes with the Galatry profile to achieve the highest precision.

TJ11   **4:35 – 4:50**

RECENT PROGRESS IN DEVELOPING A COMMERCIAL FIBER-LOOP CAVITY RINGDOWN SYSTEM

BRIAN SILLER, RYAN MATZ, HELEN WAECHTER, *Research and Development, Tiger Optics LLC, Warrington, PA, USA.*

High purity and precisely mixed liquid solutions are important to a variety of industrial processes, but sensors for such solutions often have significant drawbacks such as the need for regular calibration and the inability to continuously make real-time measurements. For some specialty liquids, such as cryogenic liquids or caustic solutions used in the semiconductor industry, direct sensors for composition and contamination don't exist at all, and indirect methods must be used instead.

Fiber-loop cavity ring-down spectroscopy (FL-CRDS) can provide an ideal solution for many challenging applications. Since fibers are resistant to chemicals and extreme temperatures, a sensor based on FL-CRDS can be used in environments where other techniques and sensors can't work.

In a FL-CRDS instrument, a laser is coupled into a loop of fiber, and a small amount of light is extracted from the loop to a detector with each pass. Spliced into the loop is a sensing element that allows the evanescent field of the light otherwise confined within the fiber core to interact with the surrounding environment. Results will be presented for detection of contaminants in liquids with several types of sensing elements: fiber tapers, side-polished fibers, and core-only fibers; each with a variety of geometries. Sampling systems for both continuous flow of small samples and for monitoring of static sample baths will also be presented.

TJ12 **4:52 – 5:07**

DOPPLER BROADENING THERMOMETRY BASED ON CAVITY RING-DOWN SPECTROSCOPY

SHUI-MING HU, CUNFENG CHENG, JIN WANG, YAN TAN, YU ROBERT SUN, AN-WEN LIU, *Hefei National Laboratory for Physical Science at Microscale, University of Science and Technology of China, Hefei, China*; JIN-TAO ZHANG, *National Institute of Metrology, Beijing, China*.

A Doppler broadening thermometry is implemented using a laser-locked cavity ring-down spectrometer[a][b] combined with a temperature-stabilized sample cell. The temperature fluctuation of the gas sample cell is kept below 1 mK for hours. The probing laser is frequency locked at a longitudinal mode of a Fabry-Pérot interferometer made of ultra-low-expansion glass, and the spectral scan is implemented by scanning the sideband produced by an electro-optic modulator. As a result, a kHz precision has been maintained during the measurement of the spectrum of 10 GHz wide. A ro-vibrational line of C_2H_2 is measured at sample pressures of a few Pa. Using a pair of mirrors with a reflectivity of 0.99997 at 787 nm, we are able to detect absorption line profiles with a signal-to-noise ratio of 10^5. Fitting of the recorded spectra allows us to determine the Doppler width with a statistical uncertainty of 10 ppm. Further improvements on the experimental reproducibility and investigations on the collision effects will probably lead to an optical determination of the Boltzmann constant with an uncertainty of a few ppm.

---

[a]H. Pan, C.-F. Cheng, Y. R. Sun, B. Gao, A.-W. Liu, S.-M. Hu, "Laser-locked, continuously tunable high resolution cavity ring-down spectrometer," Rev. Sci. Instrum. **82**, 103110 (2011)

[b]Y. R. Sun, H. Pan, C.-F. Cheng, A.-W. Liu, J.-T. Zhang, S.-M. Hu, "Application of cavity ring-down spectroscopy to the Boltzmann constant determination," Opt. Express, **19**, 19993 (2011)

TJ13 **5:09 – 5:24**

MID-IR SUB-DOPPLER ERESOLUTION SPECTROMETER USING AN ENHANCED-CAVITY ABSORPTION CELL COUPLED WITH A WIDE BEAM

MASASHI ABE, KANA IWAKUNI, SHO OKUBO[a], HIROYUKI SASADA, *Department of Physics, Faculty of Science and Technology, Keio University, Yokohama, Japan*.

We have introduced a wide-beam-coupled enhanced-cavity absorption cell (ECAC) into a 3-μm difference-frequency generation spectrometer in order to reduce transit-time broadening of Lamb dips. It contains concave and convex mirrors with a curvature radius of ± 7 m separated by 37.5 cm, has a finesse of 770, and is coupled with a Gaussian beam having a $1/e^2$ radius of 1.9 mm at beam waist. The spectrometer is applied to record sub-Doppler resolution spectra of the ν_3 band of CH_4 and the ν_1 and ν_4 bands of CH_3D, and the transit-time broadening is estimated 30 kHz for these molecules. The observed Lamb dips are about 80 kHz (HWHM) wide, which is one third of those recorded using another ECAC coupled with a $1/e^2$ radius of 0.7 mm at the beam waist. Some A_1-A_2 splittings of the low J levels for CH_3D are first resolved, and the absolute transition frequencies are determined with a relative uncertainty of 10^{-9}.

---

[a]Present address: National Metrology Institute of Japan (NMIJ), Ntional Institute of Advanced Industrial Science and Technology (AIST), Tsukuba, Japan

TJ14   **5:26 – 5:41**

QUANTITATIVE ABSORPTION AND KINETIC STUDIES OF TRANSIENT SPECIES USING GAS PHASE OPTICAL CALORIMETRY

DMITRY G. MELNIK, *Department of Chemistry and Biochemistry, The Ohio State University, Columbus, OH, USA.*

Quantitative measurements of the absorption cross-sections and reaction rates constants of free radicals by spectroscopic means requires the knowledge of the absolute concentration of the target species. We have demonstrated earlier[a] that such information can be retrieved from absorption measurements of the well-known "reporter" molecule, co-produced in radical synthesis. This method is limited to photochemical protocols allowing for production of "reporters" stochiometrically with the target species. This limitation can be overcome by use of the optical calorimetry (OC) which measures heat signatures of a photochemical protocol. These heat signatures are directly related to the amount of species produced and the thermochemical data of the reactants and stable products whose accuracy is usually substantially higher than that of the absorption data for prospective "reporters". The implementation of the OC method presented in this talk is based on the measurements of the frequency shift of the resonances due to the change in the optical density of the reactiove sample within a Fabry-Perot cavity caused by deposition of heat from the absorbed photolysis beam and subsequent chemical reactions. Preliminary results will be presented and future development of this experimental technique will be discussed.

[a]D. Melnik, R. Chhantyal-Pun and T. A. Miller, *J. Phys. Chem. A*, **114**, 11583, (2010)

TJ15   **5:43 – 5:58**

COLLINEAR TWO-COLOR SATURATION SPECTROSCOPY IN CN A-X (1-0) AND (2-0) BANDS

DAMIEN FORTHOMME, C. McRAVEN, TREVOR SEARS[a], GREGORY HALL, *Department of Chemistry, Brookhaven National Laboratory, Upton, NY, USA.*

Hyperfine-resolved saturation spectra were measured for a selection of low and medium J rotational lines in the $A^2\Pi - X^2\Sigma^+$ system of CN using two copropagating laser beams tuned to transitions in the (2–0) and (1–0) bands. A bleach laser was amplitude modulated and fixed in frequency near the center of a rotational line of the (2–0) vibrational band, while a probe laser was frequency-modulated and scanned across selected lines of the (1–0) vibrational band, sharing a common lower state with the bleach laser. Locking the probe laser with a tunable radio frequency offset to a cavity that tracks the slowly drifting bleach laser greatly improved the quality of the double-resonance saturation signals, by stabilizing the relative frequency of the two beams. The sub-Doppler resonances were fit with Lorentzian line shapes having a typical full-width at half maximum of 2–3 MHz. The hyperfine spectra observed depend on the hyperfine structure within both rovibronic transitions excited, permitting the determination of hyperfine molecular constants in the $\nu = 2$ state and the refinement of previously published values in the $\nu = 1$ state. Four nuclear magnetic dipole and two electric quadrupole hyperfine constants were determined for each of the upper states from a fit with a weighted root mean squared error of 0.5 MHz. The vibrational dependence of these constants is weak or negligible.

Acknowledgements: Work at Brookhaven National Laboratory was carried out under contract No. DE-AC02-98CH10886 with the U.S. Department of Energy and supported by its Office of Basic Energy Sciences, Division of Chemical Sciences, Geosciences and Biosciences.

[a]also: Department of Chemistry, Stony Brook University, Stony Brook, New York, 11794

TK. Metal containing
Tuesday, June 17, 2014 – 1:30 PM
Room: B102 Chemical and Life Sciences

Chair: Manori Perera, Illinois Wesleyan University, Bloomington, IL, USA

TK01   1:30 – 1:45

LASER SPECTROSCOPY OF THE $C^2\Pi$ (41242 cm^{-1}) AND $^2\Delta$ (42192 cm^{-1}) STATES OF MAGNESIUM HYDRIDE

NICHOLAS CARON, DENNIS TOKARYK, *Department of Physics, University of New Brunswick, Fredericton, NB, Canada*; ALLAN G. ADAM, *Department of Chemistry, University of New Brunswick, Fredericton, NB, Canada.*

Two laser-induced fluorescence techniques were used to access the high-lying $C^2\Pi$ state (41242 cm^{-1}) in the metal-bearing molecule MgH. First, we directly excited the $C^2\Pi$ state from the $X^2\Sigma^+$ ground state with UV light produced by frequency-doubling pulsed visible laser light. Second, a resonant two-photon experiment was performed with different laser beams tuned to the $A^2\Pi$ - $X^2\Sigma^+$ and $C^2\Pi$ - $A^2\Pi$ transitions. Term energies for levels up to J = 11.5 were fit to a $^2\Pi$ Hamiltonian to yield rotational parameters B = 6.1045(10) cm^{-1}, D = 3.176(86)×10^{-4} cm^{-1}, A = 3.843(60) cm^{-1}, and p = -2.653(88) ×10^{-2} cm^{-1}. We also observed levels of a $^2\Delta$ state (42192 cm^{-1}) up to J = 4.5 with the two-photon technique. Guntch previously observed the $^2\Delta$ - $A^2\Pi$ spectrum in 1937, but assigned the upper state as $^2\Sigma^-$. The molecular parameters determined for the $^2\Delta$ state are B = 6.2861(23) cm^{-1}and A = -0.168(17) cm^{-1}.

TK02   1:47 – 2:02

VIBRONIC PERTURBATIONS IN THE ELECTRONIC SPECTRUM OF MgC

PHALGUN LOLUR, RICHARD DAWES, *Department of Chemistry, Missouri University of Science and Technology, Rolla, MO, USA*; MICHAEL HEAVEN, *Department of Chemistry, Emory University, Atlanta, GA, USA.*

Accurate studies of the ground and low lying excited triplet pi-states of the covalently bonded alkaline earth metal carbides have been of interest to both theoreticians and experimentalists in the past few decades to understand their bonding. Diatomic beryllium carbide (BeC), which is valence iso-electronic with MgC, was probed by laser ablation and jet cooling techniques producing rotationally resolved data reported in a previous study.[1] Dynamically weighted MRCI calculations were used to construct adiabatic potential energy curves for the ground and the four lowest triplet pi-states up to 50,000 cm^{-1}. From these, diabatic potentials and couplings were obtained and used to compute vibronic levels for the four interacting states. Here we apply the same methodology to MgC and examine the similarities and differences between the two systems. Results show significantly different bonding characteristics for the pi-states of MgC when compared to BeC.

References: 1- B. J. Barker et al. J. Chem. Phys. 137, 214313 (2012).

TK03   2:04 – 2:19

ELECTRONIC TRANSITIONS OF SCANDIUM MONOXIDE

NA WANG, YUK WAI NG, ALLAN S.C. CHEUNG, *Department of Chemistry, The University of Hong Kong, Hong Kong, Hong Kong.*

The electronic transition spectrum of scandium monoxide (ScO) in the spectral region between 290 and 311 nm has been recorded using laser ablation/reaction free-jet expansion and laser induced fluorescence spectroscopy. The ScO molecule was produced by reacting laser-ablated yttrium atoms with O_2 gas seeded in argon. Twenty transition bands were observed and thirteen of them have been selected for further studied by optical-optical double resonance (OODR) spectroscopy. Higher-lying electronic excited states have been reached via the intermediate $B^2\Sigma^+$ state from the ground $X^2\Sigma^+$ state. Our analysis indicates that the upper states conform to Hund's case (c) coupling scheme, the sub-states analyzed so far have Ω = 0.5 and 1.5. In addition, for the first time, a quartet state namely $^4\Sigma^+$ state has been identified experimentally. Least squares fits of the measured rotational lines yield molecular constants for the newly identified excited states.

144

TK04   2:21–2:36

ELECTRONIC TRANSITIONS OF SCANDIUM MONOPHOSPHIDE

NA WANG, KIU FUNG NG, <u>ALLAN S.C. CHEUNG</u>, *Department of Chemistry, The University of Hong Kong, Hong Kong, Hong Kong.*

The electronic transition spectrum of scandium monophosphide (ScP) molecule in the visible region between 480 nm and 550 nm has been recorded and analyzed using laser ablation/reaction free-jet expansion and laser induced fluorescence spectroscopy. The ScP molecule was produced by reacting laser - ablated scandium atoms with PH_3 seeded in argon. Ten vibrational bands were recorded and they are identified to be belonging to three electronic transition systems, namely the $[19.1]^1\Pi_1$ - $X^1\Sigma$ transition, $[19.5]^1\Sigma$ - $X^1\Sigma$ transition, and $[19.8]$ $^3\Pi_i$ - $X^3\Pi_i$ transition. ScP has been determined to have a $X^1\Sigma$ ground state in our earlier work. A least squares fit of the measured rotational lines yielded molecular constants for the single and the triplet electronic states. Theoretical calculation results have been used to help the assignment of the observed electronic states.

TK05   2:38–2:48

PERTURBATION ANALYSIS OF THE (0,0) BAND OF THE $A^2\Pi_{3/2}$ - $X^2\Sigma^+$ TRANSITION IN ZrN

KAITLIN A WOMACK, TAYLOR DAHMS, <u>LEAH C O'BRIEN</u>, *Department of Chemistry, Southern Illinois University, Edwardsville, IL, USA*; JAMES J O'BRIEN, *Chemistry and Biochemistry, University of Missouri, St. Louis, MO, USA.*

The (0,0) band of the $A^2\Pi_{3/2}$ - $X^2\Sigma^+$ transition of ZrN is known to be perturbed. Both homogeneous and heterogeneous perturbations are observed in the spectrum. A recent high-level ab initio calculation has helped to identify several possible perturbing states. PGOPHER is used to analyze the interactions. ZrN was produced in a hollow cathode sputter source, and the spectrum was recorded in emission by a high resolution FT spectrometer. Results of the analysis will be presented.

TK06   2:50–3:05

ANALYSIS OF A NEW ELECTRONIC TRANSITION OF MoO IN THE NEAR-INFRARED

<u>JACK C HARMS</u>, KAITLIN A WOMACK, LEAH C O'BRIEN, *Department of Chemistry, Southern Illinois University, Edwardsville, IL, USA.*

A new electronic transition of molybdenum monoxide, MoO, was observed near 6735 cm^{-1}. Based upon recent ab initio calculations, the transition is tentatively assigned as the (0,0) band of the $c^3\Pi - b^3\Sigma^-$ transition. PGOPHER was used to analyze the line positions observed in the spectrum. MoO molecules were produced in a Mo-lined hollow cathode sputter source, and the spectrum was recorded in emission using the high resolution spectrometer associated with the McMath-Pierce Solar telescope at Kitt Peak, AZ. Results of the analysis will be presented.

TK07   3:07–3:22

HIGH RESOLUTION LASER SPECTROSCOPY OF RHENIUM CARBIDE

ALLAN G. ADAM, <u>RYAN M. HALL</u>, *Department of Chemistry, University of New Brunswick, Fredericton, NB, Canada*; COLAN LINTON, DENNIS TOKARYK, *Department of Physics, University of New Brunswick, Fredericton, NB, Canada.*

The first spectroscopic study of rhenium carbide, ReC, has been performed using both low and high resolution techniques to collect rotationally resolved electronic spectra from 420 to 500nm. Laser-induced fluorescence (LIF), and dispersed fluorescence (DF) techniques were employed. ReC was formed in our laser ablation molecular jet apparatus by ablating a rhenium target rod in the presence of 1% methane in helium. The low resolution spectrum identified four bands of an electronic system belonging to ReC, three of which have been studied so far. Extensive hyperfine structure composed of six hyperfine components was observed in the high resolution spectrum, as well as a clear distinction between the ^{187}ReC and ^{185}ReC isotopologues. The data seems consistent with a $^4\Pi$ - $^4\Sigma^-$ transition, as was predicted before experimentation[a].Dispersed fluorescence spectra allowed us to determine the ground state vibrational frequency (ω_e''=994.4 ± 0.3 cm^{-1}), and to identify a low-lying electronically excited state at T_e''=1118.4 ± 0.4 $cm^{-1}$with a vibrational frequency of ω_e''=984 ± 2 cm^{-1}.

---

[a]Personal communication, F. Grein, University of New Brunswick

TK08 3:24 – 3:39

OBSERVATION OF A NEW $^2\Sigma^+$ - $^2\Sigma^+$ TRANSITION OF PtF BY INTRACAVITY LASER ABSORPTION SPECTROSCOPY

TAYLOR DAHMS, <u>LEAH C O'BRIEN</u>, KAITLIN A WOMACK, *Department of Chemistry, Southern Illinois University, Edwardsville, IL, USA*; JAMES J O'BRIEN, *Chemistry and Biochemistry, University of Missouri, St. Louis, MO, USA.*

A new electronic transition of PtF was observed by intracavity laser absorption spectroscopy. Based on the theoretical calculations of the electronic structure, this spectrum is assigned as the $^2\Sigma^+$ - $^2\Sigma^+$ transition. PtF molecules were produced in a Pt-lined hollow cathode sputter source in Ar with a trace of SF_6. Results of the analysis will be presented.

Intermission

TK09 3:56 – 4:11

OPTICAL STARK SPECTROSCOPY OF GOLD CHROLRIDE

<u>RUOHAN ZHANG</u>, TIMOTHY STEIMLE, *Department of Chemistry and Biochemistry, Arizona State University, Tempe, AZ, USA.*

The bonding and electrostatic properties of gold containing molecules are highly influenced by relativistic effects and electron correlation [a]. Hence it is difficult to predict those properties via electron structure calculation, and such calculation are guided by experimental observations. Here we report on the $A(\Omega = 1) - X^1\Sigma^+$ and $B(\Omega = 0) - X^1\Sigma^+$ bands of AuCl, which have been previously recorded at Doppler limited resolution [b]. A cold molecular beam sample was generated and the bands were recorded at high resolution (FWHM =35 MHz) using laser excitation spectroscopy, both field-free and in the presence of a static electric field. An improved set of spectroscopic parameters for the $A(\Omega = 1)$ and $B(\Omega = 0)$ states were obtained. The Stark induced shifts were analyzed to determine the permanent electric dipole moments for the X, A, and B states. A comparison with AuF [c] and theory will be made.

[a] P. Pyykko; *Angew Chem. Int* **43** 4412, 2004.

[b] L. C. O'Brien, A. L. Elliott, and M. Dulick; *J. Mol. Spectrosc* **194** 124, 1999.

[c] T. C. Steimle, R. Zhang, C. Qin and T. D. Varberg; *J. Phys. Chem. A* **117**(46) 11739,2013.

TK10 4:13 – 4:28

OBSERVATION OF TWO $\Omega=0^+$ EXCITED ELECTRONIC STATES IN JET-COOLED LaH

<u>SURESH YARLAGADDA</u>, *Atomic and Molecular Physics Division, Homi Bhabha National Institute, Bhabha Atomic Research Centre, Mumbai,400085, Maharastra, India*; SHEO MUKUND, *Atomic and Molecular Physics , Bhabha Atomic Research center, Mumbai, Maharastra, India*; SOUMEN BHATTACHARYYA, *Atomic and Molecular Physics Division, Bhabha Atomic Research Centre, Mumbai,400085, Maharastra, India*; SANJAY G. NAKHATE, *Atomic and Molecular Physics , Bhabha Atomic Research center, Mumbai, Maharastra, India.*

Lanthanum hydride (LaH) molecules were produced in a pulsed supersonic molecular beam by reaction of laser produced lanthanum metal plasma with $\sim$2% ammonia seeded in helium gas. Rotationally resolved laser-induced fluorescence excitation bands involving two excited electronic states, originating from ν=0 of ground state $X^1\Sigma^+$ were observed with band origins at 21970.71(1) and 22100.31(3) cm^{-1}. Rotational analysis confirmed these transitions as $\Omega=0^+$- $X^1\Sigma^+$. The rotational lines up to J"=6 were observed in the 21970.71 cm^{-1} band. However, for the 22100.31 cm^{-1} band, rotational lines up to J"=12 in the P-branch and J" up to 10 in the R-branch was observed. The molecular constants for both the excited states were determined by fitting rotational line wavenumbers to an effective Hamiltonian operator for $\Omega=0^+$ and $X^1\Sigma^+$ state using the Pgopher program.

TK11   4:30 – 4:45

THE SUBMILLIMETER SPECTRUM OF UO

JENNIFER HOLT, CHRISTOPHER F. NEESE, FRANK C. DE LUCIA, *Department of Physics, The Ohio State University, Columbus, OH, USA*; IVAN MEDVEDEV, *Department of Physics, Wright State University, Dayton, OH, USA*; MICHAEL HEAVEN, *Department of Chemistry, Emory University, Atlanta, GA, USA*.

Gaseous ^{238}UO was prepared in a high temperature furnace, and rotational spectra were measured in the 500–550 GHz and 590–650 GHz ranges. Transitions with J from 24 to 32 were observed in the $\Omega = 4$ ground state and low lying vibronic states with $\Omega = 3$ to $\Omega = 5$. Spectra were modeled using Dunham theory for diatomic molecules.

TK12   4:47 – 5:02

APPLICATION OF TWO DIMENSIONAL FLOURESCENCE SPECTROSCOPY TO TRANSITION METAL CLUSTERS.

DAMIAN L KOKKIN, TIMOTHY STEIMLE, *Department of Chemistry and Biochemistry, Arizona State University, Tempe, AZ, USA*.

Determining the physical properties (bond lengths, angles, dipole moments, etc) of transition metal oxides and dioxides is relevant to catalysis, high temperature chemistry, materials science and astrophysics. Analysis of optical spectra is a convenient method for extraction of physical properties, but can be difficult because of the density of electronic states and in the case of the dioxides, presence of both the oxide and superoxide forms. Here we demonstrate the application of two dimensional fluorescence spectroscopy[a] for aiding in the assignment and analysis. Particular attention will be paid to the spectroscopy of first row transition metal monoxides and dioxides of Nickel, NiO and NiO_2, and Manganese, MnO. Furthermore, the application of this technique to discovering the spectrum of other transition metal systems such as Metal-dicarbides will be outlined.

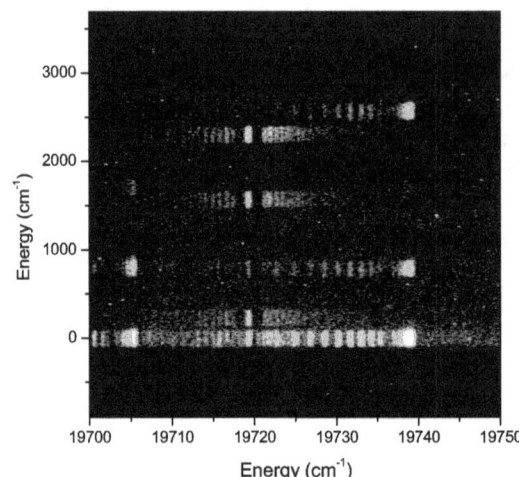

[a] N.J. Reilly, T.W. Schmidt, S.H. Kable, *J. Phys. Chem. A.*, 110(45), 12355-12359, 2006

TK13   5:04 – 5:19

FOURIER TRANSFORM INFRARED EMISSION SPECTRA OF MgF_2

DANIEL J. FROHMAN, PETER F. BERNATH, *Department of Chemistry and Biochemistry, Old Dominion University, Norfolk, VA, USA*; JACEK KOPUT, *Department of Chemistry, Adam Mickiewicz University, Poznan, Poland*.

High resolution infrared emission spectra of hot MgF_2 in the 700 to 1300 cm^{-1} region have been recorded. The molecules were generated by heating solid MgF_2 to 1675 °C. Four vibrational bands were rotationally analyzed yielding band origins and rotational constants. Observed bands are: 001-000 ($\Sigma_u^+ - \Sigma_g^+$), $01^11 - 01^10$ ($\Pi_g - \Pi_u$), 02^21 (f parity) $- 02^20$ (f parity) ($\Delta_u - \Delta_g$), and $03^31 - 03^30$ ($\Phi_g - \Phi_u$). High level *ab initio* calculations were essential in making assignments and in helping to fit the data. The $\Delta_u - \Delta_g$ band was only observed for f-parity because the e-parity is significantly perturbed by l-resonance.

TK14   **5:21 – 5:26**

MICROWAVE FREQUENCY TRANSITIONS REQUIRING LASER ABLATED URANIUM METAL DISCOVERED US-
ING CHIRP-PULSE FOURIER TRANSFORM SPECTROSCOPY

B. E. LONG, *Department of Chemistry, Wesleyan University, Middletown, CT, USA*; S. A. COOKE, *Natural and Social Science, Purchase College SUNY, Purchase, NY, USA.*

A rod of depleted uranium metal (mp = 1,132° C) has been ablated with the fundamental operating frequency of a Nd:YAG laser. The resulting ablation plume of uranium was then mixed with argon gas and expanded between the transmit/receive horn antennae of a chirp-pulse Fourier transform microwave spectrometer. The recorded spectra show nine strong transitions which are not present when the laser is not used in the experimental procedure. A series of experiments in which the backing gas conditions were altered provides evidence that the nine observed transitions are carried by the same species. Should the transitions be from one species it is most likely an asymmetric top. The transitions persist even when ultra-pure argon is used as the sole backing gas. The oxide coating of the uranium metal likely provides a source of oxygen and, presently, the "top" candidate for the unknown molecule is UO_3, which is known to have C_{2v} symmetry. Double resonance experiments are planned to aid transition assignments. A plausible explanation for an elusive assignment to date is the presence of pseudo-rotation.

TK15   **5:28 – 5:43**

ANION PHOTOELECTRON SPECTROSCOPIC STUDIES OF THE $NbC_4H_4^-$, $NbC_6H_6^-$ AND $NbC_6H_4^-$
PRODUCTS OF FLOW TUBE REACTIONS OF NIOBIUM WITH BUTADIENE

MELISSA A. BAUDHUIN, PRAVEENKUMAR BOOPALACHANDRAN, D. ALEX SCHNEPPER, DOREEN LEOPOLD, *Chemistry Department, University of Minnesota, Minneapolis, MN, USA*; STEPHEN R MILLER, *Chemistry Deparment, Gustavus Adolphus College, St. Peter, MN, USA.*

We report mass spectra, 488 nm anion photoelectron spectra, and density functional theory (DFT) calculations of organometallic complexes produced by flow tube reactions of niobium with butadiene (C_4H_6), and compare these results with those obtained upon reactions with ethylene (C_2H_4). In the C_4H_6 experiments, $NbC_4H_4^-$ is the most abundant product anion, indicating loss of H_2 upon reaction with Nb. DFT analysis of the vibrationally-resolved photoelectron spectrum indicates that the $^3A'$ anion incorporates a five-membered Nb-C_4 ring in which the Nb atom lies outside the C_4 plane. The electron affinity of the corresponding neutral molecule ($^2A'$) is measured to be 0.997 ± 0.006 eV. Upon reaction with C_2H_4, at least one additional isomer of $NbC_4H_4^-$ is produced, giving rise to broad spectral features at higher electron binding energies. Reactions with C_4H_6 also yield relatively small amounts of the $NbC_6H_6^-$ and $NbC_6H_4^-$ product anions, indicating C-C bond activation in addition to dehydrogenation. The former anion displays the $^3A_1$, C_{6v} Nb-benzene π-complex structure previously observed upon reaction with C_2H_4. The $NbC_6H_4^-$ anion produced upon reaction with C_4H_6 yields at least two vibrationally-resolved photodetachment transitions. DFT calculations performed to date suggest that the lower electron binding energy transition, which indicates an electron affinity of 1.110 ± 0.008 eV for the corresponding neutral complex, is due to the $^4B_2 \leftarrow {}^3B_2$ detachment from a planar, C_{2v} Nb-benzyne anion.

TK16   **5:45 – 5:55**

THEORETICAL STUDY OF M^+—RG_2 (M^+ = CA — RA; RG = HE — RN)

ANNA ANDREJEVA, *School of Chemistry, University of Nottingham, Nottingham, United Kingdom*; ADRIAN GARDNER, *Department of Chemistry, Emory University, Atlanta, GA, USA*; JACK B GRANEEK, *Center for Free-Electron Laser Science, Max Planck Institut für Kernphysik, Hamburg, Germany*; RICHARD PLOWRIGHT, *School of Chemistry, University of Nottingham, Nottingham, United Kingdom*; WILLIAM BRECKENRIDGE, *Department of Chemistry, The University of Utah, Salt Lake City, UT, USA*; TIM WRIGHT, *School of Chemistry, University of Nottingham, Nottingham, United Kingdom.*

Ab initio calculations were employed to investigate M^+—RG_2 species. Trends in binding energies, D_e, bond lengths, and bond angles are discussed and rationalised by analysing the electronic wavefunction. Mulliken, natural population, and atoms-in-molecules (AIM) population analyses are performed. It is found that some complexes are linear whereas others are bent. Those results are discussed in terms of hybridization and the various interactions present in these species. The work is a continuation from a previously published study[a] where Group 2 Be^+—RG_2 and Mg^+—RG_2 species are investigated and compared with Group 1 Li^+—RG_2 and Na^+—RG_2 species.

[a] A. Andrejeva, A. M. Gardner, J. B. Graneek, R. J. Plowright, W. H. Breckenridge and T. G. Wright, J. Phys. Chem. A., 117, 13578 (2013)

WA. Plenary
Wednesday, June 18, 2014 – 8:30 AM
Room: Theater Lincoln Hall

Chair: Dale Van Harlingen, University of Illinois, Urbana, IL, USA

WA01 8:30 – 9:10

BROADBAND ROTATIONAL SPECTROSCOPY

BROOKS PATE, *Department of Chemistry, The University of Virginia, Charlottesville, VA, USA.*

The past decade has seen several major technology advances in electronics operating at microwave frequencies making it possible to develop a new generation of spectrometers for molecular rotational spectroscopy. High-speed digital electronics, both arbitrary waveform generators and digitizers, continue on a Moore's Law-like development cycle that started around 1993 with device bandwidth doubling about every 36 months. These enabling technologies were the key to designing chirped-pulse Fourier transform microwave (CP-FTMW) spectrometers which offer significant sensitivity enhancements for broadband spectrum acquisition in molecular rotational spectroscopy. A special feature of the chirped-pulse spectrometer design is that it is easily implemented at low frequency (below 8 GHz) where Balle-Flygare type spectrometers with Fabry-Perot cavity designs become technologically challenging due to the mirror size requirements. The capabilities of CP-FTMW spectrometers for studies of molecular structure will be illustrated by the collaborative research effort we have been a part of to determine the structures of water clusters – a project which has identified clusters up to the pentadecamer. A second technology trend that impacts molecular rotational spectroscopy is the development of high power, solid state sources in the mm-wave/THz regions. Results from the field of mm-wave chirped-pulse Fourier transform spectroscopy will be described with an emphasis on new problems in chemical dynamics and analytical chemistry that these methods can tackle. The third (and potentially most important) technological trend is the reduction of microwave components to chip level using monolithic microwave integrated circuits (MMIC) – a technology driven by an enormous mass market in communications. Some recent advances in rotational spectrometer designs that incorporate low-cost components will be highlighted. The challenge to the high-resolution spectroscopy community – as posed by Frank De Lucia last year at the final meeting in Columbus – is what problems can we solve when real, fully capable spectrometers become essentially free to build?

WA02 9:15 – 9:55

DECELERATION AND TRAPPING OF COLD FREE RADICALS BY PULSED MAGNETIC FIELDS

TAKAMASA MOMOSE, *Department of Chemistry, University of British Columbia, Vancouver, BC, Canada.*

The study of cold and ultracold molecules is a rapidly growing interdisciplinary research field. The application of translationally cold molecules includes ultra-high resolution spectroscopy, tests of fundamental symmetry, coherent control, and investigation of cold and ultra cold chemistry. However, producing and trapping translationally cold molecules is still challenging. Cold free radicals are of great interest in relation to interstellar chemistry. In our laboratory, we have constructed a Zeeman decelerator for the deceleration of supersonic molecular beams of free radicals. Every paramagnetic molecule has a magnetic dipole moment, and therefore manipulation of the translational motion of free radicals is possible using inhomogeneous magnetic fields. Our decelerator consists of a series of solenoid coils, which provides periodic inhomogeneous magnetic fields of up to 7 T along the molecular beam axis. Rapid modulation of the field intensity removes the kinetic energy of paramagnetic species via the Zeeman effect. With this Zeeman decelerator, we have successfully decelerated supersonic beams of free radicals such as CH_3. The temperature of molecules thus created is low enough to trap them in an anti-Helmholtz type magnetic trap. Resonance-enhanced multiphoton ionization spectroscopy is used to confirm that radicals with specific rotational states are confined in the magnetic trap for more than several hundred micro seconds. One of the advantages of the Zeeman decelerator is it can decelerate and trap the rotational ground state of any paramagnetic molecule. We will discuss details of our Zeeman molecular decelerator, and possible applications including the study of high resolution spectroscopy and cold reactive collisions of free radicals below sub Kelvin temperatures. The work is supported by CFI funds for the Canadian Centre for Research on Ultra-Cold Systems (CRUCS) at UBC.

Intermission

WA03   **10:15 – 10:55**

CHIRAL MOLECULES REVISITED BY BROADBAND MICROWAVE SPECTROSCOPY

MELANIE SCHNELL, *CoCoMol, Max-Planck-Institut für Struktur und Dynamik der Materie, Hamburg, Germany.*

Chiral molecules have fascinated chemists for more than 150 years. While their physical properties are to a very good approximation identical, the two enantiomers of a chiral molecule can have completely different (bio)chemical activities. For example, the right-handed enantiomer of carvone smells of spearmint while the left-handed one smells of caraway. In addition, the active components of many drugs are of one specific handedness, such as in the case of ibuprofen. However, in nature as well as in pharmaceutical applications, chiral molecules often exist in mixtures with other chiral molecules. The analysis of these complex mixtures to identify the molecular components, to determine which enantiomers are present, and to measure the enantiomeric excesses (ee) remains a challenging task for analytical chemistry, despite its importance for modern drug development.

We present here a new method of differentiating enantiomers of chiral molecules in the gas phase based on broadband rotational spectroscopy [a, b]. The phase of the acquired signal bares the signature of the enantiomer, as it depends upon the combined quantity, $\mu_a\mu_b\mu_c$, which is of opposite sign between enantiomers. It thus also provides information on the absolute configuration of the particular enantiomer. Furthermore, the signal amplitude is proportional to the ee. A significant advantage of our technique is its inherent mixture compatibility due to the fingerprint-like character of rotational spectra. In this contribution, we will introduce the technique and present our latest results on chiral molecule spectroscopy and enantiomer differentiation.

[a] D. Patterson, M. Schnell, J.M. Doyle, *Nature* 497 (2013) 475-477

[b] V.A. Shubert, D. Schmitz, D. Patterson, J.M. Doyle, M. Schnell, *Angewandte Chemie International Edition* 53 (2014) 1152-1155

WA04   **11:00 – 11:40**

HIGH-RESOLUTION SPECTROSCOPIC STUDIES OF REACTION INTERMEDIATES RELEVANT TO ATMOSPHERIC CHEMISTRY

YASUKI ENDO, *Department of Basic Science, The University of Tokyo, Tokyo, Japan.*

We have been studying short lived reaction intermediates and complexes containing short lived species by high-resolution spectroscopic means. Laser induced fluorescence spectrosocpy with resolutions up to 0.02 cm^{-1} is used for observations of electronic transitions, and Fourier-transform microwave (FTMW) spectroscopy and FTMW-mm-wave double resonance spectroscopy are used for observations of pure rotational spectra. Both of the methods are combined with supersonic jet sytems equipped with pulse discharge nozzles to produce short lived species.

Last several years, we are concentrating on observations of short lived oxygen bearing species and their complexes especially with water. Such species are considred to be important in atmospheric chemistry, since chemistry in the earth's atmosphere mainly proceeds as oxidation reactions of trace species existing in the atmosphere, where various oxygen bearing reaction intermediates are playing important roles. Detections by high-resolution spectroscopy are expected to provide valuable information on these species. Furthermore, importance of molecular complexes containing reaction intermediates with water in atmospheric chemistry has been discussed recently, where detections of such species are highly required since quite limited experimental information has been obtained so far.

In the talk, results of a number of such species, either monomers or complexes, studied recently in our laboratory will be reviewed.

WF. Mini-symposium: Astronomical Molecular Spectroscopy in the Age of ALMA
Wednesday, June 18, 2014 – 1:30 PM
Room: 116 Roger Adams Lab

Chair: Cécile Favre, University of Michigan, Ann Arbor, MI, USA

WF01   1:30 – 1:45

THE GBT PRIMOS PROGRAM: 7 YEARS OF ASTRONOMICAL DISCOVERY

JOANNA F. CORBY, *Department of Astronomy, University of Virginia, Charlottesville, VA, USA*; BRETT A. McGUIRE, *Division of Chemistry and Chemical Engineering, California Institute of Technology, Pasadena, CA, USA*; MIKE HOLLIS, *Astrochemistry, NASA Goddard Space Flight Center, Greenbelt, MD, USA*; FRANK J LOVAS, *Sensor Science Division, National Institute of Standards and Technology, Gaithersburg, MD, USA*; PHILIP JEWELL, ANTHONY REMIJAN, *ALMA, National Radio Astronomy Observatory, Charlottesville, VA, USA.*

The GBT PRebiotic Interstellar MOlecule Survey (PRIMOS) towards Sgr B2N is the deepest, most complete spectral line survey in the range of 300MHz - 49 GHz. PRIMOS enables astronomers, chemists, and biologists to test theories of molecular formation, the origins of organic chemistry and the molecular complexity and physical and kinematic structure of material in our Galaxy. To date, PRIMOS data have resulted in 14 refereed publications since 2007, demonstrating the power of centimeter wave spectroscopy for detecting new organic species and revealing the significance of non-LTE effects including maser amplification in the cm-wave spectra of organic molecules. The survey has additionally advertised molecular astrophysics in public lectures, summer undergraduate diversity programs, and high school student projects. While the GBT is the only telescope in the world capable of conducting the PRIMOS Survey, PRIMOS data couples with newly available broad-bandwidth telescopes including the Jansky Very Large Array and ALMA. Synergistic observations with ALMA will be necessary to fully characterize the spectra of molecular material and determine excitation mechanisms leading to observed line radiation. This presentation provides an overview of the PRIMOS program, highlights PRIMOS science, and describes how the entire astronomical community can obtain the data for their own research.

WF02   1:47 – 2:02

A LOOK AT NITRILE CHEMISTRY IN SGR B2(N) USING THE COMBINED POWER OF THE GBT AND THE VLA

AMANDA STEBER, DANIEL P. ZALESKI, NATHAN A SEIFERT, JUSTIN NEILL, MATT MUCKLE, BROOKS PATE, *Department of Chemistry, The University of Virginia, Charlottesville, VA, USA*; JOANNA F. CORBY, *Department of Astronomy, University of Virginia, Charlottesville, VA, USA*; ANTHONY REMIJAN, *ALMA, National Radio Astronomy Observatory, Charlottesville, VA, USA.*

Nitriles form the most prolific family of molecules known in the ISM, and laboratory work shows that radical-driven chemistry can account for the formation of a diverse set of nitrile and imine molecules. Broadband reaction screening of nitrile chemistry in a pulsed discharge nozzle coupled to a chirped-pulse Fourier transform rotational spectrometer has enabled detections of several new interstellar species including E- and Z-ethanimine[a] and E-cyanomethanimine[b]. The detections were made by direct comparisons of laboratory broadband rotational spectra with the Robert C. Byrd Green Bank Telescope (GBT) PRebiotic Interstellar MOlecule Survery (PRIMOS) survey towards Sgr B2(N), the most chemically complex interstellar region known. In order to probe nitrile chemistry in Sgr B2, we targeted low energy rotational transitions in the 18-21 GHz range of several nitriles with the Karl G. Jansky Very Large Array (VLA) at ∼1 arcsecond resolution. The data indicate that most nitriles and nitrile derivatives are co-spatial with shell shaped continuum features thought to be expanding ionization fronts. The CH2CN radical and imine species in particular are NOT associated with the hot core known as the "Large Molecule Heimat", where most large organic molecules are thought to reside. This result suggests radical driven nitrile chemistry may be promoted by near-UV radiation in moderate density regions of molecular clouds, and the data will be useful for evaluating possible formation mechanisms.

---

[a]R.A. Loomis et al. *Ap. J. L.* **765**(L9), 2013.
[b]D.P. Zaleski et al. *Ap. J. L.* **765**(L10), 2013.

WF03   **2:04 – 2:19**

METHANIMINE AT HIGH SPATIAL RESOLUTION IN SGR B2: IMPLICATIONS FOR THE FORMATION OF CYANOMETHANIMINE

JOANNA F. CORBY, *Department of Astronomy, University of Virginia, Charlottesville, VA, USA*; AMANDA STEBER, NATHAN A SEIFERT, CRISTOBAL PEREZ, *Department of Chemistry, The University of Virginia, Charlottesville, VA, USA*; ANTHONY REMIJAN, *ALMA, National Radio Astronomy Observatory, Charlottesville, VA, USA*; BROOKS PATE, *Department of Chemistry, The University of Virginia, Charlottesville, VA, USA*.

Two transitions of methanimine (CH2NH) have been mapped towards Sgr B2 using the Jansky Very Large Array (VLA) with 1.5 arcsecond resolution. The two targeted transitions are both between low-lying energy states at similar frequencies, yet one appears in absorption whereas the other is in emission with the same line profile. The VLA data reveals that the spatial distributions of the two transitions match and that they are NOT associated with the hot core toward Sgr B2(N). As compared to other molecular lines observed towards Sgr B2 at centimeter wavelengths, the CH2NH emission line is highly uncharacteristic, and the transitions exhibits non-thermal effects implying a population inversion. We discuss the non-thermal excitation of CH2NH, observed spatial distributions, and implications for the chemistry in Sgr B2. Specifically, CH2NH may be important for the formation of the recently detected species E-cyanomethanimine [1] and of the Z- and N- conformers of cyanomethanimine. Laboratory work by Balucani et al [2] indicates that reactions between the CN radical and olefins (with a carbon-carbon double bond) may proceed without a barrier, potentially making the reaction CH2NH + CN → HCNHCN viable in the interstellar medium.

1 Zaleski, D.P., et al. 2013, ApJL, 765, L10 2 Balucani, N., et al. 2000, ApJ, 545, 892

WF04   **2:21 – 2:36**

CH$^+$ AND SH$^+$ ABSORPTION SPECTROSCOPY WITH HERSCHEL: PROBING THE TURBULENT DISSIPATION IN THE DIFFUSE ISM.

BENJAMIN GODARD, EDITH G. FALGARONE, *LERMA, Observatoire de Paris, Paris, France*; GUILLAUME PINEAU DES FORÊTS, *IAS, Université Paris-sud, Orsay, France*; MARYVONNE GERIN, PIERRE LESAFFRE, *LERMA, Observatoire de Paris, Paris, France*; D. A. NEUFELD, *Department of Physics and Astronomy, Johns Hopkins University, Baltimore, MD, USA*; FRANÇOIS LEVRIER, *LERMA, Observatoire de Paris, Paris, France*.

Because it is predominantly heated by the UV radiation field, the diffuse interstellar medium (ISM) has long been thought to behave like a photo-dissociation region (PDR). Yet, for the last 30 years, absorption spectroscopy has revealed a gas with a chemical richness that was unexpected from the sole predictions of PDR-type models. This problem has recently been deepened by the observations of large abundances of small hydrides using the Herschel/HIFI instrument. Since their production pathways are protected by highly endo-energetic reactions, it has been proposed that several of these species are nothing else but a signature of another powerful energy source, such as the dissipation of magnetized turbulence[1].

Among all the molecules detected by Herschel, CH$^+$ and SH$^+$ are a unique couple[2] because the energies involved in their formation are particularly large (4640 K and 9860 K respectively). Their presence in the cold diffuse ISM is therefore much more than a chemical riddle : it is rooted in the physics of the diffuse ISM, the intermittency of the turbulent cascade and the rate of its dissipation, and it connects with the broader issues of star formation and galaxy evolution.

The informations inferred from the absorption spectra were investigated[3] in the framework of the TDR (Turbulent Dissipation Regions) model which follows the dynamical and chemical evolutions of the gas in intermittent regions of turbulent dissipation. By comparing the predictions of the TDR model with multiwavelength observations of seven atomic and molecular species (C$^+$, CH$^+$, SH$^+$, H, H$_2$, HCO$^+$ and CO) we were able, for the first time, to measure six essential properties of the interstellar turbulence: (1) the dissipation rate and (2) how it varies across the Galactic disk, (3) the size of the dissipative structures and (4) their lifetime, and (5) the dominant dissipative process (ion-neutral friction).

[1]B. Godard, E. Falgarone, and G. Pineau des Forêts, 2009, A&A, 495, 847 ; [2]B. Godard, E. Falgarone, M. Gerin, et al., 2012, A&A, 540, A87 ; [3]B. Godard, E. Falgarone, and G. Pineau des Forêts, 2014, A&A, accepted

152

WF05   **2:38 – 2:53**

MEASUREMENT OF THE LOWEST MILLIMETER-WAVE TRANSITION FREQUENCY OF THE CH RADICAL

STEFAN TRUPPE, *Centre for Cold Matter, Blackett Laboratory, Imperial College London, London, United Kingdom*; RICHARD JAMES HENDRICKS, *Department of Physics, Imperial College London, London, United Kingdom*; ED HINDS, MICHAEL TARBUTT, *Centre for Cold Matter, Blackett Laboratory, Imperial College London, London, United Kingdom.*

The CH radical is an important constituent of stellar atmospheres, interstellar gas clouds and is of fundamental importance to interstellar chemistry. Furthermore, it offers a sensitive way to test the hypothesis that fundamental constants measured on earth may differ from those observed in other parts of the universe[a]. Here, we present a measurement of the lowest millimeter-wave transition of CH, near 535 GHz, with an accuracy of 0.6 kHz[b], an improvement of nearly two orders of magnitude compared to the previous best rest frequencies. We drive the millimeter-wave transitions using the 54th harmonic of a frequency synthesizer phase-locked to a 10 MHz GPS frequency reference. Using ALMA this transition has recently been observed in the absorber PKS 1830-211 at a redshift of $z = 0.89$[c]. As pointed out by de Nijs et al.[d] a very robust and sensitive means to search for variations in fundamental constants could be obtained by observing the lowest millimeter-wave transition of CH along with the two Λ-doublets at 3.3 and 0.7 GHz, all from the same interstellar gas cloud.

[a] S. Truppe et al., *Nature Communications* **4**, 2600, 2013
[b] S. Truppe et al., *The Astrophysical Journal* **780**, 71, 2014
[c] S. Muller, *private communication*, 2013
[d] de Nijs et al., *Physical Review A* **86**, 032501, 2012

WF06   **2:55 – 3:05**

LABORATORY CHARACTERIZATION AND ASTRONOMICAL DETECTION OF THE NITROSYLIUM ION, NO$^+$

STEPHANE BAILLEUX, *Laboratoire PhLAM, Université de Lille 1, Villeneuve de Ascq, France*; E. A. ALEKSEEV, *Radiospectrometry Department, Institute of Radio Astronomy of NASU, Kharkov, Ukraine*; JOSE CERNICHARO, BELÉN TERCERO, *Departamento de Astrofísica, Centro de Astrobiología CAB, CSIC-INTA, Madrid, Spain*; ASUNCION FUENTE, RAFAEL BACHILLER, *Observatorio Astronómico Nacional, Alcalá de Henares, Spain*; EVELYNE ROUEFF, *Laboratoire de l'Univers et de ses Théories (Luth), Observatoire de Paris-Meudon, Paris, France*; MARYVONNE GERIN, *LERMA, Observatoire de Paris, Paris, France*; SANDRA TREVIÑO-MORALES, *Astronomy and Science Support, Instituto de Radio Astronoma Milimétrica (IRAM), Granada, Spain*; NURIA MARCELINO, *ALMA, National Radio Astronomy Observatory, Charlottesville, VA, USA*; BERTRAND LEFLOCH, *Institut de Planétologie et d'Astrophysique de Grenoble (IPAG), UJF-Grenoble / CNRS-INSU, Grenoble, France.*

We report the discovery for the first time in space of the nitrosylium ion, NO$^+$. The observations were performed towards the cold dense core Barnard 1-b. The identification of the $J = 2 \leftarrow 1$ line is supported by new laboratory measurements of rotational lines of NO$^+$ in the ground vibrational state up to the $J = 8 \leftarrow 7$ transition (953207.189 MHz).

The ion was produced in a magnetically extended negative glow discharge in NO. Vibrational excitation of the ion was high enough to measure rotational lines up to $v = 2$. A few transitions of ^{15}NO$^+$ were also measured ($v = 0, 1$). All known rotational and ro-vibrational frequencies of this close-shell ion were included in an isotopically invariant analysis.

In Barnard 1-b, the observed line profile of NO$^+$ exhibits two velocity components at 6.5 and 7.5 km s^{-1}, with column densities of 1.5×10^{12} and 6.5×10^{11} cm^{-2}, respectively. New observations of NO and HNO have been performed and allowed to estimate the following abundance ratios: $X(\mathrm{NO})/X(\mathrm{NO}^+) \approx 511$, and $X(\mathrm{HNO})/X(\mathrm{NO}^+) \approx 1$. The chemistry of NO$^+$ has been investigated by means of a time-dependent gas phase model which includes an updated chemical network according to recent experimental studies. The predicted abundance for NO$^+$ and NO is found to be consistent with the observations. However, that of HNO relative to NO is too high. No satisfactory chemical paths have been found to explain the observed low abundance of HNO.

THE SEARCH FOR l-C_3H^+ (B11244) IN MORE THAN 40 ASTRONOMICAL SOURCES

BRETT A. McGUIRE, BRANDON CARROLL, *Division of Chemistry and Chemical Engineering, California Institute of Technology, Pasadena, CA, USA*; JAMES SANDERS, SUSANNA L. WIDICUS WEAVER, *Department of Chemistry, Emory University, Atlanta, GA, USA*; GEOFFREY BLAKE, *Division of Chemistry and Chemical Engineering, California Institute of Technology, Pasadena, CA, USA*; ANTHONY REMIJAN, *ALMA, National Radio Astronomy Observatory, Charlottesville, VA, USA.*

In 2012, Pety et al. (*A&A*, 548, A68) reported the detection of a series of transitions from 90 to 270 GHz arising from a molecular carrier (B11244) which they attributed to the l-C_3H^+ cation, a species never-before seen in the laboratory. Theoretical work, however, suggested the anion, C_3H^-, was a more likely carrier (Fortenberry et al. 2013, *ApJ*, 772, 39). We conducted several observational studies to examine these possibilities, and concluded l-C_3H^+ was supported by the observational evidence, a conclusion which has recently been confirmed by laboratory work (Brünken et al. 2014, *ApJL*, 783, L4). Here, we present a body of observational results compiled in our search for l-C_3H^+ toward more than 40 sources. Despite spanning a wide range of environments, including hot molecular cores, cold cores, PDRs, and HH objects, we find definitive evidence for l-C_3H^+ in only three sources. We will discuss the implications of the apparent scarcity of this molecule. What is so special about these specific regions that favors the formation of this molecule, and in turn, what can l-C_3H^+ tell us about the physical and chemical conditions within these environments? Interferometric observations with ALMA are the ideal path forward for answering these questions, and we discuss what these observations will tell us about l-C_3H^+ and the unique environments in which it is present.

THE DISTRIBUTION OF ASTRONOMICAL ALDEHYDES - THE CASE FOR EXTENDED EMISSION OF ACETALDEHYDE (CH_3CHO).

ANDREW BURKHARDT, *Department of Astronomy, University of Virginia, Charlottesville, VA, USA*; RYAN LOOMIS, *Department of Astronomy, Harvard University, Cambridge, MA, USA*; NIKLAUS DOLLHOPF, JOANNA F. CORBY, *Department of Astronomy, University of Virginia, Charlottesville, VA, USA*; ANTHONY REMIJAN, *ALMA, National Radio Astronomy Observatory, Charlottesville, VA, USA.*

With the advent of new broadband spectral line interferometric observations, we can now begin to fully characterize the spectra and distribution of complex organic molecules that have been largely ignored since their original detections using single dish telescopes. First detected in 1973, acetaldehyde (CH_3CHO) has been detected in numerous sources including TMC-1, Sgr B2(N) and Orion KL (Gottlieb et al 1973; Mathews et al. 1984; Johansson et al. 1991); yet its distribution within these sources is still not well known. Unlike a number of other molecules observed in these regions, acetaldehyde is not observed to be concentrated in hot core regions toward Sgr B2(N), but to have an extended distribution, a trait shared by other aldehydes (Hollis et al. 2001; Chengalur and Kanekar, 2003). An extended distribution may indicate formation through gas phase ion molecule reactions, or that the distribution is a result of non-thermal processes liberating the molecule off grain surfaces. Meanwhile, a compact distribution may indicate warm grain surface chemistry with subsequent desorption by thermal processes. Spatial maps will also help determine abundance correlations with other related molecules such as formic acid, aiding in the investigation of formation routes. In this talk, we present multiple transition maps of acetaldehyde toward Orion KL using both CARMA and the ALMA Band 6 Science Verification data which show evidence of an extended distribution of acetaldehyde, suggesting a similar formation chemistry in Orion KL as suggested by Chengular and Kanekar (2003) towards Sgr B2(N). In addition, spatial correlations to other molecules in the region will be shown, possibly suggesting a common formation chemistry for some aldehydes.

154

THE SEARCH FOR A COMPLEX MOLECULE IN A SELECTED HOT CORE REGION: A RIGOROUS ATTEMPT TO CONFIRM TRANS-ETHYL METHYL ETHER TOWARD W51 E1/E2

BRANDON CARROLL, BRETT A. McGUIRE, *Division of Chemistry and Chemical Engineering, California Institute of Technology, Pasadena, CA, USA*; ALDO J. APPONI, *Department of Chemistry and Astronomy, University of Arizona, Tuscon, AZ, USA*; LUCY ZIURYS, *Department of Astronomy, University of Arizona, Tucson, AZ, USA*; GEOFFREY BLAKE, *Division of Chemistry and Chemical Engineering, California Institute of Technology, Pasadena, CA, USA*; ANTHONY REMIJAN, *ALMA, National Radio Astronomy Observatory, Charlottesville, VA, USA.*

An extensive search has been conducted to confirm transitions of *trans*-ethyl methyl ether (tEME), ($C_2H_5OCH_3$), toward the high mass star forming region W51 e1/e2 using the 12 m Telescope of the Arizona Radio Observatory (ARO) at 2 mm and 3 mm wavelengths. Typical peak to peak noise levels for the present observations of W51e1/e2 were between 10 mK to 30 mK, indicating an upper limit of the tEME column density of $\leq 1.5 \times 10^{15}$ cm^{-2}, this would make tEME at least a factor 2 times less abundant than dimethyl ether (CH_3OCH_3) toward W51 e1/e2. We have also performed an extensive search for this species toward the high mass star forming region Sgr B2(N-LMH) with the NRAO 100 m Green Bank Telescope (GBT). No transitions of tEME were detected and we were able to set an upper limit to the tEME column density of $\leq 4 \times 10^{14}$ cm^{-2} toward Sgr B2(N-LMH). We will discuss these observations in the context of detecting large complex organic species toward star forming regions with next generation telescopes such as ALMA.

Intermission

SUBMILLIMETER WAVE SPECTROSCOPY OF ACETYL ISOCYANATE : $CH_3C(O)NCO$

L. MARGULÈS, R. A. MOTIYENKO, *Laboratoire PhLAM, UMR 8523 CNRS - Université Lille 1, Villeneuve d'Ascq, France*; J.-C. GUILLEMIN, *Institut des Sciences Chimiques de Rennes, UMR 6226 CNRS - Université de Rennes 1, Rennes, France*; BELÉN TERCERO, JOSE CERNICHARO, *Departamento de Astrofísica, Centro de Astrobiología CAB, CSIC-INTA, Madrid, Spain*; ATEF JABRI, ISABELLE KLEINER, *Laboratoire Interuniversitaire des Systèmes Atmosphériques (LISA), CNRS et Universités Paris Est et Paris Diderot, Créteil, France*; V. ILYUSHIN, *Radiospectrometry Department, Institute of Radio Astronomy of NASU, Kharkov, Ukraine.*

Except isocyanic acid detected in the ISM since 1972[a], the organo isocyanate derivatives are poorly studied in the millimeter wave domain. This lack of data could be the reason of their non detection in the ISM up to now. We decided to investigate the $C_3H_3NO_2$ isomer: acetyl isocyanate. Previously measured up to 40 GHz [b], the cis-conformer exhibits internal rotation motion with a medium barrier value of 360 cm^{-1}. The trans conformer conformer is calculated to have an energy of 12.55 kJ.mol^{-1} (1060 cm^{-1}) higher than the cis one[c] and is not studied here. The measurements were performed in Lille with our solid state devices spectrometer up to 500 GHz. The sample was found to have a poor stability and reacts fastly with metal parts. We should repeat measurements using a flow and a pyrex cell in order to have satisfactory signal to noise ratio. The analysis was performed with RAM36 code[d] which used the Rho Axis Method. The first results and its searche in ORION will be presented.

This work was supported by the CNES and the Action sur Projets de l'INSU, PCMI. This work was also done under Ukrainian-French CNRS-PICS 6051 project and ANR-13-BS05-0008-02 IMOLABS

[a]Snyder, L. E.; and Buhl, D.*Astrophys. J.* **177**, (1972) 619

[b]Landsberg, B.M.; and Iqbal, K.*J.C.S. Faraday II* **76**, (1980) 1208

[c]Uchida, Y.; Toyoda, M.; Kuze, N.; and Sakaizumi, T.*J. Mol. Spectrosc.* **256**, (2009) 163

[d]Ilyushin, V.V. et al;*J. Mol. Spectrosc.* **259**, (2010) 26

WF11 4:30 – 4:45

LABORATORY CHARACTERIZATION AND ASTROPHYSICAL DETECTION IN ORION KL OF HIGHER EXCITED VIBRATIONAL STATES OF VINYL CYANIDE

ALICIA LÓPEZ, BELÉN TERCERO, JOSE CERNICHARO, *Departamento de Astrofísica, Centro de Astro-biología CAB, CSIC-INTA, Madrid, Spain*; ZBIGNIEW KISIEL, LECH PSZCZÓŁKOWSKI, *ON2, Institute of Physics, Polish Academy of Sciences, Warszawa, Poland*; CELINA BERMÚDEZ, JOSÉ L. ALONSO, *Grupo de Espectroscopia Molecular, Lab. de Espectroscopia y Bioespectroscopia, Unidad Asociada CSIC, Universidad de Valladolid, Valladolid, Spain*; IVAN MEDVEDEV, *Department of Physics, Wright State University, Dayton, OH, USA*; CHRISTOPHER F. NEESE, *Department of Physics, The Ohio State University, Columbus, OH, USA*; BRIAN DROUIN, ADAM M DALY, *Jet Propulsion Laboratory, California Institute of Technology, Pasadena, CA, USA*; NURIA MARCELINO, *ALMA, National Radio Astronomy Observatory, Charlottesville, VA, USA*; SERENA VITI, HANNAH CALCUTT, *Department of Physics and Astronomy, University College London, London, IX, United Kingdom.*

Vinyl cyanide (acrylonitrile, $H_2C{=}CHC{\equiv}N$) is an interstellar molecule that was classified as a 'weed' since transitions in its isotopic species and vibrationally excited states have already been detected and need to be accounted for in searches for complex organic molecules. Presently we extend the systematic analysis of the laboratory rotational spectrum of vinyl cyanide to 9 new excited vibrational states with vibrational energies above $550\ cm^{-1}$ (785K). The spectroscopic analysis is based on the broadband 50-1900 GHz spectrum combined from results from the participating spectroscopic laboratories and covering a total of 1235 GHz.

The studied states come in the form of polyads of perturbing vibrational states, and such perturbations also affect the strong, low-K_a transitions used for astrophysical detection. It is therefore crucial to account for such effects in order to produce reliable linelists. The experimental data for three new polyads were fitted to experimental accuracy using Coriolis and Fermi perturbation models. Multiple transitions in the lowest of these polyads (and in other excited vibrational states and isotopic species of vinyl cyanide) were detected in the millimetre survey of the Orion-KL Nebula made with the IRAM 30-m radiotelescope.

WF12 4:47 – 5:02

LABORATORY AND ASTRONOMICAL DISCOVERY OF HYDROMAGNESIUM ISOCYANIDE

CARLOS CABEZAS, ISABEL PEÑA, SANTIAGO MATA, JOSÉ L. ALONSO, *Grupo de Espectroscopia Molecular, Lab. de Espectroscopia y Bioespectroscopia, Unidad Asociada CSIC, Universidad de Valladolid, Valladolid, Spain*; JOSE CERNICHARO, MARCELINO AGÚNDEZ, *Departamento de Astrofísica, Centro de Astrobiología CAB, CSIC-INTA, Madrid, Spain*; MICHEL GUÉLIN, *Institut de Radioastronomie Millimétrique, Observatoire de Paris, Paris, France.*

We report on the detection of hydromagnesium isocyanide, HMgNC, in the laboratory and in the carbon-rich evolved star IRC+10216. The J = 1-0 and J = 2-1 lines were observed in our microwave laboratory equipment in Valladolid with a spectral accuracy of 3 kHz. The hyperfine structure produced by the nitrogen atom was resolved for both transitions. Four rotational lines of this species, J = 8-7, J = 10-9, J = 12-11, and J = 13-12, have been detected toward IRC+10216. First results for another metal bearing species are also reported.

WF13   5:04 – 5:19

SPECTROSCOPIC CHARACTERIZATION AND DETECTION OF ETHYL MERCAPTAN IN ORION

LUCIE KOLESNIKOVÁ, ADAM M DALY, JOSÉ L. ALONSO, *Grupo de Espectroscopia Molecular, Lab. de Espectroscopia y Bioespectroscopia, Unidad Asociada CSIC, Universidad de Valladolid, Valladolid, Spain*; BELÉN TERCERO, JOSE CERNICHARO, *Departamento de Astrofísica, Centro de Astrobiología CAB, CSIC-INTA, Madrid, Spain*; BRI GORDON, STEVEN SHIPMAN, *Department of Chemistry, New College of Florida, Sarasota, FL, USA.*

The rotational spectrum of ethyl mercaptan, CH_3CH_2SH, has been measured in the microwave, millimeter- and submillimeter-wave regions from 8 to 880 GHz and more than 2800 distinct transition frequencies have been assigned for the *gauche-* and *trans-*conformers. Very precise values of the spectroscopic constants allowed the detection of the *gauche-*CH_3CH_2SH towards Orion KL.[a] 77 unblended or slightly blended lines plus no missing transitions in the range 80 – 280 GHz support this identification. $Trans$–CH_3CH_2SH has been detected tentatively.

[a]L. Kolesniková, B. Tercero, J. Cernicharo, A. M. Daly, J. L. Alonso, B. P. Gordon, S. Shipman, *Astrophys. J. Lett.* **2014**, accepted.

WF14   5:21 – 5:31

METHODS FOR DETECTION OF FAMILIES OF MOLECULES IN THE INTERSTELLAR MEDIUM

GLEN LANGSTON, *Astronomy, National Science Foundation, Arlington, VA, USA.*

We present a high velocity resolution (0.04 km/sec) molecular line survey of the Taurus Molecular Cloud in the frequency range 39 to 48 GHz with NSF's Robert C. Byrd Green Bank telescope (GBT). The observing method and data reduction process are outlined. We describe the method of obtaining the calibrated, averaged spectral line data online.

The RMS survey sensitivity was slightly different for each 200MHz frequency band, and ranged from 0.02 to 0.15 K (T_B) for the different bands.

A large number of molecular lines are detected, most of which have previously been associated with already known interstellar molecules. We present a summary processes to combine a number of lines of molecular species in order to identify new species.

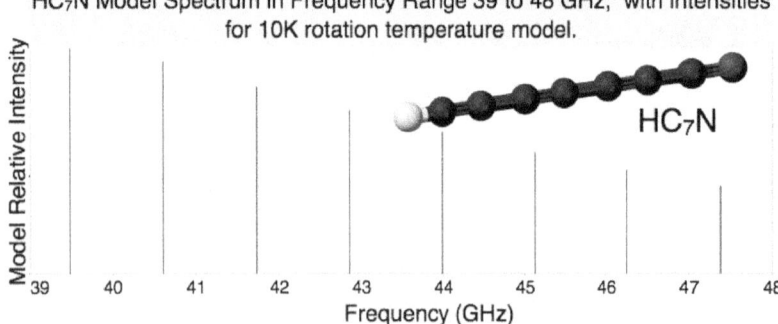

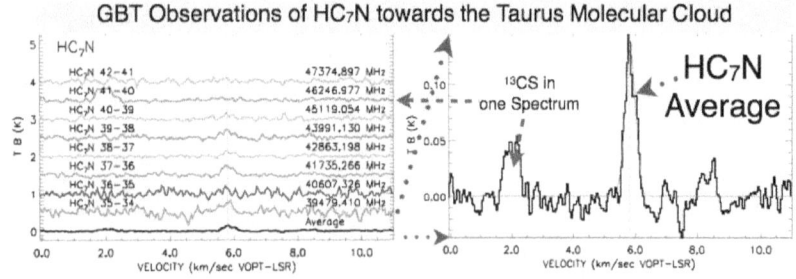

WF15 5:33–5:48

INVESTIGATING THE "MINIMUM ENERGY PRINCIPLE" IN SEARCHES FOR NEW MOLECULAR SPECIES - THE CASE OF H_2C_3O ISOMERS

RYAN LOOMIS, *Department of Astronomy, Harvard University, Cambridge, MA, USA*; AMY ROBERTSON, *Department of Astronomy, University of Arizona, Tucson, AZ, USA*; CHELEN JOHNSON, *Breck School, Golden Valley, MN, USA*; SAMANTHA BLAIR, *Department of Astronomy, University of California, Berkeley, CA, USA*; ANTHONY REMIJAN, *ALMA, National Radio Astronomy Observatory, Charlottesville, VA, USA.*

Next generation interferometers offer new possibilities of extracting information about the history and conditions of interstellar regions through analysis of molecular abundance ratios, which act as chemical fingerprints. To fully utilize these new capabilities, however, complete molecular inventories of interstellar regions must be established and the relative contributions of different chemical processes must be determined. Current theories on molecule formation will therefore be guiding forces in future observing strategies. Recently, Lattelais et al. have interpreted aggregated observations of molecular isomers to suggest that there exists a "minimum energy principle", such that molecular formation will favor more stable molecular isomers for thermodynamic reasons.

To test the predictive use of this model, we have fully characterized the spectra of the three isomers of C_3H_2O toward the well known molecular region Sgr B2. Evidence for the detection of the isomers cyclopropenone (c-C_3H_2O) and propynal (HCCCHO) is presented, along with evidence for the non-detection of the lowest zero-point energy isomer, propadienone (CH_2CCO). We interpret these observations as evidence that chemical formation pathways, which may be under kinetic control, have a more pronounced effect on final isomer abundances than thermodynamic effects such as the minimum energy principle (Lattelais et al. 2009, ApJ 696, L133). These results suggest that as ALMA opens up new space for molecular discovery, focus should be placed on molecules and molecular reactions under kinetic control, and we briefly discuss the possible applications of isomeric analysis in the era of broadband interferometry.

WF16 5:50–6:05

OBSERVING ORGANIC MOLECULES IN INTERSTELLAR GASES: NON EQUILIBRIUM EXCITATION.

LAURENT WIESENFELD, ALEXANDRE FAURE, *Institut de Planétologie et d'Astrophysique de Grenoble (IPAG), UJF-Grenoble / CNRS-INSU, Grenoble, France*; ANTHONY REMIJAN, *ALMA, National Radio Astronomy Observatory, Charlottesville, VA, USA*; KRZYSZTOF SZALEWICZ, *Department of Physics and Astronomy, University of Delaware, Newark, DE, USA.*

In order to observe quantitatively organic molecules in interstellar gas, it is necessary to understand the relative importance of photonic and collisional excitations. In order to do so, collisional excitation transfer rates have to be computed. We undertook several such studies, in particular for H_2CO and $HCOOCH_3$. Both species are observed in many astrochemical environments, including star-forming regions. We found that those two molecules behave in their low-lying rotational levels in an opposite way.

For cis methyl-formate, a non-equilibrium radiative transfer treatment of rotational lines is performed, using a new set of theoretical collisional rate coefficients. These coefficients have been computed in the temperature range 5 to 30 K by combining coupled-channel scattering calculations with a high accuracy potential energy surface for $HCOOCH_3$ – He. The results are compared to observations toward the Sagittarius B2(N) molecular cloud. A total of 2080 low-lying transitions of methyl formate, with upper levels below 25 K, were treated. These lines are found to probe a cold (30 K), moderately dense ($n \sim 10^4$ cm^{-3}) interstellar gas. In addition, our calculations indicate that all detected emission lines with a frequency below 30 GHz are collisionally pumped weak masers amplifying the background of Sgr B2(N). This result demonstrates the generality of the inversion mechanism for the low-lying transitions of methyl formate.

For formaldehyde, we performed a similar non-equilibrium treatment, with H_2 as the collisional partner, thanks to the accurate H_2CO – H_2 potential energy surface . We found very different energy transfer rates for collisions with para-H_2 ($J=0$) and ortho-H_2 ($J=1$). The well-known absorption against the cosmological background of the $1_{11} \to 1_{01}$ line is shown to depend critically on the difference of behaviour between para and ortho-H_2, for a wide range of H_2 density. [a]

[a]We thank the CNRS-PCMI French national program for continuous support and the CHESS Herschel KP program for travel supports. Discussions with C. Ceccarelli, P. Hily-Blant and S. Maret are acknowledged.

WG. Mini-symposium: Beyond the Mass-to-Charge Ratio: Spectroscopic Probes of the Structures of Ions

Wednesday, June 18, 2014 – 1:30 PM

Room: 100 Noyes Laboratory

Chair: **Jaime A. Stearns**, Air Force Research Laboratory, Kirtland AFB, NM, USA

WG01   *INVITED TALK*   1:30 – 2:00

INFRARED SPECTRA OF PROTONATED AROMATIC HYDROCARBONS AND THEIR NEUTRAL COUNTERPARTS IN SOLID *PARA*-HYDROGEN

MOHAMMED BAHOU, *Applied Chemistry, National Chiao Tung University, Hsinchu, Taiwan*; YU-JONG WU, *Molecular Science, National Synchrotron Radiation Research Center, Hsinchu, Taiwan*; YUAN-PERN LEE, *Applied Chemistry, National Chiao Tung University, Hsinchu, Taiwan*.

Protonated polycyclic aromatic hydrocarbons (H^+PAH) have been reported to have infrared (IR) bands at wavenumbers near those of unidentified infrared (UIR) emission bands from interstellar objects. However, recording IR spectra of H^+PAH in laboratories is challenging. Two spectral methods have been employed previously to yield IR spectra of H^+PAH. One employs IR multiphoton dissociation (IRMPD) of H^+PAH, but the bands are broad and red-shifted. [a] Another measures the single-photon IR photodissociation action spectrum of cold H^+PAH tagged with a weakly bound ligand, such as Ar, but application of this method to large PAH is difficult.[b] A new method for investigating IR spectra of H^+PAH and their neutral counterparts was developed using electron bombardment during p-H_2 matrix deposition. With this technique, we have recorded IR absorption spectra of protonated forms of benzene ($C_6H_7^+$), naphthalene (1- and 2-$C_{10}H_9^+$), pyrene (1-$C_{16}H_{11}^+$), coronene (1-$C_{24}H_{13}^+$), and their neutrals.[c] The significant superiority of the spectra thus recorded to those with the Ar-tagging and IRMPD methods is demonstrated. The narrow widths of the lines enabled us to distinguish clearly between isomers 1-$C_{10}H_9^+$ and 2-$C_{10}H_9^+$; 2-$C_{10}H_9^+$ was unstable and converted to 1-$C_{10}H_9^+$ in less than 30 min. A survey of these experimental results shows that three major lines in the 7-9 μm region are red-shifted from 7.19, 7.45, and 8.13 μm of 1-$C_{16}H_{11}^+$ to 7.37, 7.53, and 8.21 μm of 1-$C_{24}H_{13}^+$, showing the direction towards the UIR bands near 7.6, 7.8, and 8.6 μm. In contrast, the line at 11.5 μm for 1-$C_{16}H_{11}^+$ is blue-shifted to 11.4 μm for 1-$C_{24}H_{13}^+$, showing the direction toward the UIR band near 11.2 μm. Other examples will be presented if time permits.

---

[a]O. Dopfer, *PAHs and the Universe*, **46**, 103 (2011).

[b]A. M. Ricks, G. E. Douberly, M. A. Duncan, *Astrophys. J.* **702**, 301 (2009).

[c]M. Bahou, Y.-J. Wu, Y.-P. Lee, *J. Chem. Phys.* **136**, 154304 (2012); M. Bahou, Y.-J. Wu, Y.-P. Lee, *Phys. Chem. Chem. Phys.* **15**, 1907 (2013); M. Bahou, Y.-J. Wu, Y.-P. Lee, *J. Phys. Chem. Lett.* **4**, 1989 (2013); M. Bahou, Y.-J. Wu, Y.-P. Lee, *Angew. Chem. Int. Ed.* **53**, 1021 (2014).

WG02 2:05 – 2:20

VISIBLE PHOTODISSOCIATION SPECTRA OF THE 1-METHYL AND 2-METHYLNAPHTHALENE CATIONS: LASER SPECTROSCOPY AND THEORETICAL SIMULATIONS

HELA FRIHA, GERALDINE FERAUD[a], CYRIL FALVO, PASCAL PARNEIX, THOMAS PINO, PHILIPPE BRECHIGNAC, *Institut des Sciences Moléculaires d'Orsay, Université Paris-Sud, Orsay, France*; TYLER TROY, TIMOTHY SCHMIDT, *School of Chemistry, The University of Sydney, Sydney, NSW, Australia*; ZOUBEIDA DHAOUADI, *LSAMA, University Tunis El Manar, Tunis, Tunisia.*

Naphthalene (Np) and its methylated derivatives (1-Me-Np and 2-Me-Np) are prototype molecules for spectroscopists as first members of the polycyclic aromatic hydrocarbons (PAHs) family. High resolution studies are capable to explore the details of the internal rotation of the methyl group. Although this was achieved in neutral PAHs[b], the task is not the same in cations. Me-Np cations have been probed by resonance-enhanced multiphoton dissociation[c], showing only very broad and unresolved spectra, while absorption in argon matrix revealed more resolved vibronic bands[d].

The electronic absorption gas phase spectra of 1-Me-Np$^+$ and 2-Me-Np$^+$ were measured using an Ar-tagging technique. In both cases, a band system was observed in the visible range and assigned to the $D_2 \leftarrow D_0$ transition. The 1-Me-Np$^+$ absorption bands revealed a red shift of 808 cm^{-1}, relative to Np$^+$ (14 906 cm^{-1})[e] , while for 2-Me-Np$^+$ a blue shift of 226 cm^{-1} was found. A short vibrational progression was also observed. Moreover, insights into the internal rotation motion of the CH$_3$ were inferred, although intrinsic broadening due to intramolecular relaxation was present. These measurements were supported by detailed quantum chemical calculations that allowed exploration of the potential energy curves, along with a complete simulation of the harmonic FC factors using the cumulant Gaussian fluctuations formalism, extended to include the internal rotation.

[a]presently at University of Marseille (PIIM), France
[b]see for instance Baba et al, *J.Phys.Chem.A*, **2009**, 113, 2366
[c]Dunbar et al, *J. Am. Chem. Soc.* **1976**, 98, 7994-7999 and *J.Phys.Chem.* **1985**, 89, 3617
[d]Andrews et al, *J.Phys.Chem.* **1982**, 86, 2916
[e]Pino et al, *J. Chem. Phys.* **1999**, 111, 7337-7347

WG03 2:22 – 2:37

STRUCTURE AND ELECTRONIC PROPERTIES OF IONIZED PAH CLUSTERS [a]

CHRISTINE JOBLIN, DAMIAN L KOKKIN, HASSAN SABBAH, ANTHONY BONNAMY, *IRAP, Université de Toulouse 3 - CNRS, Toulouse, France*; LEO DONTOT, MATHIAS RAPACIOLI, AUDE SIMON, FERNAND SPIEGELMAN, *LCPQ, Université de Toulouse 3 - CNRS, Toulouse, France*; PASCAL PARNEIX, THOMAS PINO, OLIVIER PIRALI, CYRIL FALVO, ANTONIO GAMBOA, PHILIPPE BRECHIGNAC, *Institut des Sciences Moléculaires d'Orsay, Université Paris-Sud, Orsay, France*; GUSTAVO A. GARCIA, LAURENT NAHON, *DESIRS beamline, Synchrotron SOLEIL, Gif-sur-Yvette, France.*

Polycyclic aromatic hydrocarbon (PAH) clusters have been proposed as candidates for evaporating very small grains that are revealed by their mid-IR emission at the surface of UV-irradiated clouds in interstellar space[b]. This suggestion is a motivation for further characterization of the properties of these clusters in particular when they are ionized. We have used a molecular beam coupled to the photoelectron-photoion coincidence spectrometer DELICIOUS II/ III [c] at the VUV beamline DESIRS of the synchrotron SOLEIL to characterize the electronic properties of cationic coronene ($C_{24}H_{12}$) and pyrene ($C_{16}H_{10}$) clusters up to the pentamer and heptamer, respectively. These experimental results are analysed in the light of electronic structure calculations. Simulations of the properties of ionized PAH clusters are faced with the difficulty of describing charge delocalization in these large systems. We will show that recent developments combining a Density Functional Tight Binding method with Configuration Interaction scheme[d] is successful in simulating the ionization potential, which gives strong confidence into the predicted structures for these PAH clusters. We will also present current effort to study charge transfer states by performing complementary measurements with the PIRENEA ion trap set-up.

[a]Joint ANR project GASPARIM, ANR-10-BLAN-501
[b]M. Rapacioli, C. Joblin and P. Boissel *Astron. & Astrophys.* **429** (2005), 193-204.
[c]G. Garcia, H. Soldi-Lose and L. Nahon *Rev. Sci. Instrum.* **80** (2009), 023102; G. Garcia, B. Cunha de Miranda, M. Tia, S. Daly, L. Nahon, *Rev. Sci. Instrum.* **84** (2013), 053112
[d]M. Rapacioli, A. Simon, L. Dontot and F. Spiegelman *Phys. Status Solidi B* **249** (2) (2012), 245-258; L. Dontot, M. Rapacioli and F. Spiegelman (2014) *submitted*

WG04

ABSORPTIONS IN THE VISIBLE OF PROTONATED PYRENE COLLISIONALLY COOLED TO 15 K

C. A. RICE, FRANCOIS XAVIER HARDY, OLIVER GAUSE, JOHN P. MAIER, *Department of Chemistry, University of Basel, Basel, Switzerland.*

Protonated polycyclic hydrocarbons have been added to the list of suggested carriers of the diffuse interstellar absorptions. To test this proposition requires laboratory spectra measured under interstellar conditions, in particular with the rotational and vibrational degrees of freedom equilibrated to low temperatures. This has been achieved for protonated pyrene with absorption bands in the visible, using an ion trap and collisional cooling to ≈ 15 K. A two-photon excitation-dissociation scheme was employed to record the $(1)\,^1A' \leftarrow X\,^1A'$ electronic spectrum on around 10^5 ions per duty cycle. The origin band of the absorption spectrum of this relatively large polycyclic aromatic species with 27 atoms is located at 4858.86 Å. Two further comparably intense spectral features are present at 4834.48 and 4809.32 Å. This is one of the largest protonated aromatics studied in the gas phase and compared to astonomical observations; however, it is not a carrier of known diffuse interstellar bands.

WG05

ULTRAVIOLET PHOTODISSOCIATION ACTION SPECTROSCOPY OF PROTONATED AZABENZENES

CHRISTOPHER S. HANSEN, *School of Chemistry, University of Wollongong, Wollongong, New South Wales, Australia*; STEPHEN J. BLANKSBY, *Central Analytical Research Facility, Queensland University of Technology, Brisbane, Queensland, Australia*; EVAN BIESKE, *School of Chemistry, The University of Melbourne, Melbourne, Victoria, Australia*; JEFFREY R. REIMERS, *University of Technology Sydney, School of Physics and Materials Science, Broadway, New South Wales, Australia*; ADAM J. TREVITT, *School of Chemistry, University of Wollongong, Wollongong, New South Wales, Australia.*

Azabenzenes are derivatives of benzene containing between one and six nitrogen atoms. Protonated azabenzenes are the fundamental building blocks of many biomolecules, charge-transfer dyes, ionic liquids and fluorescent tags. However, despite their ubiquity, there exists limited spectroscopic data that reveals the structure, behaviour and stability of these systems in their excited states. For the case of pyridinium ($C_5H_5N\text{-}H^+$), the simplest azabenzene, the electronic spectroscopy is complicated by short excited state lifetimes, efficient non-radiative deactivation methods and limited fluorescence. Ultraviolet (UV) photodissociation (PD) action spectroscopy[a] provides new insight into the spectroscopic details, excited state behaviour and photodissociation processes of a series of protonated azabenzenes including pyridinium, diazeniums and their substituted derivatives.

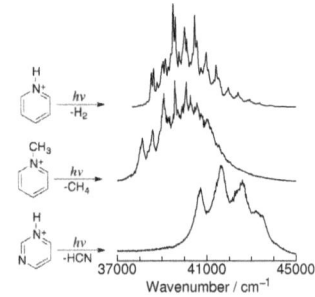

The room-temperature UV PD action spectra, often exhibiting vibronic detail,[b] will be presented alongside PD mass spectra and the kinetic data from structurally-diagnostic ion-molecule reaction kinetics. Analysis of the spectra, with the aid of quantum chemical calculations, reveal that many azabenzenes prefer a non-planar excited state geometry reminiscent of the structures encountered in 'channel 3'-like deactivation of aromatics. The normal modes active in this isomerization contribute largely to the spectroscopy of the *N*-pyridinium ion as they build upon totally-symmetric vibronic transitions leading to repeating sets of closely-spaced spectral features.

[a]Hansen, C.S. *et al.*; J. Am. Soc. Mass Spectrom. 24:932-940 (2013) [b]Hansen, C.S. *et al.*; J. Phys. Chem. A 117:10839-10846 (2013)

WG06 3:13 – 3:28

SIMULTANEOUS COUNTER-ION CO-DEPOSITION: A TECHNIQUE ENABLING MATRIX ISOLATION SPEC-
TROSCOPY STUDIES USING LOW-ENERGY BEAMS OF MASS-SELECTED IONS[a]

RYAN M LUDWIG, DAVID T MOORE, *Chemistry Dept., Lehigh University, Bethlehem, PA, USA.*

Matrix isolation spectroscopy was first developed in Pimentel's group during the 1950's to facilitate spectroscopic studies of transient species.[b] Cryogenic matrices of condensed rare gases provide an inert chemical environment with facile energy dissipation and are transparent at all wavelengths longer than vacuum UV, making them ideal for studying labile and reactive species such as radicals, weakly bound complexes, and ions. Since frozen rare gases are poor electrolytes, studies of ions require near-equal populations of anions and cations in order to stabilize the number densities required for spectroscopic experiments. Many techniques for generation of ions for using in matrix isolation studies satisfy this criterion intrinsically, however when ion beams generated in external sources are deposited, the counter-ions typically arise via secondary processes that are at best loosely controlled.[c] It has long been recognized that it would be desirable to stabilize deposition of mass-selected ions generated in an external source using simultaneous co-deposition of a beam of counter-ions, however previous attempts to achieve this have been reported as unsuccessful.[d] The Moore group at Lehigh has demonstrated successful experiments of this type, using mass-selected anions generated from a metal cluster source, co-deposited with a balanced current of cations generated in a separate electron ionization source. This talk will focus on the details of the technique, and present some results from proof-of-concept studies on anionic copper carbonyl complexes formed in argon matrices following co-deposition of Cu^- with Ar^+ or Kr^+.

[a]Funding support from NSF CAREER Award CHE-0955637 is gratefully acknowledged.
[b]Whittle et al., J. Chem. Phys. 22, p.1943 (1954); Becker et al., J. Chem. Phys. 25, p.224 (1956).
[c]Godbout et al., J. Chem. Phys. 96, p.2892 (1996).
[d]Sabo et al., Appl. Spectrosc. 45, p. 535 (1991).

WG07 3:30 – 3:45

CONTROLLED NEUTRALIZATION OF ANIONS IN CRYOGENIC MATRICES BY NEAR-THRESHOLD PHOTODE-
TACHMENT[a]

RYAN M LUDWIG, DAVID T MOORE, *Chemistry Dept., Lehigh University, Bethlehem, PA, USA.*

Using matrix isolation FTIR, we have observed the formation of anionic copper carbonyl complexes $[Cu(CO)_n]^-$ (n=1-3) following co-deposition of Cu^- and counter-cations (Ar^+ or Kr^+) into argon matrices doped with CO.[b] The infrared bands have been previously assigned in argon matrix studies employing laser ablation, however they were quite weak compared to the bands for the corresponding neutral species.[c] In the current study, when the deposition is carried out in fully darkened conditions at 10 K with high CO concentrations (1-2%), only the bands for the anionic complexes are observed initially via FTIR. However, upon mild irradiation with broadband visible light, the anionic bands are rapidly depleted, with concomitant appearance of bands corresponding to neutral copper carbonyl complexes. This photo-triggered neutralization is attributed to photodetachment of electrons from the anions, which then "flow" through the solid argon matrix to recombine in the matrix with non-adjacent trapping sites. This mechanism is supported by the appearance of a new band near 1515 cm^{-1}, assigned to the $(CO)_2^-$ species in argon.[d] The wavelength dependence of the photodetachment will be discussed in detail, although preliminary indications are that the thresholds for the copper carbonyls, which are normally in the infrared,[e] are shifted into the visible region of the spectrum in argon matrices. This likely occurs because the conduction band of solid argon is known to lie about 1 eV above the vacuum level,[f] and thus the electron must have at least this much energy in order to escape into the matrix and find a trapping site.

[a]Funding support from NSF CAREER Award CHE-0955637 is gratefully acknowledged
[b]Ryan M. Ludwig and David T. Moore, J. Chem. Phys. 139, 244202 (2013).
[c]Zhou, M.; Andrews, L., J. Chem. Phys. 111, 4548 (1999).
[d]Thompson, W.E.; Jacox, M.E.; J. Chem. Phys. 91, 735 (1991).
[e]Stanzel, J. et al.; Collect. Czech. Chem. Comm. 72, 1 (2007).
[f]Harbich, W. et al.; Phys. Rev. B. 76, 104306 (2007).

Intermission

WG08

ANOMALOUS BEHAVIOR OBSERVED UPON ANNEALING AND PHOTODETACHMENT OF ANIONIC COPPER CARBONYL CLUSTERS IN ARGON MATRICES[a]

RYAN M LUDWIG, <u>DAVID T MOORE</u>, *Chemistry Dept., Lehigh University, Bethlehem, PA, USA.*

Using matrix isolation FTIR, we have observed the formation of anionic copper carbonyl complexes $[Cu(CO)_n]^-$ (n=1-3) following co-deposition of Cu^- and counter-cations (Ar^+ or Kr^+) into argon matrices doped with CO.[b] When the deposition is carried out at 20 K, weak bands corresponding to the neutral copper carbonyl complexes $Cu(CO)_n$ (n=1-3) are also observed, and these grow in steadily as the matrix is annealed up to 30 K. This is in contrast to what is observed at 10 K (c.f. ISMS 2014 abstract #P631), where no appreciable neutral bands are observed, and indicates that some neutralization occurs during the formation of the complexes in the 20 K matrix. In addition, sharp peaks not previously observed grow in around the anionic bands upon annealing to 30 K; this is somewhat odd, since annealing typically simplifies the spectra of matrix samples as kinetically trapped metastable species relax to more stable forms. In this case, higher-resolution (0.125 cm^{-1}) spectra reveal considerable new fine structure, with 5 and 20 peaks appearing in the regions of the mono- and tricarbonyl anions, respectively, each of which nominally has but a single IR-active CO-stretching mode. These new features are tentatively assigned (at least in part) to electric-field-induced splitting arising from long-range interactions with cationic species in the matrix. A second anomalous feature of these spectra is that, upon photodetachment, several new bands are observed in the region of the neutral copper carbonyl species. Upon annealing these bands then disappear, with concomitant growth of the expected neutral bands. This behavior raises the exciting possibility that these transient bands represent metastable "vertical detachment products", where the neutral species has been kinetically trapped by the matrix in the geometry of the anion. Evidence supporting this interpretation will be presented.

[a]Funding support from NSF CAREER Award CHE-0955637 is gratefully acknowledged
[b]Ryan M. Ludwig and David T. Moore, J. Chem. Phys. 139, 244202 (2013).

WG09

CARBON DIOXIDE CLUSTERS AND COPPER COMPLEXES FORMED IN ARGON MATRICES[a]

<u>MICHAEL E. GOODRICH</u>, DAVID T MOORE, *Chemistry Dept., Lehigh University, Bethlehem, PA, USA.*

Neutral and anionic clusters containing carbon monoxide and carbon dioxide molecules were observed in FTIR matrix isolation spectroscopy experiments following co-deposition of Cu^- and Ar^+ in argon matrices doped with moderate amounts (0.1-1%) of CO_2 and/or CO. Following deposition at 10 K with 0.5% CO_2, peaks for $CuCO_2^-$ and $Cu(CO_2)_2^-$ were observed, along with a small peak for CO_2^- and several bands assigned to neutral $(CO_2)_x$ clusters. Upon annealing to 20 K, a new peak appears at 1857 cm^{-1}, which is assigned to the $C_2O_4^-$ ion, based on previous work in neon matrices.[b] When CO is added as a dopant (0.5 %) to the matrix gas mix, formation of the anionic copper CO_2 complexes is suppressed, and neutral CO-CO_2 complexes are observed in the spectra, as are bands corresponding to $C_2O_3^-$, based on previous studies.[c] Interestingly, the copper carbonyl bands typically observed for equivalent deposition conditions in the absence of CO_2 are strongly suppressed in these spectra. The implication is that complexation with the CO_2 molecules, which are far more abundant in the matrix, inhibits the CO molecules from "finding" the Cu^- centers during the matrix formation process. (c.f. ISMS 2014 abstract #P616)

[a]Funding support from NSF CAREER Award CHE-0955637 is gratefully acknowledged
[b]Zhou, M.F.; Andrews, L.; J. Chem. Phys. 110, 6820 (1999).
[c]Zhou, M.F.; et al., J. Chem. Phys. 112, 7089 (2000).

WG10 4:36 – 4:51

THEORETICAL HIGH-RESOLUTION SPECTROSCOPY BEYOND CCSD(T): THE INTERSTELLAR ANIONS CN^-, CCH^-, C_3N^-, AND C_4H^-

PETER BOTSCHWINA, BENJAMIN SCHRÖDER, PETER SEBALD, RAINER OSWALD, *Institute of Physical Chemistry, Georg-August-Universität Göttingen, Göttingen, Germany.*

Using extended coupled cluster methods well beyond fc-CCSD(T), spectroscopic properties of several molecular anions of astrochemical interest have been calculated. Excellent agreement with MW data is observed for CN^-, CCH^-, C_3N^-, and C_4H^- and accurate equilibrium structures are presented for all four species. The results for CCH^- are superior to recent theoretical results of Huang and Lee[a] and confirm the quality of our earlier predictions.[b] The new calculations predict $\nu_1 = 3209.3$ cm^{-1}, ν_2 (band origin) $= 510.0$ cm^{-1}, and $\nu_3 = 1804.4$ cm^{-1}, estimated errors not exceeding 1 cm^{-1}.

[a] X. Huang, T. J. Lee *J. Chem. Phys.* **2009**, *131*, 104301.
[b] M. Mladenović, P. Botschwina, P. Sebald, S. Carter, *Theor. Chem. Acc.* **1998**, *100*, 134.

WG11 4:53 – 5:08

HIGH-J ROTATIONAL LINES OF HCO^+ AND ITS ISOTOPOLOGUES MEASURED BY USING EVENSON-TYPE TUNABLE FIR SPECTROMETER

RYO OISHI, TATSUYA MIYAMOTO, MARI SUZUKI, YOSHIKI MORIWAKI, FUSAKAZU MATSUSHIMA, *Department of Physics, University of Toyama, Toyama, Japan*; TAKAYOSHI AMANO, *Department of Chemistry, University of Waterloo, Waterloo, ON, Canada.*

Frequencies of high-J rotational lines of HCO^+ and DCO^+ have been measured precisely by using an Evenson-type tunable far-infrared spectrometer in Toyama. The spectrometer (sometimes called TuFIR in short) is based on synthesizing terahertz radiation from two mid-infrared CO_2 laser lines and one microwave source. The HCO^+ ions are produced by discharging a CO, H_2 (or D_2), and Ar mixture in an extended negative glow discharge cell cooled with liquid nitrogen. The low-J rotational lines have been extensively studied. Information on the higher-J levels have been obtained by observing the vibration-rotation bands. More recently high precision infrared measurements have been made by observing the Lamb dips.[a] We focus our efforts to measuring the high-J rotational lines. Currently we have observed the lines $J + 1 \leftarrow J$ (J=7-19, 21) for HCO^+, and $J + 1 \leftarrow J$ (J=11, 13-14, 16-24) for DCO^+. An additional isotopologue, $H^{13}CO^+$ is now under investigation. From the analysis of the intensity of each rotational line, we estimate the rotational temperature to be as low as 120K. Apparently, due to this low temperature, it seems to be difficult to extend the measurements to yet higher-J lines.

[a] B. M. Siller, J. N. Hodges, A. J. Perry, and B. J. McCall, *J. Phys. Chem. A.* **117**, 10034(2013)

WG12   5:10–5:25

MID-INFRARED NICE-OHMS SPECTROMETER FOR THE STUDY OF COLD MOLECULAR IONS

MICHAEL PORAMBO, JESSICA PEARSON, COURTNEY TALICSKA, *Department of Chemistry, University of Illinois at Urbana-Champaign, Urbana, IL, USA*; BENJAMIN J. McCALL, *Departments of Chemistry and Astronomy, University of Illinois at Urbana-Champaign, Urbana, IL, USA.*

Molecular ions are involved in the chemistry of many interesting systems, such as the atmosphere, combustion reactions, and the interstellar medium. Challenging aspects of studying molecular ions spectroscopically include producing ions in enough abundance and, for large or fluxional ions, overcoming the problem of quantum dilution at high vibrational and rotational temperature. Furthermore, highly precise transition frequencies are needed to answer many questions involving molecular ions, such as the presence of specific candidate ions in the interstellar medium. To address these challenges, we have constructed a mid-infrared spectrometer that uses a difference frequency generation (DFG) light source to probe cooled molecular ions produced in a continuous supersonic expansion discharge source. The cooling of the ions achieved through supersonic expansion mitigates the problem of quantum dilution. High sensitivity to detect the 10^{12} ions per cm^3 produced is accomplished through the use of noise-immune cavity-enhanced optical heterodyne molecular spectroscopy (NICE-OHMS) as a detection technique. Finally, an optical frequency comb is used to measure the transition frequencies of molecular ions precisely. This talk will present the current status of the instrument and preliminary studies to optimize and characterize its performance. Initial studies of room temperature methane allowed us to verify the use of NICE-OHMS for inferring rotational temperature of a molecular sample through Boltzmann plot analysis. Spectroscopy of H_3^+ and HN_2^+ extended this temperature verification to molecular ions. Future work on H_2CO^+, with the goal of determining its rovibrational transitions to a precision on the order of 1 MHz to aid in astronomical detection, will also be presented.

WG13   5:27–5:42

RIGIDITY OF THE MOLECULAR ION H_5^+

CSABA FÁBRI, *Laboratory of Physical Chemistry, ETH Zurich, Zürich, Switzerland*; JÁNOS SARKA, ATTILA CSÁSZÁR, *Research Group on Complex Chemical Systems, MTA-ELTE, Budapest, Hungary.*

The in-house fourth-age quantum chemical code GENIUSH is used for the variational determination of rotational-vibrational energy levels corresponding to reduced- and full-dimensional models of H_5^+, a molecular ion exhibiting several strongly coupled large-amplitude motions and strong coupling between its vibrations and rotations. The quantum chemical computations are supplemented with simplified modeling efforts, including one- and two-dimensional exactly solvable models. These models help to understand the peculiar rovibrational energy-level structure computed for H_5^+ and its deuterated isotopologues. It is shown that while a 1D "active torsion" model provides proper rovibrational levels when compared to the full, 9D treatment, models excluding the torsion have limited physical significance due to the extremely strong coupling between the torsion and one of the rotations. The role the choice of the internal coordinates and the embedding of the rotational axes play in obtaining converged rovibrational results is discussed in detail. The structure of the rovibrational energy levels of H_5^+ proves that it is useful to view this ion as a prototypical astructural molecule: the rotational and vibrational level spacings are of the same order of magnitude and the level structure drastically deviates from that computed via perturbed rigid-rotor and harmonic-oscillator models.

WG14   5:44–5:59

FULL DIMENSIONAL POTENTIALS, DIPOLE MOMENT SURFACES AND (RO)VIBRATIONAL CALCULATIONS FOR H_5^+, H_7^+ AND HOCO

JOEL BOWMAN, STUART CARTER, YIMIN WANG, *Department of Chemistry, Emory University, Atlanta, GA, USA.*

I will describe progress in the first-principles calculations of "line-list" ro-vibrational spectra of H_5^+, D_5^+ and HOCO and low-resolution spectrum of H_7^+, along with insights into the internal motions of these species. The calculations make use of full-dimensional ab initio potential and dipole moment surfaces and the code "MULTIMODE".

WH. Mini-symposium: Spectroscopy in Kinetics and Dynamics

Wednesday, June 18, 2014 – 1:30 PM

Room: B102 Chemical and Life Sciences

Chair: Toshinori Suzuki, Kyoto University, Kyoto, Japan

WH01   ***INVITED TALK***   1:30 – 2:00

RADICALLY DIFFERENT KINETICS AT LOW TEMPERATURES

IAN SIMS, *Institut de Physique de Rennes, Université de Rennes 1, Rennes, France.*

The use of the CRESU (Cinétique de Réaction en Ecoulement Supersonique Uniforme, or Reaction Kinetics in Uniform Supersonic Flow) technique coupled with pulsed laser photochemical kinetics methods has shown that reactions involving radicals can be very rapid at temperatures down to 10 K or below. The results have had a major impact in astrochemistry and planetology, as well as proving an exacting test for theory.[a] The technique has also been applied to the formation of transient complexes of interest both in atmospheric chemistry[b] and combustion.[c]

Until now, all of the chemical reactions studied in this way have taken place on attractive potential energy surfaces with no overall barrier to reaction. The $F + H_2 \rightarrow HF + H$ reaction does possess a substantial energetic barrier ($\cong$ 800 K), and might therefore be expected to slow to a negligible rate at very low temperatures. In fact, this H-atom abstraction reaction does take place efficiently at low temperatures due entirely to tunneling. I will report direct experimental measurements of the rate of this reaction down to a temperature of 11 K, in remarkable agreement with state-of-the-art quantum reactive scattering calculations by François Lique (Université du Havre) and Millard Alexander (University of Maryland). [d]

It is thought that long chain cyanopolyyne molecules $H(C_2)_nCN$ may play an important role in the formation of the orange haze layer in Titan's atmosphere. The longest carbon chain molecule observed in interstellar space, $HC_{11}N$, is also a member of this series. I will present new results, obtained in collaboration with Jean-Claude Guillemin (Ecole de Chimie de Rennes) and Stephen Klippenstein (Argonne National Labs), on reactions of C_2H, CN[e] and C_3N radicals (using a new LIF scheme by Hoshina and Endo[f]) which contribute to the low temperature formation of (cyano)polyynes.

---

[a]H. Sabbah, L. Biennier, I. R. Sims, Y. Georgievskii, S. J. Klippenstein, I. W. M. Smith, Science **317**, 102 (2007).

[b]S. D. Le Picard, M. Tizniti, A. Canosa, I. R. Sims, I. W. M. Smith, Science **328**, 1258 (2010).

[c]H. Sabbah, L. Biennier, S. J. Klippenstein, I. R. Sims, B. R. Rowe, J. Phys. Chem. Lett. **1**, 2962 (2010).

[d]M. Tizniti, S. D. Le Picard, F. Lique, C. Berteloite, A. Canosa, M. H. Alexander, I. R. Sims, Nature Chemistry **6**, 141 (2014).

[e]S. Cheikh Sid Ely, S. B. Morales, J. C. Guillemin, S. J. Klippenstein, I. R. Sims, J. Phys. Chem. A **117**, 12155 (2013).

[f]K. Hoshina, Y. Endo, J. Chem. Phys. **127**, 184304 (2007).

WH02   2:05 – 2:20

H-ATOM REACTION KINETICS IN SOLID PARAHYDROGEN FOLLOWED BY RAPID SCAN FTIR

DAVID T. ANDERSON, *Department of Chemistry, University of Wyoming, Laramie, WY, USA.*

Reactions of migrating H-atoms in parahydrogen (pH_2) matrices with trapped molecular species provide a relatively unexplored yet well-established experimental method to study the kinetics and mechanisms of atom tunneling reactions in the 1.5 to 5 K temperature range. My group has now completed a series of experimental studies on the kinetics of reactions of H-atoms with HCOOH, CH_3OH, and N_2O which all show a pronounced inverse temperature dependence over this small temperature range.[a] Conversely, the analogous H-atom reaction with NO displays a more standard Arrhenius behavior. In this talk, I will present a brief summary of these results with the objective of developing a predictive understanding of the kinetics of these H-atom tunneling reactions. I will also emphasize the advantages of following the kinetics using rapid scan FTIR.

---

[a]Fredrick M. Mutunga, Shelby E. Follett, and David T. Anderson, *J. Chem. Phys.* **139**, 151104 (2013).

166

2:22–2:37

TO TUNNEL OR NOT TO TUNNEL, PROTON TRANSFER IS THE QUESTION.

KATHRYN CHEW, DEACON NEMCHICK, PATRICK VACCARO, *Department of Chemistry, Yale University, New Haven, CT, USA.*

The transduction of protons between donor and acceptor sites, as mediated by the action of adjoining hydrogen bonds, represents one of the most ubiquitous of chemical transformations. While the basic mechanisms underlying such phenomena often can be ascribed to simple acid-base chemistry, the putative roles of selective nuclear and electronic displacements should not be discounted, especially when the presence of a sizeable potential barrier impedes classical hydron-migration pathways. The vibrational and isotopic specificity of hindered intramolecular proton transfer taking place within the ground ($\tilde{X}^1A_1$) and the lowest-lying excited ($\tilde{A}^1B_2$ ($\pi^*\pi$)) electronic states of the prototypical tropolone (TrOH) system has been probed by implementing multiple-color variants of resonant four-wave mixing (RFWM) spectroscopy, with polarization-resolved detection allowing for the extraction of quantitative rotation-tunneling information. The marked dependence of unimolecular dynamics on the extent and the type of excitation deposited into TrOH internal degrees of freedom will be discussed. Experimentally observed trends and propensities for tunneling-mediated reactivity will be interpreted through use of accompanying quantum-chemical calculations.

WH04 2:39–2:54

SPECTROSCOPIC AND KINETIC STUDIES OF ATMOSPHERIC FREE RADICALS

ELIZABETH FOREMAN, YITIEN JOU, KARA KAPNAS, *Chemistry, University of California, Irvine, Irvine, CA, USA*; CRAIG MURRAY, *Department of Chemistry, The University of California, Irvine, CA, USA.*

Photo-induced radical chemistry is crucial to atmospheric processes, namely: oxidation, particulate matter formation, and climate change. The combination of transient absorption and pulsed cavity ring-down spectroscopies is used to study weak electronic or overtone transitions of trace gas-phase species and investigate the kinetic and photochemical properties of important transient atmospheric radicals.

WH05 2:56–3:11

RADICAL INTERMEDIATES IN THE ADDITION OF OH TO PROPENE: PHOTOLYTIC PRECURSORS AND ANGULAR MOMENTUM EFFECTS

MATTHEW D BRYNTESON, CARRIE C WOMACK, RYAN S BOOTH, *Department of Chemistry, The University of Chicago, Chicago, IL, USA*; SHIH -H LEE, *Molecular Science, National Synchrotron Radiation Research Center, Hsinchu, Taiwan*; JIM J LIN, *Institute of Atomic and Molecular Sciences, Academia Sinica, Taipei, Taiwan*; LAURIE BUTLER, *Department of Chemistry, The University of Chicago, Chicago, IL, USA.*

We investigate the photolytic production of two radical intermediates in the reaction of OH with propene, one from addition of the hydroxyl radical to the terminal carbon and the other from addition to the center carbon. In a collision-free environment, we photodissociate a mixture of 1-bromo-2-propanol and 2-bromo-1-propanol at 193 nm to produce these radical intermediates. Using a velocity map imaging apparatus, we measured the speed distribution of the recoiling bromine atoms, yielding the distribution of kinetic energies of the nascent C_3H_6OH radicals + Br. Resolving the velocity distributions of $Br(^2P_{1/2})$ and $Br(^2P_{3/2})$ separately with 2+1 REMPI allows us to determine the total (vibrational + rotational) internal energy distribution in the nascent radicals. Using an impulsive model to estimate the rotational energy imparted to the nascent C_3H_6OH radicals, we predict the percentage of radicals having vibrational energy above and below the lowest dissociation barrier, that to OH + propene; it accurately predicts the measured velocity distribution of the stable C_3H_6OH radicals. In addition, we use photofragment translational spectroscopy to detect several dissociation products of the unstable C_3H_6OH radicals: OH + propene, methyl + acetaldehyde, and ethyl + formaldehyde. We also use the angular momenta of the unstable radicals to estimate the energy partitioned to relative kinetic energy when they dissociate to OH + propene, which agrees very well with the data.

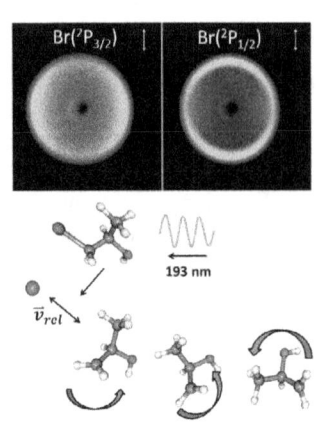

WH06   **3:13 – 3:28**

STATE-RESOLVED AND STATE-TO-STATE PHOTODISSOCIATION STUDY OF CO_2 BY TWO-COLOR VUV-VUV LASER PUMP-PROBE METHOD

ZHOU LU, YIH-CHUNG CHANG, WILLIAM M. JACKSON, CHEUK-YIU NG, *Department of Chemistry, The University of California, Davis, CA, USA.*

CO_2 is known as a strong contributor to the greenhouse effect, and its concentration in the atmosphere increases annually. Photodissociation of CO_2 is considered an important photochemical sink of CO_2 molecules which could ultimately limit the increase of CO_2 concentration in the atmosphere. Since CO_2 molecules have negligibly small absorption from the visible region down to about 200 nm, photodissociation studies of CO_2 in the vacuum ultraviolet (VUV) region below 200 nm are of great importance in understanding the photochemical decomposition processes of CO_2 molecules. State-to-state photodissociation of CO_2 has been investigated by employing two independent VUV lasers and the time-slice velocity-map-imaging-photoion (VMI-PI) method. The spin-allowed photoproduct channels, $CO(X^1\Sigma^+) + O(^1D)$, $CO(X^1\Sigma^+) + O(^1S)$, $CO(a^3\Pi) + O(^3P_J)$, and $C(^3P_J) + O_2(X^3\Sigma_g^-)$, and the spin-forbidden photoproduct channel, $CO(X^1\Sigma^+) + O(^3P_J)$, were directly observed from the time-slice VMI-PI images. The angular anisotropic parameters were evaluated, allowing us to estimate the lifetimes for the formations of these dissociation channels. To the authors' knowledge, the current CO_2 photodissociation studies show for the first time that all of the energetically available photodissociation channels are formed in the CO_2 photoexcitation energy of interest.

Intermission

WH07   **3:45 – 4:00**

SUBMILLIMETER MEASUREMENTS OF THE CRIEGEE INTERMEDIATE, CH_2OO, IN THE GAS PHASE

ADAM M DALY, BRIAN DROUIN, SHANSHAN YU, *Jet Propulsion Laboratory, California Institute of Technology, Pasadena, CA, USA.*

High frequency pure rotational transitions of the simplest Criegee intermediate, methylene peroxide (CH2OO), have been measured up to 1 THz. The data provide the most accurate spectroscopic parameters for the ground vibrational state. The molecule was produced in a flow cell with a DC discharge of CH_2I_2, O_2 and Ar. Doppler limited measurements were made in the frequency ranges 220-320, 580-680 and 970-1080 GHz at room temperature. A total of 211 transitions were measured and added to four transitions measured in the microwave to obtain a fit up to $J^{max} = 49$ and $K^{max} = 14$ of the a-dipole spectrum.

WH08   **4:02 – 4:17**

MEASURING RATE CONSTANTS FOR REACTIONS OF THE SIMPLEST CRIEGEE INTERMEDIATE CH_2OO BY MONITORING THE OH RADICAL

YINGDI LIU, KYLE D BAYES, STANLEY P. SANDER, *Jet Propulsion Laboratory, California Institute of Technology, Pasadena, CA, USA.*

Criegee radicals are important atmospheric intermediates formed from ozonolysis of alkenes. It potentially contributes to the atmospheric oxidation cycle mainly by generating OH radicals through unimolecular decomposition. In this work, we focus on studying the unimolecular decomposition reaction of the smallest Criegee intermediate (CH_2OO), which was generated by reacting CH_2I with O_2. While generating the CH_2OO molecule by reacting CH_2I with O_2, significant amounts of the OH radical were observed by laser-induced fluorescence. The addition of molecules known to react with CH_2OO increased the observed decay rates of the OH signal. Using the OH signals as a proxy for the CH_2OO concentration, the rate constant for the reaction of hexafluoroacetone with CH_2OO was determined. The rate constant for the reaction of SO_2 with CH_2OO showed no pressure dependence over the range of 50 to 200 Torr. This work provides the direct experimental evidence for the unimolecular decomposition of CH_2OO, and possible mechanisms of CH_2OO have been investigated by this multidimensional study.

WH09   4:19 – 4:34

EFFECTS OF REACTANT ROTATIONAL EXCITATION ON Cl + CH_4 / CHD_3 REACTIONS

HUILIN PAN, *Institute of Atomic and Molecular Sciences, Academia Sinica, Taipei, Taiwan*; FENGYAN WANG, *Fudan University, Department of Chemistry, Shanghai,, China*; YUAN CHENG, JUI-SAN LIN, KOPIN LIU, *Institute of Atomic and Molecular Sciences, Academia Sinica, Taipei, Taiwan.*

Effects of reactant rotation, which can disentangle the stereodynamical properties of chemical reactivity, are of great importance in understanding and controlling the steric effects in chemical reactions. Here, we report the rotational quantum-state control of a late-barrier reaction of chlorine atom with CH_4 and CHD_3 in a crossed molecular beam experiment. Experimental results demonstrate that, in both reactions, the more detailed product translational and angular distributions are essentially the same for different rotational states of the vibrationally excited CH_4 and CHD_3 reactants. Yet, the integral cross sections show strong dependence on the reactant rotational excitation, suggesting that the reactivity diversity arises from the anisotropic interactions enroute to the reaction barrier. More detailed analysis indicates that the effects of reactant rotation do not derive from the rotational-energy effects or long-range forces, rather are the result of short-range forces in the transition state region. Exactly how the transition-state properties, e.g., the barrier location and the tightness of barrier, influence the rotational reactivity diversity, however, remains unclear. Further investigations are on-going to gain deeper insights.

WH10   4:36 – 4:51

ROTATIONAL ENERGY TRANSFER AND DEPOLARIZATION IN RARE GAS + CN ($\tilde{X}$, v=0) COLLISIONS

GREGORY HALL, DAMIEN FORTHOMME, TREVOR SEARS[a], *Department of Chemistry, Brookhaven National Laboratory, Upton, NY, USA.*

Rotational energy transfer and depolarization rates have been determined for fine-structure-selected rotational levels of the vibrationless ground state of CN ($\tilde{X}$ $^2\Sigma^+$) radicals. Collisions with Ar, He and the photolytic precursor, CH_3COCN have been studied with polarized transient frequency modulation (FM) absorption spectroscopy. Transient FM signals were recorded as a function of probe laser detuning across Doppler-broadened lines of the $\tilde{A} - \tilde{X}$ (1–0) band, monitoring the saturation recovery kinetics after a tunable dye laser pulse selectively bleached the probed CN rotational level. The refilling kinetics of a hole in an otherwise thermalized state distribution is identical to the hypothetical decay of the corresponding isolated level in an empty manifold, even to the extent of displaying hyperfine quantum beats in the hole alignment. The Doppler-resolved kinetics reflect a competition between the speed-dependent rotational energy transfer rates, which tend to cool the velocity distribution of the hole, and velocity-changing collisions, which tend to rethermalize the velocity distribution. The observations are of relevance to speed-dependent effects in pressure broadening, but measured under Doppler-limited pressure conditions. Elastic depolarization contributes significantly to the observed signals at low rotational states, negligibly so at high J. A strongly J-dependent contribution to the relaxation kinetics due to small amounts of the photolytic precursor cannot be neglected when extracting the rare-gas-dependent rate coefficients from the observed kinetics. Some qualitative differences are thereby found with previously published energy transfer studies on He or Ar + CN ($\tilde{X}$, v=2) by Fei, et al. [*J. Chem. Phys.* **100**, 1190 (1994); *Chem. Phys. Lett.* **232**, 547 (1995)].

Acknowledgments: Work at Brookhaven National Laboratory was carried out under Contract No. DE-AC02-98CH10886 with the U.S. Department of Energy and supported by its Office of Basic Energy Sciences, Division of Chemical Sciences, Geosciences and Biosciences.

---

[a]and Department of Chemistry, Stony Brook University, Stony Brook, NY, USA

WH11

THE VIBRATIONALLY DRIVEN H-ATOM ABSTRACTION FROM METHANE BY BROMINE RADICALS

ETHAN VOLPA, *Department of Chemistry, University of Wisconsin–Madison, Madison, WI, USA*; ANDREW BERKE, *Department of Chemistry, Indiana University, Bloomington, IN, USA*; FLEMING CRIM, *Department of Chemistry, The Univeristy of Wisconsin, Madison, WI, USA.*

In an effort to understand how the Polanyi rules can be extended from radical-diatomic molecule reactions to radical-polyatomic molecule reactions, members of the bimolecular gas-phase dynamics community have often used H-atom abstraction from methane (or one of its isotopologues) by X (where X=H, F or Cl) as a model system. Each of these model reactions can be separately characterized by both the height and location of the potential energy barrier along the reaction coordinate. Currently, we are working to understand the gas-phase dynamics of H-atom abstraction from CH_4 when X=Br. In this iteration of the model system, the abstraction barrier is located very late along the reaction coordinate and is quite high by comparison to other studied systems. This leads to some surprising dynamical effects in the X=Br system that we have not seen in other systems studied thus far.

WH12

STUDYING OZONOLYSIS REACTIONS OF 2-BUTENES USING CAVITY RING-DOWN SPECTROSCOPY

LIMING WANG, *School of Chemistry and Chemical Engineering, South China University of Technology, Guangzhou, China*; YINGDI LIU, *Jet Propulsion Laboratory, California Institute of Technology, Pasadena, CA, USA*; MIXTLI CAMPOS-PINEDA, CHAD PRIEST, *Department of Chemistry, University of California, Riverside, CA, USA*; JINGSONG ZHANG, *Department of Chemistry and Air Pollution Research Center, University of California, Riverside, CA, USA.*

Ozonolysis reactions of alkenes are important oxidation pathways of alkenes in the atmosphere, and they are also significant sources of tropospheric hydroxyl radicals. In this work, ozone reactions with *trans-* and *cis*-2-butene are studied using cavity ring-down spectroscopy (CRDS). Vinoxy (CH_2CHO) radical, a proposed co-product of OH from dissociation of Criegee intermediates following the primary ozonolysis of 2-butenes, is directly observed. The vinoxy formation is found to decrease with increasing pressure. Formaldehyde (HCHO), a side-product in the ozonolysis reactions, is also monitored. Chemical kinetic modeling has been performed to illustrate the reaction mechanisms and to quantify the reactive intermediate and product yields.

WH13

COLLISIONALLY-MEDIATED SINGLET-TRIPLET CROSSING IN $\tilde{a}^1A_1$ CH$_2$ REVISITED: (010) COUPLING

ANH T. LE, GREGORY HALL, TREVOR SEARS[a], *Department of Chemistry, Brookhaven National Laboratory, Upton, NY, USA.*

Methylene, CH$_2$, possesses a ground $\tilde{X}^3B_1$ ground electronic state and an excited $\tilde{a}^1A_1$ state only 3150cm^{-1} higher in energy. The collision-induced singlet-triplet crossing in the gaseous mixtures is important in determining overall reaction rates and chemical behavior. Accidental near-degeneracies between rotational levels of the singlet state and the vibrationally excited triplet state result in a few gateway rotational levels that mediate collision-induced intersystem crossing. The mixed states can be recognized and quantified by deperturbation, knowing the zero-order singlet and triplet energy levels. Hyperfine structure can be used as alternative indicator of singlet-triplet mixing. Non-zero mixing will induce hyperfine splittings intermediate between the unresolved hyperfine structure of pure singlet and the resolvable ($\approx$50MHz) splittings of pure triplet, arising from the ($\mathbf{I}\cdot\mathbf{S}$) interaction in the ortho states, where nuclear spin I=1[b]. Collision-induced intersystem crossing rates from the (010) state are comparable to those for (000)[c], yet the identities and characters of the presumed gateway states are unknown. A new spectrometer is under construction to investigate triplet mixing rotational levels of $\tilde{a}^1A_1$(010) by sub-Doppler measurements of perturbation-induced hyperfine splittings. Their observation will permit the identification of gateway states and quantification of the degree of triplet contamination of the singlet wavefunction. Progress in the measurements and the analysis of rotational energy transfer in (010) will be reported.

Acknowledgments: Work at Brookhaven National Laboratory was carried out under Contract No. DE-AC02-98CH10886 with the U.S. Department of Energy and supported by its Office of Basic Energy Sciences, Division of Chemical Sciences, Geosciences and Biosciences.

---

[a]also: *Department of Chemistry, Stony Brook University, Stony Brook, New York 11794*
[b]C.-H. Chang, G. E. Hall, T. J. Sears, *J. Chem. Phys* **133**, 144310(2010)
[c]G. E. Hall, A. V. Komissarov, and T. J. Sears, *J. Phys. Chem. A* **108** 7922-7927 (2004)

WH14

THEORETICAL INVESTIGATION OF THE UV-VIS PHOTODISSOCIATION DYNAMICS OF Ar$_n$(BRCN$^-$)

BERNICE OPOKU-AGYEMAN, *Department of Chemistry and Biochemistry, The Ohio State University, Columbus, OH, USA*; JULIA H LEHMAN, AMANDA CASE, CARL LINEBERGER, *Department of Chemistry and Biochemistry, JILA - University of Colorado, Boulder, CO, USA*; ANNE B McCOY, *Department of Chemistry and Biochemistry, The Ohio State University, Columbus, OH, USA.*

We present the results of quantum dynamics studies of photodissociation of BrCN$^-$ following electronic excitation to states that dissociate to Br$^-$ + CN and Br* + CN$^-$. The electronic structure of BrCN$^-$ was evaluated with MRCI-SO/aug-cc-pVTZ at a fixed CN distance of 1.18 Å. The calculations were used to evaluate the two-dimensional potential surfaces for the ground and excited states as functions of the Br-CN(center of mass) distances, R, and angles, θ, between CN and R. A diabatic model developed for the two relevant excited states shows a dramatic change in the electronic character of the states near the BrCN$^-$ geometry when $\theta \leq \pi/2$. The quantum dynamics studies on the bare BrCN$^-$ were carried out by exciting wave packets of six vibrational states of BrCN$^-$ that are thermally populated at 150K onto each of the two excited states. Upon excitations of the wave packets onto the state accessible in the visible region, 51% Br$^-$ and 49% CN$^-$ photoproducts were calculated. Similar calculations in the UV region yielded 56% Br$^-$ and 44% CN$^-$ photoproducts. Experimentally, when BrCN$^-$ is excited in the visible region, 60% Br$^-$ and 40% CN$^-$ photoproducts are obtained while 80% Br$^-$ and 20% CN$^-$ yields were obtained in the UV region. Further dynamics studies of BrCN$^-$ solvated in argon will be carried out with varying Ar$_n$(BrCN$^-$) cluster sizes.

WI. Comparing theory and experiment

Wednesday, June 18, 2014 – 1:30 PM

Room: 112 Chemistry Annex

Chair: Richard Dawes, University of Missouri - Rolla, Rolla, MO, USA

WI01   1:30 – 1:45

ROVIBRATIONAL STATES OF HBF^+ AND HCO^+ ISOTOPOLOGUES UP TO HIGH J: THEORY AND EXPERIMENT

PETER BOTSCHWINA, PETER SEBALD, BENJAMIN SCHRÖDER, *Institute of Physical Chemistry, Georg-August-Universität Göttingen, Göttingen, Germany*; KENTAROU KAWAGUCHI, *Graduate School of Natural Science and Technology , Okayama University, Okayama, Japan*; TAKAYOSHI AMANO, *Department of Chemistry, University of Waterloo, Waterloo, ON, Canada.*

Near-equilibrium potential energy surfaces for HBF^+ and HCO^+, obtained from high-level calculations beyond fc-CCSD(T), are employed in variational calculations for many rovibrational states of various isotopologues. Calculated effective spectroscopic parameters are in excellent agreement with available experimental data and many predictions are being made, also for line intensities of HBF^+ and HCO^+ isotopologues. Combining a difference frequency system with glow discharge and a discharge modulation scheme, six and seven lines of the ν_1 bands for $H^{11}BF^+$ and $H^{10}BF^+$, respectively, were observed. Together with data obtained from microwave spectroscopy, the spectroscopic constants of the ν_1 states could be derived through least-squares fitting.

WI02   1:47 – 2:02

A COMPARISON OF THE METHODS OF STUDYING THE SPECTRA OF THE AsH_2 RADICAL

GEOFFREY DUXBURY, *Department of Physics, University of Strathclyde, Glasgow, United Kingdom*; ALEXANDER ALIJAH, *GSMA - Champagne Ardennne, Université de Reims, Reims Cedex 2, France.*

The first studies of the $^2A_1$-$^2B_1$ electronic band system of the AsH_2 and AsD_2 radicals were made at Sheffield University in the period from 1966 to 1968 by Dixon, Duxbury and Lamberton using flash photolysis of arsine and deuterated arsine. The bands have a complex rotational structure associated with that of an asymmetric rotor. Band centres of the 0,v_2,0-0,0,0 progression were identified for v_2'=0 tp v_2'=5, although only the structure of the bands from v_2'=1 to 3 was analysed in detail. After a long time interval in 1986 a low resolution emission spectrum of AsH_2 was recorded by NI et al. However, it was not until 2007 that He and Clouthier studied the electronic transition of jet-cooled AsH_2 using laser induced fluorescence and wavelength-resolved emission. Following on from this in 2009 Zhao and colleagues recorded absorption spectra of the AsH_2 radical by cavity ringdown spectroscopy. Finally in 2012 Grimminger and Clouthier recorded the equivalent transitions in AsD_2 and AsHD. They also carried out ab initio calculations. By comparing the recent spectroscopic results with those of Dixon et al, we wish to show the complementarity of the different methods for understanding the behaviour of AsH_2 and AsD_2 radicals.

EFFECTS OF SPIN-ORBIT COUPLING ON THE SPIN-ROTATION INTERACTION IN THE AsH$_2$ RADICAL

GEOFFREY DUXBURY, *Department of Physics, University of Strathclyde, Glasgow, United Kingdom*; ALEXANDER ALIJAH, *GSMA - Champagne Ardennne, Université de Reims, Reims Cedex 2, France.*

The occurence of predissociation in the electronic spectrum of AsH$_2$ is very dependent upon the magnitude of the spin-orbit coupling parameter of the central atom. Making use of Table 5.6 in "The Spectra and Dynamics of Diatomic Molecules, ELSEVIER" by H. Lefebvre-Brion and R.W. Field, it is possible to appreciate the rapid rate of increase of the spin-orbit constants associated with the heavy central atom in the di-hydrides NH$_2$, PH$_2$ and AsH$_2$. The spin-orbit constants range from 42.7 cm^{-1} for NH$_2$, to 191.3 cm^{-1} for PH$_2$, and 1178 cm^{-1} for AsH$_2$.

The effects of spin-orbit coupling may be seen in a plot of the separation of the central $^{R}Q_{0,9}$ and $^{P}Q_{1,N}$ sub-bands as the value of v$_2$' increases from 0 to 5. As the value of v$_2$' increases beyond 2 the spectrum becomes more and more fuzzy as the effects of predissociation become more obvious. This means that unlike the example of the behaviour of PH$_2$, where the vibronic level pattern can be followed below and above the barrier to linearity, in AsH$_2$ and AsD$_2$ the absorption spectrum becomes completely diffuse below the barrier to linearity in the A ^{2}A$_1$ state. The change in the magnitude of the doublet splittings as v$_2$' increases may be seen in the plots of the doublet splittings showing the spin-uncoupling as a result of the increase of overall rotation. In the absorption spectrum of SbH$_2$, recorded in 1967 by T. Barrow in the Chemistry Department at Sheffield University, all the absorption features showed the effects of predissociation, consistent with a spin-orbit constant of 2834 cm^{-1} for the central atom of SbH$_2$.

AB INITIO STUDY OF ION-PAIR STATES OF THE IODINE MOLECULE

VADIM A ALEKSEEV, *Department of Physics, Saint-Petersburg State University, St. Petersburg, Russia.*

Ion-pair states of the I$_2$ molecule have been the subject of many experimentals studies and to date all 18 states correlating with I$^+$(^{3}P$_{J=2,1,0}$, ^{1}D$_2$) + I$^-$(^{1}S$_0$) asymptotes are known from experiment. This contribution reports on *ab initio* study of the I$_2$ molecule with an emphasis on the ion-pair states. As an illustration, Figure shows experimental and calculated potentials of the ion-pair states correlating with I$^+$(^{3}P$_2$) + I$^-$(^{1}S$_0$) (energy is relative to I (^{2}P$_{3/2}$) + I (^{2}P$_{3/2}$) asymptote).

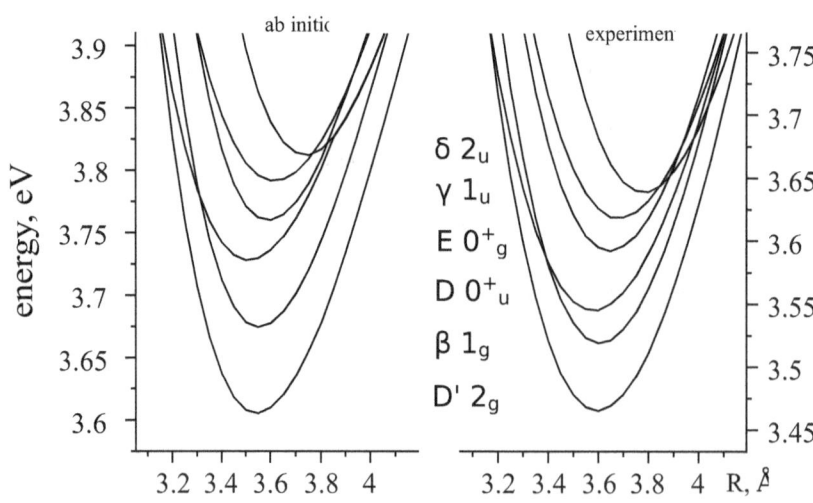

WI05   2:38 – 2:53

EXPERIMENTAL AND THEORETICAL STUDIES ON THE ELECTRONIC ABSORPTION SPECTRA OF QUINOLINE CARBOXALDEHYDES

MUSTAFA KUMRU, MUSTAFA KOCADEMIR, HAIDAR M ALFANDA, *Department of Physics, Fatih University, Istanbul, Turkey.*

We have investigated electronic spectra of some quinoline carboxaldehydes dissolved in water by UV-vis measurements in the range of 190-1100 nm and by theoretical calculations with density functional theory (DFT). The geometrical structures of the quinoline carboxaldehydes have been obtained at the B3LYP/6-311++G(d,p) level, while the electronic absorption spectra have been simulated in water by using time-dependent DFT at the same level. Theoretical and experimental spectra agree to each other very well.

Keywords: Electronic spectra, Quinoline carboxaldehydes, DFT

[1] V. Kucuk, A. Altun, M. Kumru, Spectrochim. Acta Part A 85 (2012)9298.
[2] M. Kumru, V. Kucuk, T. Bardakci, Spectrochim. Acta Part A 90(2012)2834.
[3] M. Kumru, V. Kucuk, M. Kocademir, Spectrochim. Acta Part A, 96 (2012) 242251.
[4] M. Kumru, V. Kucuk, P. Akyürek, Spectrochim. Acta Part A, 113 (2013) 72–79.

WI06   2:55 – 3:10

POLARIZED MATRIX INFRARED SPECTRA OF CYCLOPENTADIENONE - AN IMPORTANT REACTIVE INTERMEDIATE IN COMBUSTION AND BIOMASS PYROLYSIS

THOMAS ORMOND, BARNEY ELLISON, *Department of Chemistry and Biochemistry, University of Colorado, Boulder, CO, USA*; JOHN F. STANTON, *Department of Chemistry, The University of Texas, Austin, TX, USA.*

A detailed vibrational analysis of the infrared spectra of cyclopentadienone ($C_5H_4=O$ and $C_5D_4=O$) in rare gas matrices has been carried out. Ab initio coupled-cluster anharmonic force field calculations were used to guide the assignments. Flash pyrolysis of o-phenylene sulfite ($C_6H_4O_2SO$ and $C_6D_4O_2SO$) was used to provide a molecular beam of cyclopentadienone entrained in the rare gas carrier. The beam was interrogated with time-of-flight photoionization mass spectrometry (TOF-PIMS), confirming the clean, intense production of $C_5H_4=O$. Matrix isolation infrared spectroscopy was coupled with 355 nm polarized UV for photo-orientation and linear dichroism experiments to determine the symmetries of the vibrations.

Intermission

WI07   3:27 – 3:42

CONSISTENT ASSIGNMENT OF THE VIBRATIONS OF MONOHALOSUBSTITUTED BENZENES

JOE HARRIS, ANNA ANDREJEVA, WILLIAM DUNCAN TUTTLE, *School of Chemistry, University of Nottingham, Nottingham, United Kingdom*; IGOR PUGLIESI, CHRISTIAN SCHRIEVER, *Fakultät für Physik, Ludwig-Maximilians-Universität, München, Deutschland*; TIM WRIGHT, *School of Chemistry, University of Nottingham, Nottingham, United Kingdom.*

When substituted benzenes become a focus of a spectroscopic study there are various well known vibrational labelling schemes present,[a,b] however it was shown in recent works the description of monohalobenzene vibrations in terms of benzene modes (ie. Wilson notation) is questionable in some cases.[c,d] A new scheme is presented which uses the motions of monofluorobenzene vibrations as a basis for labelling vibrational assignments of monosubstituted benzenes.[d] The scheme has been successfully applied to the ground and excited states of toluene and its deuterated-methyl group isotopologue.[e,f] Here we present the application of the scheme to fluorobenzene and its fully deuterated analogue. One-colour resonance-enhanced multiphoton ionization (REMPI) spectroscopy was employed in order to characterise the fluorobenzene and fluorobenzene-d_5 excited state.

[a]E. B. Wilson Jr., Phys. Rev., 45, 706 (1934)
[b]G .Varsanyi, Assignments of the Vibrational Spectra of Seven Hundred Benzene Derivatives,Wiley, New York, 1974, Vol. I and II
[c]I. Pugliesi, N. C. Tonge and M. C. R. Cockett, J. Chem. Phys., 129, 104303 (2008)
[d]A. M. Gardner and T. G. Wright, J. Chem. Phys., 135,114305 (2011)
[e]A. M. Gardner, A. M. Green, V. M. Tame-Reyes, V. H. K. Wilton and T. G. Wright, 138, 134303 (2013)
[f]A. M. Gardner, A. M. Green, V. M. Tame-Reyes, K. L. Reid, J. A. Davies, V. H. K. Wilton and T. G. Wright, manuscript accepted

WI08 3:44 – 3:54

CONSISTENT ASSIGNMENT OF THE VIBRATIONS OF CHLORO- AND BROMOBENZENE MOLECULES AND THEIR DEUTERATED ANALOGUES

ANNA ANDREJEVA, JOE HARRIS, WILLIAM DUNCAN TUTTLE, TIM WRIGHT, *School of Chemistry, University of Nottingham, Nottingham, United Kingdom.*

The consistency of the labelling and assignments of the vibrational frequencies of the chloro- and bromo- monosubstituted benzene molecules is investigated in their first electronically excited states (S_1). The assignments given utilise a recent nomenclature[a] discussed in a previous talk, allowing the ring-localised vibrations to be compared straightforwardly across different monohalosubstituted benzenes. For the S_1 state, one-colour resonance-enhanced multiphoton ionization (REMPI) spectroscopy was employed. The assignments of the frequencies include previous work but also the calculated wavenumbers for both fully hydrogenated monohalosubstituted benzenes ($-h_5$) and the deuterated isotopologues ($-d_5$) employing time-dependent density functional theory (TDDFT).

---

[a] A. M. Gardner and T. G. Wright, J. Chem. Phys., 135, 114305 (2011)

WI09 3:56 – 4:11

ACCURATE ANHARMONIC IR SPECTRA FROM INTEGRATED CC/DFT APPROACH

VINCENZO BARONE, *Scuola Normale Superiore, Scuola Normale Superiore, Pisa, Italy*; MALGORZATA BICZYSKO, JULIEN BLOINO, *Istituto di Chimica dei Composti OrganoMetallici (ICCOM-CNR), UOS di Pisa, Consiglio Nazionale delle Ricerche, Pisa, Italy*; IVAN CARNIMEO, *Scuola Normale Superiore, Scuola Normale Superiore, Pisa, Italy*; CRISTINA PUZZARINI, *Dep. Chemistry 'Giacomo Ciamician', University of Bologna, Bologna, Italy.*

The recent implementation of the computation of infrared (IR) intensities beyond the double harmonic approximation [1] paved the route to routine calculations of infrared spectra for a wide set of molecular systems. Contrary to common beliefs, second-order perturbation theory is able to deliver results of high accuracy provided that anharmonic resonances are properly managed [1,2].

It has been already shown for several small closed- and open shell molecular systems that the differences between coupled cluster (CC) and DFT anharmonic wavenumbers are mainly due to the harmonic terms, paving the route to introduce effective yet accurate hybrid CC/DFT schemes [2]. In this work we present that hybrid CC/DFT models can be applied also to the IR intensities leading to the simulation of highly accurate fully anharmonic IR spectra for medium-size molecules, including ones of atmospheric interest, showing in all cases good agreement with experiment even in the spectral ranges where non-fundamental transitions are predominant[3].

[1] J. Bloino and V. Barone, J. Chem. Phys. 136, 124108 (2012) [2] V. Barone, M. Biczysko, J. Bloino, Phys. Chem. Chem. Phys., 16, 1759-1787 (2014) [3] I. Carnimeo, C. Puzzarini, N. Tasinato, P. Stoppa, A. P. Charmet, M. Biczysko, C. Cappelli and V. Barone, J. Chem. Phys., 139, 074310 (2013)

WI10

A FIRST-PRINCIPLES MODEL OF FERMI RESONANCE IN THE ALKYL CH STRETCH REGION: APPLICATION TO HYDRONAPHTHALENES, INDANES, AND CYCLOHEXANE

EDWIN SIBERT, *Department of Chemistry, The Univeristy of Wisconsin, Madison, WI, USA*; NATHANAEL KIDWELL, TIMOTHY S. ZWIER, *Department of Chemistry, Purdue University, West Lafayette, IN, USA.*

The infrared (IR) spectroscopy of the alkyl CH stretch region (2750-3000 cm^{-1}) of a series of bicyclic hydrocarbons and free radicals has been studied under supersonic expansion cooling in the gas phase, and compared with a theoretical model that describes the local mode stretch-bend Fermi resonance interactions. The double resonance method of fluorescence-dip infrared (FDIR) spectroscopy was used on the stable molecules 1,2-dihydronaphthalene, 1,4-dihydronaphthalene, tetralin, indene, and indane using the S_0-S_1 origin transition as a monitor of transitions. Resonant ion-dip infrared (RIDIR) spectra were recorded for the trihydronaphthyl (THN) and inden-2-yl methyl (I2M) radicals. The previously developed model Hamiltonian [J. Chem. Phys. **138** 064308 (2013)] incorporates cubic stretch-bend coupling with parameters obtained from density functional theory methods. Full dimensional calculations are compared to reduced dimensional Hamiltonian results in which anharmonic CH streches and CH$_2$ scissor modes are Fermi coupled. Excellent agreement between theoretical results is found. Scale factors of select terms in the reduced dimensional Hamiltonian, obtained by fitting the theoretical Hamiltonian predictions to the experimental spectra, are found to be similar to previous work. The resulting Hamiltonian predicts successfully all the major spectral features considered in this study. A simplified model is introduced in which the CH$_2$ groups are decoupled. This model enables the assignment of many of the spectral features. The model results are extended to describe the CH stretch spectrum of the chair and twist-boat conformers of cyclohexane. The chair conformer is used to illustrate the shortcomings of the CH$_2$ coupling model.

WI11

THE ROLE OF ELECTRICAL ANHARMONICITY IN DETERMINING INTENSITY IN THE 2100 cm^{-1} REGION OF THE WATER SPECTURM

ANNE B McCOY, *Department of Chemistry and Biochemistry, The Ohio State University, Columbus, OH, USA.*

The origins of the intensity of the feature in the spectrum of liquid water near 2100 cm^{-1} are investigated through calculations of the spectra of water clusters. The spectra of thirteen water clusters with six or fewer water molecules are calculated using low-order expansions of the potential and dipole surfaces in internal and normal mode coordinates. The results are dissected to interpret the origins of the intensity in 2100 cm^{-1} region of the spectra. It is found that the intensity near 2100 cm^{-1} can be attributed to combination bands involving the HOH bend and intermolecular vibrations that break the hydrogen bonding network. Based on the robustness of a feature near 2100 cm^{-1} in all of the water clusters studied, it is inferred that a similar assignment can be made for the spectrum of liquid water. When the origins of the intensity near 2100 cm^{-1} is analyzed from calculations that used internal coordinates to expand the potential and dipole surfaces, the leading contribution to the intensity is attributed to large second derivatives of the dipole moment with respect to the internal coordinates that are excited, or electrical anharmonicity. This picture changes if the derivatives of the potential and dipole surfaces are taken with respect to normal modes. In this representation, the second derivatives of the dipole moment are often vanishingly small, while the mixed third and fourth derivatives of the potential become quite large. Based on this result, mechanical anharmonicity appears to be responsible for the intensity in the 2100 cm^{-1} region. This strong dependence of the interpretation of the origins of the intensity in the 2100 cm^{-1} region of the water spectrum is also investigated and discussed.

WI12 4:47 – 4:57

A PADE APPROXIMATION OF THE TRANSFORMED TRANSITION MOMENT OPERATOR OF THE ν_2 BAND OF WATER VAPOR

OLEG V. EGOROV, *Quantum Electronics and Photonics, Tomsk State University, Tomsk, Russia.*

Centrifugal and other intramolecular interactions have a significant influence on the line strength of the light asymmetric rotors with C_{2v} symmetry group. A previously proposed method[a], based on the expansion of the transformed dipole moment with the terms up to degree 2 in the total angular momentum J, is widely used now in the high resolution molecular spectroscopy of water vapor. As the perturbation theory in the mentioned method is performed by Taylor series the best fitting is occurred within the certain ranges of quantum numbers of spectral lines.

For modern experimental spectra of water vapor, including high-temperature ones, many spectral lines with various values of quantum numbers are identified. In this case the direct using of the scheme[a] doesn't lead to satisfaction describing of intramolecular effects in line strengths of the ν_2 band of water vapor[b]. A semiemperical obtained Pade approximation of the transformed transition moment operator of B-type bands has been applied by us for treating of the recent experimental spectrum of water vapor in the ν_2 band[c]. Finally, the considerable improvement has been found: the value of the RMS deviation for the spectral lines with $\Delta K = +\text{-}1$ can be decreased from 8.94 +-0.42% to 2.65+-0.12%.

---

[a]Mol.Phys., 1976, Vol. 32, No. 2, 523-537.
[b]J.Opt.Soc.Am.B, 1991, Vol. 8, No. 11, 2236-2255.
[c]J.Mol.Spectrosc., 2011, Vol. 265, 59-68.

WI13 4:59 – 5:14

VIBRATIONAL SPECTROSCOPY AND GAS-PHASE THERMOCHEMISTRY OF THE MODEL DIPEPTIDE N-ACETYL GLYCINE METHYL AMIDE

CHRISTOPHER LEAVITT, *Department of Chemistry, University of Georgia, Athens, GA, USA*; PAUL RASTON, *School of Chemistry and Physics, The University of Adelaide, Adelaide, South Australia, Australia*; GRANT MOODY, CAITLYNE SHIRLEY, GARY DOUBERLY, *Department of Chemistry, University of Georgia, Athens, GA, USA.*

The structure-function relationship in proteins is widely recognized, motivating numerous investigations of isolated neutral and ionic polypeptides that generally employ conformation specific, multidimensional UV and IR spectroscopies. This data taken in conjunction with computed harmonic frequencies has provided a snapshot of the underlying molecular physics at play in many polypeptides, but few experiments have been able to probe the energetics of these systems. In this study, we use vibrational spectroscopy to measure the gas-phase enthalpy change for isomerization between two conformations of the dipeptide N-acetyl glycine methyl amide (NAGMA). A two-stage oven source is implemented producing a gas-phase equilibrium distribution of NAGMA molecules that is flash frozen upon pickup by He nanodroplets. Using polarization spectroscopy, the IR spectrum is assigned to a mixture of two conformers having intramolecular hydrogen bonds made up of either five- or seven-membered rings, C_5 and C_7, respectively. The interconversion enthalpy, obtained from the van't Hoff relation, is 4.52 ± 0.12 kJ/mol for isomerization from the C_7 to the C_5-conformer. This experimental measurement is compared to computations employing a broad range of theoretical methods.

EXPERIMENTAL AND THEORETICAL STUDIES OF THE PURE ROTATIONAL SPECTRA OF LEAD HALIDES: PbF AND PbCl

SPENCER NORMAN, RICHARD DAWES, G. S. GRUBBS II, *Department of Chemistry, Missouri University of Science and Technology, Rolla, MO, USA*; S. A. COOKE, *Natural and Social Science, Purchase College SUNY, Purchase, NY, USA*; B. E. LONG, *Department of Chemistry, Wesleyan University, Middletown, CT, USA*; CHRIS DEWBERRY, *Chemistry Department, University of Minnesota, Minneapolis, MN, USA*.

The pure rotational spectrum of lead monochloride, PbCl, has been measured and analyzed using chirped pulse and cavity Fourier transform microwave (CP-FTMW and FTMW) spectrometers equipped with an ablation source. Refined parameters of an effective Hamiltonian including fine and hyperfine interactions similar to those previously reported by Fink et al. [1] were determined. Dynamically-weighted, explicitly-correlated MRCI-F12 calculations [2] were performed for both PbF and the valence isoelectronic PbCl to predict potential energy curves (PEC). Spin-orbit coupling was included in the calculations, which is known to split the $X_1\,^2\Pi_{1/2}$ and $X_2\,^2\Pi_{3/2}$ components of the ground electronic state by roughly 8280 cm^{-1} in both lead halide systems. Calculated rotational levels were obtained using the PECs and compared with experiment including previously published results for PbF [3].

References: 1- K. Ziebarth, K. D. Setzer, O. Shestakov,1 and E. H. Fink, J. Mol. Spec. 191, 108 (1998). 2- B. J. Barker et al. J. Chem. Phys. 137, 214313 (2012). 3- R. J. Mawhorter et al. Phys. Rev. A 84, 022508 (2011).

WJ. Structure determination
Wednesday, June 18, 2014 – 1:30 PM
Room: 274 Medical Sciences Building

Chair: Isabel Peña, Universidad de Valladolid, Valladolid, Spain

---

WJ01   1:30 – 1:45

MILLIMETER-WAVE STUDIES OF THE ISOTOPOLOGUES OF IZnCH$_3$(X^{1}A$_1$): GEOMETRIC PARAMETERS AND EVIDENCE FOR ZINC INSERTION

MATTHEW BUCCHINO, *Department of Chemistry and Astronomy, University of Arizona, Tuscon, AZ, USA*; JUSTIN YOUNG, PHILLIP SHERIDAN, *Department of Chemistry and Biochemistry, Canisius College, Buffalo, NY, USA*; LUCY ZIURYS, *Department of Astronomy, University of Arizona, Tucson, AZ, USA*.

The laboratory detection of gas-phase IZnCH$_3$ (X^{1}A$_1$), using millimeter-wave direct absorption methods, was reported previously. This work has been extended by the measurement of the pure rotational spectrum of several isotopolgues: I^{64}ZnCH$_3$, I^{66}ZnCH$_3$, I^{64}ZnCD$_3$, and I^{64}Zn^{13}CH$_3$. These species were all created by the reaction of zinc vapor with CH$_3$I, CD$_3$I, or ^{13}CH$_3$I in the presence of a DC discharge. The zinc isotopolgues were observed in natural abundance. Rotational transitions in the range 256–293 GHz (J = 109 ← 108 to J = 132 ← 131, for K = 0 to 6) have been recorded for each species. From these measurements, an r$_0$ structure has been determined. This structure was found to be in good agreement with previous DFT calculations. Interestingly, the 110.2° Zn – C – H bond angle of IZnCH$_3$ is identical to that of the hydrogen substituted zinc insertion complex HZnCH$_3$ (X^{1}A$_1$). These data are further evidence that IZnCH$_3$ is not created by the generation of free radical fragments, but by the direct insertion of atomic zinc into the C – I bond of iodomethane.

WJ02   1:47 – 2:02

LABORATORY DETECTION OF ClZnCH$_3$ (X^{1}A$_1$): FURTHER EVIDENCE FOR ZINC INSERTION

MATTHEW BUCCHINO, *Department of Chemistry and Astronomy, University of Arizona, Tuscon, AZ, USA*; LUCY ZIURYS, *Department of Astronomy, University of Arizona, Tucson, AZ, USA*.

The pure rotational spectrum of methylzinc chloride, ClZnCH$_3$ (X^{1}A$_1$), has been recorded in the gas phase using direct absorption spectroscopic techniques. ClZnCH$_3$ was synthesized by the reaction of zinc vapor, generated in a Broida-type oven, with chloromethane in the presence of a DC discharge. Rotational transitions of the main isotopologue, ^{35}Cl^{64}ZnCH$_3$, were measured in the frequency range of 260–297 GHz. The presence of clear K-ladder structure (K = 0–6) indicates that the species is a symmetric top with C$_{3V}$ symmetry. ClZnCH$_3$ appears to be formed by the oxidative addition of atomic zinc to chloromethane. Searches for the ^{37}Cl, ^{66}Zn, ^{13}C, and D substituted isotopologues are currently in progress. The geometries of ClZnCH$_3$ and IZnCH$_3$ will be compared with reactivity trends for halogen-substituted organometallic reagents.

WJ03   2:04 – 2:19

THE SIMPLEST CRIEGEE INTERMEDIATE (H$_2$C = O–O): EQUILIBRIUM STRUCTURE AND POSSIBLE FORMATION FROM ATMOSPHERIC LIGHTNING

MICHAEL C McCARTHY, *Atomic and Molecular Physics, Harvard-Smithsonian Center for Astrophysics, Cambridge, MA, USA*; LAN CHENG, *Department of Chemistry, The University of Texas, Austin, TX, USA*; KYLE N CRABTREE, OSCAR MARTINEZ JR., *Atomic and Molecular Physics, Harvard-Smithsonian Center for Astrophysics, Cambridge, MA, USA*; THANH LAM NGUYEN, *Department of Chemistry, The University of Texas, Austin, TX, USA*; CARRIE WOMACK, *Department of Chemistry, MIT, Cambridge, MA, USA*; JOHN F. STANTON, *Department of Chemistry, The University of Texas, Austin, TX, USA*.

Fourier transform microwave spectroscopy in combination with double-resonance techniques has been used to detect the rotational spectra of all five singly-substituted isotopic species of H$_2$C = O–O, the simplest Criegee intermediate. By correcting the rotational constants of these species and those of four others previously reported by Nakajima and Endo (*J. Chem. Phys.* **39**, 101103, 2013) for zero-point vibrational motion calculated theoretically, a highly precise equilibrium structure is reported for this important atmospheric intermediate. In contrast to the production method employed by most other groups, which has emphasized the use of halogenated percursors, we find that H$_2$C = O–O is produced in good yield and fairly selectively by passing a mixture of methane and excess molecular oxygen through an electrical discharge. For this reason H$_2$C = O–O may be produced in the direct vicinity of atmospheric lightning.

WJ04   2:21 – 2:36

MILLIMETER AND SUBMILLIMETER SPECTROSCOPIC STUDIES OF HO_3

LUYAO ZOU, SUSANNA L. WIDICUS WEAVER, *Department of Chemistry, Emory University, Atlanta, GA, USA.*

HO_3 is a radical species of atmospheric and astrophysical chemical importance. While the microwave spectrum of this radical is known, its rotational spectrum has not been measured at higher frequencies. Studies of HO_3 in the millimeter and submillimeter ranges would provide the information needed to further refine its structural determination, as well as provide additional spectroscopic guidance for observational searches. In order to study the rotational spectrum of HO_3 at higher frequencies, we have coupled a pulsed supersonic expansion discharge source with a multipass direct absorption millimeter and submillimeter spectrometer. Initial experiments focused on the optimization of the HO_2 radical, which has a known rotational spectrum and is the intermediate species in the production of HO_3. Searches for lines of HO_3 are currently underway. The spectrometer design, results of the HO_2 studies, and initial results for HO_3 will be presented.

WJ05   2:38 – 2:53

MILLIMETER-WAVE SPECTROSCOPY OF HYDRAZOIC ACID (HN_3)

BRENT K. AMBERGER, BRIAN J. ESSELMAN, R. CLAUDE WOODS, ROBERT J. McMAHON, *Department of Chemistry, The Univeristy of Wisconsin, Madison, WI, USA.*

The rotational spectra for hydrazoic acid (HN_3), its isotopologues, and its vibrational satellites have been reexamined using millimeter-wave rotational spectroscopy in the range of 240-360 GHz. Treating sodium azide (NaN_3) or the commercially available singly ^{15}N-labeled NaN_3 with phosphoric acid or deuterated phosphoric acid yielded 6 different isotopologues. From these samples, we were also able to observe all of the isotopologues containing one additional ^{15}N at natural abundance. In total, we assigned rotational transitions to 14 different species; only $H^{15}N_3$ and $D^{15}N_3$ were not accessible. With the large number of rotational constants determined for these isotopologues, an excellent equilibrium structure determination was performed with CFOUR's xrefit routine. This structure shows a bent azide sub-unit, and is in excellent agreement with the geometry optimization performed at the CCSD(T)/ANO2 level of theory. The Coriolis perturbation of the ground and first two vibrationally excited states of HN_3 will also be discussed.

WJ06   2:55 – 3:10

PHOSPHORUS AND SILICON ANALOGS OF ISOCYANIC ACID:
FOURIER-TRANSFORM MICROWAVE SPECTROSCOPY OF HPCO AND HNSiO

SVEN THORWIRTH, *I. Physikalisches Institut, Universität zu Köln, Köln, Germany;* VALERIO LATTANZI, *Dep. Chemistry 'Giacomo Ciamician', University of Bologna, Bologna, Italy;* MICHAEL C McCARTHY, *Atomic and Molecular Physics, Harvard-Smithsonian Center for Astrophysics, Cambridge, MA, USA.*

By means of Fourier transform microwave spectroscopy of a supersonic jet, the pure rotational spectra of two second-row analogs to isocyanic acid, HNCO, have been observed for the first time. The phosphorus and silicon analogs HPCO and HNSiO, respectively, were observed by their fundamental a-type rotational transitions ($\mu_{a(HPCO)}$= 0.45 D, $\mu_{a(HNSiO)}$= 2.10 D) in the centimeter wave range from 10 to 32 GHz through discharges of appropriate precursor gases highly diluted in neon. Spectroscopic searches and identification were based on predictions from high-level quantum-chemical calculations at the CCSD(T) level of theory in combination with large basis sets. Excellent agreement between experimental and calculated molecular parameters is found. In case of HPCO, the ^{13}C isotopic species was also observed.

Since both the stem compound HNCO and its sulphur analog HNCS are known to be present in space, and because also a sizable number of phosphorus and silicon-bearing species were detected there, both compounds are plausible targets for future radio astronomical searches using sensitive radio astronomical instrumentation.

Intermission

180

WJ07   3:27 – 3:42

BOND ANGLES AROUND A TETRAVALENT CENTRAL ATOM

ROBERT KARL BOHN, *Department of Chemistry, University of Connecticut, Storrs, CT, USA.*

There are several practical algorithms for building molecular geometries based on bond lengths, bond angles, and torsional angles. There seem to be few discussions of the effect changing one angle has on the remaining bond angles depending upon local symmetry. For example, in methane, CH_4, the H-C-H bond angles are all tetrahedral, i.e., $\alpha = 109.4712206...$ deg. If one considers CH_3F, a molecule with C_{3v} symmetry, how are the H-C-F bond angles related to the H-C-H bond angles? This study derives the bond angle relationships for a 4-bonded central atom such as a saturated C atom. For a 4-bonded central atom (6 bond angles) the possible local point group symmetries are $T_d(0)$, $D_{2d}(1)$, $C_{3v}(1)$, $C_{2v}(1)$, $D_2(2)$, $C_2(3)$, $C_s(3)$, and $C_1(4)$. The numbers in parentheses are the degrees of freedom, i.e., the number of angles which can be assigned arbitrary values with the remaining angles fixed by symmetry. Analytical formulas relating the bond angles for each of the eight possible symmetries are derived. Also, formulas have been derived for the five possible symmetries of a planar 4-bonded atom, $D_{4h}(0)$, $D_{2h}(1)$, C_{2v}(pendant, 1), C_{2v}(trapezoid, 2), and $C_s(3)$; the three possible structures of planar 3-bonded atoms, $(D_{3h}(0)$, $C_{2v}(1)$, and $C_s(2)$; the three possible symmetries of pyramidal 3-bonded atoms, $C_{3v}(1)$, $C_s(2)$, and $C_1(3)$; and the trivial case of 2-bonded atoms, $D_{\infty h}(0)$ and C_{2v}(bent 1). There are also six distinct 4-bonded central atom structures with all the bonds directed into a hemisphere, $C_{4v}(1)$, $C_{2v}(2)$, $C_2(3)$, C_s(trapezoid 3), C_s(pendant 3), and $C_1(4)$, geometries rarely seen in molecules.

WJ08   3:44 – 3:59

THE CHIRPED PULSE AND CAVITY FOURIER TRANSFORM MICROWAVE (CP-FTMW AND FTMW) SPECTRUM OF BROMOPERFLUOROACETONE

NICHOLAS FORCE, DAVID JOSEPH GILLCRIST, CASSANDRA C. HURLEY, FRANK E MARSHALL, NICHOLAS A. PAYTON, THOMAS D. PERSINGER, N. E. SHREVE, G. S. GRUBBS II, *Department of Chemistry, Missouri University of Science and Technology, Rolla, MO, USA.*

The microwave spectrum of the molecule bromoperfluoroacetone has been measured on a newly constructed CP-FTMW spectrometer along with a FTMW spectrometer relocated from Oxford University to Missouri S&T. Rotational constants, centrifugal distortion parameters, and nuclear quadrupole coupling constants will be discussed. Comparisons to the previously studied halogen analogues perfluoroacetone[a] and chloroperfluoroacetone[b] will be discussed.

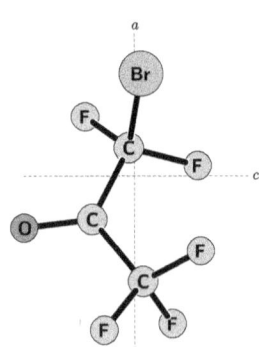

[a]J.-U. Grabow, N. Heineking, and W. Stahl, *Z. Naturforsch.* **46a** (1991) 229.
[b]G. Kadiwar, C. T. Dewberry, G. S. Grubbs II and S. A. Cooke, Talk **RH11**, 65th International Symposium on Molecular Spectroscopy (2010).

WJ09

PERFLUOROBUTYRIC ACID AND ITS MONOHYDRATE: A CHIRPED PULSE AND CAVITY BASED FOURIER TRANSFORM MICROWAVE SPECTROSCOPIC STUDY

JAVIX THOMAS, *Department of Chemistry, University of Alberta, Edmonton, AB, Canada*; AGAPITO SERRATO III, <u>WEI LIN</u>, *Department of Chemistry, University of Texas, Brownsville, TX, USA*; WOLFGANG JAEGER, YUNJIE XU, *Department of Chemistry, University of Alberta, Edmonton, AB, Canada.*

Perfluorobutyric acid (PFBA) is highly soluble in water and is a molecule of environmental importance. Rotational spectra of PFBA and its monohydrate were studied using a broadband chirped pulse and a narrow band cavity based Fourier transform microwave spectrometers and high level ab initio calculations. Extensive conformational search was performed for both the acid and its monohydrate at the MP2/6-311++G(2d,p) level of theory. Two and three conformers were predicted for PFBA and its monohydrate, respectively. One set of rotational transitions of PFBA and its mono-hydrate in each case was observed and assigned. Based on the broadband spectra obtained, one can confidently conclude that only one dominate conformer exists in each case. The orientation of the hydroxyl group in PFBA was determined using isotopic analysis. Comparison of the observed transition intensities and the calculated electric dipole moment components allowed one to identify the most stable monohydrate conformation which takes on the insertion hydrogen-bonding topology. Comparison to the shorter chain analogues, i.e. trifluoroacetic acid, perfluoropropionic acid, and their monohydrates, was made to elucidate the general trend in their conformational preference and binding topologies.

WJ10

ANALYSIS OF THE ROTATIONAL STRUCTURE IN THE HIGH-RESOLUTION IR SPECTRUM OF *trans*-HEXATRIENE-3-d_1

<u>NORMAN C. CRAIG</u>, YIHUI CHEN, *Department of Chemistry and Biochemistry, Oberlin College, Oberlin, OH, USA*; THOMAS BLAKE, *Chemical Physics, Pacific Northwest National Laboratory, Richland, WA, USA.*

For use in determining the semiexperimental structure of *trans*-hexatriene, its 3-d_1 isotopologue has been synthesized and the high-resolution (0.0015 cm^{-1}) IR spectrum has been recorded. The rotational structure in four C-type bands has been analyzed. These bands are for ν_{26} at 997.4, ν_{28} at 908.8, ν_{29} at 902.2, and ν_{32} at 678.6 cm^{-1}, which are all out-of-plane modes. Ground state rotational constants are $A_0 = 0.7952226(8)$, $B_0 = 0.0446149(7)$, and $C_0 = 0.0422661(4)$ cm^{-1}. The inertial defect is –0.2009 amu Å^2, which confirms planarity for this molecular species. The ultimate intent of this investigation is to evaluate the degree to which the "C=C" bonds are lengthened and the sp^2–sp^2 "C–C" bonds are shortened in comparison with localized bonds and with butadiene.

WJ11

MOLECULAR STRUCTURE OF THE PHENYL RADICAL (C_6H_5)

KYLE N CRABTREE, OSCAR MARTINEZ JR., *Atomic and Molecular Physics, Harvard-Smithsonian Center for Astrophysics, Cambridge, MA, USA*; <u>JOHN F. STANTON</u>, *Department of Chemistry, The University of Texas, Austin, TX, USA*; MICHAEL C McCARTHY, *Atomic and Molecular Physics, Harvard-Smithsonian Center for Astrophysics, Cambridge, MA, USA.*

Phenyl (C_6H_5) is the prototypical aryl radical, and is thought to be a critical intermediate for soot formation in combustion environments and a precursor to the formation of PAHs in space. Despite the fundamental importance of this species, its molecular structure has never been measured experimentally. We have measured the rotational spectra of C_6H_5 and all of its singly-substituted D and ^{13}C isotopologues in the 170–190 and 250–270 GHz bands in a positive column discharge of dilute benzene (or $C_6H_5D/^{13}CC_5H_6$) in Ar. The experimentally determined rotational constants were combined with zero point vibrational and electronic corrections calculated at the CCSD(T) level of theory to derive the semi-experimental equilibrium structure (r_e^{emp}).

WJ12 4:52 – 5:07

EQUILIBRIUM STRUCTURE OF PIPERIDINE

JEAN DEMAISON, *Laboratoire PhLAM, Université de Lille 1, Villeneuve de Ascq, France*; HEINZ DIETER RUDOLPH, *Department of Chemistry, Universitaet Ulm, Ulm, Germany*; NORMAN C. CRAIG, *Department of Chemistry and Biochemistry, Oberlin College, Oberlin, OH, USA*; PATRICIA ECIJA, EMILIO J. COCINERO, *Physical Chemistry Department, Universidad del País Vasco, Bilbao, Spain*; ALBERTO LESARRI, *Department Quimica Fisica y Quimica Inorganica, Universidad de Valladolid, Valladolid, Spain*.

In order to determine an accurate equilibrium structure for the equatorial form of piperidine, $C_5H_{11}N$, microwave transitions and ground state rotational constants are reported for the ^{13}C and ^{15}N isotopologues measured in natural abundance. These rotational constants along with those of the parent and ND species were used together with vibration-rotation constants obtained from a cubic force fields calculated at the B3LYP/cc-pVTZ and MP2/cc-pVTZ levels of theory to derive a semiexperimental (SE) equilibrium structure. However, the rotational constants of the CD deuterated species are not available, and there are many small Cartesian coordinates. Furthermore, piperidine is an oblate top. Quite generally, there is a large rotation of the principal axis system upon isotopic substitution in oblate tops that may considerably reduce the accuracy of the SE structure. For these reasons, the mixed estimation method was used. In this method, internal coordinates from good-quality quantum chemical calculations (with appropriate uncertainties) are fitted simultaneously with moments of inertia of the full set of isotopologues. In order to have predicates as accurate as possible, the geometry was optimized at the MP2 and CCSD(T) levels of theory using Dunning basis sets up to quadruple-zeta quality. This combination of experimental rotational constants and high-level ab initio calculations allowed us to determine an accurate equilibrium structure.

WJ13 5:09 – 5:24

MICROWAVE SPECTRA AND MOLECULAR GEOMETRIES OF BENZONITRILE AND PENTAFLUOROBENZONITRILE

MAHDI KAMAEE, JENNIFER VAN WIJNGAARDEN, *Department of Chemistry, University of Manitoba, Winnipeg, MB, Canada*.

The ground state rotational spectra of benzonitrile (BN) and pentafluorobenzonitrile (PFBN) were investigated using Balle-Flygare Fourier transform microwave (FTMW) spectroscopy in the region between 5 and 25 GHz. In addition to the parent species, transitions due to the five ^{13}C isotopologues were measured in natural abundance and used to calculate relevant geometric parameters of the ring backbone of the two compounds. The experimental results compare well with the equilibrium structure determined via *ab initio* theory (MP2/6-311++G(2d,2p)). To better understand the effect of fluorination on the ring moiety, the electronic environments around the nitrogen atom in both species were compared through a Townes-Dailey-type analysis of the observed ^{14}N quadrupole hyperfine structure.

WJ14

MICROWAVE SPECTROSCOPY OF MONOTERPENES OF ATMOSPHERIC INTEREST: α-PINENE, β-PINENE, AND NOPINONE

JUAN-RAMON AVILES MORENO, ELIAS NEEMAN, T. R. HUET, *Laboratoire PhLAM, UMR8523 CNRS - Université Lille 1, Villeneuve d'Ascq, France.*

Several monoterpenes and terpenoids are biogenic volatile organic compounds which are emitted in the atmosphere, and react with OH, O_3, NO_x, etc. to give rise to several oxydation and degradation products. Spectroscopic information on these atmospheric species are still very scarse. Meanwhile we have demonstrated that combining quantum calculations to microwave spectroscopy led to the unambiguous characterization of the most stable conformers for perillaldehyde, [a] limonene and carvone. [b] This information can be used to subsequently model accurately the vibrational signature for atmospheric purposes. [c]

We have recorded the pure rotational spectra of α-pinene and β-pinene ($C_{10}H_{16}$), and of nopinone ($C_9H_{14}O$), using the MB-FTMW spectrometer of Lille, in the 2-20 GHz range at temperatures varying between 340 and 380 K. For these three bicyclic molecules only one conformer can be observed, and the rotational structure was observed up to J, K_a = 8, 3 ; 8, 4 ; 8, 5, respectively. All the spectra were modeled with a semi-rigid rotor Hamiltonian and fitted to obtain a rms value better than 5 kHz using a-, b- and c- type transitions.

All the experimental results were supported by several quantum calculations performed at different levels of theory (DFT and ab initio). In particular no experimental evidence of internal rotation motion was found (methyl groups), in good agreement with the calculated barriers.

Support from the French Laboratoire d'Excellence CaPPA (Chemical and Physical Properties of the Atmosphere) through contract ANR-10-LABX-0005 of the Programme d'Investissements d'Avenir is acknowledged.

[a] J.-R. Aviles Moreno, F. Partal Urena, J.-J. Lopez Gonzalez and T. R. Huet, *Chem. Phys. Lett.* **473** (2009) 17

[b] J.-R. Aviles Moreno, T. R. Huet, F. Partal Urena, J.-J. Lopez Gonzalez, *Struct. Chem.* **24** (2013) 1163

[c] T. R. Huet, J.-R. Aviles Moreno, O. Pirali, M. Tudorie, F. Partal Urena, J.-J. Lopez Gonzalez, *JQSRT.* **113** (2012) 1261

RA. Mini-symposium: Astronomical Molecular Spectroscopy in the Age of ALMA

Thursday, June 19, 2014 – 8:30 AM

Room: 116 Roger Adams Lab

Chair: Anthony Remijan, NRAO, Charlottesville, VA, USA

RA01   8:30 – 8:45

IMPROVED INFRASTUCTURE FOR CDMS AND JPL MOLECULAR SPECTROSCOPY CATALOGUES

CHRISTIAN ENDRES, STEPHAN SCHLEMMER, *I. Physikalisches Institut, Universität zu Köln, Köln, Germany*; BRIAN DROUIN, JOHN PEARSON, *Jet Propulsion Laboratory, California Institute of Technology, Pasadena, CA, USA*; HOLGER S. P. MÜLLER, P. SCHILKE, JÜRGEN STUTZKI, *I. Physikalisches Institut, Universität zu Köln, Köln, Germany.*

Over the past years a new infrastructure for atomic and molecular databases has been developed within the framework of the Virtual Atomic and Molecular Data Centre (VAMDC). Standards for the representation of atomic and molecular data as well as a set of protocols have been established which allow now to retrieve data from various databases through one portal and to combine the data easily. Apart from spectroscopic databases such as the Cologne Database for Molecular Spectroscopy (CDMS), the Jet Propulsion Laboratory microwave, millimeter and submillimeter spectral line catalogue (JPL) and the HITRAN database, various databases on molecular collisions (BASECOL, KIDA) and reactions (UMIST) are connected. Together with other groups within the VAMDC consortium we are working on common user tools to simplify the access for new customers and to tailor data requests for users with specified needs. This comprises in particular tools to support the analysis of complex observational data obtained with the ALMA telescope.

In this presentation requests to CDMS and JPL will be used to explain the basic concepts and the tools which are provided by VAMDC. In addition a new portal to CDMS will be presented which has a number of new features, in particular meaningful quantum numbers, references linked to data points, access to state energies and improved documentation. Fit files are accessible for download and queries to other databases are possible.

RA02   8:47 – 9:02

HIGHLY ACCURATE QUANTUM-CHEMICAL CALCULATIONS FOR THE INTERSTELLAR MOLECULES C_3 AND l-C_3H^+

PETER BOTSCHWINA, BENJAMIN SCHRÖDER, CHRISTOPHER STEIN, PETER SEBALD, RAINER OSWALD, *Institute of Physical Chemistry, Georg-August-Universität Göttingen, Göttingen, Germany.*

Composite potential energy surfaces with coupled-cluster contributions up to CCSDTQP were constructed for C_3 and l-C_3H^+ and used in the calculation of spectroscopic properties. The use of very large AO basis sets and the consideration of higher-order correlation beyond CCSD(T) is of utmost importance for C_3 in order to arrive at quantitative spectroscopic data. The first detection of l-C_3H^+ in the interstellar medium was reported by Pety et al.,[a] who attributed 9 radio lines observed in the horsehead photodissociation region to that species. That assignment was questioned by the recent theoretical work of Huang et al.[b] However, our more accurate calculations are well in support of the original assignment. The calculated ground-state rotational constant is $B_0 = 11248$ MHz, only 0.03% off from the radio astronomical value of 11244.9512 ± 0.0015 MHz. The ratio of centrifugal distortion constants $D_0(\text{exp.})/D_e(\text{theor.})$ of 1.8 is quite large, but reasonable in comparison with C_3O^c and C_3.

---

[a]J. Pety, P. Gratier, V. Guzmán, E. Roueff, M. Gerin et al., *Astron. Astrophys.* **2012**, *A68*, 1-8.
[b]X. Huang, R. C. Fortenberry, T. J. Lee, *Astrophys. J. Lett.* **2013**, *768:L25*, 1-5.
[c]P. Botschwina, R. Oswald, *J. Chem. Phys.* **2008**, *129*, 044305.

RA03 9:04 – 9:19

TERAHERTZ MEASUREMENTS OF HOT HYDRONIUM IONS (H$_3$O$^+$) WITH AN EXTENDED NEGATIVE GLOW DISCHARGE

SHANSHAN YU, <u>JOHN PEARSON</u>, *Jet Propulsion Laboratory, California Institute of Technology, Pasadena, CA, USA.*

Terahertz absorption spectroscopy was employed to detect the ground-state inversion transitions of the hydronium ion. The highly excited ions were created with an extended negative glow discharge through a gas mixture of 1 mtorr of H$_2$O, 2 mtorr of H$_2$, and 12 mtorr of Ar, which allowed observation of transitions with J and K up to 12. In total forty seven transitions were measured in the 0.9–2.0 THz region and twenty two of them were observed for the first time. The experimental uncertainties range from 100 to 300 kHz, which are much better than those of 0.3–1.2 MHz reported in previous work. Differences up to 25.6 MHz were found between the observed positions and the catalog values that have been used for Herschel data analysis of observations towards Sagittarius B2(N), NGC 4418 and Arp 220.[ab] The new and improved measurements were fit to experimental accuracies with an updated Hamiltonian; and better H$_3$O$^+$ predictions are reported to support the proper analysis of astronomical observations by high-resolution spectroscopy telescopes, such as Herschel,[c] SOFIA, and ALMA.

[a]Lis et al., "Hot, metastable hydronium ion in the Galactic center: formation pumping in X-ray-irradiated gas?", *Phil. Trans. R. Soc. A* **370**, 5162 (2012).

[b]González-Alfonso et al., "Excited OH$^+$, H$_2$O$^+$ and H$_3$O$^+$ in NGC 4418 and Arp 220", *Astrophys. Astronom* **550**, A25 (2013).

[c]Lis et al., "Widespread rotaional-hot hydronium in the Galactic Interstellar Medium", *Astrophys. J.*, submitted (2014).

RA04 9:21 – 9:36

TORSION-ROTATION-VIBRATION EFFECTS IN THE v$_{20}$, 2v$_{21}$, 2v$_{13}$ AND v$_{21}$+v$_{13}$ STATES OF CH$_3$CH$_2$CN

<u>ADAM M DALY</u>, JOHN PEARSON, SHANSHAN YU, BRIAN DROUIN, *Jet Propulsion Laboratory, California Institute of Technology, Pasadena, CA, USA*; CELINA BERMÚDEZ, JOSÉ L. ALONSO, *Grupo de Espectroscopia Molecular, Lab. de Espectroscopia y Bioespectroscopia, Unidad Asociada CSIC, Universidad de Valladolid, Valladolid, Spain.*

Ethyl cyanide, CH$_3$CH$_2$CN, is a highly abundant molecule in hot cores associated with massive star formation where temperatures often approach 200 K. Astrophysicists would like to use the many thousands of observed lines to evaluate thermal equilibrium, temperature distributions, heating sources, and radiative pumping effects. In spite of a recent partial success in characterizing the v$_{20}$ and v$_{12}$ vibrational states[a], many aspects of the spectroscopy of the v$_{20}$ state are not adequately characterized. Torsional splittings in the b-type spectrum of v$_{20}$ are typically a few MHz and many a-type transitions also show resolved torsional splittings, both are incompatible with the expected 1200 cm^{-1} barrier to internal rotation in a v$_t$ = 0 state. Additionally all K values above 2 show some obvious perturbations. The three states that lie just above v$_{20}$ are 2v$_{21}$, 2v$_{13}$ and v$_{21}$ + v$_{13}$. It has been determined that v$_{20}$ interacts weakly with both 2v$_{21}$ and 2v$_{13}$ and that 2v$_{21}$ interacts weakly with 2v$_{13}$, in spite of their common symmetry and very close proximity. However, all the interactions of v$_{21}$ + v$_{13}$ appear to be very strong, making assignments of the combination band particularly problematic. The numerous interactions result in wide spread anomalous torsional splittings. These splittings provide valuable insight into the nature of the interactions, however without a reasonable model, assignment of A or E to a torsional component is far from obvious. There remains no reasonable quantum mechanical description of how to proceed with a torsion-rotation-vibration analysis involving large and small amplitude motions. We present what is known and unknown in this quartet of CH$_3$CH$_2$CN states.

[a]Daly, A. M., Bermúdez, C., López, A., et al., 768, 1, ApJ, 2013

186

RA05   9:38 – 9:53

LABORATORY MEASUREMENTS IN SUPPORT OF ASTRONOMICAL OBSERVATIONS: ROTATIONAL SPECTROSCOPY UP THE THz REGION

GABRIELE CAZZOLI, CRISTINA PUZZARINI, *Dep. Chemistry 'Giacomo Ciamician', University of Bologna, Bologna, Italy.*

The identification of a large variety of chemical compounds in space as well as planetary atmospheres has been made possible through the spectral signatures of astronomical bodies. For their interpretation, laboratory measurements are essential and rotational spectroscopy plays a leading role. Recent missions, such as the Herschel Space Observatory and the Stratospheric Observatory for Infrared Astronomy (SOFIA), have pointed out the need for precise and accurate frequency measurements and spectroscopic parameters in the submillimeter-wave range. The higher spectral resolution that can be exploited by the Atacama Large Millimeter Array (ALMA) is expected to play a key role in the detection of complex species as well as rare isotopologues. In the present contribution, the millimeter-/submillimeter-wave spectrometer working at the University of Bologna and its applications in the field of astronomical observations are presented. The focus is here on the accuracy of the retrieved transition frequencies of neutral as well as ionic species.

RA06   9:55 – 10:10

THE COMPLETE, TEMPERATURE RESOLVED SPECTRUM OF METHANOL BETWEEN 214 AND 265 GHZ

JAMES McMILLAN, CHRISTOPHER F. NEESE, SARAH FORTMAN, FRANK C. DE LUCIA, *Department of Physics, The Ohio State University, Columbus, OH, USA.*

We have studied methanol, one of the so-called 'astronomical weeds', in the 215–265 GHz band. We have gathered a set of intensity calibrated, complete, experimental, and temperature resolved spectra from across the temperature range of 240–389 K. A number of low lying transitions, including the $\nu_t = 3$, have not been produced by available catalogs. Using our previously reported method of analysis[a] we were able generate a line list that contains lower state energies and linestrengths, for all of the observed lines in the band. This line list includes those lines which have no quantum mechanical assignment.

In addition to this line list we provide a point by point method capable of generating the complete spectrum at an arbitrary temperature. The sensitivity of the point by point analysis is such that we are able to identify lines which would not have manifest in a single scan across the band. The consequence has been to reveal not only a number of new methanol lines, but also trace amounts of contaminants. We show how the intensities from the contaminants can be removed with indiscernible impact on the signal from methanol. To do this we use the point by point results from our previous studies of these contaminants. The efficacy of this process serves as strong proof of concept for usage of our point by point results on the problem of the weeds. The success of this approach for dealing with the weeds has also previously been reported[b].

[a] S. Fortman et al. 2014 ApJ 782 75
[b] S. Fortman et al. 2012 J. Mol. Spectrosc. 782, 75

Intermission

RA07   10:27 – 10:42

AN ANALYSIS OF THE ROTATIONAL SPECTRUM OF ACETONITRILE (CH$_3$CN) IN EXCITED VIBRATIONAL STATES

CHRISTOPHER F. NEESE, JAMES McMILLAN, SARAH FORTMAN, FRANK C. DE LUCIA, *Department of Physics, The Ohio State University, Columbus, OH, USA.*

Acetonitrile (CH$_3$CN) is a well-known interstellar molecule whose vibrationally excited states need to be accounted for in searches for new molecules in the interstellar medium. To help catalog such 'weed' molecules, we have developed a technique that involves recording complete spectra over a range of astrophysically significant temperatures. With such a data set, we can experimentally measure the line strengths and lower state energies of unassigned lines in the spectrum.

In this talk we will present the ongoing analysis of complete temperature resolved spectra in the 215–265 GHz and 570–650 GHz regions. We have been able to assign many vibrationally hot lines from this data and a room temperature data set spanning 165–700 GHz. To date, we have assigned lines from most of the vibrational states below ν_6 at 1448 cm^{-1}.

RA08

THE MILLIMETER- AND SUBMILLIMETER-WAVE SPECTRUM OF PROPENAL

ADAM M DALY, CELINA BERMÚDEZ, LUCIE KOLESNIKOVÁ, JOSÉ L. ALONSO, *Grupo de Espectroscopia Molecular, Lab. de Espectroscopia y Bioespectroscopia, Unidad Asociada CSIC, Universidad de Valladolid, Valladolid, Spain.*

The detection of propenal (acrolein) in Sgr B2 is based in two transitions detected with the Green-Bank Telescope,[a] operating in the frequency range 18-26 GHz. A combination of Stark modulated microwave spectroscopy and frequency modulated millimeter- and submillimeter-wave spectroscopy has been used to record the rotational spectrum of propenal from 26 to 675 GHz in order to provide accurate rotational parameters for the identification of new lines of propenal in the interstellar medium. Ground state and the vibrational excited states below 700 cm^{-1} of trans-propenal (ν_{18}, $2\nu_{18}$, ν_{13}, $3\nu_{18}$, $\nu_{18}+\nu_{13}$, ν_{12}, ν_{17}, $4\nu_{18}$, $2\nu_{18}+\nu_{13}$, $2\nu_{13}$) have been analyzed in a global fit taking in to account Coriolis and Fermi interactions. Spectroscopic constants for all three ^{13}C trans-isotopologes in natural abundance as well as ground state spectroscopic constants for cis-propenal are also reported.

---

[a] J. M. Hollis, P. R. Jewell, F. J. Lovas, A. Remijan, H. Møllendal *Astrophys. J.* **610**, *L21-L24*, (2010)

RA09

A COMPREHENSIVE INTENSITY STUDY OF THE ν_4 TORSIONAL BAND OF ETHANE

JALAL NOROOZ OLIAEE, NASSER MOAZZEN-AHMADI, *Department of Physics and Astronomy, University of Calgary, Calgary, AB, Canada*; IRVING OZIER, *Department of Physics and Astronomy, University of British Columbia, Vancouver, BC, Canada*; KEEYOON SUNG, *Jet Propulsion Laboratory, Science Division, California Institute of Technology, Pasadena, CA, USA*; TIMOTHY J CRAWFORD, LINDA BROWN, *Jet Propulsion Laboratory, California Institute of Technology, Pasadena, CA, USA*; EDWARD H WISHNOW, *Space Sciences Laboratory and Department of Physics, University of California, Berkeley, CA, USA*; V. MALATHY DEVI, *Department of Physics, College of William and Mary, Williamsburg, VA, USA.*

The torsional spectrum of C_2H_6 has been investigated from 220 to 330 cm^{-1} to measure the intensity of the fundamental and the first torsional hot band needed for atmospheric studies of Titan. Several spectra were measured at resolutions of 0.01 and 0.02 cm^{-1} using the JPL Bruker IFS-125 coupled to a coolable multi-pass absorption cell originally developed at University of British Columbia.[a] Spectra were recorded at several temperatures from 293 K to 166 K, with the lower temperatures relevant to the stratosphere of Titan. Because this spectrum is very weak, a long absorption path of 52 m was used along with substantial sample pressures from 35 to 255 Torr. Intensities were analysed using a quantum mechanical model reported previously.[b] The torsional fundamental of C_2H_6 is observed in the CIRS spectra of Titan. Line parameters for the torsional bands are required for accurate characterization of spectral features of Titan's far-infrared region. The current study should lead to a better understanding of the methane cycle in planetary atmospheres and permit the identification of the other molecular features in the CIRS data.[c]

---

[a] E. H. Wishnow, A. Leung, and H. P. Gush, Rev. Sci. Instr., 70, 23 (1999).

[b] N. Moazzen-Ahmadi, A.R.W. McKellar, J.W.C. Johns, and I.Ozier, J. Chem. Phys. 97, 3981 (1992).

[c] Research described in this paper was performed, in part, at the Jet Propulsion Laboratory, California Institute of Technology under contracts and cooperative agreements with the NASA.The data were obtained using NASA's OPR Grant awarded to the College of William and Mary. The research conducted at the University of Calgary is supported by the Canadian Space Agency.

RA10 11:18 – 11:33

ISOTOPE SELECTIVE PHOTODISSOCIATION OF N_2 BY THE INTERSTELLAR RADIATION FIELD AND COSMIC RAYS

ALAN HEAYS, EWINE VAN DISHOECK, *Leiden Observatory, University of Leiden, Leiden, Netherlands*; RUUD VISSER, *Department of Astronomy, University of Michigan, Ann Arbor, MI, USA*; ROLAND GREDEL, *MPIA, Max Planck Institute for Astronomy, Heidelberg, Germany*; WIM UBACHS, *Department of Physics and Astronomy, VU University , Amsterdam, Netherlands*; BRENTON R LEWIS, STEPHEN GIBSON, *Research School of Physics and Engineering, Australian National University, Canberra, ACT, Australia.*

Photodissociation of $^{14}N_2$ and $^{14}N^{15}N$ occurs in interstellar clouds, protoplanetary discs, (exo)planetary atmospheres, and other environments due to ultraviolet radiation originating from stellar sources and the presence of cosmic rays. We study this process in detail in search of an explanation for the observed non-elemental ratios of N isotopologues observed in solar system bodies and in molecular clouds.

High-resolution theoretical photodissociation cross sections of N_2 and competing UV-absorbing species are used to calculate the isotope-selective shielding of N_2 in photochemical models of a diffuse interstellar cloud and protoplanetary disk. An enhancement of the atomic $^{15}N/^{14}N$ ratio over the elemental value is obtained due to the self-shielding of external radiation at an extinction of about $A_v = 1$ mag, and leads to a similar mass fractionation in daughter species. This effect is larger where assumed grain growth has reduced the opacity of dust to ultraviolet radiation.

The cosmic-ray induced dissociation of N_2 is calculated from a high-resolution model of H_2 emission and is found to depend sensitively on details of the emission and photodissociation cross sections.

RA11 11:35 – 11:50

THEORETICAL STUDY ON VIBRONIC INTERACTIONS AND PHOTOPHYSICS OF LOW-LYING EXCITED ELECTRONIC STATES OF POLYCYCLIC AROMATIC HYDROCARBONS

NAGAPRASAD REDDY SAMALA, S. MAHAPATRA, *School of Chemistry, University of Hyderabad, Hyderabad, Andhra Pradesh, India.*

Polycyclic aromatic hydrocarbons (PAHs), in particular, their radical cation (PAH$^+$), have long been postulated to be the important molecular species in connection with the spectroscopic observations[a] in the interstellar medium. Motivated by numerous important observations by stellar as well as laboratory spectroscopists, we undertook detailed quantum mechanical studies of the structure and dynamics of electronically excited PAH$^+$ in an attempt to establish possible synergism with the recorded data[b]. In this study, we focus on the quantum chemistry and dynamics of the doublet ground (X) and low-lying excited (A, B and C) electronic states of the radical cation of tetracene (Tn), pentacene (Pn), and hexacene (Hn) molecule. This study is aimed to unravel photostability, spectroscopy, and time-dependent dynamics of their excited electronic states. In order to proceed with the theoretical investigations, we construct suitable multistate and multimode Hamiltonian for these systems with the aid of extensive ab initio calculations of their electronic energy surfaces. The diabatic coupling surfaces are derived from the calculated adiabatic electronic energies. First principles nuclear dynamics calculations are then carried out employing the constructed Hamiltonians and with the aid of time-independent and time-dependent quantum mechanical methods[c]. We compared our theoretical results with available photoelectron spectroscopy, zero kinetic energy photoelectron (ZEKE) spectroscopy and matrix isolation spectroscopy (MIS) results. A peak at 8650 Å in the B state spectrum of Tn$^+$ is in good agreement with the DIB at 8648 Å observed by Salama et al. Similarly in Pn$^+$, a peak at 8350 Å can be correlated to the DIB at 8321 Å observed by Salama et al.

[a]J. Zhang et al., J. Chem. Phys., 128,104301 (2008).; F. Salama, Origins of Life Evol. Biosphere, 28, 349 (1998).; F. Salama et al., Astrophys. J., 526, 265 (1999).; J. Szczepanski et al., Chem. Phys. Lett., 232, 221 (1995).; J. Szczepanskiet al., Chem. Phys. Lett., 245, 539 (1995).; J. Zhang et al., Astrophys. J., 715, 485 (2010).
[b]V. Sivaranjana Reddy et al., Phys. Rev. Lett, 104, 111102 (2010).; S. Ghanta et al., Phys. Chem. Chem. Phys. 13, 14523.; & 13, 14531 (2011).
[c]S. Nagaprasad Reddy et al., J. Phys. Chem. A., 117, 8737 (2013).

RB. Clusters/Complexes
Thursday, June 19, 2014 – 8:30 AM
Room: 100 Noyes Laboratory

Chair: Bob McKellar, National Research Council of Canada, Ottawa, ON, Canada

---

RB01   8:30 – 8:45

INFRARED SPECTROSCOPY OF $C_6D_6 - Rg_n$ (n=1,2)

<u>JOBIN GEORGE</u>, *Department of Physics and Astronomy, University of Calgary, Calgary, AB, Canada*; MAHDI YOUSEFI, *Physics and Astronomy/Institute for Quantum Science and Technology, University of Calgary, Calgary, AB, Canada*; MOJTABA REZAEI, *Department of Physics and Astronomy, University of Calgary, Calgary, AB, Canada*; BOB McKELLAR, *Steacie Laboratory, National Research Council of Canada, Ottawa, ON, Canada*; NASSER MOAZZEN-AHMADI, *Physics and Astronomy/Institute for Quantum Science and Technology, University of Calgary, Calgary, AB, Canada*.

Benzene-noble gas complexes were one of the earliest topics of interest in spectroscopic investigation of van der Waals (vdW) complexes. Smalley et al.[a] observed $C_6H_6-(He)_{1,2}$ vdW complexes in the late 1970s by means of electronic spectroscopy. A recent study on the same species was done by M. Hayashi et al.[b] at higher resolution (250 MHz). Here, we present the infrared observation of $C_6D_6 - Rg_n$ (n=1,2) with the rare gas being He, Ne, or Ar, in the regions of ν_{12} fundamental band of C_6D_6 ($\sim$2289 cm^{-1}) and the $\nu_2 + \nu_{13}$ combination band ($\sim$2275 cm^{-1}) which are coupled by a Fermi resonance. The spectra were observed at a resolution of 60 MHz using a tunable optical parametric oscillator to probe a pulsed supersonic-jet expansion from a slit nozzle. In the case of $C_6D_6 - Rg$ dimers, the spectra were assigned to a symmetric top with C_{6v} symmetry with the rare gas atom being located on the C_6 symmetry axis. To observe $C_6D_6 - Rg_2$ trimers, the nozzle was cooled using a closed-cycle methanol refrigerator and the spectra were simulated with a rotational temperature of 1.3K. The spectra of the $C_6D_6 - Rg_2$ trimers were in agreement with a D_{6h} symmetry structure, where the rare gas atoms are positioned above and below the C_6D_6 plane. Data analysis and observation are presently ongoing.

---

[a]S. M. Beck, M. G. Liverman, D. L. Monts and R. E. Smalley, J. Chem. Phys. **70**, 232 (1979).

[b]M. Hayashi, Y. Ohshima , Chem. Phys. **419**, 131 (2013).

RB02   8:47 – 9:02

SUB-DOPPLER ELECTRONIC SPECTRUM OF THE BENZENE–D_2 COMPLEX

MASATO HAYASHI, <u>YASUHIRO OHSHIMA</u>, *Department of Photomolecular Science, Institute for Molecular Science, Okazaki, Japan.*

Excitation spectrum of the benzene–D_2 van der Waals (vdW) complex in the vicinity of the $S_1 \leftarrow S_0\ 6_0^1$ vibronic transition of the monomer was recorded by utilizing mass-selective two-color resonance-enhanced two-photon ionization. Extensive adiabatic cooling with the rotational temperature of < 0.5 K was conducted by the high-pressure pulsed expansion, and sub-Doppler resolution yielding the line width of 250 MHz was realized in a collimated molecular beam by employing Fourier-transform-limited ultraviolet pulses for the excitation.[a] In contrast to our previous study on the benzene–H_2 complex,[b] the weaker binding *ortho* nuclear-spin isomer, correlating to the $j = 0$ state of a freely rotating D_2, was observed in addition to the stronger binding *para* isomer (with $j = 1$), by using a gas sample of normal D_2. Three and two vibronic bands involving vdW-mode excitation were observed for the *para* and *ortho* isomers, respectively. By comparing the present results with those of the benzene–H_2 complex, we made unambiguous assignments on the vdW modes involved in each observed band, and obtained complete sets of vibrational frequencies of all the three vdW modes for the both H_2 and D_2 isotopomers in the $S_1\ 6^1$ manifold. One of the vdW frequency correlates to the splitting between the $m = 0$ and ±1 sublevels in the $j = 1$ state of a freely rotating H_2/D_2 molecule, and the potential barrier for the hindered internal rotation has been evaluated to be ca. 60 cm^{-1} from the values. Ratio of the vdW frequencies between the H_2 and D_2 species deviate significantly from the value for the harmonic vibration (i.e., $\sqrt{2} \approx 1.4$), indicating substantial anharmonic character of the vdW modes in the complex.

---

[a]M. Hayashi and Y. Ohshima, *Chem. Phys.* **419**, 131–137 (2013).

[b]M. Hayashi and Y. Ohshima, *J. Phys. Chem. A* **117**, 9819–9830 (2013).

RB03   9:04 – 9:19

ISOMER-SPECIFIC IR SPECTROSCOPY OF BENZENE-(WATER)$_N$ CLUSTERS WITH N=1-8: NEW INSIGHTS FROM THE WATER BEND FUNDAMENTALS AND ISOTOPICALLY SUBSTITUTED CLUSTERS

RYOJI KUSAKA, *Chemistry, Hiroshima University, Higashi-Hiroshima, Japan*; PATRICK S. WALSH, TIMOTHY S. ZWIER, *Department of Chemistry, Purdue University, West Lafayette, IN, USA.*

This talk will focus on the isomer-specific IR spectra of benzene-(water)n (BWn) clusters with n = 1-8, returning to a topic studied by our group[a] some 20 years ago, but now with higher resolution (OH stretch region), with inclusion of data from isotopically substituted clusters, and with extension into the HOH bending mode region. Spectra are recorded using resonant ion-dip infrared spectroscopy, an IR-UV double resonance method. Isomer-specific IR spectra in the regions of OH, OD stretches and HOH, HOD bend of benzene-H_2O, -D_2O, -HOD, -(H_2O)$_2$, -(D_2O)$_2$, -HOD-DOD were recorded in order to investigate in greater detail the intermolecular potential energy surface between water and benzene. These spectra show strong combination bands in addition to the OH/OD stretch fundamentals arising from large-amplitude "tumbling" and tunneling along internal rotation and torsion coordinates of water(s) on the surface of benzene. Interestingly, the number of extra bands and spectral patterns change dramatically depending on cluster size, the kind of deuterated isomer, and the spectral region probed. In larger clusters with n=3-8, the water HOH bending region is explored for the first time. The prominent bending mode transitions in BW1-8 are spread over a relatively small range (1610-1660 cm^{-1}), and shift with cluster size in a way that reflects the known structural changes that accompany the increase in size. By comparison of experiment with calculation, it is possible to assign the experimentally observed 1614 cm^{-1} transition of BW1 and 1615 cm^{-1} of BW2 bands to the π-bound water molecule. The 1620-1660 cm^{-1} bands of BW3-8 are due to water molecules that can be categorized as single-acceptor, single-donor (AD) hydrogen-bonded waters. In the case of single-acceptor, double-donor (ADD) water molecules, which are expected to be seen from BW6,[a] they show higher-frequency bending vibrations and weaker IR intensity, which would correspond to very weakly observed bands in 1660-1750 cm^{-1} for BW6-8.

[a]R. N. Pribble and T. S. Zwier, Science, 1994, 265, 75-79.

RB04   9:21 – 9:36

MICROWAVE SPECTRUM, STRUCTURE, AND INTERNAL DYNAMICS OF THE PYRIDINE - ACETYLENE WEAKLY BOUND COMPLEX

BECCA MACKENZIE, CHRIS DEWBERRY, *Chemistry Department, University of Minnesota, Minneapolis, MN, USA*; EMMA JARRETT, ANTHONY LEGON, *School of Chemistry, University of Bristol, Bristol, United Kingdom*; KEN LEOPOLD, *Chemistry Department, University of Minnesota, Minneapolis, MN, USA.*

A-type rotational spectra of the weakly bound complex formed from pyridine and acetylene are reported. Contrary to expectation based on the symmetric structure of HCCH$\cdots$NH$_3$, the acetylene moiety in HCCH$\cdots$NC$_5$H$_5$ does not lie along the symmetry axis of the pyridine. Rotational and ^{14}N hyperfine constants instead indicate that, while the complex is indeed planar with an acetylenic hydrogen directed toward the nitrogen, the HCCH axis forms an angle of $\sim$23° with the C_2 axis of the pyridine. Spectra of HCCH$\cdots$NC$_5$H$_5$, HCCD$\cdots$NC$_5$H$_5$, DCCH$\cdots$NC$_5$H$_5$, and DCCD$\cdots$NC$_5$H$_5$ are all doubled, revealing the existence of a pair of low energy states. In light of the bent structure, this suggests a tunneling motion through a barrier at the C_{2v} configuration. Because the splitting persists in the singly deuterated species, we conclude that the motion does not involve interchange of the acetylenic hydrogens. Single ^{13}C substitution in either the ortho or meta positions of the pyridine eliminates the doubling and gives rise to separate sets of spectra for which the rotational constants are well predicted by a bent geometry. In this case, the two sets correspond to distinct species in which the ^{13}C is either on the same or the opposite side as the acetylene. This further suggests that the doubling observed with unsubstituted pyridine arises from wagging of the acetylene, as such a tunneling motion is expected to be quenched when the pyridine is rendered asymmetric. The bent structure of the system may arise due to a secondary hydrogen bonding interaction between the ortho hydrogens of the pyridine and the π system of the acetylene.

ROTATIONAL SPECTRA OF ADDUCTS OF PYRIDINE WITH METHANE AND ITS HALIDES

QIAN GOU, LORENZO SPADA, *Dep. Chemistry 'Giacomo Ciamician', University of Bologna, Bologna, Italy*; MONTSERRAT VALLEJO-LÓPEZ, ALBERTO LESARRI, *Department Quimica Fisica y Quimica Inorganica, Universidad de Valladolid, Valladolid, Spain*; EMILIO J. COCINERO, *Physical Chemistry Department, Universidad del País Vasco, Bilbao, Spain*; WALTHER CAMINATI, *Dep. Chemistry 'Giacomo Ciamician', University of Bologna, Bologna, Italy*.

The rotational spectra of the 1:1 complexes of pyridine (PYR) with methane and its halides have been observed and assigned using pulsed jet Fourier transform microwave technique. Depending on the nature of the chemical species linked to PYR, π or σ type complexes have been observed, in relation to the PYR interaction sites, that is the σ system of the aromatic ring, or the n orbital of the nitrogen atom. In the case of PYR-CF_4, where the two subunits are held together by a $CF_3 \bullet\bullet N$ halogen bond, with the top undergoing a free rotation with respect to PYR [1]. Instead, a C-Cl$\bullet\bullet$N halogen bond characterizes PYR-CF_3Cl. In PYR with CHF_3 [2], CH_3F and CH_2F_2 two kind of weak hydrogen bonds, C-H$\bullet\bullet$N and C-H$\bullet\bullet$F, have been found to connect the two constituent units. The barriers to the internal rotations of the CHF_3 and of the CH_3F groups have been determined from the A-E splittings of the rotational transitions. The rotational spectrum of the adduct PYR-CH_4 shows that CH_4 links to an aromatic molecule through a C H$\bullet\bullet\pi$ weak hydrogen bond. The shape and the internal dynamic behavior of this complex are very similar to that of the van der Waals complexes involving aromatic molecules with rare gases, which would suggest classifying CH_4, in relation to its ability to form molecular complexes with aromatic molecules, as a pseudo rare gas.

References:

[1] A. Maris, L. B. Favero, B. Velino, W. Caminati, J. Phys. Chem. A, 2013, 117, 11289.

[2] L. B. Favero, B. M. Giuliano, A. Maris, S. Melandri, P. Ottaviani, B. Velino, W. Caminati, Chem. Eur. J. 2010, 16, 1761.

Intermission

REACTIVE PATHWAYS IN THE CHLOROBENZENE-AMMONIA DIMER CATION RADICAL: NEW INSIGHTS FROM EXPERIMENT AND THEORY

SILVER NYAMBO, BRANDON UHLER, AIMABLE KALUME, LLOYD MUZANGWA, SCOTT REID, *Department of Chemistry, Marquette University, Milwaukee, WI, USA*.

Previously, we have studied non-covalent interactions in mono-halogenated benzene clusters using mass selected resonant 2-photon ionization methods. We have extended our studies by investigating the interaction between these mono-halobenzenes with a prototypical N atom donor (NH_3). Thus, we have obtained electronic spectra of PhX$\cdots$(NH_3)n (X=F, Cl, Br and n=1,2....) complexes in the region of the PhX monomer S_0-S_1 ($\pi\pi^*$) transition. Here we are mainly focusing on PhCl$\cdots$$NH_3$ dimer. We found that upon ionization of the dimer, three reactive pathways of the [PhCl$\cdots$$NH_3$]$^+$· have been evidenced. The primary pathway is the Cl atom elimination, previously evidenced. The second and third pathways, HCl elimination and H atom elimination are identified for the first time in the R2PI studies of the dimer. Electronic spectra obtained for the three pathways shows that they originate from a common precursor. The reactive pathways in this system were extensively characterized computationally. We used DFT and post-Hartree Fock electronic structure calculations, Frank-Condon analysis to support our experimental findings. The results were consistent with previous direct ab initio molecular dynamics calculations, we found two nearly iso-energetic Wheland intermediates which lie significantly lower in energy than the initially formed dimer cation radical [PhCl$\cdots$$NH_3$]$^+$·.

RB07 10:27 – 10:42

PHOTOIONIZATION INDUCED INTERMOLECULAR PROTON TRANSFER IN THE CH...O HYDROGEN BONDED CYCLOPENTANONE DIMER IN THE GAS PHASE

ARUP K. GHOSH, TAPAS CHAKRABORTY, *Physical Chemistry, Indian Association for the Cultivation of Science, Kolkata, India.*

Although α-cleavage is the preferred reaction channel upon photoionization of small carbonyl compounds, we present experimental and theoretical evidences that in a jet-cooled CH...O hydrogen bonded dimeric complex, enolization and intermolecular proton transfer turns out to be one of the prominent channels of cyclopentanone when photoionized at the threshold by a two-photon ionization scheme. On changing the ionization wavelength of the nanosecond laser pulses, we show that the fragmentation channels of dimer as well as monomer are immensely altered for photoionizations by the 2nd (532 nm), 3rd (355 nm)and 4th (266 nm) harmonics of a Nd: YAG laser. The observed changes are interpreted with the energetic predictions of electronic structure calculations.

RB08 10:44 – 10:59

MICROWAVE SPECTRUM OF HYDROGEN BONDED HEXAFLUOROISOPROPANOL●●●WATER COMPLEX

ABHISHEK SHAHI, ELANGANNAN ARUNAN, *Department of Inorganic and Physical Chemistry, Indian Institute of Science, Bangalore, India.*

Stabilizing α-helical structure of protein and dissolving a hard to dissolve polymer, polythene terphthalete, are some of the unique properties of the organic solvent Hexafluoroisopropanol (HFIP). After determining the complete microwave spectrum of HFIP monomer[a], we have recorded the spectrum of HFIP●●●H_2O complex. Ab initio calculations were used to optimize three different possible structures. The global minimum, structure 1, had HFIP as proton donor. Another promising structure, Structure 2, has been obtained from a molecular dynamic study[b]. A total of 46 observed lines have been fitted well for obtaining the rotational and distortion constants within experimental uncertainty. The observed rotational constants are A = 1134.53898(77) MHz, B = 989.67594(44) MHz and C = 705.26602(20) MHz. Interestingly, the rotational constants of structure 1, structure 2 and experiments were very close. Experimentally observed distortion constants were close to structure 1. $b-type$ transitions were stronger than $c-type$ which is also consistent with the calculated dipole moment components of structure 1. Calculations predict a non-zero a-dipole moment but experimentally $a-type$ transitions were absent. Microwave spectra of two of the deuterium isotopologues of this complex i.e. HFIP●●●D_2O (30 transitions) and HFIP●●●HOD (33 transitions) have been also observed. Search for other isotopologues are in progress. To characterize the nature of hydrogen bonding, Atoms in Molecules and Natural Bond Orbital theoretical analysis have been done. Experimental structure and these theoretical analyses indicate that the hydrogen bonding in HFIP●●●H_2O complex is stronger than that in water dimer.

[a] A. Shahi and E. Arunan, Talk number RK16, 68th International Symposium on Molecular Spectroscopy 2013, Ohio, USA.
[b] Yamaguchi, T.; Imura, S.; Kai, T.; Yoshida, K. Zeitschrift für Naturforsch. A 2013, 68a, 145.

RB09　　　　　　　　　　　　　　　　　　　　　　　　　　　　　**11:01 – 11:16**

HYDROGEN BOUND COMPLEXES WITH TROPOLONE: BINDING MOTIFS, BARRIER HEIGHTS, AND THE SEARCH FOR BIFURCATING SYSTEMS

DEACON NEMCHICK, KATHRYN CHEW, PATRICK VACCARO, *Department of Chemistry, Yale University, New Haven, CT, USA.*

The potentially frustrated transfer of a proton between the hydroxylic (proton-donating) and ketonic (proton-accepting) oxygen atom centers in tropolone (TrOH) long has served as a model system for the study of coherent (symmetrical) proton-transfer events. A litany of hydrogen-bound complexes [TrOH · X_n] can be formed *in situ* by docking amphoteric ligands onto the TrOH substrate under supersonic free-jet expansion conditions. Binary ($n = 1$) and higher order ($n = 2, 3, \ldots$) complexes formed with formic acid, hydrogen fluoride, acetic acid and propiolic acid (X = FA, HF, AA, and PA) have been synthesized and interrogated using a variety of spectroscopic probes built upon the intense $\tilde{A}^1B_2 - \tilde{X}^1A_1$ ($\pi^* \leftarrow \pi$) near-ultraviolet absorption system of bare tropolone, thereby providing vibronically resolved information through combined use of laser-induced fluorescence (LIF), dispersed fluorescence (DF), fluorescence hole-burning (FHB), and stimulated emission pumping (SEP) methods. Experimental results reveal the propensity for binary complexes to adopt a higher-energy external binding motif (ligand attached to the seven membered aromatic ring) over the energetically preferred internal form (ligand bound to the O−H···O reaction center), where the latter cleft-bound species can undergo unique symmetric (coherent) double proton-transfer reactions. These findings will be discussed in light of supporting quantum-chemical calculations.

RB10　　　　　　　　　　　　　　　　　　　　　　　　　　　　　**11:18 – 11:33**

INTERMOLECULAR INTERACTIONS BETWEEN FORMALDEHYDE AND DIMETHYL ETHER AND BETWEEN FORMALDEHYDE ABD DIMETHYL SULFIDE IN THE COMPLEX, INVESTIGATED BY FOURIER TRANSFORM MICROWAVE SPECTROSCOPY AND AB INITIO CALCULATIONS

YOSHIYUKI KAWASHIMA, YOSHIO TATAMITANI, YOSHIHIRO OSAMURA, *Applied Chemistry, Kanagawa Institute of Technology, Atsugi, Japan*; EIZI HIROTA, *The Central Office, The Graduate University for Advanced Studies, Hayama, Kanagawa, Japan.*

The ground-state rotational spectra of the formaldehyde-dimethyl ether (H_2CO − DME) and the formaldehyde-dimethyl sulfide (H_2CO − DMS) complexes have been studied by Fourier transform microwave spectroscopy, and a-type and c-type transitions have been observed and assigned for the H_2CO as well as D_2CO species of both the complexes. In the case of the H_2CO − DME, doublets were observed with the splitting of a few 100 kHz, whereas no such splitting was found for the H_2CO − DMS. The observed rotational constants were analyzed to conclude a C_s geometry, where DME and DMS were bound to H_2CO by the two C-H(DME/DMS)—O(H_2CO) and the two O(DME)/S(DMS)—H-C(H_2CO) hydrogen bonds. The distances R_{cm} between the centers of mass of the constituents were determined to be 3.102 and 3.200 Åfor the two complexes, respectively. Both the H_2CO − DME and H_2CO − DMS complexes were shown by the natural bond orbital (NBO) analysis, to have strong charge transfer interactions between the constituents, as expected from the strong electron-accepting character of the H_2CO.

RB11　　　　　　　　　　　　　　　　　　　　　　　　　　　　　**11:35 – 11:50**

A VIBRATIONAL MODEL FOR ACCURATE DETERMINATION OF OSCILLATOR STRENGTHS IN HYDROGEN BONDED COMPLEXES

KASPER MACKEPRANG, HENRIK KJÆRGAARD, *Department of Chemistry, University of Copenhagen, Copenhagen, Denmark.*

Equilibrium constants of the formation of gas-phase hydrogen bonded molecular complexes have been determined using a method combining experimental and theoretically simulated vibrational spectra. The equilibrium constants are determined from the monomer pressures and the pressure of the complex. The pressures of the monomers are measured experimentally, but the pressure of the complex is too low to detect. However, a combination of an experimentally measured integrated absorbance from Fourier-Transform Infrared spectra and a theoretically calculated oscillator strength allows us to estimate the pressure of the complex and determine the equilibrium constant of the complex formation. We have developed a vibrational model to improve the accuracy of the calculated oscillator strengths, which is used to determine the pressures of the molecular complexes, thus improving the accuracy of the equilibrium constants. Our model and results obtained using this model will be presented and discussed

194

RB12 **11:52 – 12:07**

EFFECT OF SUBSTITUENTS IN ALCOHOL-AMINE COMPLEXES

<u>ANNE SCHOU HANSEN</u>, LIN DU, HENRIK KJÆRGAARD, *Department of Chemistry, University of Copenhagen, Copenhagen, Denmark.*

A series of alcohol-amine complexes have been investigated to gain physical insight into the effect on the hydrogen bond strength as different substituents are attached. The series of complexes investigated are shown in the figure, where R_1 = CH_3, CH_3CH_2 or CF_3CH_2 and R_2 = H or CH_3. To estimate the hydrogen bond strength, redshifts of the OH-stretching transition frequency upon complexation were measured using gas phase Fourier Transform InfraRed (FTIR) spectroscopy. Equilibrium constants for the formation of the complexes were also determined, exploiting a combination of a calculated oscillator strength and the measured integrated absorbance of the fundamental OH-stretching and second overtone NH-stretching transitions.

RC. Theory and Computation
Thursday, June 19, 2014 – 8:30 AM
Room: B102 Chemical and Life Sciences

Chair: So Hirata, University of Illinois, Urbana, IL, USA

RC01   8:30–8:45

SIMULTANEOUS EVALUATION OF MULTIPLE ROTATIONALLY EXCITED STATES OF FLOPPY MOLECULES USING DIFFUSION MONTE CARLO

ANNE B McCOY, JASON E. FORD, MELANIE L. MARLETT, *Department of Chemistry and Biochemistry, The Ohio State University, Columbus, OH, USA*; ANDREW S. PETIT, *Department of Chemistry, University of Pennsylvania, Philadelphia, PA, USA.*

In this work, an extension to diffusion Monte Carlo (DMC) is proposed, allowing for the simultaneous calculation of the energy and wave function of multiple rotationally excited states of floppy molecules.[a] The total wave function is expanded into a set of Dirac δ-functions called walkers, while the rotational portion of the wave function is expanded in a symmetric top basis set. Each walker is given a rotational state vector containing coefficients for all states of interest. The positions of the atoms and the coefficients in the state vector evolve according to the split operator approximation of the quantum propagator. The method was benchmarked by comparing calculated rotation-vibration energies for H_3^+, H_2D^+, and H_3O^+ to experimental values. For low to moderate values of J, the resulting energies are within the statistical uncertainty of the calculation. Rotation-vibration coupling is captured through flexibility introduced in the form of the vibrational wave function. This coupling is found to increase with increasing J-values. Based on the success achieved through these systems, the method was applied to CH_5^+ and its deuterated isotopologues for $v = 0$, $J \geq 10$. Based on these calculations, the energy level structure of CH_5^+ is found to resemble that for a of a spherical top, and excitations up to $J = 10$ displayed insignificant rotation-vibration coupling. Extensions of this approach that explicitly account for vibrations will also be discussed.

[a] A. S. Petit, J. E. Ford and A. B. McCoy, J. Phys. Chem. A, in press, K. D. Jordan Festschrift, DOI: 10.1021/jp408821a

RC02   8:47–9:02

NUMERICALLY EXACT CALCULATION OF ROVIBRATIONAL LEVELS OF Cl^-H_2O

XIAO-GANG WANG, TUCKER CARRINGTON, *Department of Chemistry, Queen's University, Kingston, ON, Canada.*

Large amplitude vibrations of Van der Waals clusters are important because they reveal large regions of a potential energy surface (PES). To calculate spectra of Van der Waals clusters it is common to use an adiabatic approximation. When coupling between intra- and inter-molecular coordinates is important non-adiabatic coupling cannot be neglected and it is therefore critical to develop and test theoretical methods that couple both types of coordinates. We have developed new product basis and contracted basis Lanczos methods for Van der Waals complexes and tested them by computing rovibrational energy levels of Cl^-H_2O. The new product basis is made of functions of the inter-monomer distance, Wigner functions that depend on Euler angles specifying the orientation of H_2O with respect to a frame attached to the inter-monomer Jacobi vector, basis functions for H_2O vibration, and Wigner functions that depend on Euler angles specifying the orientation of the inter-monomer Jacobi vector with respect to a space-fixed frame. An advantage of this product basis is that it can be used to make an efficient contracted basis by replacing the vibrational basis functions for the monomer with monomer vibrational wavefunctions. Due to weak coupling between intra- and inter-molecular coordinates, only a few tens of monomer vibrational wavefunctions are necessary. The validity of the two new methods is established by comparing energy levels with benchmark rovibrational levels obtained with polyspherical coordinates and spherical harmonic type basis functions. For all bases, product structure is exploited to calculate eigenvalues with the Lanczos algorithm.

For Cl^-H_2O, we are able, for the first time, to compute accurate splittings due to tunnelling between the two equivalent C_s minima. We use the PES of Rheinecker and Bowman (RB).[a] Our results are in good agreement with experiment for the five fundamental bands observed.[b]

[a] J. Rheinecker and J. M. Bowman, J. Chem. Phys. **124** 131102 (2006); J. Rheinecker and J. M. Bowman, J. Chem. Phys. **125** 133206 (2006)
[b] S. Horvath, A. B. McCoy, B. M. Elliott, G. H. Weddle, J. R. Roscioli, and M. A. Johnson J. Phys. Chem. A **114** 1556 (2010)

RC03 9:04 – 9:19

AUTOMATIC GENERATION OF ANALYTIC EQUATIONS FOR VIBRATIONAL AND ROVIBRATIONAL CONSTANTS FROM FOURTH-ORDER VIBRATIONAL PERTURBATION THEORY

DEVIN A. MATTHEWS, JUSTIN Z GONG, *Department of Chemistry and Biochemistry, The University of Texas, Austin, TX, USA*; JOHN F. STANTON, *Department of Chemistry, The University of Texas, Austin, TX, USA.*

The derivation of analytic expressions for vibrational and rovibrational constants, for example the anharmonicity constants χ_{ij} and the vibration-rotation interaction constants α_r^B, from second-order vibrational perturbation theory (VPT2) can be accomplished with pen and paper and some practice. However, the corresponding quantities from fourth-order perturbation theory (VPT4) are considerably more complex, with the only known derivations by hand extensively using many layers of complicated intermediates and for rotational quantities requiring specialization to orthorhombic cases or the form of Watson's reduced Hamiltonian. We present an automatic computer program for generating these expressions with full generality based on the adaptation of an existing numerical program based on the sum-over-states representation of the energy to a computer algebra context. The measures taken to produce well-simplified and factored expressions in an efficient manner are discussed, as well as the framework for automatically checking the correctness of the generated equations.

RC04 9:21 – 9:36

ROVIBRATIONAL CONSTANTS FROM FOURTH-ORDER PERTURBATION THEORY AND THE RELATIONSHIP TO THE CONTACT TRANSFORMATION APPROACH

DEVIN A. MATTHEWS, JUSTIN Z GONG, *Department of Chemistry and Biochemistry, The University of Texas, Austin, TX, USA*; JOHN F. STANTON, *Department of Chemistry, The University of Texas, Austin, TX, USA.*

The nuclear Schrödinger equation forms the basis for vibrational perturbation theory. However, the canonical "$\hat{H}\Psi = E\Psi$" form is shown to be insufficient for computing elements of the rotational effective Hamiltonian and ultimately the various rovibrational and vibrationally-averaged rotational molecular properties. The necessary modifications are shown to be a natural result of derivation of the Schrödinger equation in a non-commutative algebra. The resulting equation is also compared to the contact transformation approach (CVVPT) and the two are shown to be equivalent (as is well known in the pure vibrational case). Lastly, equations for various rovibrational constants using this corrected approach and fourth-order perturbation theory (VPT4) are presented and compared to the available literature results.

RC05 9:38 – 9:53

APPLICATION OF FOURTH ORDER VIBRATIONAL PERTURBATION THEORY WITH ANALYTIC HARTREE-FOCK FORCE FIELDS

JUSTIN Z GONG, DEVIN A. MATTHEWS, *Department of Chemistry and Biochemistry, The University of Texas, Austin, TX, USA*; JOHN F. STANTON, *Department of Chemistry, The University of Texas, Austin, TX, USA.*

Fourth-Order Rayleigh-Schrodinger Perturbation Theory (VPT4) is applied to a series of small molecules. The quality of results have been shown to be heavily dependent on the quality of the quintic and sextic force constants used and that numerical sextic force constants converge poorly and are unreliable for VPT4. Using analytic Hartree-Fock force constants, it is shown that these analytic higher-order force constants are comparable to corresponding force constants from numerical calculations at a higher level of theory. Calculations show that analytic Hartree-Fock sextic force constants are reliable and can provide good results with Fourth-Order Rayleigh-Schrodinger Perturbation Theory.

RC06   9:55 – 10:05

ROTATIONAL SPECTRUM OF SO_3 AND THEORETICAL EVIDENCE FOR THE FORMATION OF ROTATIONAL ENERGY LEVEL CLUSTERS IN ITS VIBRATIONAL GROUND STATE

DANIEL S UNDERWOOD, *Department of Physics and Astronomy, University College London, Gower Street, London WC1E 6BT, United Kingdom*; SERGEI N. YURCHENKO, *Department of Physics and Astronomy, University College London, Gower Street, London WC1E 6BT, United Kingdom*; JONATHAN TENNYSON, *Department of Physics and Astronomy, University College London, London, IX, United Kingdom*; PER JENSEN, *Faculty of Mathematics and Natural Sciences, University of Wuppertal, Wuppertal, Germany.*

The structure of the purely rotational spectrum of sulphur trioxide SO_3 is investigated using a new synthetic line list. The list combines line positions from an empirical model with line intensities determined, in the form of Einstein coefficients, from variationally computed ro-vibrational wavefunctions in conjunction with an *ab initio* dipole moment surface. The empirical model providing the line positions involves an effective, Watsonian-type rotational Hamiltonian with literature parameter values resulting from least-squares fittings to observed transition frequencies. The formation of so-called rotational energy clusters at high rotational excitation are investigated. The SO_3 molecule is planar at equilibrium and exhibits a unique type of rotational-energy clustering associated with unusual stabilization axes perpendicular to the S–O bonds. This behaviour is characterized theoretically in the J range from 100 through 250. The wavefunctions for these cluster states are analysed, and the results are compared to those of a classical analysis in terms of the rotational-energy-surface formalism.

RC07   10:07 – 10:22

CONSISTENT ASSIGNMENT OF THE VIBRATIONS OF PARA DISUBSTITUTED BENZENE MOLECULES

ANNA ANDREJEVA, ALISON JASMIN LEE, TIM WRIGHT, *School of Chemistry, University of Nottingham, Nottingham, United Kingdom.*

When disubstituted benzene molecules are considered the relative position of the substituents must be defined. The three possible forms are ortho-, meta-, and para- where the latter is investigated in this work by consideration of the effect para positioned substituents will have on the vibrational modes. The consistency of the labelling and assignment of the vibrational frequencies of the para disubstituted benzene molecules is investigated in their ground states (S_0) and first electronically excited states (S_1). The work extends a previously published nomenclature where ring-localised vibrations are compared straightforwardly across different monosubstituted benzene species and given the label M_i.[a,b,c] The assignments of the frequencies include previous work but also the calculated wavenumbers for both hydrogenated disubstituted benzenes (-h_4) and the deuterated isotopologues (-d_4) employing density functional theory (DFT) and time-dependent density functional theory (TDDFT).

[a] A. M. Gardner and T. G. Wright, J. Chem. Phys., 135, 114305 (2011)

[b] A. M. Gardner, A. M. Green, V. M. Tame-Reyes, V. H. K. Wilton and T. G. Wright, 138, 134303 (2013)

[c] A. M. Gardner, A. M. Green, V. M. Tame-Reyes, K. L. Reid, J. A. Davies, V. H. K. Wilton and T. G. Wright, manuscript accepted

Intermission

198

ANHARMONIC VIBRATIONAL SPECTROSCOPY ON METAL TRANSITION COMPLEXES

CAMILLE LATOUCHE, *Scuola Normale Superiore, Scuola Normale Superiore, Pisa, Italy*; JULIEN BLOINO, *Istituto di Chimica dei Composti OrganoMetallici (ICCOM-CNR), UOS di Pisa, Consiglio Nazionale delle Ricerche, Pisa, Italy*; VINCENZO BARONE, *Scuola Normale Superiore, Scuola Normale Superiore, Pisa, Italy.*

Advances in hardware performance and the availability of efficient and reliable computational models have made possible the application of computational spectroscopy to ever larger molecular systems. The systematic interpretation of experimental data and the full characterization of complex molecules can then be facilitated. Focusing on vibrational spectroscopy, several approaches have been proposed to simulate spectra beyond the double harmonic approximation, so that more details become available. However, a routine use of such tools requires the preliminary definition of a valid protocol with the most appropriate combination of electronic structure and nuclear calculation models. Several benchmark of anharmonic calculations frequency have been realized on organic molecules. Nevertheless, benchmarks of organometallics or inorganic metal complexes at this level are strongly lacking despite the interest of these systems due to their strong emission and vibrational properties.

Herein we report the benchmark study realized with anharmonic calculations on simple metal complexes, along with some pilot applications on systems of direct technological or biological interest.

FREE O-H ANHARMONIC STRETCHING MOTIONS IN $H^+(CH_3OH)_{1-3}$ WITH/WITHOUT ATTACHED ARGON.

HSIAO-HAN CHUANG, JER-LAI KUO, KAITO TAKAHASHI, *Institute of Atomic and Molecular Sciences, Academia Sinica, Taipei, Taiwan*; ASUKA FUJII, *Department of Chemistry, Tohoku University, Sendai, Japan.*

The free O-H stretching modes of $H^+(CH_3OH)_{1-3}$ clusters with/without argon have been unraveled by combining Infrared Pre-Dissociation (IR-PD) spectra and multidimensional normal mode analysis by Density Functional Theory (DFT) methods. Experimental IR-PD spectra have shown sharp peaks between 3200 and 3800 cm^{-1}, and these peaks shift when the size of cluster is varied, as well as increasing the number of argon attachment. We benchmarked several DFT and ab initio methods and the results shown that the free O-H stretching modes can well described by using a few selected normal mode coordinates with standard DFT methods.

UNFOLDING THE QUANTUM NATURE OF PROTON BOUND SYMMETRIC DIMERS OF $(MeOH)_2H^+$ AND $(Me_2O)_2H^+$: A THEORETICAL STUDY

JAKE ACEDERA TAN, JER-LAI KUO, *Institute of Atomic and Molecular Sciences, Academia Sinica, Taipei, Taiwan.*

A proton under a tug of war between two competing Lewis bases is a common motif in biological systems and proton transfer processes[a – b]. Over the past decades, model compounds for such motifs can be prepared by delicate stoichiometric control of salt solutions [c]. Unfortunately, condensed phase studies, which aims to identify the key vibrational signatures are complicated to analyze. As a result, gas-phase studies do provide promising insights on the behavior of the shared proton. This study attempts to understand the quantum nature of the shared proton under theoretical paradigms. Proton bound symmetric dimers of $(MeOH)_2H^+$ and $(Me_2O)_2H^+$ are chosen as the model compounds. The simulation is performed using Density Functional Theory (DFT) at the B3LYP level with 6-311+G(d,p) as the basis set. It was found out that stretching mode of shared proton couples with several other normal modes and its corresponding oscillator strength do distribute to other normal modes.

[a]J.R. Roscioli, L.R. McCunn and M.A. Johnson. Science 2007, 316, 249
[b]T.E. DeCoursey. Physiol. Rev., 2003, 83, 475
[c]E.S. Stoyanov. Psys. Chem. Phys., 2000,2,1137

RC11 **11:30 – 11:45**

ANHARMONIC IR SPECTRA OF BIOMOLECULES: NUCLEOBASES AND THEIR OLIGOMERS

VINCENZO BARONE, *Scuola Normale Superiore, Scuola Normale Superiore, Pisa, Italy*; MALGORZATA BICZYSKO, JULIEN BLOINO, *Istituto di Chimica dei Composti OrganoMetallici (ICCOM-CNR), UOS di Pisa, Consiglio Nazionale delle Ricerche, Pisa, Italy*; IVAN CARNIMEO, TERESA FORNARO, *Scuola Normale Superiore, Scuola Normale Superiore, Pisa, Italy*.

Computational spectroscopy techniques have become in the last years effective means to predict and characterize spectra, such as infrared, for molecular systems of increasing dimensions with account for different environments. We are actively developing a comprehensive and robust computational protocol, set within a perturbative vibrational framework [1], aimed at a quantitative reproduction of the spectra of biomolecules.

In order to model the vibrational spectra of weakly bound molecular complexes, dispersion interactions should be taken into proper account. In this work, we present critical assessment of dispersion-corrected DFT approaches for anharmonic vibrational frequency calculations. It is shown that fully anharmonic IR spectra, simulated through full and reduced-dimensionality generalized second-order vibrational perturbation theory (GVPT2)[1] with the potential energy surfaces computed with the B3LYP-D3 approach, may be used to interpret experimental data of nucleobases and their complexes[2] by the direct comparison of experimental IR spectra with their theoretical anharmonic counterpart, taking into account also overtones and combination bands.

[1] V. Barone, M. Biczysko, J. Bloino, Phys. Chem. Chem. Phys., 2014,16, 1759-1787

[2] T. Fornaro, M. Biczysko, S. Monti, V. Barone , Phys. Chem. Chem. Phys., 2014, DOI: 10.1039/C3CP54724H

RC12 **11:47 – 12:02**

ACCURATE CHARACTERIZATION OF THE PEPTIDE LINKAGE IN THE GAS PHASE: A JOINT QUANTUM-CHEMICAL AND ROTATIONAL SPECTROSCOPY STUDY OF THE GLYCINE DIPEPTIDE ANALOGUE

CRISTINA PUZZARINI, *Dep. Chemistry 'Giacomo Ciamician', University of Bologna, Bologna, Italy*; MALGORZATA BICZYSKO, VINCENZO BARONE, *Scuola Normale Superiore, Scuola Normale Superiore, Pisa, Italy*; LAURA LARGO, ISABEL PEÑA, CARLOS CABEZAS, JOSÉ L. ALONSO, *Departmento Química Física/ Grupo de Espectroscopía Molecular, Universidad de Valladolid, Valladolid, Spain*.

Accurate structures of aminoacids in the gas phase have been obtained by joint microwave and quantum-chemical investigations. However, the structure and conformational behavior of α-aminoacids once incorporated into peptide chains are completely different and have not yet been characterized with the same accuracy. To fill this gap, we present here an accurate characterization of the simplest dipeptide analogue (N-acetylglycinamide) involving peptidic bonds. State-of-the-art quantum-chemical computations are complemented by a comprehensive study of the rotational spectrum using a combination of Fourier transform microwave spectroscopy with laser ablation. The coexistence of the C_7 and C_5 conformers has been proved and energetically as well as spectroscopically characterized. This joint theoretical-experimental investigation demonstrated the feasibility of obtaining accurate structures for flexible small biomolecules, thus paving the route to the elucidation of the inherent behavior of peptides.

RD. Fundamental interest

Thursday, June 19, 2014 – 8:30 AM

Room: 112 Chemistry Annex

Chair: Ming-Wei Chen, University of Illinois, Urbana, IL, USA

RD01   8:30 – 8:45

HALO NUCLEIC MOLECULES: MOLECULES FORMED FROM AT LEAST ONE ATOM WITH A HALO NUCLEUS. EMPHASIS ON $^{11,11}Li_2$ ALONG WITH OTHER EXOTIC ISOTOPOLOGUES.

NIKESH S. DATTANI, STASZEK WELSH, *Physical and Theoretical Chemistry Laboratory, Oxford University, Oxford, United Kingdom.*

Atoms whose nuclei have an exotic number of nucleons can have a 'core nucleus' surrounded by a 'halo' formed by a nucleon orbiting the core nucleus. For example, due to the two halo neutrons orbiting the core nucleus of ^{11}Li, its nucleus has a cross section that is roughly the same size as that of ^{208}Pb. Halo nucleic atoms have been studied extensively both in theory and in experiments, however halo nucleic molecules have not been studied in either. We first show, using HeH^+, BeH, and MgH as examples, that with measurements of any two isotopologues of a molecule, we can determine crucial properties of a third isotopologue well within spectroscopic accuracy. We then use the extremely precise empirical information available[a,b,c,d] for the low-lying states of $^{6,6}Li_2$, $^{6,7}Li_2$ and $^{7,7}Li_2$ to predict potentials and various properties of the halo nucleic molecule $^{11,11}Li_2$, along with isotopologues containing ^3Li, ^4Li, ^5Li, ^8Li, ^9Li, ^{10}Li, and ^{12}Li. We believe that our predictions of the ro-vibrational energies are reliable for experiments for the first detection of a halo nucleic molecule.

[a]R. J. Le Roy, N. S. Dattani, J. A. Coxon, A. J. Ross, P. Crozet, C. Linton, *J. Chem. Phys.* **131**, 204309 (2009).

[b]N. S. Dattani, R. J. Le Roy, *J. Mol. Spec.* **268**, 199-210 (2011).

[c]M. Semczuk, X. Li, W. Gunton, M. Haw, N. S. Dattani, J. Witz, A. Mills, D. J. Jones, K. W. Madison, *Phys. Rev. A* **87**, 052505 (2013)

[d]W. Gunton, M. Semczuk, N. S. Dattani, K. W. Madison, *Phys. Rev. A* **88**, 062510 (2013)

RD02   8:47 – 9:02

QED CORRECTION FOR H_3^+

LORENZO LODI, OLEG POLYANSKY, JONATHAN TENNYSON, *Department of Physics and Astronomy, University College London, London, IX, United Kingdom*; ALEXANDER ALIJAH, *GSMA - Champagne Ardennne, Université de Reims, Reims Cedex 2, France*; NIKOLAY FEDOROVICH ZOBOV, *Microwave Spectroscopy, Institute of Applied Physics, Nizhny Novgorod, Russia.*

A quantum electrodynamics (QED) correction surface for the simplest polyatomic and polyelectronic system H_3^+ is computed using an approximate procedure. This surface is used to calculate the shifts to vibration-rotation energy levels due to QED; such shifts have a magnitude of up to 0.25 cm^{-1} for vibrational levels up to $15\,000$ cm^{-1} and are expected to have an accuracy of about 0.02 $^{-1}$. Combining the new H_3^+ QED correction surface with existing highly accurate Born-Oppenheimer (BO), relativistic and adiabatic components suggests that deviations of the resulting *ab initio* energy levels from observed ones are largely due to non-adiabatic effects.

RD03

CALCULATING POTENTIAL ENERGY CURVES WITH QUANTUM MONTE CARLO

ANDREW D POWELL, RICHARD DAWES, *Department of Chemistry, Missouri University of Science and Technology, Rolla, MO, USA.*

Quantum Monte Carlo (QMC) is a computational technique that can be applied to the electronic Schrödinger equation for molecules. QMC methods such as Variational Monte Carlo (VMC) and Diffusion Monte Carlo (DMC) have demonstrated the capability of capturing large fractions of the correlation energy, thus suggesting their possible use for high-accuracy quantum chemistry calculations. QMC methods scale particularly well with respect to parallelization making them an attractive consideration in anticipation of next-generation computing architectures which will involve massive parallelization with millions of cores. Due to the statistical nature of the approach, in contrast to standard quantum chemistry methods, uncertainties (error-bars) are associated with each calculated energy. This study focuses on the cost, feasibility and practical application of calculating potential energy curves for small molecules with QMC methods. Trial wave functions were constructed with the multi-configurational self-consistent field (MCSCF) method from GAMESS-US.[1] The CASINO Monte Carlo quantum chemistry package [2] was used for all of the DMC calculations. An overview of our progress in this direction will be given.

References: M. W. Schmidt et al. J. Comput. Chem. 14, 1347 (1993). R. J. Needs et al. J. Phys.: Condensed Matter 22, 023201 (2010).

RD04

MILLIMETER-WAVE SPECTROSCOPY OF S_2Cl_2: A CANDIDATE MOLECULE FOR THE DETECTION OF *ORTHO-PARA* TRANSITION

ZEINAB TAFTI DEHGHANI, ASAO MIZOGUCHI, HIDETO KANAMORI, *Department of Physics, Tokyo Institute of Technology, Tokyo, Japan.*

S_2Cl_2 is a candidate molecule for the detection of *ortho-para* transition, because the Cl atoms on the skewed position from the rotational principle axes give large off-diagonal terms in the quadrupole interaction, which can mix *ortho* and *para* states. In order to estimate the *ortho-para* mixing in a hyperfine-resolved rotational state, pure rotational transitions were measured by millimeter-wave spectroscopy using two different experimental set-ups. The transitions from the term value around 20 K was measured with a supersonic jet and those around 200 K were measured with a dry ice cooled gas cell. Hundreds of peaks were assigned for the naturally abundant $S_2{}^{35}Cl_2$ and $S_2{}^{35}Cl^{37}Cl$ isotopic species, and the rotational molecular constants including the fourth-order and sixth-order centrifugal distortion constants were determined. The hyperfine structures were partly resolved in some Q-branch transitions and those spectral patterns were well reproduced with the hyperfine constants determined by the previous FTMW spectroscopy in the cm-wave region[a]. With the new molecular constants determined in this study and the previous hyperfine constants, it becomes possible to predict a more reliable *ortho-para* mixing ratio and to narrow down the possible candidate transitions in the mm-wave region for the detection of *ortho-para* transition.

---

[a]Mizoguchi et al., J. Mol. Spectrosc. 250,86-97(2008)

202

RD05   9:38 – 9:53

HIGH-RESOLUTION INFRARED SPECTROSCOPY OF CUBANE, C_8H_8

VINCENT BOUDON, *Laboratoire ICB, CNRS/Université de Bourgogne, DIJON, France*; OLIVIER PIRALI, SÉBASTIEN GRUET, *AILES beamline, Synchrotron SOLEIL, Saint Aubin, France*; LUCIA D'ACCOLTI, CATERINA FUSCO, COSIMO ANNESE, *Dipartimento di Chimica, Università di Bari A. Moro, Bari, Italy.*

Carbon-cage molecules have generated a considerable interest from both experimental and theoretical point of views. We recently performed a high-resolution study of adamantane ($C_{10}H_{16}$), the smallest hydrocarbon cage belonging to the diamandoid family. There exist another family of hydrocarbon cages with additional interesting chemical properties: the so-called Platonic hydrocarbons that comprise dodecahedrane ($C_{20}H_{20}$) and cubane (C_8H_8). Both possess C–C bond angles that deviate from the tetrahedral angle ($109.8°$) of the sp^3 hybridized form of carbon. This generates a considerable strain in the molecule. Cubane itself has the highest density of all hydrocarbons (1.29 g/cm^3). This makes it able to store larges amounts of energy, although the molecule is fully stable. Up to now, only one high-resolution study of cubane has been performed on a few bands [2].

We report here a new wide-range high-resolution study of the infrared spectrum of cubane. The sample was synthesized in Bari upon decarboxylation of 1,4-cubanedicarboxylic acid thanks to the improved synthesis of literature [3]; its ^{1}H and ^{13}C NMR, FTIR, and mass spectrometry agreed with reported data [4]. Several spectra have been recorded at the AILES beamline of the SOLEIL French synchrotron facility. They cover the 800 to 3100 cm^{-1} region. Besides the three infrared-active fundamentals (ν_{10}, ν_{11} and ν_{12}), we could record many combination bands, all of them displaying a well-resolved octahedral rotational structure. We present here a preliminary analysis of some of the recorded bands, performed thanks the SPVIEW and XTDS software, based on the tensrorial formalism developed in the Dijon group [5].

[1] O. Pirali, V. Boudon, J. Oomens, M. Vervloet, *J. Chem. Phys.*, **136**, 024310 (2012).

[2] A. S. Pine, A. G. Maki, A. G. Robiette, B. J. Krohn, J. K. G. Watson, Th. Urbanek, *J. Am. Chem. Soc.*, **106**, 891–897 (1984).

[3] P. E. Eaton, N. Nordari, J. Tsanaktsidis, P. S. Upadhyaya, *Synthesis*, **1**, 501, (1995).

[4] E. W. Della, P. T. Hine, H. K. Patney, *J. Org. Chem.*, **42**, 2940 (1978).

[5] Ch. Wenger, V. Boudon, M. Rotger, M. Sanzharov and J.-P. Champion, *J. Mol. Spectrosc.*, **251** 102–113 (2008).

Intermission

RD06   10:10 – 10:25

MICROWAVE SPECTROSCOPY AND MOLECULAR STRUCTURE OF ISONITROSYL HYDROXIDE (HOON)

KYLE N CRABTREE, *Atomic and Molecular Physics, Harvard-Smithsonian Center for Astrophysics, Cambridge, MA, USA*; MARAT R TALIPOV, *Department of Chemistry, Marquette University, Milwaukee, WI, USA*; GERARD O'CONNOR, *School of Chemistry, University of Sydney, Sydney, NSW, Australia*; OSCAR MARTINEZ JR., *Atomic and Molecular Physics, Harvard-Smithsonian Center for Astrophysics, Cambridge, MA, USA*; SERGEY L KHURSAN, *Institute of Organic Chemistry, Ufa Scientific Centre, Russian Academy of Sciences, Ufa, Russia*; MICHAEL C McCARTHY, *Atomic and Molecular Physics, Harvard-Smithsonian Center for Astrophysics, Cambridge, MA, USA.*

Nitrous acid (HONO) is an important member of the atmospheric nitrogen cycle whose chemistry involves a variety of gas-phase, photochemical, and heterogeneous processes. Among its formation pathways in the atmosphere is the ternary association of hydroxyl (OH) with nitric oxide (NO), but the formation of the isonitrosyl hydroxide (HOON) isomer has largely been ignored owing to early theoretical studies that questioned its stability. Guided by new high-level ab initio calculations, we have detected the rotational spectrum of *trans*-HOON in an electrical discharge of a dilute mixture of NO and water vapor by a combination of Fourier transform microwave spectroscopy and double resonance methods. No evidence for the *cis* isomer was found in any of our spectroscopic surveys between 15.4–17.0 GHz. A semi-experimental equilibrium structure for *trans*-HOON has been derived to high precision from isotopic substitution (DOON, H^{18}OON, HO^{18}ON, HOO^{15}N) along with zero-point vibrational corrections calculated at the CCSD(T)/aug-cc-pVTZ level of theory. Most notably, the central O–O bond in *trans*-HOON is found to be 1.9149 ± 0.0005 Å in length, which is the longest known O–O bond in a molecule (nearly 20% longer than the analogous bond in the HOOO radical).

RD07

HIGH RESOLUTION JET-COOLED INFRARED ABSORPTION SPECTRA OF FORMIC ACID DIMER: A REINVESTIGATION OF THE FERMI-TRIAD SYSTEM IN THE C-O STRETCHING REGION

CHUANXI DUAN, *College of Physical Science and Technology, Central China Normal University, Wuhan, China.*

High resolution jet-cooled absorption spectra of the formic acid dimer $(HCOOH)_2$ have been measured in the C-O stretching region at 1215-1240 cm^{-1} using a rapid-scan tunable diode laser spectrometer. Three vibrational bands of $(HCOOH)_2$ have been assigned unambiguously. They were interpreted as the Fermi-triad system consisting of the ν_{22} fundamental band and two combination bands in a previous low-resolution study [F. Ito, Chem. Phys. Lett. 447, 202(2007)]. The spectral coverage in the high-resolution study of the middle band [M. Ortlieb and M. Havenith, J. Phys. Chem. A. 111, 7355(2007)] were extended. These three vibrational bands were analyzed together using a standard rigid rotor Watson A-reduced Hamiltonian without explicit consideration of the perturbation among three vibrationally excited states. The perturbed energies for three vibrationally excited states are 1219.71637(20), 1225.34666(15), and 1233.95863(17) cm^{-1}, respectively.

RD08

BROADBAND MID-INFRARED COMB-RESOLVED FOURIER TRANSFORM SPECTROSCOPY

ANDREW MILLS, KEVIN LEE, CHRISTIAN MOHR, JIE JIANG, MARTIN FERMANN, *Laser Research, IMRA AMERICA, Inc, Ann Arbor, MI, USA*; PIOTR MASLOWSKI, *Institute of Physics, Faculty of Physics, Astronomy and Informatics, Nicolaus Copernicus University, Torun, Poland.*

We report on a comb-resolved, broadband, direct-comb spectroscopy system in the mid-IR and its application to the detection of trace gases and molecular line shape analysis. By coupling an optical parametric oscillator (OPO), a 100 m multipass cell, and a high-resolution Fourier transform spectrometer (FTS), sensitive, comb-resolved broadband spectroscopy of dilute gases is possible. The OPO has radiation output at 3.1-3.7 and 4.5-5.5 μm. The laser repetition rate is scanned to arbitrary values with 1 Hz accuracy around 417 MHz. The comb-resolved spectrum is produced with an absolute frequency axis depending only on the RF reference (in this case a GPS disciplined oscillator), stable to 1 part in 10^9. The minimum detectable absorption is 1.6×10^{-6} $cm^{-1}Hz^{-1/2}$.

The operating range of the experimental setup enables access to strong fundamental transitions of numerous molecular species for applications based on trace gas detection such as environmental monitoring, industrial gas calibration or medical application of human breath analysis. In addition to these capabilities, we show the application for careful line shape analysis of argon-broadened CO band spectra around 4.7 μm. Fits of the obtained spectra clearly illustrate the discrepancy between the measured spectra and the Voigt profile (VP), indicating the need to include effects such as Dicke narrowing and the speed-dependence of the collisional width and shift in the line shape model, as was shown in previous cw-laser studies.[a] In contrast to cw-laser based experiments, in this case the entire spectrum (~ 250 cm^{-1}) covering the whole P and R branches can be measured in 16 s with 417 MHz resolution, decreasing the acquisition time by orders of magnitude. The parallel acquisition allows collection of multiple lines simultaneously, removing the correlation of possible temperature and pressure drifts. While cw-systems are capable of measuring spectra with higher precision,[b] this demonstration opens the door for fast, massively parallel line shape parameters retrieval combined with analysis reaching beyond the VP and with absolute frequency calibration delivered by frequency combs.

[a] R. Wehr et al. *J. Mol. Spec.* **235** 54-68 (2003)
[b] A. Cygan, et al. *Eur. Phys. J. Special Topics* **222** 2119-2142 (2013)

RD09

SINGLE MOLECULE RAMAN SPECTROSCOPY UNDER HIGH PRESSURE

YUANXI FU, DANA DLOTT, *Department of Chemistry, University of Illinois at Urbana-Champaign, Urbana, IL, USA.*

Pressure effects on surface-enhanced Raman scattering spectra of Rhdoamine 6G adsorbed on silver nanoparticle surfaces was studied using a confocal Raman microscope. Colloidal silver nanoparticles were treated with Rhodamine 6G (R6G) and its isotopically substituted partner, R6G-d4. Mixed isotopomers let us identify single-molecule spectra, since multiple-molecule spectra would show vibrational transitions from both species. The nanoparticles were embedded into a poly vinyl alcohol film, and loaded into a diamond anvil cell for the high-pressure Raman scattering measurement. Argon was the pressure medium. Ambient pressure Raman scattering spectra showed few single-molecule spectra. At moderately high pressure (1GPa), a surprising effect was observed. The number of sites with observable spectra decreased dramatically, and most of the spectra that could be observed were due to single molecules. The effects of high pressure suppressed the multiple-molecule Raman sites, leaving only the single-molecule sites to be observed.

RD10

THERMAL EXPANSION INVESTIGATIONS OF SINGLE WALLED CARBON NANOTUBES BY RAMAN SPECTROSCOPY AND MOLECULAR DYNAMICS SIMULATIONS

DANIEL CASIMIR, RAUL F GARCIA-SANCHEZ, PRABHAKAR MISRA, *Department of Physics and Astronomy, Howard University, Washington, DC, USA.*

The mechanical properties of nano sized materials differ significantly from the predicted behavior of their bulk macroscopic counterparts. The former tend to be stronger, more malleable and exhibit greater flexibility. The thermal properties of materials have also been shown to be altered significantly after having been shrunken to nanometer dimensions. The nano material and thermo mechanical quantity this study is based on in the examination of this unique behavior is the linear thermal expansion of single walled carbon nanotubes respectively. Single walled carbon nanotubes are hollow cylindrical tubes of sp^2 hybridized carbon atoms with the majority of samples having diameters on the order 1nm, with lengths ranging from 1um to sometimes a centimeter. The foundational material in the mathematical description of these objects' properties (mechanical, electrical, etc) is the planar graphene sheet of graphite. Hence, due to both single walled carbon nanotubes and graphite sharing intra planar, σ-bonded, hexagonally arranged carbon atoms in their structural make up, the latter material is usually used as the conceptual bulk parent material in the transition from the nanometer length scale to the macroscopic regime. The mean linear thermal expansion coefficient is the ratio of the change in unit length in response to a 1 degree Celsius rise in temperature. The "true" value of this quantity is obtained in the theoretical limit of a vanishing temperature range ΔT in the ratio stated above. This simply stated thermo mechanical quantity is structurally sensitive, making it very important in studying any phase transitions in crystalline systems, the intermolecular binding interactions among lattice units and a host of other properties intrinsic in the understanding of the crystalline material of interest (Single walled carbon nanotubes in this case). In conclusion, there is much incentive in examining the somewhat controversial variation in the behavior and quoted values of the thermal expansion of these quasi one dimensional objects. In this study we examine this important property of single walled carbon nanotubes using Resonant Raman Spectroscopy and Molecular Dynamics Simulation based on the Adaptive Intermolecular Reactive Empirical Bond Order potential. The latter is a well established potential that is well suited to modeling chemical reactions in condensed hydrocarbons.

RD11   **11:20 – 11:25**

SPECTROSCOPIC INVESTIGATION OF THE EFFECTS OF ENVIRONMENT ON NEWLY DEVELOPED EMISSIVE MATERIALS

LOUIS E. McNAMARA, NATHAN I HAMMER, *Chemistry and Biochemistry, University of Mississippi, Oxford, MS, USA*; HEMALI RATHNAYAKE, *Chemistry, Western Kentucky University, Bowling Green, OH, USA*; KIETH HOLLIS, *Chemistry, Mississippi State University, Starkville, MS, USA*; JARED DELCAMP, *Chemistry, University of Mississippi, Oxford, MS, USA*.

A comprehensive spectroscopic analysis of recently synthesized novel emissive materials, such as perylene diimide nanostructures, pincer complexes, and newly developed dyes, provides insight into how to modify these materials to be better suited for applications in photovoltaics and photodiodes. Properties of interest in this study include fluorescence emission, fluorescence lifetime, and quantum yield.

Tracking how the photophysics of a compound change as different environments are introduced to the system helps to develop a better understanding of the fundamental photophysical properties of the material. Both solid phase and samples in solution are examined on the bulk and single molecule level.

RD12   **11:27 – 11:37**

NQR STUDY OF LATTICS DYNAMICS in $RbIO_3$ and $KBrO_3$

IGOR VERTEGEL, EUGENY CHESNOKOV, ALEXANDER BARABASH, ALEXANDER OVCHARENKO, *Department of optics and spectroscopy of crystals, Institute of Physics National Academy of Sciences, Kiev, Ukraine*; OLEG PONKRATENKO, *Department of physics, Institute for Nuclear Research, Kiev, Ukraine*; IVAN VERTEGEL, *The Faculty of Mechanics and Mathematics, National Taras Shevchenko University, Kiev, Ukraine*.

The temperature dependence of the I^{127} NQR frequencies for $RbIO_3$ and $KBrO_3$ in the temperature region 4,2-100K has been investigated. The calculation was made by the least-square method by program[a] using the Bayer-Kushida relation [b]. It is assumed that in these crystals the electric field gradient effectively is averaged by the two libration modes polarized along the x and y axes of the molecule (the two-frequency model of the average field gradient at the resonance nucleus site). The calculated values of the parameters ν_1=92+-2 sm^{-1}, ν_2=175+-3 sm^{-1}, ν_1'=120+-3 sm^{-1}, ν_2'=173+-3 sm^{-1} are compared with the symmetry and the position bands translational vibrations of crystals observed in the IR and Raman spectra at 100K.

[a]Rene Brun and Fons Rademakers "ROOT-An object oriented Data Analysis Framework", Proceedings AiHENP 96 Workshop, Lausanne, Sep. 1996, Nucl. Inst. and Meth. in Phys. Res A 389 (1997), 81-86.
[b]G.K. Semin, T.A. Babushkina, G.G. Jacobson. Nuclear Quadrupole Resonance in chemistry, Wiley, New York (1975)

RD13   **11:39 – 11:49**

INVESTIGATION OF THE PECULIARITIES OF THE ABSORBED WATER ON TiO_2 BY NMR-RELAXATION METHOD

IGOR VERTEGEL, EUGENY CHESNOKOV, ALEXANDER OVCHARENKO, VLADIMIR GAIVORONSKY, *Department of optics and spectroscopy of crystals, Institute of Physics National Academy of Sciences, Kiev, Ukraine*; IVAN VERTEGEL, *The Faculty of Mechanics and Mathematics, National Taras Shevchenko University, Kiev, Ukraine*.

The temperature dependence of the spin-lattice relaxation time (SLR) for protons of absorbed water on the surface of the TiO_2 micro-sphere in the temperature range 170-400K has been investigated. It was shown that in investigated temperature range there are two mechanisms of SLR: low-temperature (170-280K) and high-temperature (300-400K). The obtained data evidence of the fact that two types absorbed molecules of water on the TiO_2 surface exists: more active with E_{a1}=3.9kJ/mol and less mobile with E_{a2}=6.7 kJ/mol. This can be explained by different position of absorbed molecular groups of water on the surface of the TiO_2 micro-sphere.

RE. Chirped pulse
Thursday, June 19, 2014 – 8:30 AM
Room: 274 Medical Sciences Building

Chair: Brooks Pate, The University of Virginia, Charlottesville, VA, USA

RE01   8:30 – 8:45

A LOW-COST CHIRPED-PULSE FOURIER TRANSFORM MICROWAVE SPECTROMETER FOR UNDERGRADUATE PHYSICAL CHEMISTRY LAB

BRANDON CARROLL, IAN FINNERAN, GEOFFREY BLAKE, *Division of Chemistry and Chemical Engineering, California Institute of Technology, Pasadena, CA, USA.*

We present the design and construction of a simple and low-cost waveguide chirped pulse Fourier transform microwave (CP-FTMW) spectrometer suitable for gas-phase rotational spectroscopy experiments in undergraduate physical chemistry labs as well as graduate level research. The spectrometer operates with modest bandwidth, using phased locked loop (PLL) microwave sources and a direct digital synthesis (DDS) chirp source, making it an affordable for undergraduate labs. The performance of the instrument is benchmarked by acquiring the pure rotational spectrum of the J = 1 - 0 transition OCS and its isotopologues from 11-12.5 GHz.

RE02   8:47 – 8:57

A CHIRPED-PULSE FOURIER TRANSFORM SPECTROMETER OPERATING FROM 110 TO 170 GHZ

LAUREN E. BERNIER, STEVEN SHIPMAN, *Department of Chemistry, New College of Florida, Sarasota, FL, USA.*

A chirped-pulse Fourier transform spectrometer operating from 110 - 170 GHz was constructed. The design of this spectrometer is directly adapted from that of the 260 - 295 GHz chirped-pulse spectrometer built by Steber and co-workers at the University of Virginia[a]. In this instrument, an arbitrary waveform generator (AWG) produces a chirped pulse which is frequency shifted to a range between 9.2 and 14.1 GHz and then multiplied by a factor of 12 via an active multiplier chain to a range between 110 and 170 GHz. As in the Pate lab design, the AWG also serves as a local oscillator (LO) source; this LO is multiplied and used to downconvert the molecular emission, allowing it to be collected by a 40 GS/s digitizer. Benchmark measurements were taken for methanol at room temperature, and details of the instrument's performance will be discussed.

---

[a]A.L. Steber, B.J. Harris, J.L. Neill, and B.H. Pate, J. Mol. Spectrosc., 280, 3 (2012)

RE03   8:59 – 9:14

SPECTRAL TAXONOMY: A SEMI-AUTOMATED COMBINATION OF CHIRPED-PULSE AND CAVITY FOURIER TRANSFORM MICROWAVE SPECTROSCOPY

KYLE N CRABTREE, MICHAEL C McCARTHY, *Atomic and Molecular Physics, Harvard-Smithsonian Center for Astrophysics, Cambridge, MA, USA.*

Chirped-pulse Fourier transform microwave spectroscopy (CP-FTMW) has proven to be a powerful tool for broadband spectral surveys in the cm-wave band. In conjunction with a non-specific production source, such as an electrical discharge, new, unexpected molecules can be detected by their rotational spectra provided that they can be disentangled from other species that may be present. As an example, we have recently used a CP-FTMW spectrometer operating in the 8–18 GHz band to detect and identify two new silicon nitrides, HSiNSi and H_3SiNSi, in a discharge of dilute silane and nitrogen, although neither species had been the subject of prior experimental or theoretical study. However, of the $\sim$100 lines that are observed in this plasma, only $\sim$20 have been assigned to known species.

To further investigate unassigned lines in CP-FTMW spectra, we take advantage of the higher sensitivity of a traditional cavity FTMW spectrometer to rapidly perform follow-up assays in an approach we call "spectral taxonomy." Lines are classified according to whether their intensities are significantly altered by, for instance, turning off the discharge, applying a magnetic field, or removing a precursor gas; lines that show the same behavior for all tests may arise from a common carrier. After taxonometric classification, lines within each group are exhaustively tested with double resonance methods in an attempt to establish linkages which would identify lines arising from a shared quantum state and give clues as to the structure of the carrier. Using newly-designed control software for our cavity spectrometer, this entire procedure can be performed with minimal human intervention.

RE04   9:16 – 9:31

BUFFER GAS COOLED MOLECULE SOURCE FOR CPMMW SPECTROSCOPY

YAN ZHOU, DAVID GRIMES, <u>TIMOTHY J BARNUM</u>, ETHAN KLEIN, ROBERT W FIELD, *Department of Chemistry, MIT, Cambridge, MA, USA.*

We have built a new molecular beam source that implements 20 K Neon buffer gas cooling for the study of the spectra of small molecules. In particular, laser ablation of BaF_2 pellets has been optimized to produce a molecular beam of BaF with a number density more than 100 times greater than what we have previously obtained from a typical Smalley-type photoablation supersonic beam source. Moreover, the forward beam velocity of 150 m/s in our apparatus represents an approximate 10-fold reduction, improving spectroscopic resolution from 500 kHz to better than 50 kHz at 100 GHz in a chirped-pulse millimeter-wave experiment in which resolution is limited by Doppler broadening. Novel improvements in our buffer gas source and advantages for CPmmW spectroscopy studies will be discussed. We thank David Patterson, John Barry, John Doyle, and David DeMille for help in the design of our source.

RE05   9:33 – 9:48

DIRECT OBSERVATION OF RYDBERG-RYDBERG TRANSITIONS VIA CPMMW SPECTROSCOPY

<u>YAN ZHOU</u>, DAVID GRIMES, ETHAN KLEIN, TIMOTHY J BARNUM, ROBERT W FIELD, *Department of Chemistry, MIT, Cambridge, MA, USA.*

Rydberg-Rydberg transitions of Ca atoms are *directly* observed by chirped-pulse millimeter-wave spectroscopy, which is a form of broadband, high-resolution, free induction decay-detected (FID) spectroscopy with accurate relative intensities. A new setup, a 20 K Neon buffer gas cooled molecular beam system, has been constructed and tested in our lab. The number density of our target molecules, BaF, is increased by a factor of >100 relative to a Smalley-type laser ablation supersonic beam source. In addition, the laboratory frame velocity is decreased by factor 10, which improves our spectroscopic resolution to better than 50 kHz FWHM at 100 GHz. The improved molecular beam source opens the door to an extension of the CPmmW spectroscopy from atomic Rydberg states to molecular Rydberg states. I expect to present preliminary data from "pure electronic" spectra of BaF Rydberg molecules. We expect to produce 10^8 state-selected core-nonpenetrating Rydberg molecules in a single pulse of a laser-laser-mm-wave triple resonance excitation sequence.

RE06   9:50 – 10:05

TOWARD THE USE OF RYDBERG STATES FOR STATE-SELECTIVE PRODUCTION OF MOLECULAR IONS

<u>DAVID GRIMES</u>, TIMOTHY J BARNUM, STEPHEN COY, ROBERT W FIELD, *Department of Chemistry, MIT, Cambridge, MA, USA.*

The usual simplified view of Rydberg states of molecules as consisting of a single Rydberg electron loosely bound to a molecular ion core in a well-defined rotation-vibration state suggests an attractive possibility for state-selective production of molecular ions. A Rydberg electron excited above the energy of the ground state of the ion core will spontaneously autoionize, leaving behind a molecular ion. The autoionizing states are of strongly mixed character due to the ubiquitous nonadiabatic interactions between Rydberg series associated with different states of the ion core. Using our complete Multichannel Quantum Defect Theory (MQDT) fit model for CaF, we have predicted the locations and strengths of special autoionizing resonances that decay into a single rotation-vibration state of a molecular ion. Few molecules are as well characterized as CaF, nor as elegantly simple. We additionally describe the use of core nonpenetrating states as a general method to produce an ensemble of molecular ions in a single, selectable quantum state.

RE07   10:07 – 10:22

SELECTIVE POPULATION OF MOLECULAR CORE NONPENETRATING RYDBERG STATES

DAVID GRIMES, YAN ZHOU, TIMOTHY J BARNUM, ETHAN KLEIN, ROBERT W FIELD, *Department of Chemistry, MIT, Cambridge, MA, USA.*

Core nonpenetrating Rydberg states of molecules are a neglected state of matter. They could have a variety of uses, notably state-selective production of molecular ions. Due to the $l(l+1)/r^2$ centrifugal barrier that prevents the Rydberg electron in high-l states from penetrating inside of the ion core, the electron is essentially uncoupled from the ion core, and the system becomes atom-like with long lifetimes, an "almost good" l quantum number, and electronic transitions that follow $\Delta J^+=0$ and $\Delta v^+=0$ propensity rules. However, in most molecules access to these states, via a sequence of $\Delta l=+1$ transitions from low-n*, low-l states, is blocked by the necessity to traverse the "zone of death," in which nonradiative decay mechanisms are prohibitively fast. We exploit Chirped Pulse millimeter-Wave (CPmmW) spectroscopy to efficiently excite Ca atoms and BaF molecules to core nonpenetrating states in the absence of nonradiative decay mechanisms. A universal method for preparing core nonpenetrating Rydberg states of molecules, which combines CPmmW spectroscopy with STImulated Raman Adiabatic Passage (STIRAP), will be discussed.

Intermission

RE08   10:39 – 10:54

THE PURE ROTATIONAL SPECTRUM OF PbI FROM BROADBAND ROTATIONAL SPECTROSCOPY

DANIEL P. ZALESKI, *School of Chemistry, Newcastle University, Newcastle-upon-Tyne, United Kingdom*; HANSJOCHEN KÖCKERT, *School of Chemistry, Newcastle University, Newcastle upon Tyne, United Kingdom*; SUSANNA LOUISE STEPHENS, NICK WALKER, *School of Chemistry, Newcastle University, Newcastle-upon-Tyne, United Kingdom*; LISA-MARIA DICKENS, COREY EVANS, *Department of Chemistry, University of Leicester, Leicester, United Kingdom.*

The pure rotational spectrum of the open-shelled diatomic PbI is presented. The deep averaged broadband spectrum has allowed for characterization of several rotational spectra belonging to various lead isotopologues, as well as multiple vibrationally excited states of PbI. Being an opened-shelled diatomic, Λ-type doubling was observed. Analysis of the diatomic's spectral constants and potential energy surface will be described. These results will also be compared to early results from SnI.

RE09   10:56 – 11:06

WAVEGUIDE CHIRPED-PULSE FOURIER TRANSFORM MICROWAVE SPECTROSCOPY OF ALLYL BROMIDE

MORGAN N McCABE, STEVEN SHIPMAN, *Department of Chemistry, New College of Florida, Sarasota, FL, USA.*

The rotational spectrum of allyl bromide was recorded from 8.7 to 26.5 GHz at -20 °C with a waveguide chirped-pulse Fourier transform microwave spectrometer. The rotational spectrum of allyl bromide has been previously studied by Niide and coworkers. [a,b] However, previous assignments of this spectrum only extended to J = 12 and K_a = 1. Newly acquired data from our spectrometer has allowed us to extend the previous work to higher values of J and K_a, leading to significant improvements in the distortion constants in particular. Comparisons between the spectra and conformational preferences of the allyl halides will also be discussed.

[a] Y. Niide, M, Takano,T. Satoh, and Y. Sasada J. Mol. Spectrosc., 63, 108(1976)
[b] Niide, Yuzuru, J. Sci. Hiroshima Univ., Ser. A, 48, 1(1984)

RE10 11:08 – 11:18

CHIRPED-PULSE FOURIER TRANSFORM MICROWAVE SPECTROSCOPY OF 3-METHOXYPROPYLAMINE

MORGAN N McCABE, STEVEN SHIPMAN, *Department of Chemistry, New College of Florida, Sarasota, FL, USA*; SEAN ARNOLD, J. CHASE CHEWNING, MIRANDA SMITH, GORDON BROWN, *Department of Science and Mathematics, Coker College, Hartsville, SC, USA.*

The rotational spectrum of 3-methoxypropylamine was collected from 8.0 to 18.5 GHz with the Coker College chirped-pulse FTMW molecular beam spectrometer. Ab initio predictions using the B3LYP-D3 dispersion-corrected density functional gave high quality starting geometries, enabling us to quickly assign the spectrum of the lowest energy conformer, which has a g'gt configuration (moving from the amine end to the methoxy end of the molecule). Attempts were also made to collect the spectrum of this molecule in the room-temperature waveguide instrument at New College, but these attempts were unsuccessful as the molecule rapidly reacts with the copper walls of the waveguide to produce ammonia.

RE11 11:20 – 11:35

THE ROTATIONAL SPECTRA, STRUCTURES, AND CHLORINE NUCLEAR ELECTRIC QUADRUPOLE COUPLING CONSTANTS FOR A FAMILY OF THREE HALOGENATED CYCLIC ALKENES, $C_nF_{2n-4}Cl_2$: n = 4, 5, AND 6

B. E. LONG, E. A. ARSENAULT, LUCAS HANSEN, *Department of Chemistry, Wesleyan University, Middletown, CT, USA*; S. A. COOKE, *Natural and Social Science, Purchase College SUNY, Purchase, NY, USA.*

Microwave spectra for a family of three halogenated cyclic alkenes, $C_nF_{2n-4}Cl_2$; n = 4, 5, and 6, have been observed and analyzed. Rotational constants, quartic centrifugal distortion constants, and nuclear electric quadrupole coupling constants for a total of 9 isotopologues have been reported. These molecules are near oblate tops containing two quadrupolar nuclei. Data was first obtained via a chirp pulse Fourier transform microwave (FTMW) spectrometer, and then further analyzed with a Balle-Flygare type FTMW spectrometer to determine the chlorine nuclear electric quadrupole coupling constants. The spectroscopic parameters for all 9 species will be presented and compared.

RE12 11:37 – 11:52

CHIRPED-PULSE FOURIER TRANSFORM MICROWAVE SPECTROSCOPY OF 2-CHLORO-3-FLUOROPYRIDINE AND 2-CHLORO-6-FLUOROPYRIDINE

SEAN ARNOLD, J. CHASE CHEWNING, GORDON BROWN, *Department of Science and Mathematics, Coker College, Hartsville, SC, USA.*

The pure rotational spectra of 2-chloro-3-fluoropyridine and 2-chloro-6-fluoropyridine were measured on a chirped-pulsed Fourier transform microwave (CP-FTMW) spectrometer in the 8 – 18.5 GHz frequency range. The spectra were analyzed to find the rotational constants of the molecules for both the ^{35}Cl and the ^{37}Cl isotopologues. The measured rotational transitions exhibit hyperfine splitting, from which the nuclear quadrupole coupling constants have been assigned. The rotational constants and the nuclear quadrupole coupling constants have been compared to *ab initio* calculations performed using the Gaussian 03W software package.

RF. Mini-symposium: Astronomical Molecular Spectroscopy in the Age of ALMA

Thursday, June 19, 2014 – 1:30 PM

Room: 116 Roger Adams Lab

Chair: Leslie Looney, University of Illinois, Urbana, IL, USA

RF01   ***INVITED TALK***   1:30 – 2:00

OBSERVATIONS OF VOLATILE SPECIES IN PROTOPLANETARY DISKS

GEOFFREY BLAKE, *Division of Chemistry and Chemical Engineering, California Institute of Technology, Pasadena, CA, USA.*

This talk will review recent highlights from early shared risk observations of the rotational emission lines from small molecules in the protoplanetary disks around young Sun-like stars. Particular emphasis will be placed on the synergy of the ALMA observations of the outer disk with high resolution spectroscopy from the ground and space (Herschel, Spitzer) at infrared through THz wavelenghts, and on observational constraints that can be placed on the location of key condensation fronts such as the (water) snow line.

RF02   2:05 – 2:15

THE FIRST UNBIASED RADIO EMISSION LINE SURVEY OF THE PROTOPLANETARY DISK ORBITING LKCA 15

KRISTINA MARIE PUNZI, JOEL H KASTNER, *School of Physics and Astronomy, Rochester Institute of Technology, Rochester, NY, USA*; PIERRE HILY-BLANT, THIERRY FORVEILLE, *Institut de Planétologie et d'Astrophysique de Grenoble (IPAG), UJF-Grenoble / CNRS-INSU, Grenoble, France*; G G SACCO, *Arcetri Observatory, INAF, Florence, Italy.*

We have conducted the first comprehensive mm-wave molecular emission line survey of the circumstellar disk orbiting the nearby, pre-main sequence (T Tauri) star LkCa 15 (D = 140 pc). The outer disk is chemically rich, with numerous previous detections of molecular emission lines revealing a significant gas mass. The disk around this young ($\sim$3-5 Myr), actively accreting solar analog likely hosts a young protoplanet (LkCa 15b) within its central cavity. Hence, LkCa 15 is an excellent target for an unbiased radio spectroscopic survey intended to produce a full census of the detectable molecular species within an evolved, protoplanetary disk. Our survey of LkCa 15 was conducted with the Institute de Radioastronomie Millimétrique (IRAM) 30 meter telescope over the 1.1-1.4 mm wavelength range. The survey includes detections of the three most abundant CO isotopologues (^{12}CO, ^{13}CO, and $C^{18}O$) which facilitate estimates of the spatially integrated CO emission line optical depths, and complete coverage of the hyperfine line complexes of CN and C_2H that provide diagnostics of excitation and opacity for these species. This work demonstrates the value of comprehensive single-dish line surveys in guiding future high resolution interferometric imaging by ALMA of protoplanetary disks orbiting T Tauri stars.

RF03 2:17 – 2:32

NEAR-INFRARED SPECTROSCOPY OF SIMPLE ORGANIC MOLECULES IN THE GV TAU N PROTOPLANETARY DISK

ERIKA GIBB, *Physics & Astronomy, University of Missouri St. Louis, St. Louis, MO, USA.*

T Tauri stars are low mass young stars that may serve as analogs to the early solar system. Observations of organic molecules in the protoplanetary disks surrounding T Tauri stars are important for characterizing the chemical and physical processes that lead to planet formation. We used NIRSPEC on Keck 2 to perform a high resolution ($\lambda/\Delta\lambda \sim 25{,}000$) L-band survey of T Tauri star GV Tau N, a nearly edge-on young star in the L1524 molecular cloud. The nearly edge-on orientation is rare but necessary to sample the disk in absorption, rather than the more common emission line measurements. GV Tau N is one of only two sources for which HCN and C_2H_2 have been reported in absorption (Gibb et al. 2007; Doppmann et al. 2008). More recently, we reported the first detection of methane, CH_4 (Gibb & Horne 2013). The rotational temperatures are relatively high, implying that HCN, C_2H_2, CH_4, and water originate in the warm molecular layer of the inner protoplanetary disk. Differences in rotational temperature for different molecules suggest that the absorbing column for each molecule samples a different radial distribution.

Doppmann, G. W., Najita, K. R., & Carr, J. S. 2008, ApJ, 685, 298
Gibb, E. L., Van Brunt, K. A., Brittain, S. D., & Rettig, T. W. 2007, ApJ, 660, 1572
Gibb, E. L., Horne, D. 2013, ApJ, 776, L28

E.L.G. was supported by NSF Astronomy grant AST-0908230 and NASA Exobiology grant NNX07AK38G.

RF04 2:34 – 2:49

NEAR-INFRARED SPECTROSCOPIC STUDY OF AA TAU: WATER AND OH OBSERVATIONS

LOGAN RYAN BROWN, ERIKA GIBB, *Physics & Astronomy, University of Missouri St. Louis, St. Louis, MO, USA.*

To understand our own solar origins, we must investigate the composition of the protoplanetary disk from which the solar system formed. To infer this, we study analogs to the early solar system called T Tauri stars. These objects are low-mass, pre-main sequence stars surrounded by circumstellar disks of material from which planets are believed to form. We present high-resolution ($\lambda/\Delta\lambda \sim 25{,}000$), near-infrared spectroscopic data from the T Tauri star AA Tau using NIRSPEC at the Keck II telescope, located on Mauna Kea, HI, taken in 2009 and 2010. AA Tau has a close to edge-on geometry, with an inclination of $70° \pm 10°$ (Donati et al. 2010). Objects must have a nearly edge-on inclination for the disk to be sampled via absorption line spectroscopy. We observed strong absorption lines of both water and OH to which a spectroscopic model was fit in order for us to determine column density and rotational temperature. These near-infrared observations complement the work being done with ALMA, allowing us to probe the inner most disk regions and the chemistry contained within while ALMA primarily samples and is most sensitive to the outer disk.

PHYSICS AND CHEMISTRY IN UV ILLUMINATED REGIONS: THE HORSEHEAD CASE

VIVIANA V. GUZMAN, *SSP, Harvard-Smithsonian Center for Astrophysics, Cambridge, MA, USA*; JÉRÔME PETY, *Observatoire de Paris, IRAM, Grenoble, France*; PIERRE GRATIER, *LAB, Université de Bordeaux, Bordeaux, France*; JAVIER GOICOECHEA, *Departamento de Astrofísica, Centro de Astrobiología CAB, CSIC-INTA, Madrid, Spain*; MARYVONNE GERIN, *LERMA, Observatoire de Paris, Paris, France*; EVELYNE ROUEFF, *LUTH, Observatoire de Paris-Meudon, Paris, France*.

Molecular lines are used to trace the structure of the interstellar medium and the physical conditions of the gas in different environments, from protoplanetary disks to high-z galaxies. To fully benefit from the diagnostic power of molecular lines, the formation and destruction paths of the molecules, including the interplay between gas-phase and grain surface chemistry, must be quantitatively understood. Well-defined sets of observations of simple template sources are key to benchmark the theoretical models. In this context the PDR of the Horsehead mane is a particularly interesting case because it has a simple geometry (almost 1D, viewed edge-on) and the density profile across the PDR is well constrained.

In this talk, I will summarize our recent results on the ISM physics and chemistry in the Horsehead, from a complete and unbiased line survey at 1, 2 and 3mm performed with the IRAM-30m telescope, where approximately 30 species (plus their isotopologues) are detected with up to 7 atoms. I will show the importance of the interplay between the solid and gas phase chemistry in the formation of (complex) organic molecules, like H_2CO, CH_3OH, and CH_3CN, which reveal that photo-desorption of ices is an efficient mechanism to release molecules into the gas phase. The case of CH_3CN is especially surprising, as it is 40 times more abundant in the warm ($T_{kin} \sim 60$ K) UV-illuminated edge of the nebula, than in the shielded and colder ($T_{kin} \sim 20$ K) inner layers. I will show that complex molecules, such as HCOOH, CH_2CO, CH_3CHO, and CH_3CCH are easily detected in the PDR. I will also discuss new diagnostics of the UV-illuminated gas, like CF^+ (for which we recently resolved its hyperfine structure for the first time), which is observable from the ground, and we propose it can be used as a proxy of C^+. I will finish by reporting the first detection of a new molecule, recently confirmed to be the small hydrocarbon C_3H^+, which shows that photo-erosion of PAHs is needed to explain the enhanced abundance of other small hydrocarbons, like C_3H and C_3H_2.

WHAT MOLECULAR ABUNDANCES CAN TELL US ABOUT THE DYNAMICS OF STAR FORMATION

KONSTANTINOS TASSIS, *Department of Physics, University of Crete, Heraklion, Greece*; KAREN WILLACY, HAROLD W YORKE, NEAL J TURNER, *Jet Propulsion Laboratory, Science Division, California Institute of Technology, Pasadena, CA, USA*.

Molecular clouds are the sites where new stars form. Spectroscopic observations of different molecular species in these clouds can provide invaluable information regarding the dynamical evolution of star forming sites: first, they provide direct dynamical information (velocities as a function of density); second, they reveal the abundance of various molecules, which in turn depends on the chemodynamical evolutionary stage and history of the observed region. However, the connection between theoretical models of cloud dynamics and astronomical molecular spectroscopy is far from straight forward. The chemistry and dynamics of the clouds are interlinked, and various parameters such as the cloud temperature and its initial elemental abundances affect theoretical predictions, resulting in large model degeneracies: radically different dynamical models can often result in similar molecular abundances. In this talk, I will discuss first results from a massive effort undertaken to overcome this problem. By coupling non-equilibrium chemistry with a large array of different dynamical models of molecular cloud evolution, we are looking for these molecular line observables that are least affected by varying parameters and model degeneracies, and can be used to drastically constrain the possible dynamical histories of observed star-forming regions. To this end, we have studied a variety of dynamical models describing the evolution of pre- stellar molecular cloud cores (the initial phase of star formation) that cover the entire spectrum of proposed mechanisms, including pure hydrodynamical collapse and magnetically mediated collapse at various levels of importance of the magnetic field in the cloud dynamics. These models have been coupled to a network of chemical reactions that follow the relative abundances for ~ 100 molecular species, by solving the non- equilibrium chemical reactions for the first time simultaneously with the dynamical equations. I will present highlights from the results of this work, including newly proposed observables with maximal potential for discrimination between different models of cloud evolution and star formation. These results are especially timely as ALMA is able to measure many of these quantities and contribute to the resolution of long-standing questions in star formation, such as the timescale of pre-stellar core evolution, and the relative importance of magnetic field and turbulence in their dynamics.

RF07   **3:25 – 3:40**

PROBING THE CHEMISTRY AND DYNAMICS OF HOT MOLECULAR CORES USING HIGHLY EXCITED CYANOPOLYYNIC TRANSITIONS

ROBERT JOHN LOUGHNANE, *Centro de Radioastronomía y Astrofísica, Universidad Nacional Autónoma de México, Morelia, Michoacán, Mexico*; FRANÇOIS LIQUE, *NRS-Université du Havre, Laboratoire Ondes et Milieux complexes, Le Havre, France*; NAVTEJ SINGH, *Centre for Astronomy, National University of Ireland, Galway, Republic of Ireland*; STAN KURTZ, *Centro de Radioastronomía y Astrofísica, Universidad Nacional Autónoma de México, Morelia, Michoacán, Mexico*.

A hyperfine line fitting program is presented, which decomposes an observed rotational transition into its individual hyperfine components. The fit is optimized by the use of the Levenberg-Marquardt algorithm (for non-linear fitting) or Caruana's algorithm (linearization of the Gaussian function). From the optimal fit, various parameters from the decomposed components are derived such as the linewidth dispersion, peak brightness temperature and peak position in velocity units. The closeness in frequency units of two neighbouring hyperfine components within a rotational transition spectrum allows the derivation of a more credible estimate of the optical depth for the observed source of emission. Effective smoothing of the data subsequent to the fitting procedure greatly reduces the perceived error in the determination of various physical conditions of the observed region. The technique has been employed in observations of massive hot molecular cores (HMCs), considered to be the birthplace of high mass stars. In particular, observations of the cyanopolyynes HC_3N and HC_5N, each of which include a quadrupole hyperfine structure, as well as methyl cyanide, CH_3CN, have been analysed with this technique and modelled with a radiative transfer code incorporating non-LTE conditions, in order to derive abundances and column densities for a total of 10 HMCs and 5 massive cores. Using these derived parameters for each core, we have been able to test the time-dependent chemical models presented for these species by Chapman et al. (2009) and thus verify the suitability of their usefulness as "chemical clocks" by which to constrain the ages of the observed objects. In addition to this work, a detailed study of the magnetic hyperfine structure of a selection of inversion transitions of NH_3 is presented. As part of the continuing preparatory work for Herschel, SOFIA and, in particular, ALMA - improved rest frequencies for this commonly used kinetic temperature detecting species in star-forming cores will be forthcoming.

RF08   **3:42 – 3:57**

THE CO AND SIO PROTOSTELLAR OUTFLOWS OF 30 PROTOSTARS

DOMINIQUE M. SEGURA-COX, LESLIE LOONEY, *Department of Astronomy, University of Illinois at Urbana-Champaign, Urbana, IL, USA*.

We present a study of outflows driven by $\sim$30 protostars using CARMA 1.3 mm observations ($\sim$2.5$''$ resolution) of CO ($J = 2 \rightarrow 1$) and SiO ($J = 5 \rightarrow 4$). The sample is taken from the CARMA protostellar polarization survey. The polarization measurements revealed that the magnetic fields are either randomly oriented or preferentially oriented perpendicular to the molecular outflow axis. To probe the properties of the outflow itself, we calculate the mass, momentum, and kinetic energy of each outflow. To examine outflow evolution, we measure the opening angles of each outflow and examine how these vary with age determined from the bolometric temperature. When the protostellar targets are multiple systems, we attempt to disentangle the outflows originating from each component in order to study the system in detail. Observations of molecules such as CO and SiO are key to detecting outflows and determining their properties, which in turn helps constrain the details of the underlying physics driving outflows and how the outflows relate to their driving protostars.

Intermission

RF09 4:14 – 4:29

SHOCKS AND MOLECULES IN PROTOSTELLAR OUTFLOWS

HÉCTOR ARCE, *Astronomy Department, Yale University, New Haven, CT, USA.*

As protostars form through the gravitational infall of material from their parent molecular cloud, they power energetic bipolar outflows that interact with the surrounding medium. Protostellar outflows are important to the chemical evolution of star forming regions, as the shocks produced by the interaction of the high-velocity protostellar wind and the ambient cloud can heat the surrounding medium and trigger chemical and physical processes that would otherwise not take place in a quiescent molecular cloud. Protostellar outflows, are therefore a great laboratory to study shock physics and shock-induced chemistry. I will present results from millimeter-wave observations of a small sample of outflow shocks. The spectra show clear evidence of the existence of complex organic molecules (e.g., methyl formate, ethanol, acetaldehyde) and high abundance of certain simple molecules (e.g., HCO^+, HCN, H_2O) in outflows. Results indicate that, most likely, the complex species formed on the surface of grains and were then ejected from the grain mantles by the shock. Spectral surveys of shocked regions using ALMA could therefore be used to probe the composition of dust in molecular clouds. Our results demonstrate that outflows modify the chemical composition of the surrounding gaseous environment and that this needs to be considered when using certain species to study active star forming regions.

RF10 4:31 – 4:46

DETECTION, IDENTIFICATION AND CORRELATION OF COMPLEX ORGANIC MOLECULES IN 32 INTERSTELLAR CLOUDS USING SUBMM OBSERVATIONS

NADINE WEHRES, SHIYA WANG, MARY RADHUBER, JAMES SANDERS, JAY A KROLL, JACOB LAAS, BRIAN HAYS, TREVOR CROSS, *Department of Chemistry, Emory University, Atlanta, GA, USA*; D. C. LIS, *Cahill Center for Astronomy and Astrophysics, California Institute of Technology, Pasadena, CA, USA*; ERIC HERBST, *Department of Chemistry, The University of Virginia, Charlottesville, VA, USA*; SUSANNA L. WIDICUS WEAVER, *Department of Chemistry, Emory University, Atlanta, GA, USA.*

We present spectral line surveys of 32 galactic sources using the Caltech Submillimeter Observatory (CSO) and the HIFI instrument on the Herschel Space Observatory. This study covers the 220 – 265 GHz frequency range using the CSO, as well as higher frequencies, 645 – 676 GHz and 1.14 THz – 1.19 THz using the HIFI instrument. Deconvolution of the double sideband spectra was performed using the CLASS program and the Herschel/HIFI pipeline. Analysis of the data sets was performed using Global Optimization and Broadband Analysis Software for Interstellar Chemistry (GOBASIC), a new software program developed by our group for the evaluation and study of large astronomical spectroscopic data sets.

Initial analysis has focused on 12 complex organic molecules that can be used to trace grain-surface and gas-phase chemical processing in the interstellar medium. GOBASIC was used to determine molecular column densities and rotational temperatures. This information is being used to study correlations between molecular abundances within a given source, and source-to-source correlations for a given molecule, with the ultimate goal of determining which molecules can be used as clocks of the star-formation process. The spectra and the results of this initial analysis will be presented.

RF11 4:48 – 5:03

LIGHTING THE DARK MOLECULAR GAS USING THE MID INFRARED H_2 ROTATIONAL LINES

ADITYA TOGI, JD SMITH, *Physics and Astronomy, University of Toledo, Toledo, OH, USA.*

The knowledge of molecular gas distribution is necessary to understand star formation in galaxies[a]. The molecular gas content of galaxies must be inferred using indirect tracers since H_2 which forms a major component of molecular gas in galaxies is not observable under typical conditions of interstellar medium[b]. Physical processes causing enhancement and reduction of these tracers can cause misleading estimates of the molecular gas content in galaxies. We have devised a new method to measure molecular gas mass using quadrupole rotational lines of H_2 found in the mid infrared spectra of various types of galaxies. We apply our model to derive the amount of molecular gas even in low metallicity galaxies where indirect tracers are unable to estimate the dark molecular gas mass[c].

[a]Bigiel, F., Leroy, A., Walter, F., et al. 2008, The Astronomical Journal, 136, 2846 (2008)

[b]Solomon, P. M., Rivolo, A. R., Barett, J., and Yahil, A. The Astrophysical Journal, 319, 730 (1987)

[c]Wolfire, M. G., Hollenbach, D., and McKee, C. F. The Astrophysical Journal, 716, 1191 (2010)

RF12   5:05 – 5:20

COMPARATIVE CHEMISTRY IN PLANETARY NEBULAE: THE ROLE OF THE CARBON TO OXYGEN RATIO

JESSICA L EDWARDS, *Department of Chemistry and Biochemistry, Department of Astronomy, The University of Arizona, Tucson, AZ, USA*; LUCY ZIURYS, *Department of Chemistry and Biochemistry, University of Arizona, Tucson, AZ, USA.*

While the vast majority of stars in our galaxy will go through the Planetary Nebula (PN) stage near the end of their lives, these objects are not very well understood both physically and chemically. It has long been thought that nebular age is a major factor in determining the chemical content of PNe, but our recent studies have shown this is not likely the case. Millimeter-wave observations of planetary nebulae using the telescopes of the Arizona Radio Observatory have shown that the molecular content of these PNe appears to vary with the C to O ratio of the progenitor star. For example, SO, SO_2, and SiO, molecules that are typically seen in oxygen-righ AGB stars and supergiants, were all detected in the oxygen-rich M2-48. Previous studies of the young, carbon-rich object NGC 7027 show the presence of multiple carbon containing molecules like HC_3N, CCH, and C_3H_2. For NGC 6537, where C/O is close to 1, SO and CCH were identified, indicating an S-type progenitor star. The molecular content of PNe that are carbon-rich, oxygen-rich, and have a C/O ratio close to unity will be compared. The chemical relationship with respect to age will be discussed, as well as implications that the molecular content of the remnant circumstellar shell and its C/O ratio is the main contributing factor behind chemical variation in PNe.

RF13   5:22 – 5:37

MOLECULAR ABUNDANCES IN THE CIRCUMSTELLAR ENVELOPE OF OXYGEN-RICH SUPERGIANT VY CANIS MAJORIS

JESSICA L EDWARDS, *Department of Chemistry and Biochemistry, Department of Astronomy, The University of Arizona, Tucson, AZ, USA*; LUCY ZIURYS, *Department of Chemistry and Biochemistry, University of Arizona, Tucson, AZ, USA.*

A complete set of molecular abundances have been established for the Oxygen-rich circumstellar envelope (CSE) surrounding the supergiant star VY Canis Majoris (VY CMa). These data were obtained from The Arizona Radio Observatory (ARO) 1-mm spectral line survey of this object using the ARO Sub-millimeter Telescope (SMT), as well as complimentary transitions taken with the ARO 12-meter. The non-LTE radiative transfer code ESCAPADE has been used to obtain the molecular abundances and distributions in VY CMa, including modeling of the various asymmetric outflow geometries in this source. For example, SO and SO_2 were determined to arise from five distinct outflows, four of which are asymmetric with respect to the central star. Abundances of these two sulfur-bearing molecules range from 3 x 10^{-8} - 2.5 x 10^{-7} for the various outflows. Similar results will be presented for molecules like CS, SiS, HCN, and SiO, as well as more exotic species like NS, PO, AlO, and AlOH. The molecular abundances between the various outflows will be compared and implications for supergiant chemistry will be discussed.

RF14   5:39 – 5:49

ROTATIONALLY EXCITED H_2 IN THE MAGELLANIC CLOUDS

RUI XUE, TONY WONG, *Department of Astronomy, University of Illinois at Urbana-Champaign, Urbana, IL, USA.*

We have performed a systematic analysis of excited-state (up to $J=5$) H_2 Lyman-Werner absorption lines using archival spectra in the *FUSE* Magellanic Clouds Legacy Project. The H_2 column densities at different ground state J-levels and the Doppler broadening parameter b are determined for both Magellanic and Galactic components along each line of sight. Combining the results with previously measured total gas column densities of HI and H_2, we derive the H_2 excitation temperature, volume density, and local UV field strength for the absorbing gas. The physical and chemical properties of the absorbers are compared with Galactic samples, and also used to test predictions from multiple-phase ISM equilibrium models. Finally, we compare the absorbing gas from the Magellanic Clouds with its larger-scale ISM environment as revealed in previous surveys of gas and dust emission, extending our results from the UV data measured along moderately reddened sight lines to more dense gas detected in emission.

RF15   **5:51 – 6:06**

DOES PLASMA STRUCTURE INFLUENCE MOLECULE FORMATION AND RADIATION CHARACTERISTICS?

DAAN C SCHRAM, *Physics, Eindhoven University of Technology, Eindhoven, Netherlands.*

In this presentation several consequences of the plasma state with spatial structures will be discussed. The basic thought is that a concentration of neutral hydrogen gas in combination with weak magnetic fields will cause elongated plasmas with higher density, which will expand along the field lines. As a consequence the plasma is more concentrated in filamentary structures, the charge density is higher and electron temperature can be higher as well. Clusters will be more charged, which will influence the transport of radicals and charge. The electron density life time may not be limited by dissociative recombination. A comparison is made of the effects of such a stratified medium as compared to a homogeneous plasma for dark clouds conditions. The effect of a structured medium could be that more radicals (as H) are formed and more molecules are formed. Also finite fields and currents may have influence on plasma kinetics.

RG. Mini-symposium: Beyond the Mass-to-Charge Ratio: Spectroscopic Probes of the Structures of Ions

Thursday, June 19, 2014 – 1:30 PM

Room: 100 Noyes Laboratory

Chair: Etienne Garand, University of Wisconsin, Madison, WI, USA

RG01   ***INVITED TALK***   **1:30 – 2:00**

ISOMER-SPECIFIC IR^{2}MS^2 SPECTROSCOPY OF PROTONATED WATER CLUSTERS

__KNUT R. ASMIS__, *Wilhelm-Ostwald-Institut für Physikalische und Theoretische Chemie , Universität Leipzig, Leipzig, Germany.*

Understanding how protons are hydrated remains an important and challenging research area. The anomalously high proton mobility of water, for example, can be explained by a periodic isomerization between the Eigen and Zundel binding motifs, $H_3O^+(aq)$ and $H_5O_2^+(aq)$, respectively, even though the detailed mechanism is considerably more complex and not completely understood. These rapidly interconverting structures from the condensed phase can be stabilized, isolated and studied in the gas phase in the form of protonated water clusters.

Infrared photodissociation (IRPD) spectroscopy serves as a powerful tool for studying the structure of gas phase clusters. However, the contribution of multiple isomers to the IRPD spectrum can complicate the assignment. Here, results on the isomer-specific IR/IR double resonance (IR^{2}MS^2) spectroscopy of the protonated water clusters $H^+(H_2O)_n \cdot H_2$ with $n = 5$-10 are reported. IR^{2}MS^2 spectra are measured in the spectral region of the free and hydrogen-bonded OH-stretching vibrations (2880-3850 cm^{-1}) and assigned on the basis of a comparison to the results of electronic structure calculations. For the protonated water hexamer, it is demonstrated that combining the radiation from an IR free electron laser with that from a widely tunable table-top IR laser allows extending this technique across nearly the complete IR region (260-3900 cm^{-1}). *Ab initio* molecular dynamics calculations qualitatively recover the IR spectra of the two isomers for $n = 6$ and allow identifying characteristic hydrogen-bond stretching bands below 400 cm^{-1}.

RG02   **2:05 – 2:20**

A THEORETICAL STUDY ON THE STRUCTURAL EVOLUTION OF IONIZED WATER CLUSTERS, $(H_2O)_n^+$, n$\leq$ 3 ~ 8

EN-PING LU, YING-CHENG LI, __JER-LAI KUO__, *Institute of Atomic and Molecular Sciences, Academia Sinica, Taipei, Taiwan*; MING-KANG TSAI, *Department of Chemistry, National Taiwan Normal University, Taipei, Taiwan.*

Based on similairity between vibrational spectra of $H^+(H_2O)_n$ and $(H_2O)_n^+$, it has been suggestted that the structure of $(H_2O)_n^+$ can be understood as $H^+(H_2O)_{n-1}OH$.[a] Dominating isomers for n$\leq$5 have been analyzed in detail based on free OH stretching modes. In this work, we perform extensive search for local minima structures for n=5~8 with first-principles methods. More than 600 stable isomers were located for n=8. Based on these structures, we analyze the structural evolution, assign the OH stretching regions and the solvation of OH radical in $(H_2O)_n^+$.

---

[a]K. Mizuse, J.-L Kuo, A. Fujii, Chem. Sci. 2011, 2, 868, K. Mizuse, A. Fujii, J. Phys. Chem. A 2013, 117, 929

RG03 2:22–2:37

MULTIDIMENTIONAL NORMAL MODE CALCULATIONS FOR THE OH VIBRATIONAL SPECTRA OF $(H_2O)_3^+$, $(H_2O)_3^+$Ar, $H^+(H_2O)_3$, AND $H^+(H_2O)_3$Ar

YING-CHENG LI, HSIAO-HAN CHUANG, JAKE ACEDERA TAN, KAITO TAKAHASHI, JER-LAI KUO, *Institute of Atomic and Molecular Sciences, Academia Sinica, Taipei, Taiwan.*

Recent experimental observations of $(H_2O)_3^+$, $(H_2O)_3^+$Ar, $H^+(H_2O)_3$, and $H^+(H_2O)_3$Ar clusters in the region 1400-3800 cm^{-1} show that the OH stretching vibration has distinct characteristics.[a][b][c] Multidimensional normal mode calculations were carried out for OH stretching vibrations in the 1200-4000 cm^{-1} photon energy range. The potential energy and dipole surfaces were evaluated by using first-principles methods. By comparing the calculated frequencies and intensities of OH stretching vibration with experimental spectra, we found that the assignment of OH strecthing of H_3O^+ moiety and free OH strectching vibration have resonable agreement with experimental data.

[a]Jeffrey M. Headrick, Eric G. Diken, Richard S. Walters, Nathan I. Hammer, Richard A. Christie, Jun Cui, Evgeniy M. Myshakin, Michael A. Duncan, Mark A. Johnson, Kenneth D. Jordan, Science, 2005, **17**, 1765.
[b]Kenta Mizuse, Jer-Lai Kuo and Asuka Fujii, Chem. Sci., 2011, **2**, 868.
[c]Kenta Mizuse and Asuka Fujii, J. Phys. Chem. A, 2013, **117**, 929.

RG04 2:39–2:54

COMPUTATIONAL FRAMEWORK FOR ANALYSIS OF HYDROGEN BONDING IN THE OH STRETCH REGION

LAURA C. DZUGAN, ANNE B McCOY, *Department of Chemistry and Biochemistry, The Ohio State University, Columbus, OH, USA.*

There are two types of bands in the OH stretch region of the vibrational spectra of hydrogen-bonded complexes; narrow peaks due to isolated OH stretches and a broadened feature reflecting the OH stretches involved in strong hydrogen bonding. This second region can be as wide as several hundred wavenumbers and is shifted to the red of the narrow peaks. In this work we focus on $(CaOH)^+\cdot(H_2O)_n$ and $(MgOH)^+\cdot(H_2O)_n$ systems.[1] When n<4, the spectra are characterized by only the narrow peaks near 3700 cm^{-1}. When n≥4, there is an additional band that is several hundred cm^{-1} wide, which is attributed to hydrogen bonding. This breadth arises from coupling between the OH stretches in the water molecules and the low frequency modes of the complex. To understand the broadening observed in the spectra, we have developed a computational framework in which we sample displacement geometries from the equilibrium structure based on the ground state harmonic wavefunction. Then we combine the harmonic spectra in the OH stretch region for each computed geometry to generate the spectrum for this complex. As the calculated spectra agree well with the experimental spectra, we then investigated which geometric parameters in the system are correlated to the size of the red-shift of the frequencies. The hydrogen-bonded OH stretches were found to be very sensitive to how the water molecules were arranged around the hydroxide group.

1. Leavitt, C.M.; Johnson, C.J.; Johnson, M.A. Unpublished work.

RG05 2:56–3:11

ADDING THE TEMPERATURE DIMENSION TO SIZE-SELECTED ION VIBRATIONAL PREDISSOCIATION SPECTROSCOPY: OBSERVATION OF "MELTING" IN THE PRIMARY SOLVATION SHELL OF MICROHYDRATED IODIDE CLUSTERS

OLGA GORLOVA, CONRAD WOLKE, MARK JOHNSON, *Department of Chemistry, Yale University, New Haven, CT, USA.*

Over the past decade, an intensive integrated effort involving theory and experiment have revealed the structures at play in the first hydration shell of simple ions. The water molecules generally adopt configurations in which one OH group is directed toward the ion while the other is integrated into a water network. One of the reasons this endeavor was difficult is that the three-body repulsion in this regime acts to significantly lower the effective inter-water binding energies, making the equilibrium structures much more fragile than their neutral counterparts. Here we exploit very recent advances in the temperature control of size-selected ionic clusters using cryogenic ion traps to monitor how the spectroscopic signatures of the water networks evolve as the temperature of the $I^-(H_2O)_n$ clusters is varied over the range 10 to 200 K. The breaking of the hydrogen bond interactions is observed around 150 K in the dimer. Qualitative assignments of the free, bound and ring hydrogens in the OH stretching region clarify the evolution from closed to linear hydrogen bond networks as the temperature changes. The success of this temperature programmed ion vibration predissociation (TPIVP) spectroscopy opens the way to sample large amplitude exploration of potential energy landscapes of such systems.

Intermission

RG06 3:28 – 3:43

STRUCTURES OF HYDRATED ALKALI METAL CATIONS, $M^+(H_2O)_n$Ar (M = Li, Na, K, Rb and Cs, n = 3-5), USING INFRARED PHOTODISSOCIATION SPECTROSCOPY AND THERMODYNAMIC ANALYSIS

HAOCHEN KE, CHRISTIAN VAN DER LINDE, JAMES M. LISY, *Department of Chemistry, University of Illinois at Urbana-Champaign, Urbana, IL, USA.*

Alkali metal cations play vital roles in chemical and biochemical systems. Lithium is widely used in psychiatric treatment of manic states and bipolar disorder; Sodium and potassium are essential elements, having major biological roles as electrolytes, balancing osmotic pressure on body cells and assisting the electroneurographic signal transmission; Rubidium has seen increasing usage as a supplementation for manic depression and depression treatment; Cesium doped compounds are used as essential catalysts in chemical production and organic synthesis. Since hydrated alkali metal cations are ubiquitous and the basic form of the alkali metal cations in chemical and biochemical systems, their structural and thermodynamic properties serve as the foundation for modeling more complex chemical and biochemical processes, such as ion transport and ion size-selectivity of ionophores and protein channels. By combining mass spectrometry and infrared photodissociation spectroscopy, we have characterized the structures and thermodynamic properties of the hydrated alkali metal cations, i.e. $M^+(H_2O)_n$Ar, (M = Li, Na, K, Rb and Cs, n = 3-5). Ab initio calculations and RRKM-EE (evaporative ensemble) calculations were used to assist in the spectral assignments and thermodynamic analysis. Results showed that the structures of hydrated alkali metal cations were determined predominantly by the competition between non-covalent interactions, i.e. the water—water hydrogen bonding interactions and the water—cation electrostatic interactions. This balance, however, is very delicate and small changes, i.e. different cations, different levels of hydration and different effective temperatures clearly impact the balance.

RG07 3:45 – 4:00

UNRAVELING THE ROLES OF HYDROGEN BONDING, ELECTROSTATICS, AND FERMI RESONANCES IN THE IONIC LIQUID [EMIM][BF$_4$] THROUGH CRYOGENIC ION VIBRATIONAL SPECTROSCOPY

JOSEPH FOURNIER, CONRAD WOLKE, CHRISTOPHER JOHNSON, MARK JOHNSON, *Department of Chemistry, Yale University, New Haven, CT, USA.*

The importance of hydrogen bonding in imidazolium-based ionic liquids (IL) has become a topic of vigorous debate. Red shifted features in the ring CH stretching region observed in bulk FTIR of several IL have been identified as the hydrogen bonded $C_{(2)}H$ stretch. However, recent theoretical analysis suggests that the complexity of the ring CH stretching region is a result of Fermi resonance interactions between the overtones and combination bands of the ring stretching modes with the ring CH stretches. To help clarify the role of the $C_{(2)}H$ group and the nature of the intermolecular cation-anion interactions, we report the vibrational spectra of cryogenically cooled, composition-selected ionic clusters of the prototypical IL [EMIM][BF$_4$]. We have confirmed that the CH stretching region is indeed plagued by strong Fermi resonance interactions and, therefore, have turned to isotopic and chemical substitution to determine the position and role of the $C_{(2)}H$ oscillator. The spectra are consistent with electrostatics being the dominant interaction while hydrogen bonding is not critical in this IL.

Time permitting, recent spectra on temperature controlled protonated water clusters will be discussed.

220

RG08 4:02 – 4:17

ION PAIR STRUCTURE AND PHOTODISSOCIATION DYNAMICS OF IONIC LIQUID [EMIM][TF$_2$N]

JAIME A. STEARNS, RUSSELL COOPER[a], DAVID SPORLEDER[b], *Space Vehicles Directorate, Air Force Research Lab, Kirtland AFB, NM, USA*; ALEXANDER M. ZOLOT, *Institute for Scientific Research, Boston College, Boston, MA, USA*; JERRY BOATZ, *Aerospace Systems Directorate, Air Force Research Lab, Edwards AFB, CA, USA.*

The Air Force has a pressing need to find new means of spacecraft propulsion, enabling cheaper, safer, more efficient maneuvering on orbit. Ionic liquids are a potential replacement for hydrazine in hypergolic combustion propellant systems and for xenon in electric propulsion systems. However, both applications require considerable further development, leading us to study the fundamental structural and optical properties of candidate systems. Our benchmark measurements will provide validation of theoretical models of all types, from *ab initio* methods up to codes describing full thruster plumes. Using standard supersonic jet time-of-flight spectroscopy techniques, we have measured the ultraviolet and infrared spectra of ion pairs of the only space-qualified ionic liquid, [emim][Tf$_2$N]. The ultraviolet photodissociation spectrum, though broad and essentially featureless, reveals rich underlying photodynamics involving both single- and multi-photon excitations and a wealth of interacting excited states. The infrared spectrum and MP2 calculations establish the structure as one in which the cation and anion are stacked on top of one another rather than sitting in the same plane, answering a long-standing question in this field. The complexity of the infrared spectrum and its behavior under varying jet temperatures indicates the presence of multiple conformations and likely contributions from Fermi resonance.

[a]National Research Council Associate
[b]National Research Council Associate

RG09 4:19 – 4:34

INFRARED SPECTRA OF ANIONIC COBALT-CARBON DIOXIDE CLUSTERS

BENJAMIN KNURR, J. MATHIAS WEBER, *Department of Chemistry and Biochemistry, JILA - University of Colorado, Boulder, CO, USA.*

We present infrared photodissociation spectra of Co(CO$_2$)$_n^-$ ions ($n = 3 - 11$) in the wavenumber region 1000 – 2400 cm^{-1}, interpreted with the aid of density functional theory calculations. The spectra show signatures of several structural motifs for the interaction of a Co atom and CO$_2$ ligands. The spectra are dominated by a core ion showing bidentate interaction of two CO$_2$ ligands forming C-Co and O-Co bonds. The prevalence of triplet vs singlet states and the charge distribution in the Co(CO$_2$)$_2^-$ core ion will also be discussed.

RG10 4:36 – 4:51

INFRARED PHOTODISSOCIATION SPECTROSCOPY OF DOUBLY CHARGED M(CO)$_n$ CATIONS

JON MANER, ANTONIO BRATHWAITE, NICKI REISHUS, MICHAEL A DUNCAN, *Department of Chemistry, University of Georgia, Athens, GA, USA.*

M^{2+}(CO)$_n$ (M=Ti,Cr) clusters are produced in the gas phase by laser vaporization, mass-selected in a time-of-flight mass spectrometer, and probed with infrared photodissociation spectroscopy. The mass spectra, photodissociation patterns, and IR spectra of these clusters are used to establish the coordination number of the metal dication. The IR spectra are compared with the spectra of the analogous singly charged clusters, revealing a strong blue shift in the CO stretching vibration. This blue shift is attributed to polarization of CO by M^{2+} via an ion-dipole interaction with minimal pi-backbonding.

RG11

VIBRATIONAL SPECTROSCOPY AND THEORY OF $Cu^+(CH_4)_n$ AND $Ag^+(CH_4)_n$ (n = 1 - 6)

ABDULKADIR KOCAK, <u>MUHAMMAD AFFAWN ASHRAF</u>, RICARDO B. METZ, *Department of Chemistry, University of Massachusetts, Amherst, MA, USA.*

Vibrational spectra are measured for $Cu^+(CH_4)_n$ and $Ag^+(CH_4)_n$ (n = 1 - 6) in the C-H stretching region (2500-3100 cm^{-1}) using photofragment spectroscopy. Spectra are obtained by monitoring CH_4 fragment loss following absorption of one photon (for n = 3 - 6), sequential absorption of multiple photons for $Ag^+(CH_4)_n$(n = 1 - 2) or one photon absorption by Ar-tagged $Cu^+(CH_4)_n$ (for n = 1-2). Determination of the structures of the complexes was done by comparing calculated vibrational spectra of low-lying candidate structures to the observed IR photodissociation spectra. Calculations were carried out using both the B3LYP and CAM-B3LYP hybrid density functionals with the 6-311++G(3df,3pd) basis set for Cu, C and H and the aug-cc-pVTZ basis with an effective core potential (ECP) for Ag. Calculations predict that the positions and intensities of bands in the C-H stretching region depend strongly on the coordination of the CH_4 to the metal (η^2 or η^3) and on the M-C bond length, and thus is sensitive to the number of ligands in the first and second shell. For clusters with $n \leq 4$ these ions adopt symmetrical structures with η^2 methane coordination. For copper, the larger clusters are primarily formed by addition of second-shell CH_4 to a tetrahedral core; silver primarily coordinates the 5th and 6th ligands in the first shell.

RG12

STRUCTURES, ENERGETICS, AND VIBRATIONS OF SMALL TRANSITION METAL OXIDE CLUSTERS BY HIGH-RESOLUTION ANION PHOTOELECTRON SPECTROSCOPY

<u>JONGJIN B. KIM</u>, MARISSA L. WEICHMAN, DANIEL NEUMARK, *Department of Chemistry, The University of California, Berkeley, CA, USA.*

Anion photoelectron spectroscopy has been a major tool in understanding the vibronic structure of metal oxide clusters, due to its universality and sensitivity. However, high ion temperatures and modest photoelectron energy resolutions have hampered the observation of vibrational structure. We have recently coupled our high-resolution slow photoelectron velocity-map imaging (SEVI) spectrometer to a cryogenic ion trap and a laser ablation ion source, allowing for the acquisition of photoelectron spectra of vibrationally cold metal oxide anions with a resolution down to ~ 4 cm^{-1}, limited by unresolved rotational structure. A test study of the simple d^0 group 4 MO_2 triatomic metal oxides yielded fully vibrationally-resolved spectra, allowing for reassignments of electron affinities, new measurements of vibrational fundamentals, and estimates of the anion geometries based on the observed FC structure. Studies of the corresponding Ti_2O_4 and Zr_2O_4 systems revealed vibrational progressions that allows for an unambiguous assignment of the anion isomers; previous photoelectron spectra could not distinguish the isomers based on detachment energies alone. Spectra of the VO_2^- anion identified the first three electronic states of the d^1 neutral as well as ν_1 and ν_2 vibrations in each state.

IR SPECTROSCOPY OF FIRST-ROW TRANSITION METAL CLUSTERS AND THEIR COMPLEXES WITH SIMPLE MOLECULES

D. M. KIAWI, J. BAKKER, J. OOMENS, *Institute for Molecules and Materials (IMM), Radboud University Nijmegen, Nijmegen, Netherlands*; W. J. BUMA, *Van' t Hoff Institute for Molecular Sciences, University of Amsterdam, Amsterdam, Netherlands*; L.B.F.M WATERS, *Netherlands Institute for Space Research, SRON, Utrecht, Netherlands.*

Iron is an important element in the formation of solids in space. Spectroscopic observations of interstellar iron shows that its atomic gas-phase abundance is strongly depleted with respect to that of hydrogen. In contrast, sulfur is mostly found in the gas phase in low-density regions of interstellar space, but is highly depleted in regions of star- and planet formation. Furthermore, the dominant source of sulfur in our solar system is solid FeS, as found in primitive meteorites, implying an efficient chemical pathway to convert sulphur or sulphur containing compounds into solid FeS during the (early phases of) the star formation process. We address the evolution of iron and sulfur in space on a molecular level by studying metal nanoclusters and their interaction with ligands using IR action spectroscopy. Clusters are formed through laser ablation of solid precursor materials and brought into a molecular beam environment. Complexes with ligands are obtained by directing the beam through a reaction channel containing low-pressure reactant gas. Mass-selected IR action spectra are recorded by irradiating the clusters using the Free Electron Laser for Infrared eXperiments (FELIX). Experimental spectra are then compared with DFT predictions which enables us to determine the structure of the selected cluster and its binding interactions with ligands. As part of this project, we here present IR action spectra of size-selected Fe clusters and the chemically closely related Co clusters, and their complexes with relevant ligands.

DETERMINATION OF THE STRUCTURES OF SILICON AND METAL DOPED SILICON CLUSTERS

JONATHAN T LYON, *Department of Natural Sciences, Clayton State University, Morrow, GA, USA*; ANDRE FIELICKE, *Department of Molecular Physics, Fritz-Haber-Institut der Max-Planck-Gesellschaft, Berlin, Germany*; EWALD JANSSENS, PETER LIEVENS, *Laboratory of Solid State Physics and Magnestism, Katholieke Universiteit Leuven, Leuven, Belgium.*

Strongly bound clusters are often used as convenient models for bulk material. Silicon clusters are particularly interesting due to their importance in the electronics industry. We perform experimental IR multiple photon dissociation spectroscopy in the gas-phase, which makes use of a free electron laser, and compare the results with that predicted by density functional and MP2 theory calculations. Comparison of the vibrational spectra with that predicted by theoretical calculations for several structural isomers for each cluster size leads to accurate structural assignments. Here, we present our results for silicon clusters,[a] and compare the structures with those of select transition metal doped $Si_n M$ clusters.[b] Of particular interest is the transition from exohedral to endoheral metal doped silicon clusters and how the transition size changes for different metal dopant atoms.

---

[a]Journal of Chemical Physics 2012, 136, 064301.
[b]e.g., ChemPhysChem 2014, 15, 328.

RG15 **6:01 – 6:16**

PHOTOELECTRON SPECTROSCOPY STUDY OF $[Ta_2B_6]^-$: A HEXAGONAL BIPYRAMDIAL CLUSTER

TIAN JIAN, WEILI LI, CONSTANTIN ROMANESCU, LAI-SHENG WANG, *Department of Chemistry, Brown University, Providence, RI, USA.*

It has been a long-sought goal in cluster science to discover stable atomic clusters as building blocks for cluster-assembled nanomaterials, as exemplified by the fullerenes and their subsequent bulk syntheses.[1,2] Clusters have also been considered as models to understand bulk properties, providing a bridge between molecular and solid-state chemistry.[3] Herein we report a joint photoelectron spectroscopy and theoretical study on the $[Ta_2B_6]^-$ and $[Ta_2B_6]$ clusters.[4] The photoelectron spectrum of $[Ta_2B_6]^-$ displays a simple spectral pattern and a large HOMO-LUMO gap, suggesting its high symmetry. Theoretical calculations show that both the neutral and anion are D_{6h} pyramidal. The chemical bonding analyses for $[Ta_2B_6]$ revealed the nature of the B_6 and Ta interactions and uncovered strong covalent bonding between B_6 and Ta. The D_{6h}-$[TaB_6Ta]$ gaseous cluster is reminiscent of the structural pattern in the ReB_6X_6Re core in the $[(Cp^*Re)_2B_6H_4Cl_2]$ and the TiB_6Ti motif in the newly synthesized $Ti_7Rh_4Ir_2B_8$ solid-state compound.[5,6] The current work provides an intrinsic link between a gaseous cluster and motifs for solid materials. Continued investigations of the transition-metal boron clusters may lead to the discovery of new structural motifs involving pure boron clusters for the design of novel boride materials.

Reference

[1] H.W. Kroto, J. R. Heath, S. C. OBrien, R. F. Curl, R. E. Smalley, Nature 1985, 318, 162 – 163.

[2] W. Krtschmer, L. D. Lamb, K. Fostiropoulos, D. R. Huffman, Nature 1990, 347, 354 – 358.

[3] T. P. Fehlner, J.-F. Halet, J.-Y. Saillard, Molecular Clusters: A Bridge to Solid-State Chemitry, Cambridge University Press, UK, 2007.

[4] W. L. Li, L. Xie, T. Jian, C. Romanescu, X. Huang, L.-S. Wang, Angew. Chem. Int. Ed. 2014, 126, 1312 – 1316.

[5] B. Le Guennic, H. Jiao, S. Kahlal, J.-Y. Saillard, J.-F. Halet, S. Ghosh, M. Shang, A. M. Beatty, A. L. Rheingold, T. P. Fehlner, J. Am. Chem. Soc. 2004, 126, 3203 – 3217.

[6] B. P. T. Fokwa, M. Hermus, Angew. Chem. 2012, 124, 1734 – 1737; Angew. Chem. Int. Ed. 2012, 51, 1702 – 1705.

RH. Cold/Ultra-cold/Physics
Thursday, June 19, 2014 – 1:30 PM
Room: B102 Chemical and Life Sciences

Chair: Brian DeMarco, University of Illinois, Urbana, IL, USA

RH01   1:30 – 1:45

A NOVEL METHOD TO MEASURE SPECTRA OF COLD MOLECULAR IONS

SATRAJIT CHAKRABARTY, *Department of Chemistry, The University of California, Berkeley, CA, USA*; MATHIAS HOLZ, <u>EWEN CAMPBELL</u>, AGNIVA BANERJEE, *Department of Chemistry, University of Basel, Basel, Switzerland*; DIETER GERLICH, *Institut für Physik, Technische Universität Chemnitz, Chemnitz, Germany*; JOHN P. MAIER, *Department of Chemistry, University of Basel, Basel, Switzerland.*

A universal method has been developed in our group for measuring the spectra of molecular ions in a 22-pole radio frequency trap at low temperatures. It is based on laser induced inhibition of complex growth (LIICG)[1]. At low temperatures and high number densities of buffer gas, helium attaches to ions *via* ternary association. The formation of these weakly bound complexes, however, is inhibited following resonant absorption of the bare molecular ion.

The first successful measurements have been demonstrated on the $A\,^2\Pi_u \leftarrow X\,^2\Sigma_g^+$ electronic transition of N_2^+, with some thousand N_2^+ ions, helium densities of $10^{15}\,\mathrm{cm}^{-3}$, and storage times of $1\,\mathrm{s}$. The reduction in the number of N_2^+–He complexes is the result of an interplay between excitation, radiative and collisional cooling, ternary association, and collision induced dissociation, and is explained using a kinetic model.

The method is also applicable to larger molecular species. In this case internal conversion following electronic excitation produces internally "hot" ions, reducing the attachment of helium. The technique is universal because complex formation can be impeded over a wide wavelength range.

[1] S. Chakrbarty, M. Holz, E. K. Campbell, A. Banerjee, D. Gerlich, and J. P. Maier, J. Phys. Chem. Lett. 2013, 4, 4051.

RH02   1:47 – 2:02

INFRARED LASER STARK SPECTROSCOPY AND AB INITIO COMPUTATIONS OF THE OH··· CO COMPLEX

<u>TAO LIANG</u>, *Department of Chemistry, University of Georgia, Athens, GA, USA*; PAUL RASTON, *School of Chemistry and Physics, The University of Adelaide, Adelaide, South Australia, Australia*; GARY DOUBERLY, *Department of Chemistry, University of Georgia, Athens, GA, USA.*

Following the sequential pick-up of OH and CO by helium nanodroplets, the infrared depletion spectrum is measured in the fundamental OH stretching region. Although several potentially accessible minima exist on the associated OH + CO reactive potential energy surface [e.g. J. Ma, J. Li, and H. Guo, J. Phys. Chem. Lett. 3 (2012) 2482], such as the weakly bound OH-OC dimer and the chemically bound HOCO molecule, we only observe the weakly bound OH-CO dimer. The rovibrational spectrum of this complex displays narrow ($0.02\,\mathrm{cm}^{-1}$) Lorentzian shaped peaks with spacings that are characteristic of a linear complex with unquenched electronic angular momentum, similar to what was previously observed in the gas phase [M.I. Lester, B.V. Pond, D.T. Anderson, L.B. Harding, and A.F. Wagner, J. Chem. Phys. 113 (2000) 9889]. Analogous spectra involving OD were collected, for which we also only observe the OD-CO isomer. From the Stark spectra, the dipole moments for OH-CO are determined to be 1.85(3) and 1.89(3) D for v=0 and v=1, respectively, while the analogous dipole moments for OD-CO are determined to be 1.88(8) and 1.94(5) D. The computed equilibrium ground state dipole moment at the CCSD(T)/Def2-TZVPD level of theory is 2.185 D, in disagreement with experiment. The role of vibrational averaging is investigated via the solution of a three-dimensional vibrational Schrödinger equation, which is constructed in internal bond-angle coordinates. The computed expectation value of the ground state dipole moment is in excellent agreement with experiment, indicating a floppy molecular complex.

RH03   **2:04 – 2:19**

MOLECULAR BEAM OPTICAL STARK SPECTROSCOPY OF MAGNESIUM DEUTERIDE

TIMOTHY STEIMLE, RUOHAN ZHANG, *Department of Chemistry and Biochemistry, Arizona State University, Tempe, AZ, USA*; HAILING WANG, *Physics Department , East China Normal University , Shanghai, China.*

Light polar, paramagnetic molecules, such as magnesium hydride, MgH, are attractive for slowing and trapping experiments because these molecules have both non-zero permanent electric dipole, μ_{el}, and magnetic dipole, μ_m moments. The permanent electric dipole moment is particularly relevant to Stark deceleration which depends on the ratio of the Stark shift to molecular mass. Here we report on the Stark effect in the (0,0) $A^2\Pi - X\,^2\Sigma^+$ band system of a cold molecular beam sample of magnesium deuteride, MgD. The lines associated with the lowest rotational levels are detected for the first time. The field-free spectrum was analyzed to produce an improved set of fine structure parameters for the $A^2\Pi$(v = 0) state. The observed electric field induced splittings and shifts were analyzed to produce permanent electric dipole moments, μ_{el},of 2.561(10)D and 1.34(8)D for $A^2\Pi$(v = 0) and $X^2\Sigma^+$(v=0)states, respectively. This is the first molecular beam study of MgD.

RH04   **2:21 – 2:36**

MM-WAVE SPECTROSCOPY AND DETERMINATION OF THE RADIATIVE BRANCHING RATIOS OF ^{11}BH FOR LASER COOLING EXPERIMENTS

STEFAN TRUPPE, *Centre for Cold Matter, Blackett Laboratory, Imperial College London, London, United Kingdom*; DARREN HOLLAND, RICHARD JAMES HENDRICKS, *Department of Physics, Imperial College London, London, United Kingdom*; ED HINDS, MICHAEL TARBUTT, *Centre for Cold Matter, Blackett Laboratory, Imperial College London, London, United Kingdom.*

We aim to slow a supersonic, molecular beam of ^{11}BH using a Zeeman slower and subsequently cool the molecules to sub-millikelvin temperatures in a magneto-optical trap. Most molecules are not suitable for direct laser cooling because the presence of rotational and vibrational degrees of freedom means there is no closed-cycle transition which is necessary to scatter a large number of photons. As was pointed out by Di Rosa[a] there exists a class of molecules for which the excitation of vibrational modes is suppressed due to highly diagonal Franck-Condon factors. Furthermore, Stuhl et al.[b] showed that angular momentum selection rules can be used to suppress leakage to undesired rotational states. Here we present a measurement of the radiative branching ratios of the $A^1\Pi \to X^1\Sigma$ transition in ^{11}BH - a necessary step towards subsequent laser cooling experiments. We also perform high-resolution mm-wave spectroscopy of the $J' = 1 \leftarrow J = 0$ rotational transition in the $X^1\Sigma(v = 0)$ state near 708 GHz. From this measurement we derive new, accurate hyper fine constants and compare these to theoretical descriptions. The measured branching ratios suggest that it is possible to laser cool ^{11}BH molecules close to the recoil temperature of 4 μK using three laser frequencies only.

[a]M. D. Di Rosa, *The European Physical Journal D* **31**, 395, 2004
[b]B. K. Stuhl et al., *Physical Review Letters* **101**, 243002, 2008

RH05   **2:38 – 2:53**

CHARACTERIZATION OF CaO$^+$ AND BaO$^+$ BY TWO-PHOTON IONIZATION SPECTROSCOPY

JOSHUA BARTLETT, ROBERT A. VANGUNDY, MICHAEL HEAVEN, *Department of Chemistry, Emory University, Atlanta, GA, USA.*

Reactions of laser-cooled Ca$^+$ and Ba$^+$ ions with O_2 provide pathways to the formation of cold molecular cations (CaO$^+$ and BaO$^+$) that may be further manipulated using ion trapping techniques. Spectroscopic data for these ions are needed to facilitate the characterization of internal state population distributions using highly sensitive detection schemes such as resonantly enhanced multi-photon dissociation (REMPD). Ab initio electronic structure calculations predict that both ions have $X^1\Sigma^+$ ground states, accompanied by low-lying $A^2\Pi$ states. We are currently using two-color photoionization techniques to observe the low-lying ro-vibronic states of CaO$^+$ and BaO$^+$. The neutral molecules are produced by laser ablation of the metals, combined with free-jet expansion driven by He/O_2 (0.1-0.2%) mixtures. Photoionization efficiency curves and zero kinetic energy photoelectron spectra will be reported.

RH06   2:55 – 3:10

RYDBERG SPECTROSCOPY OF ZEEMAN-DECELERATED BEAMS OF METASTABLE HELIUM MOLECULES

PAUL JANSEN, MICHAEL MOTSCH, DANIEL SPRECHER, FREDERIC MERKT, *Laboratorium für Physikalische Chemie, ETH Zurich, Zurich, Switzerland.*

Having three and four electrons, respectively, He_2^+ and He_2 represent systems for which highly accurate *ab-initio* calculations might become feasible in the near future[a]. With the goal of performing accurate measurements of the rovibrational energy-level structure of He_2^+ by Rydberg spectroscopy of He_2 and multichannel quantum-defect theory extrapolation techniques[b], we have produced samples of helium molecules in the $a^3\Sigma_u^+$ state in supersonic beams with velocities tunable down to $100\,\text{m/s}$ by combining a cryogenic supersonic-beam source with a multistage Zeeman decelerator[c]. The molecules are formed at an initial velocity of 500 m/s by striking a discharge in the pulsed expansion of helium gas from a reservoir kept at a cryogenic temperature of 10 K. Using rotationally-resolved PFI-ZEKE (pulsed-field-ionization zero-kinetic-energy) photoelectron spectroscopy, we have probed the rotational-state distribution of the molecules produced in the discharge and found vibrational levels up to $\nu'' = 2$ and rotational levels up to $N'' = 21$ to be populated. The molecular beam is coupled to a multistage Zeeman decelerator[d] that employs pulsed inhomogeneous magnetic fields to further reduce the beam velocity. By measuring the quantum-state distribution of the decelerated sample using photoelectron and photoionization spectroscopy we observed no rotational or vibrational state-selectivity of the deceleration process, but found that one of the three spin-rotation components of the He_2 $a^3\Sigma_u^+$ rotational levels is eliminated.

[a] W.-C. Tung, M. Pavanello, L. Adamowicz, *J. Chem. Phys.* **136**, 104309 (2012).

[b] D. Sprecher, J. Liu, T. Krähenmann, M. Schäfer, and F. Merkt, *J. Chem. Phys.* **140**, 064304 (2014).

[c] M. Motsch, P. Jansen, J. A. Agner, H. Schmutz, and F. Merkt, *arXiv:1401.7774*.

[d] N. Vanhaecke, U. Meier, M. Andrist, B. H. Meier, and F. Merkt, *Phys. Rev. A* **75**, 031402(R) (2007).

RH07   3:12 – 3:27

LASER COOLING THE DIATOMIC MOLECULE CaH

JOE VELASQUEZ, III, MICHAEL DI ROSA, *Physical Chemistry and Applied Spectroscopy, Los Alamos National Laboratory, Los Alamos, NM, USA.*

To laser-cool a species, a closed (or nearly closed) cycle is required to dissipate translational energy through many directed laser-photon absorption and subsequent randomly-directed spontaneous emission events. Many atoms lend themselves to such a closed-loop cooling cycle. Attaining laser-cooled molecular species is challenging because of their inherently complex internal structure, yet laser-cooling molecules could lead to studies in interesting chemical dynamics among other applications. Typically, laser-cooled atoms are assembled into molecules through photoassociation or Feschbach resonance. CaH is one of a few molecules whose internal structure is quite atom-like, allowing a nearly closed cycle without the need for many repumping lasers. We will also present our work-to-date on laser cooling this molecule. We employ traditional pulsed atomic/molecular beam techniques with a laser vaporization source to generate species with well-defined translational energies over a narrow range of velocity. In this way, we can apply laser-cooling to most species in the beam along a single dimension (the beam's axis). This project is funded by the LDRD program of the Los Alamos National Laboratory.

Intermission

RH08

VIBRATIONAL SPECTROSCOPY ON TRAPPED COLD MOLECULAR IONS

NCAMISO B KHANYILE, KENNETH R BROWN, *Department of Chemistry and Biochemistry, Georgia Institute of Technology, Atlanta, GA, USA.*

We perform vibrational spectroscopy on the $V_{0 \leftarrow 10}$ overtone of a trapped and sympathetically cooled CaH^+ molecular ion using a resonance enhanced two photon dissociation scheme. Our experiments are motivated by theoretical work that proposes comparing the vibrational overtones of CaH^+ with electronic transitions in atoms to detect possible time variation of in the mass ratio of the proton to electron [a]. Due to the nonexistence of experimental data of the transition, we start the search with a broadband femtosecond Ti:Saph laser guided by theoretical calculations [b]. Once the vibrational transition has been identified, we will move to CW lasers to perform rotationally resolved spectroscopy.

[a]M. Kajita and Y. Moriwaki, *J.Phys.B.At.Mol.Opt.Phys.*, **42**,154022(2009)
[b]Private communication

RH09

SYMPATHETIC SIDEBAND COOLING OF CaH^+

RENE RUGANGO, KENNETH R BROWN, *Department of Chemistry and Biochemistry, Georgia Institute of Technology, Atlanta, GA, USA.*

We demonstrate sympathetic Doppler cooling and our progress towards sideband cooling of a CaH^+ ion co-trapped with a Ca^+ atomic ion in a linear Paul trap. Molecular ions are generally difficult to laser cool due to a lack of closed electronic transitions as a result of vibrational and rotational states. Despite this challenge, they can be cooled indirectly through their Coulombic interaction with a fluorescent atomic ion that is being directly laser cooled. Ions are firstly Doppler cooled to get to the Lamb-Dicke regime, where the ion motion is small relative to the excitation wavelength and then sideband cooled reaching temperatures below $1 \mu K$. All the ions' axial modes (center of mass and breathing mode) and radial modes (two center of mass and two tilt modes) are addressed, and the temperature is determined by examining the ratio of sidebands.

RH10

BROADBAND OPTICAL COOLING OF AlH^+ TO THE ROTATIONAL GROUND STATE

CHRISTOPHER M. SECK, CHIEN-YU LIEN, BRIAN C. ODOM, *Physics and Astronomy, Northwestern University, Evanston, IL, USA.*

We demonstrate that a single spectrally filtered femtosecond laser, tuned to the electronic A-X transition of trapped AlH^+, can efficiently cool rotations from room temperature to the ground state. The nearly diagonal Franck-Condon-Factors between the electronic X and A states create semi-closed cycling transitions between the vibrational ground states of the X and A states. Parity-preserving electronic cycling cools to the two lowest rotational levels with a $10 \mu s$ timescale set by repeated electronic relaxation, and collection into the lowest rotational level relies upon a slower vibrational relaxation event setting the overall cooling timescale to 140 ms. The population distribution among the rotational levels is detected by $(1 + 1')$ resonance-enhanced multiphoton dissociation (REMPD) and time-of-flight mass-spectrometry (TOFMS).

228

RH11   *Post-Deadline Abstract*   **4:35 – 4:50**

PHYSICS WITH COLD MOLECULES USING BUFFER GAS COOLING: PRECISION MEASUREMENT, COLLISIONS, AND LASER COOLING

<u>NICHOLAS R HUTZLER</u>, JOHN M. DOYLE, *Department of Physics, Harvard University, Cambridge, MA, USA.*

Cryogenic buffer gas cooled beams and cells can be used to study many species, from atoms and polar molecules to biomolecules. We report on recent applications of this technique to improve the limit on the electron electric dipole moment [1], load polar molecules into a magnetic trap through optical pumping [2], perform chirally sensitive microwave spectroscopy on polyatomic molecules [3], progress towards magneto-optical trapping of polar molecules [4], and studies of atom-molecule sticking [5].

[1] The ACME Collaboration: J. Baron et al., Science 343, p. 269 (2014)
[2] B. Hemmerling et al., arXiv:1310.2669, to appear in Phys. Rev. Lett.
[3] D. Patterson, M. Schnell, & J. M. Doyle, Nature 497, p. 475 (2013)
[4] H. Lu et al., arXiv:1310.3239, to appear in New. J. Phys.
[5] J. Piskorski et al., under preparation

RH12   **4:52 – 5:07**

THE [18.1], [18.6] and [18.7] EXCITED STATES OF YTTERBIUM FLUORIDE

<u>TIMOTHY STEIMLE</u>, FANG WANG, *Department of Chemistry and Biochemistry, Arizona State University, Tempe, AZ, USA*; JOE SMALLMAN, *Department of Physics, Imperial College London, London, United Kingdom.*

The generation of a fountain of laser-cooled ytterbium fluoride, YbF, has been recently proposed[a] as a method for long coherent observation times, thereby improving the electron electric dipole moment (eEDM) measurement. Understanding the properties of the excited electronic states of YbF is essential for the development of such a scheme for laser cooling. Here we report on the measurement of the radiative lifetimes, τ, permanent electric dipole moments, μ_{el}, and magnetic g-factors for the [18.6] and [18.7] excited states of YbF. The results are compared with the previously determined values for [18.1] state[bc]. The [18.1] state is the Ω=1/2 spin-orbit component of the A $^2\Pi$(v=0) electronic state arising from the $Yb^+(4f^{14}6p\pi)F^-(2p^{14})$ configuration. The experimentally determined μ_{el}, and g-factors will be used to unravel the nature of the [18.6] and [18.7] states, which are known to be admixtures A $^2\Pi$ and an additional Ω=1/2 state of unknown electronic configuration.

[a]Tarbutt, M R; Sauer, B E; Hudson, J J; Hinds E A, New J. Phys 15, 053034, 2013.
[b]Zhuang,X; Le,A.;Steimle, T C; Bulleid, N E; Smallman, I J; Hendricks, R J; Skoff, S M ; M R; Hudson, J J; Sauer, B E; Hinds, Tarbutt, M R, PCCP, 13 19103, 2011
[c]Condylis,P C; Hudson, J J; Tarbutt, M R; Sauer, B E; Hinds E A, J. Chem. Phys. 123, 231101, 2005

RH13   **5:09 – 5:24**

HIGH-RESOLUTION MOLECULAR SPECTROSCOPY OF H_2 AT 10% THE AGE OF THE UNIVERSE; TESTING THE CONSTANCY OF PHYSICAL LAW

<u>WIM UBACHS</u>, JULIJA BAGDONAITE, MARIO DAPRA, *Department of Physics and Astronomy, VU University , Amsterdam, Netherlands*; MICHAEL T MURPHY, *Center for Astrophysics and Supercomputing, Swinburne University of Technology, Melbourne, Australia*; LEX KAPER, *Anton Pannekoek Astronomical Institute, University of Amsterdam, Amsterdam, Netherlands.*

Spectroscopy has taught us that atoms and molecules are the same now as they were in the early Universe. This means that the nature of the fundamental forces, and the chemical bonds that derive from the strength of the electromagnetic force, have not changed over cosmological history. High-resolution spectroscopy of molecules in far-distant galaxies is currently applied to tighten the constraints on the constancy of physical law, which is expressed in terms of the constancy of the fundamental constants, i.e. the fine-structure constant α and the proton-electron mass ratio μ. H_2 observable at redshifts $z = 2 - 4$, corresponding to look-back times of 10-12 billion years, is sensitive to probe a possible variation of μ. We have examined a number of high redshift objects, Q0405-443 at $z = 2.59$, Q0347-383 at $z = 3.02$, Q0528-250 at $z = 2.81$, Q2123-005 at $z = 2.06$, Q2348-011 at $z = 2.42$, and Q0642-504 at $z = 2.66$ to constrain $\Delta\mu/\mu$. Currently work is in progress to analyze spectra of additional objects Q1237+064 at $z = 2.69$, in which also CO is observed, and Q1441+272, which is the highest redshift object (at $z = 4.22$) in which H_2 is abundantly detected. From the combined studies it follows that μ varies by less than 10^{-5} for look back times of 90% of the age of the Universe.

RH14

SEARCH FOR A VARIATION OF THE PROTON-ELECTRON MASS RATIO FROM METHANOL OBSERVATIONS

WIM UBACHS, JULIJA BAGDONAITE, MARIO DAPRA, HENDRICK BETHLEM, *Department of Physics and Astronomy, VU University , Amsterdam, Netherlands*; NISSIM KANEKAR, *Tata Institute f of Fundamental Research, National Centre for Radio Astrophysics, Pune, India*; SEBASTIEN MULLER, *Onsala Space Observatory, Chalmers University of Technology, Onsala, Sweden*; CHRISTIAN HENKEL, KARL MENTEN, *Millimeter- und Submillimeter-Astronomie, Max-Planck-Institut für Radioastronomie, Bonn, NRW, Germany.*

A limit on a possible cosmological variation of the proton-to-electron mass ratio μ is derived from observation of methanol lines in the PKS1830-211 lensed galaxy at redshift $z \sim 0.89$ with the Effelsberg 100 m single-dish radio telescope (at frequencies 6.5 - 32 GHz), the IRAM 30 m telescope (at frequencies 80 - 160 GHz), and band-6 of the novel ALMA telescope array (at 260 GHz). Ten different absorption lines of CH_3OH are detected covering a wide range of sensitivity coefficients K_μ. Systematic effects of chemical segregation, excitation temperature, frequency dependence, and time variability of the back- ground source are quantified. A robust constraint of $\Delta\mu/\mu = (1.0 \pm 0.8_{stat} \pm 1.0_{syst}) \times 10^{-7}$ is derived from this large sample of lines belonging to a single molecular species. Analysis of additional observations at the E-VLA radio telescope is under way.

RH15

"SIMPLEST MOLECULE" CLARIFIES MODERN PHYSICS I. CW LASER SPACE-TIME FRAME DYNAMICS

T.C. REIMER, W. G. HARTER, *Department of Physics, The University of Arkansas, Fayetteville, AR, USA.*

Molecular spectroscopy makes very precise applications of quantum theory including GPS, BEC, and laser clocks. Now it can return the favor by shedding some light on modern physics mysteries by further unifying quantum theory and relativity.

We first ask, "What is the simplest molecule?" Hydrogen H_2 is the simplest stable molecule. *Positronium* is an electron-positron (e^+e^-)-pair. An even simpler "molecule" or "radical" is a photon-pair (γ, γ) that under certain conditions can create an (e^+e^-)-pair.

To help unravel relativistic and quantum mysteries consider CW laser beam pairs or TE-waveguides. Remarkably, their wave interference immediately gives Minkowski space-time coordinates and clearly relates eight kinds of space-time wave dilations or contractions to shifts in Doppler frequency or wavenumber.

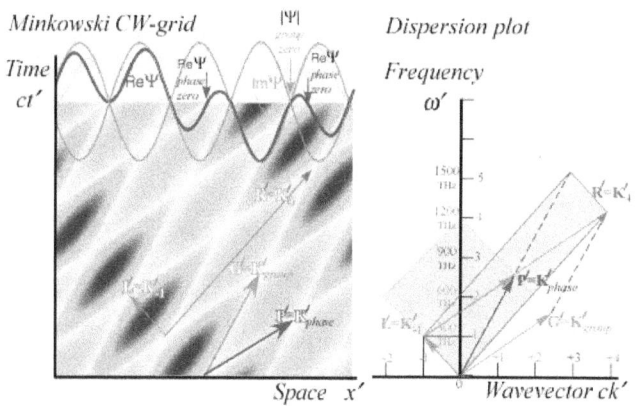

Modern physics students may find this approach significantly simplifies and clarifies relativistic physics in space-time *(x,ct)* and inverse time-space *(ω,ck)*. It resolves some mysteries surrounding super-constant *c=299,792,458m/s* by proving "Evenson's Axiom" named in honor of NIST metrologist Ken Evenson (1932-2002) whose spectroscopy established c to start a precision-renaissance in spectroscopy and GPS metrology.

The following Talk II applies this approach to relativistic quantum mechanics.

RH16

"SIMPLEST MOLECULE" CLARIFIES MODERN PHYSICS II. RELATIVISTIC QUANTUM MECHANICS

T.C. REIMER, W. G. HARTER, *Department of Physics, The University of Arkansas, Fayetteville, AR, USA.*

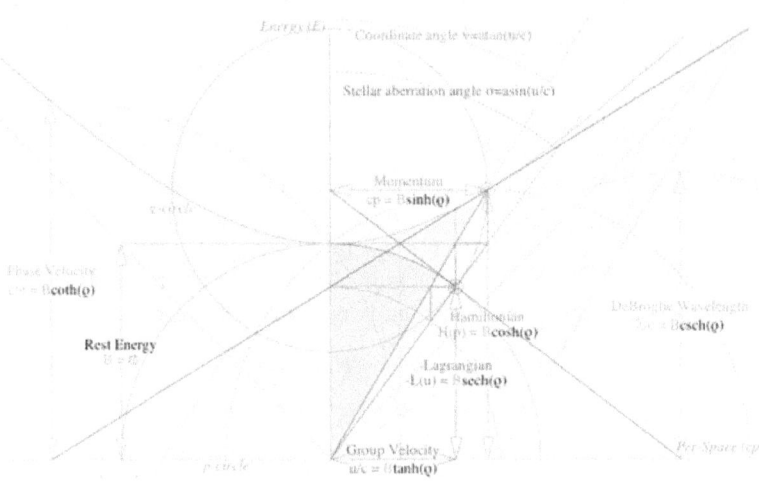

A "simplest molecule" consisting of CW-laser beam pairs helps to clarify relativity in Talk I. In spite of a seemingly massless evanescence, an optical pair also clarifies classical and quantum mechanics of relativistic matter and anti-matter.

Logical extension of *(x,ct)* and *(ω,ck)* geometry gives relativistic action functions of Hamiltonian, Lagrangian, and Poincare that may be constructed in a few ruler-and-compass steps to relate relativistic parameters for group or phase velocity, momentum, energy, rapidity, stellar aberration, Doppler shifts, and DeBroglie wavelength. This exposes hyperbolic and circular trigonometry as two sides of one coin connected by Legendre contact transforms. One is Hamiltonian-like with a longitudinal rapidity parameter ρ (log of Doppler shift). The other is Lagrange-like with a transverse angle parameter σ (stellar aberration). Optical geometry gives recoil in absorption, emission, and resonant Raman-Compton acceleration and distinguishes Einstein rest mass, Galilean momentum mass, and Newtonian effective mass. (Molecular photons appear less bullet-like and more rocket-like.) In conclusion, modern space-time physics appears as a simple result of the more self-evident Evenson's axiom: *"All colors go c."*

RI. Matrix isolation (and droplets)

Thursday, June 19, 2014 – 1:30 PM

Room: 112 Chemistry Annex

Chair: Takamasa Momose, University of British Columbia, Vancouver, BC, Canada

---

RI01   1:30 – 1:45

MARILYN JACOX: HER CONTRIBUTIONS TO MATRIX ISOLATION SPECTROSCOPY AND BEYOND

BARBARA MILLER, TERRY A. MILLER, *Department of Chemistry and Biochemistry, The Ohio State University, Columbus, OH, USA.*

This introductory talk will briefly overview the contributions of Marilyn Jacox to the technique of matrix isolation spectroscopy and her use of it to investigate the vibrational spectroscopy of a multitude of interesting molecular species. This work as well as her cataloging of the results of others on many more molecules is well known in the spectroscopic community. Much less well known is her life outside spectroscopy. The talk will conclude with some glimpses into the varied interests of her extraordinary life.

RI02   1:47 – 2:02

INFRARED ABSORPTION OF CH_3O/CD_3O RADICALS PRODUCED UPON PHOTOLYSIS OF CH_3ONO/CD_3ONO IN A p-H_2 MATRIX

YU-FANG LEE, WEI-TE CHOU, *Applied Chemistry, National Chiao Tung University, Hsinchu, Taiwan*; BRITTA JOHNSON, EDWIN SIBERT, *Department of Chemistry, The Univeristy of Wisconsin, Madison, WI, USA*; YUAN-PERN LEE, *Applied Chemistry, National Chiao Tung University, Hsinchu, Taiwan.*

The methoxy radical, CH_3O, has attracted much attention because of its important molecular structure and also as a reaction intermediate in combustion and atmospheric chemistry. Previous investigations include laser-induced fluorescence, laser magnetic resonance, and stimulated emission pumping. High-resolution infrared spectrum of jet-cooled CH_3O, produced by laser photolysis of CH_3ONO, in the C-H stretching region 2850-2940 cm^{-1} has been reported.[a] However, direct infrared absorption spectrum of CH_3O other than the C–H stretching region remains unreported. Irradiation of a p-H_2 matrix containing CH_3ONO at 3.2 K with UV light produced main features at 1365.4, 1427.5 (2_1^-, 2_1^+), 1041.8 (3_1^-), 1346.8, 1427.5, 1520.9, 1520.9 (5_1^-, 5_1^+, 5_1^-, 5_1^+), and 689.3/694.9, 945.9/951.7, 1233.5, 1235.9 cm^{-1} (6_1^-, 6_1^+, 6_1^-, 6_1^+); labels 2-6 in parentheses indicate transitions to vibrational states attributable to the umbrella, C–O stretching, CH_2 scissoring, and HCO deformation modes of CH_3O, respectively. These features appeared upon photolysis and diminished after five minutes; formation of CH_2OH was observed as CH_3O decayed. The assignments were based on comparison of observed vibrational wavenumbers with those predicted with the quadratic potential energy force field and quadratic dipole moment expansion calculated with the CCSD(T)/cc-pVTZ method.[b] Jahn-Teller and anharmonic vibrational contributions were included in the full Hamiltonian to estimate the correlation diagram connecting the harmonic eigenvalues to those of the fully coupled problem. Similarly, lines of CD_3O were observed upon UV photolysis of CD_3ONO, but became diminished within five minutes. These observations demonstrates the advantage of diminished cage effect of solid p-H_2; CH_3O and CD_3O are produced via *insitu* UV photodissociation of CH_3ONO isolated in p-H_2, but not in Ar or Ne.

---

[a]J.-X. Han, Y. G. Utkin, H.-B. Chen, L. A. Burns and R. F. Curl, *J. Chem. Phys.* **117**, 6538 (2009).
[b]J. Nagesh and E. L. Sibert III, *J. Phys. Chem. A* **116**, 3846 (2012).

RI03   2:04 – 2:19

ELECTRONIC SPECTROSCOPY OF MASS-SELECTED $C_7H_3^+$ AND C_7H_3 ISOMERS IN 6 K NEON MATRICES

JOHN P. MAIER, ARGHYA CHAKRABORTY, JAN FULARA, *Department of Chemistry, University of Basel, Basel, Switzerland.*

Three open chain isomers of $C_7H_3^+$ including a three member carbon ring in the structure were identified in a neon matrix held at 6 K using a mass-selected ion beam. Both indene and acyclic precursors were used in an ion source to produce the $C_7H_3^+$ cations which were codeposited with excess of neon to produce the matrix. The electronic absorption spectra were measured by the sensitive waveguide technique sampling around 10^{16} cm^{-3} species. Some of the cations show well-resolved fluorescence. Some neutral C_7H_3 isomers are also detected in absorption produced in the matrix by photo-bleaching of the cations. The charge of the mass-selected species is experimentally established using presence of electron scavengers, release of electrons, changing the kinetic energy during deposition and precursor dependence, whereas the structure of the isomers is inferred by comparison of the energies of the transitions with CASPT2 calculations. The assignment is supported by comparison of the vibrational frequencies inferred from the spectra and theory. The first time identification of the electronic transitions of these structurally interesting isomers provides their means for their *in situ* monitoring in hydrocarbon combustion processes and formation of aromatics therein.

RI04   2:21 – 2:36

ASSIGNING STATES IN THE JAHN-TELLER COUPLED INFRARED SPECTRA OF CH_3O AND CD_3O

BRITTA JOHNSON, EDWIN SIBERT, *Department of Chemistry, The Univeristy of Wisconsin, Madison, WI, USA.*

The ground $\tilde{X}^2E$ vibrations of the methoxy radical have intrigued both experimentalists and theorists alike due to the presence of a conical intersection at the C_{3v} molecular geometry. This conical intersection causes methoxy's vibrational spectrum to be strongly influenced by Jahn-Teller coupling, this leading to large amplitude vibrations and extensive mixing of the two lowest electronic states. This coupling combined with spin-orbit and Fermi couplings greatly complicates the assignments of states. In this talk we describe our efforts to assign the states of both CH_3O and CD_3O.

Using the potential energy force field and calculated spectra of Nagesh and Sibert[1] as a starting point, vibrational mixing is considered using various zero-order representations. When the zero-order states are the diabatic normal mode states, there is sufficient mode mixing that the normal mode quantum numbers are no longer good labels. The mixing of the zero-order states can be reduced by including additional terms in the zero-order Hamiltonian, H^o. We consider the choice of including the first order Jahn-Teller coupling between one of the three degenerate normal modes. As the rocking motion has the largest Jahn-Teller coupling, this is the coupling that is included in H^o. Although the normal mode quantum numbers of the rocking basis functions are no longer good quantum numbers, due to the Jahn-Teller induced vibronic mixing, the zero-order states can be labeled with the linear Jahn-Teller quantum numbers.[2] This work extends these ideas by considering an H^o that includes linear Jahn-Teller coupling between *two* sets of degenerate vibrations. Plots of the resulting zero-order states are presented, and the spectral transitions recently observed[3] for both CH_3O and CD_3O in a p-H_2 matrix are assigned using these basis functions. The extent of state-mixing found for the full Hamiltonian H for various choices of H^o is illustrated via the use of correlation diagrams obtained by plotting the eigenvalues of $H^o + \delta(H - H^o)$ as a function of δ where δ varies from zero to one.

[1] Nagesh, J.; Sibert, E. L. *J. Phys. Chem. A* **2012**, *116*, 3846–3855.

[2] Barckholtz, T. A.; Miller, T. A. *Int. Revs. in Phys. Chem.* **1998**, *17*, 435–524.

[3] Yu-Fang Lee, Wei-Te Chou, and Yuan-Pern Lee (private communication).

RI05   2:38 – 2:53

FTIR STUDIES OF THE PHOTOCHEMISTRY OF DEUTERATED FORMIC ACID IN A PARAHYDROGEN MATRIX

DAVID T. ANDERSON, *Department of Chemistry, University of Wyoming, Laramie, WY, USA.*

We report new FTIR studies of the 193 nm photochemistry of deuterated formic acid (DCOOD) trapped in parahydrogen matrices. In our 2011 preliminary report,[a] we showed the 193 nm *in situ* photolysis of formic acid (HCOOH) produces small amounts of HCO and HOCO and that after the laser is turned off, we observe continued slow growth in the HOCO radical for up to 10 hours after photolysis. At that time we were unsure of the detailed chemical mechanism by which the HOCO continues to grow after photolysis, but we suspected it had to do with reactions of mobile H-atoms with the HCOOH precursor that remains at significant concentrations after photolysis under these conditions. The present deuterated formic acid photolysis studies provide strong circumstantial evidence that H-atom reactions with formic acid are the source of the continued HOCO growth. Further, variable-temperature kinetic studies conducted with the deuterated formic acid sample show a strong inverse temperature dependence to the reaction kinetics. Essentially, the reaction that leads to HOCO growth only occurs at temperatures below 2.4 K. We are currently trying to model the kinetics using standard methods and the most recent analysis will be presented at the meeting.

[a]David T. Anderson, Leif O. Paulson, *66th Ohio State University International Symposium on Molecular Spectroscopy*, talk FE02 (2011).

RI06   2:55 – 3:10

STEERING H-ATOM DIFFUSION THROUGH IMPURITY-DOPED SOLID PARAHYDROGEN: THE ROLE OF DIFFERENTIAL SOLVATION ENERGIES

ROBERT HINDE, *Department of Chemistry, University of Tennessee, Knoxville, TN, USA.*

Ultraviolet irradiation of solid parahydrogen (pH_2) matrices doped with suitable H-atom precursor molecules generates H atoms *in situ* through a series of photoinitiated chemical reactions; these H atoms move through the pH_2 matrix through a quantum diffusion process that involves the tunneling-mediated process $H + H_2 \rightarrow H_2 + H$. The mobile H atoms may react chemically with other species that are also embedded in the pH_2 matrix; an investigation of the kinetics of these H-atom chemical reactions provides us with information about reaction dynamics in the pH_2 matrix environment. A recent study of the $H + N_2O \rightarrow HNNO$ reaction in solid pH_2 [Mutunga, Follett, and Anderson, J. Chem. Phys. **139**, 151104 (2013)] demonstrates that this reaction exhibits strongly non-Arrhenius behavior, proceeding at measurable rates only when the temperature of the system drops below $T \approx 2.4$ K. A molecular-level understanding of these findings requires information about how the solid pH_2 matrix environment affects the long-range entrance channel of the $H + N_2O$ reaction. Here, we carry out quantum Monte Carlo simulations of a simple model system (Ar-doped solid pH_2) to investigate matrix-induced changes to long-range interactions between H atoms and other impurities embedded in solid pH_2 matrices. Our results suggest that the pH_2 matrix creates an effective long-range repulsion between the H atom and the Ar impurity, which we explain in terms of differential solvation energies of Ar and H atoms in solid pH_2.

RI07   3:12 – 3:27

INFRARED SPECTRA AND CALCULATED BINDING ENERGIES OF γ-BUTYROLACTONE DIMERS AND TRIMERS

ERIC WILLIS, CHRIS BAUMANN, *Department of Chemistry, University of Scranton, Scranton, PA, USA.*

Infrared spectra for matrix-isolated γ-butyrolactone and γ-butyrolactone-d_6 were obtained. The carbonyl stretching mode occurs at 1803 cm^{-1} for the monomer species, 1786 cm^{-1} for the dimer species, and 1774 cm^{-1} for the trimer species (1797, 1789 and 1770 cm^{-1} for the deuterated isotopomer.) Vibrational frequencies calculated using density functional theory are in agreement with the experimental values. Density functional theory was used to calculate the structures and binding energies of γ-butyrolactone dimers and trimers. Binding energies of 55-58 kJ mol^{-1} are predicted for the dimer structures. Optimized geometries for stacked and ring trimer structures have been calculated, with predicted binding energies of up to 68 kJ mol^{-1}.

Intermission

RI08 3:44 – 3:59

OBSERVATION OF TRANSIENT SURFACE-BOUND INTERMEDIATES BY INTERFACIAL MATRIX STABILIZATION SPECTROSCOPY (IMSS)[a]

NINA K JARRAH, DAVID T MOORE, *Chemistry Dept., Lehigh University, Bethlehem, PA, USA.*

Interfacial matrix stabilization spectroscopy is a new technique based on matrix isolation spectroscopy, but where a cryogenic matrix is deposited over the top of a film sample, in order to characterize interactions between the substrate and molecular dopants. The IMSS technique harnesses the well-established ability of cryogenic matrices to trap and stabilize transient species, although in this case it is applied to intermediates relevant to heterogeneous catalysis. In this proof-of-concept study, we present data for CO and O_2 reactants binding to TiO_2 and Au/TiO_2 nanoparticle films, where in the latter case the Au nanoparticles were created by de-wetting of a 22.5 nm overlayer at 450 K. The films are first pre-saturated with CO at 40 K, then cooled to 20 K, at which point an argon matrix is deposited over the top of them. The spectra are then annealed in stages over a range of temperatures between 20 and 40 K. In all cases, the presence of the Ar matrix alters the appearance of the CO bands, revealing additional structure, such as a broad feature at 2150 cm^{-1}, which is typically attributed to CO interacting with OH groups on the TiO_2 surface, but is not observed at 40 K for these samples in the absence of the matrix. The interpretation is that the matrix induces a caging-effect that prevents molecules from desorbing from weak binding sites from which they would be "pumped away" in the vacuum chamber if the matrix were not present. Perhaps the most interesting feature of these spectra is a small but sharp band at 2112 cm^{-1} that appears ONLY when O_2 is added to the argon matrix as a dopant. This transient band grows in following annealing at 32 K, but then disappears upon annealing above 34 K, suggesting that it may correspond to a reactive intermediate. The band occurs for samples both with and without Au present on the TiO_2 surface, but shows a larger intensity in the latter case. Possible assignments for the observed band in light of previous studies from the literature will be discussed in detail.

[a]Funding support through NSF CAREER Award CHE-0955637 is gratefully acknowledged

RI09 4:01 – 4:16

TEMPERATURE DEPENDENT ANISOTROPIC LINE SPLITTING AND LINEWIDTH IN MOTIONALLY AVERAGED EPR SPECTRA OF CH_3 RADICALS IN SOLID Ar

YURIJ DMITRIEV, *Division of Plasma Physics, Atomic Physics and Astrophysics, Ioffe Physical-Technical Institute, St. Petersburg, Russia*; NIKOLAS PLOUTARCH BENETIS, *Department of Pollution Control, Technological Education Institution, TEI, Kozani, Greece.*

Rotational dynamics is one of the most fascinating features of molecular solids. This report presents EPR studies of methyl radical rotation in solid Ar, a matrix possessing no nuclear magnetic moments that facilitates recording magnetic resonance spectra of high resolution. As a result, we were able to make a thorough investigation of the lineshape temperature behavior of the A- and E-symmetry multiplets related to different rotational states of the radical. Methyl radicals in solid Ar matrix were recorderd at the liquid helium temperature 4.2 K and in the range from 9 through 31 K. Both EPR multiplets exhibited anisotropic features which were explained by a model accounting for fast rotation about the radical C_3 axis and relatively slow reorientation about the in-plane C_2 axes. We present the first observation of anisotropic splitting in the powder EPR of methyl radical E-states suggesting that the A- and E- rotational states which differ in their spatial characteristics should exhibit a varying degree of coupling to the surrounding cage in the matrix. The simulation model of the E-state doublet was tentatively based on the transition of radical dynamics from rotational jumps at low temperatures to rotational diffusion at high temperatures with a conditional border 19 K. For the molecule in A-state , temperature assisted tunneling was supposed to occur at temperatures below 16 K while temperature activated rotational diffusion was taking place at higher temperatures.[a]

[a]Y. D. acknowledges a partial financial support from Russian Foundation for Basic Research under grant 13-02-00373-a.

RI10   **4:18 – 4:33**

INFRARED SPECTROSCOPY OF ALANINE IN SOLID PARAHYDROGEN

SHIN YI TOH, YING-TUNG ANGEL WONG, PAVLE DJURICANIN, TAKAMASA MOMOSE, *Department of Chemistry, University of British Columbia, Vancouver, BC, Canada.*

Amino acids are the building blocks of biological molecules, and thus the investigation of their physical and chemical properties would allow for further understanding of their functions in biological systems. In addition, the existence of amino acids in interstellar space has been discussed for many years, but it is still under intense debate. The effect of UV radiation on amino acids is one of the keys for their search in interstellar space, where strong UV radiation exists. In this experiment, conformational compositions of alpha and beta alanine and their UV photolysis were investigated via matrix-isolation FTIR spectroscopy and quantum chemical calculations. Solid parahydrogen was used as the matrix, which provides higher resolution spectra than other noble gas matrices. We have identified several stable conformers for both alpha and beta alanine in solid parahydrogen. A clear correlation between conformational ratio and sublimation temperature was found for beta alanine. Furthermore, it was found that UV photolysis of alanine yields not only its conformational changes, but also photodissociation into a CO_2 molecule and fragments. Observed spectra and their analysis will be discussed in relation to interstellar chemistry.

RI11   **4:35 – 4:50**

ELECTRONIC ABSORPTION SPECTROSCOPY AND FRANCK-CONDON SIMULATIONS FOR HC_7H and MeC_7H

BENJAMIN C. HAENNI, *Department of Chemistry, University of Wisconsin–Madison, Madison, WI, USA*; CHRISTOPHER J. SHAFFER, *Department of Chemistry, University of Washington, Seattle, WA, USA*; JOHN F. STANTON, *Department of Chemistry, The University of Texas, Austin, TX, USA*; ROBERT J. McMAHON, *Department of Chemistry, The Univeristy of Wisconsin, Madison, WI, USA.*

Highly unsaturated carbon chains of the HC_nH family are important to the studies of combustion chemistry and the interstellar medium (ISM). Several members of this family (n=2,4,6) have been detected in the ISM by infrared spectroscopy. We have successfully matrix-isolated HC_7H and MeC_7H species and studied them using electronic absorption, FTIR, and EPR spectroscopy. The ground state potential energy surface was explored using ab initio (CCSD(T)/cc-pVTZ (fc)) methods to discover triplet minima for both species. Equation of motion coupled cluster calculations (EOM-CCSD/ANO1) of low-lying excited states allowed for Franck-Condon simulations. The comparison of the simulated spectra to the vibronic progression observed experimentally in the UV/Vis spectra permits assignment of a linear ground state triplet structure for both HC_7H and MeC_7H.

RI12   **4:52 – 5:02**

CONFORMATIONAL ANALYSIS OF R-(+)-3-METHYLCYCLOPENTANONE BY IR SPECTROSCOPY IN PARA-HYDROGEN CRYSTAL

WATHEQ AL-BASHEER, *Department of Physics, King Fahd University of Petroleum and Minerals, Dhahran, Saudi Arabia*; SHIN YI TOH, *Department of Chemistry, University of British Columbia, Vancouver, BC, Canada*; JUN MIYAZAKI, *Department of Liberal Arts and Basic Sciences, College of Industrial Technology, niho, japan*; TAKAMASA MOMOSE, *Department of Chemistry, University of British Columbia, Vancouver, BC, Canada.*

Para-hydrogen (pH_2) soft quantum crystal is an ideal isolation matrix due to its impressive intrinsic properties, i.e. its significant lattice constant, large zero-point vibration as well as its ability to repair itself of crystal defects. To investigate molecular conformation of a chiral ketone, IR spectra of R-(+)-3-methylcyclopentanone (R3MCP), hosted in pH_2 crystal, were recorded as a function of sample concentration and host pH_2 crystal temperature over the low deposition range {3.5-6.0K}. IR spectra of R3MCP in pH_2 crystal will be presented and compared against corresponding spectra in Ar matrix as well as IR spectra of the neat crystalline R3MCP at low deposition temperatures. Furthermore, density functional theory calculations of simulated IR spectra for the optimized geometries of R3MCP, equatorial-methyl and axial-methyl conformers are compared against experimental spectra for the purpose of investigating molecular conformation. Upon comparison between theoretical and experimental IR spectra, vibrational modes arising from equatorial and axial conformers have been successfully assigned and related to the individual conformer's structure.

THE CHARACTERIZATION OF GeH_2 AND GeH USING MATRIX ISOLATION INFRARED SPECTROSCOPY

JAY AMICANGELO, CHRISTOPHER BAILEY, MADELYN HOOVER, BRUCE HUFFMAN, *School of Science (Chemistry), Penn State Erie, Erie, PA, USA.*

Matrix isolation infrared spectroscopy was used to characterize the fundamental infrared peaks of the germanium hydride species GeH_2 and GeH in low temperature argon matrices that result from the vacuum-ultraviolet (VUV) photolysis of germane (GeH_4). Experiments were performed by depositing mixtures of GeH_4 with argon onto a CsI window cooled to 12 K while simultaneously photolyzing the mixture with 121 nm VUV radiation from a hydrogen resonance lamp. For GeH_2, the fundamental infrared peaks are observed at 1839.1 cm^{-1} (ν_3, antisymmetric stretch), 1816.6 cm^{-1} (ν_1, symmetric stretch) and 913.4 cm^{-1} (ν_2, bend). For GeH, the fundamental infrared stretching peak is observed at 1813.4 cm^{-1}. The assignment of the observed peaks is established by performing experiments with isotopic germane (GeD_4), by performing matrix annealing experiments (warming to 25 - 35 K and refreezing to 12 K), by performing mercury-xenon lamp matrix photolysis experiments (200 - 900 nm), and by comparison to quantum chemical calculations performed at the B3LYP and MP2 levels of theory. This work corrects what appears to be incorrect assignments made in the earlier report of Smith and Guillory[a].

[a] G. R. Smith and W. A. Guillory, J. Chem. Phys. 56, 1423 (1972)

ELECTRONIC RELAXATION PROCESSES OF TRANSITION METAL ATOMS IN HELIUM NANODROPLETS

ANDREAS KAUTSCH, FRIEDRICH LINDEBNER, MARKUS KOCH, WOLFGANG E. ERNST, *Institute of Experimental Physics, Graz University of Technology, Graz, Austria.*

Spectroscopy of doped superfluid helium nanodroplets (He_N) gives information about the influence of this cold, chemically inert, and least interacting matrix environment on the excitation and relaxation dynamics of dopant atoms and molecules [a]. We present the results from laser induced fluorescence (LIF), photoionization (PI), and mass spectroscopy of Cr [b] and Cu [c] doped He_N. From these results, we can draw a comprehensive picture of the complex behavior of such transition metal atoms in He_N upon photo-excitation. The strong Cr and Cu ground state transitions show an excitation blueshift and broadening with respect to the bare atom transitions which can be taken as indication for the solvation inside the droplet. From the originally excited states the atoms relax to energetically lower states and are ejected from the He_N. The relaxation processes include bare atom spin-forbidden transitions, which clearly bears the signature of the He_N influence. Two-color resonant two-photon ionization (2CR2PI) also shows the formation of bare atoms and small Cr-He_n and Cu-He_n clusters in their ground and metastable states [c] [d]. Currently, Cr dimer excitation studies are in progress and a brief outlook on the available results will be given.

[a] C. Callegari and W. E. Ernst, Helium Droplets as Nanocryostats for Molecular Spectroscopy - from the Vacuum Ultraviolet to the Microwave Regime, in Handbook of High-Resolution Spectroscopy, eds. M. Quack and F. Merkt, John Wiley & Sons, Chichester, 2011.
[b] A. Kautsch, M. Koch, and W. E. Ernst, J. Phys. Chem. A, 117 (2013) 9621-9625, DOI: 10.1021/jp312336m
[c] F. Lindebner, A. Kautsch, M. Koch, and W. E. Ernst, Int. J. Mass Spectrom. (2014) in press, DOI: 10.1016/j.ijms.2013.12.022
[d] M. Koch, A. Kautsch, F. Lackner, and W. E. Ernst, submitted to J. Phys. Chem. A

RI15

MOLECULAR SPECTRA OF RbSr: HELIUM DROPLET ASSISTED PREPARATION OF A DIATOMIC MOLECULE

FLORIAN LACKNER, GÜNTER KROIS, THOMAS BUCHSTEINER, JOHANN V. POTOTSCHNIG, WOLF-GANG E. ERNST, *Institute of Experimental Physics, Graz University of Technology, Graz, Austria.*

We report on the first spectroscopic investigation of the ground and excited states of RbSr. The molecules are prepared in their vibronic ground state ($X^2\Sigma^+_{1/2}, \nu'' = 0$) in a sequential pickup process[a] on the surface of helium nanodroplets, confined in a cold (0.38 K) and weakly perturbing superfluid environment[b]. Utilizing resonance-enhanced multi-photon ionization time-of-flight (REMPI-TOF) spectroscopy and laser induced fluorescence (LIF) spectroscopy our investigations cover the spectral regime of $11500\,\mathrm{cm}^{-1}$ - $23000\,\mathrm{cm}^{-1}$. The weak interaction between molecules and helium droplets causes a broadening of the observed transitions. For spectrally resolved band systems the helium droplet isolation approach facilitates the determination of molecular constants. Our assignment is assisted by theoretical calculations of potential energy curves based on a multireference configuration interaction (MRCI) approach. Several strong transitions could be identified; the most prominent spectral feature is a vibrational resolved band system at $14000\,\mathrm{cm}^{-1}$. In contrast to the excitation spectra, dispersed fluorescence (DF) spectra are not influenced by the helium environment, because the molecules leave the droplets upon photoexcitation, revealing detailed insights into the electronic structure of the free RbSr molecule.
Our experiments will aid the ongoing search for optimal pathways for the preparation of ultracold ground state RbSr molecules[c,d].

[a]G. Krois, J.V. Pototschnig, F. Lackner and W.E. Ernst, J. Phys. Chem. A, 117 (50), 13719-13731 (2013)

[b]C. Callegari and W. E. Ernst, Helium Droplets as Nanocryostats for Molecular Spectroscopy - from the Vacuum Ultraviolet to the Microwave Regime, in: Handbook of High-Resolution Spectroscopy, eds. M. Quack and F. Merkt, John Wiley & Sons, Chichester, (2011)

[c]P.S. Żuchowski, R. Guerout, and O. Dulieu, arXiv preprint arXiv:1402.0702 (2014)

[d]B. Pasquiou, A. Bayerle, S.M. Tzanova, S. Stellmer, J. Szczepkowski, M. Parigger, R. Grimm, and F. Schreck, Phys. Rev. A, 88 (2), 023601 (2013)

RI16

HELIUM NANODROPLET ISOLATION AND ROVIBRATIONAL SPECTROSCOPY OF HYDROXYMETHYLENE

CHRISTOPHER LEAVITT, CHRIS MORADI, *Department of Chemistry, University of Georgia, Athens, GA, USA*; JOHN F. STANTON, *Department of Chemistry, The University of Texas, Austin, TX, USA*; GARY DOUBERLY, *Department of Chemistry, University of Georgia, Athens, GA, USA.*

Hydroxymethylene (HCOH) and the d_1-isotopologue (HCOD) are isolated in low temperature helium nanodroplets through pyrolysis of glyoxylic acid. Transitions measured in the infrared spectra are assigned exclusively to the *trans*-conformation based on previously reported anharmonic frequency calculations (P. R. Schreiner, *et. al.* Nature 453, 906-909 and L. Kuziol, *et. al.* J. Chem. Phys. 128, 204310). For the OH(D) and CH stretches, *a*- and *b*-type lines are observed, and when taken in conjunction with CCSD(T)/cc-pVTZ computations, lower limits to the vibrational band origins are determined. The relative intensities of the *a*- and *b*-type transitions provide the orientation of the transition dipole moment in the inertial frame. Overall, the He nanodroplet data is in excellent agreement with the anharmonic frequency computations, confirming the appreciable Ar-matrix shift of the OH and OD stretches and a Fermi resonance interaction in HCOH between the ν_2 CH stretch and the $\nu_3+\nu_4$ combination band.

RJ. Large amplitude motions, internal rotation

Thursday, June 19, 2014 – 1:30 PM

Room: 274 Medical Sciences Building

Chair: Marie-Aline Martin-Drumel, Universität zu Köln, Köln, Germany

---

RJ01   1:30 – 1:45

SPECTRAL ASSIGNMENTS AND ANALYSIS OF THE GROUND STATE OF NITROMETHANE IN HIGH-RESOLUTION FTIR SYNCHROTRON SPECTRA

SYLVESTRE TWAGIRAYEZU[a], BRANT E BILLINGHURST, TIM E MAY, *EFD, Canadian Light Source Inc., Saskatoon, Saskatchewan, Canada*; MAHESH B. DAWADI, DAVID S. PERRY, *Department of Chemistry, The University of Akron, Akron, OH, USA.*

The Fourier Transform infrared spectra of CH_3NO_2, have been recorded, in the 400-950 cm^{-1} spectral region, at a resolution of 0.00096 cm^{-1}, using the Far-Infrared Beamline at Canadian Light Source. The observed spectra contain four fundamental vibrations: the NO_2 in-plane rock (475.2 cm^{-1}), the NO_2 out-of-plane rock (604.9 cm^{-1}), the NO_2 symmetric bend (657.1 cm^{-1}), and the CN-stretch (917.2 cm^{-1}). For the lowest torsional state of CN-stretch and NO_2 in-plane rock, transitions involving quantum numbers, $m'' = 0$; $J'' \leq 50$ and $K_a'' \leq 10$, have been assigned with the aid of an automated ground state combination difference program together with a traditional Loomis Wood approach[b]. Ground state combination differences derived from more than 2100 infrared transitions have been fit with the six-fold torsion-rotation program developed by Ilyushin et.al[c]. Additional sextic and octic centrifugal distortion parameters are derived for the ground vibrational state.

---

[a] Present address: Department of Chemistry, Brookhaven National Laboratory, Upton, New York, 11973, USA.
[b] C. F. Neese., *An Interactive Loomis-Wood Package, V2.0,* **56**[th],OSU Interanational Symposium on Molecular Spectroscopy (2001).
[c] V. V. Ilyushin, Z. Kisiel, L. Pszczolkowski, H. Mader, and J. T. Hougen, *J. Mol. Spectrosc.* **259**, 26, (2010).

RJ02   1:47 – 1:57

ASSIGNMENT AND ANALYSIS OF THE NO_2 IN-PLANE ROCK BAND OF NITROMETHANE RECORDED BY HIGH-RESOLUTION FTIR SYNCHROTRON SPECTROSCOPY

MAHESH B. DAWADI, DAVID S. PERRY, *Department of Chemistry, The University of Akron, Akron, OH, USA*; SYLVESTRE TWAGIRAYEZU[a], BRANT E BILLINGHURST, *EFD, Canadian Light Source Inc., Saskatoon, Saskatchewan, Canada.*

The high-resolution rotationally resolved Fourier Transform Far-infrared spectrum of the NO_2 in plane-rock band (440-510 cm^{-1}) of nitromethane (CH_3NO_2) has been recorded using the Far-Infrared Beamline at the Canadian Light Source, with a resolution of 0.00096 cm^{-1}. More than 1500 transitions lines have been assigned for $m' = 0$; $K_a' \leq 7$; $J' \leq 50$; using an automated ground state combination difference program together with the traditional Loomis Wood approach[b]. Transitions involving $m' = 0$; $K_a' \leq 7$; $J' \leq 20$; in the upper vibrational state are fit using the six-fold torsion-rotation program developed by Ilyushin et.al[c]. The torsion-rotation energy pattern in the lowest torsional state ($m' = 0$) of the upper vibrational state is similar to that of the vibrational ground state.

---

[a] Present address: Department of Chemistry, Brookhaven National Laboratory, Upton, New York, 11973, USA.
[b] C. F. Neese., *An Interactive Loomis-Wood Package, V2.0,* **56**[th],OSU Interanational Symposium on Molecular Spectroscopy (2001).
[c] V. V. Ilyushin, Z. Kisiel, L. Pszczolkowski, H. Mader, and J. T. Hougen, *M. Mol. Spectrosc.* **259**, 26, (2010).

RJ03

A MICROWAVE SPECTROSCOPIC STUDY OF METHYLATED INDOLES: INTERNAL ROTATION AND NUCLEAR QUADRUPOLE COUPLING

RANIL GURUSINGHE, MICHAEL TUBERGEN, *Department of Chemistry and Biochemistry, Kent State University, Kent, OH, USA.*

The barrier to methyl internal rotation of an asymmetric two ring system depends on the position of the methyl substitution. A cavity based Fourier transform microwave spectrometer was used to record rotational spectra of different methyl substituted indoles in the range of 10.5 - 20 GHz. About 160 hyperfine components arising from about 30 rotational transitions were assigned for each 1- and 3-methylindole. The program XIAM[a] was used to fit the rotational constants, distortion constants, nuclear quadrupole coupling constants and barrier to internal rotation to the measured transition frequencies of the A and E internal rotation states.

The best fit values for the rotational constants are A = 2651.12(2) MHz, B = 1305.266(2) MHz, C = 879.800(2) MHz for 1-methylindole and A = 2603.7224(5) MHz, B = 1268.7886(1) MHz, C = 857.8091(1) MHz for 3-methylindole. The different values observed for the barrier to internal rotation, 279.8(3) cm^{-1} for 1-methylindole and 433(1) cm^{-1} for 3-methylindole, may be due to the different rotor axis lengths and differences in local π-electron density.

Progress on the assignment of additional methylated indoles will also be presented.

[a]H. Hartwig and H. Dreizler, Z. Naturforsch, 51a, 923 - 932.

RJ04

MILLIMETER AND SUBMILLIMETER WAVE SPECTRA OF N-METHYLFORMAMIDE AND PROPIONAMIDE[a]

A. A. MESCHERYAKOV, E. A. ALEKSEEV, V. ILYUSHIN, *Radiospectrometry Department, Institute of Radio Astronomy of NASU, Kharkov, Ukraine*; R. A. MOTIYENKO, L. MARGULÈS, *Laboratoire PhLAM, UMR 8523 CNRS - Université Lille 1, Villeneuve d'Ascq, France.*

We present the rotational spectra studies of two acetamide conjugated molecules, namely, N-methylformamide (CH_3NHCHO) and propionamide ($CH_3CH_2CONH_2$). New measurements have been performed in the frequency range 50 – 150 GHz using the spectrometer in Kharkov, and in the frequency range 150 – 630 GHz using the spectrometer in Lille. The analysis of the rotational spectra of both molecules is complicated by the methyl top internal rotation and nuclear quadrupole hyperfine structure. In case of N-methylformamide the barrier to internal rotation is relatively small, $V_3 = 51.7$ cm^{-1}, whereas for propionamide the barrier is high, $V_3 = 751.9$ cm^{-1}. For propionamide the presence of the low-lying excited vibrational state (60 cm^{-1}) makes difficult the analysis within the classical rho-axis method Hamiltonian. In this case only the rotational transitions with $K_a < 10$ could be fitted within experimental accuracy. The rotational spectra of both molecules were analyzed using modified version of the RAM36 code, taking nuclear quadrupole hyperfine coupling into account. Details of the new study and problems encountered in the analysis will be discussed.

[a]Part of this work was done within the Ukrainian-French CNRS-PICS 6051 project.

RJ05 2:28–2:43

THE MILLIMETER-WAVE SPECTRUM OF VINYL ACETATE

LUCIE KOLESNIKOVÁ, ISABEL PEÑA, JOSÉ L. ALONSO, *Grupo de Espectroscopia Molecular, Lab. de Espectroscopia y Bioespectroscopia, Unidad Asociada CSIC, Universidad de Valladolid, Valladolid, Spain*; JOSE CERNICHARO, *Departamento de Astrofísica, Centro de Astrobiología CAB, CSIC-INTA, Madrid, Spain*; ISABELLE KLEINER, *CNRS et Universités Paris Est et Paris Diderot, Laboratoire Interuniversitaire des Systèmes Atmosphériques (LISA), Créteil, France.*

Recent discovery of methyl acetate in Orion KL[a] places the vinyl acetate as a potential candidate possibly present in the interstellar medium. The room-temperature rotational spectrum of vinyl acetate has been measured from 125 up to 360 GHz to provide direct frequencies to the astronomical community. Transition lines, corresponding to the most stable conformer, have been observed and assigned on the basis of the previously determined spectroscopic constants.[b] All the rotational transitions reveal the $A - E$ splitting due to the methyl internal rotation and the precise set of the spectroscopic constants obtained from the least-squares fit to a threefold barrier internal rotation Hamiltonian is reported. Additional measurements have been also made using a broadband CP-FTMW spectrometer in the region of 6–18 GHz which made possible to assign all monosubstituted ^{13}C and ^{18}O isotopic species in natural abundance and to derive the molecular structure.

[a]B. Tercero, I. Kleiner, J. Cernicharo, H. V. L. Nguyen, A. López, and G. M. Muñoz Caro, *Astrophys. J. Lett.* **2013**, *770*, 13.
[b]B. Velino, A. Maris, S. Melandri, W. Caminati, *J. Mol. Specrosc.* **2009**, *256*, 228.

RJ06 2:45–3:00

COMPARISON OF INDEPENDENTLY CALCULATED AB-INITIO NORMAL-MODE DISPLACEMENTS FOR THE THREE C-H STRETCHING VIBRATIONS OF METHANOL ALONG THE INTERNAL ROTATION PATH

LI-HONG XU, RONALD LEES, *Department of Physics, University of New Brunswick, Saint John, NB, Canada*; JON T. HOUGEN, *Sensor Science Division, National Institute of Standards and Technology, Gaithersburg, MD, USA*; JOEL BOWMAN, *Department of Chemistry, Emory University, Atlanta, GA, USA*; XINCHUAN HUANG, *SETI, NASA Ames Research Center, Moffett Field, CA, USA*; STUART CARTER, *Department of Chemistry, Emory University, Atlanta, GA, USA.*

Graphical displays of C-H stretching normal-mode coefficients from recent quantum chemical projected-frequency calculations are compared with analogous displays constructed after reexamination of results from more extensive higher-level calculations described earlier in the literature. Such comparisons confirm the facts that: (i) no geometrical phase is accumulated in these coefficients when the methyl top undergoes one complete internal-rotation revolution with respect to the frame, and (ii) some of the coefficients, when plotted against the internal rotation angle, exhibit near-cusp-like behavior at one or two angles. The connection between these graphical displays and the magnitude of "Jahn-Teller-like" and "Renner-Teller-like" torsion-vibration interaction terms in a previously reported model Hamiltonian, as well as the connection between the lack of geometric-phase accumulation in these graphs and the number of conical intersections enclosed by one full internal-rotation motion, will also be briefly discussed.

RJ07   **3:02 – 3:17**

STUDIES ON THE CONFORMATIONAL LANDSCAPE OF TERT-BUTYL ACETATE USING MICROWAVE SPEC-
TROSCOPY AND QUANTUM CHEMICAL CALCULATIONS

YUEYUE ZHAO, HALIMA MOUHIB, GUOHUA LI, WOLFGANG STAHL, *Institute for Physical Chemistry, RWTH Aachen University, Aachen, Germany*; ISABELLE KLEINER, *Laboratoire Interuniversitaire des Systèmes Atmosphériques (LISA), CNRS et Universités Paris Est et Paris Diderot, Créteil, France.*

The tert-Butyl acetate molecule was studied using a combination of quantum chemical calculations and molecular beam Fourier transform microwave spectroscopy in the 9 to 14 GHz range. Due to its rather rigid frame, the molecule possesses only two different conformers: one of C_s and one of C_1 symmetry. According to ab initio calculations, the C_s conformer is 46 kJ/mol lower in energy and is the one observed in the supersonic jet. We report on the structure and dynamics of the most abundant conformer of tert-butyl acetate, with accurate rotational and centrifugal distortion constants. Additionally, the barrier to internal rotation of the acetyl methyl group was determined. Splittings due to the internal rotation of the methyl group of up to 1.3 GHz were observed in the spectrum. Using the programs XIAM and BELGI-Cs, we determine the barrier height to be about 113 cm^{-1} and compare the molecular parameters obtained from these two codes. Additionally, the experimental rotational constants were used to validate numerous quantum chemical calculations.

This study is part of a larger project which aims at determining the lowest energy conformers of organic esters and ketones which are of interest for flavor or perfume synthetic applications[a].

---

[a]Project partly supported by the PHC PROCOPE 25059YB

RJ08   **3:19 – 3:34**

TORSION - VIBRATION COUPLING IN THE METHYL ROTOR SYSTEMS

MENG HUANG, ANNE B McCOY, TERRY A. MILLER, *Department of Chemistry and Biochemistry, The Ohio State University, Columbus, OH, USA.*

The couplings between the CH stretch and CH_3 torsion in the methyl rotor have been widely studied in methanol.[a] In this research, we are focusing on the effect of this coupling on the vibrational spectrum in the CH stretch region of the methanol cation - argon cluster, $CH_3OH^+ \cdot Ar$, and the methyl peroxy radical, $CH_3OO\cdot$. A reduced dimensional analysis including the three CH stretches and the CH_3 torsion is used to calculate the spectra. The CH stretches are treated as harmonic oscillators whose frequency depends on the torsional angle because of coupling between the CH stretch and CH_3 torsion.

The infrared spectrum of $CH_3OH^+ \cdot Ar$ cluster taken by the Duncan group[b] shows multiple peak structure in the CH stretch region. In this system, the calculation indicates the coupling between the CH_3 torsion and CH stretch is relatively strong. The multiple peak structure in the experimental spectrum can be assigned to the CH stretch fundamentals and the combination band involving the CH stretch and the CH_3 torsion. However, for the methyl peroxy radical, the calculated coupling is very weak. In the CH stretch region of the infrared spectrum of $CH_3OO\cdot$ taken by the Lee group[c], only CH stretch fundamentals with relatively broad rotational contours are observed. The broadened structure of the CH stretch fundamental is possibly caused by sequence band structure from low lying torsional levels which are well populated and shift very little from the origin band.

---

[a]D. S. Perry *J. Mol. Spectpsc.* 2009, 257, 1-10

[b]J. D. Mosley, J. W. Young, M. A. Duncan *68th International Symosium of Molecular Spectroscopy*

[c]K.-H. Hsu, Y.-P. Lee, M. Huang, T. A. Miller *68th International Symosium of Molecular Spectroscopy*

RJ09 3:36 – 3:51

ANALYSIS OF THE FAR IR SPECTRUM OF TRIMETHYLENE SULFIDE USING EVOLUTIONARY ALGORITHMS

JENNIFER VAN WIJNGAARDEN, DURELL DESMOND, *Department of Chemistry, University of Manitoba, Winnipeg, MB, Canada*; W. LEO MEERTS, *Institute for Molecules and Materials (IMM), Radboud University Nijmegen, Nijmegen, Netherlands.*

Rotationally-resolved vibrational spectra have been collected from 100-1000 cm^{-1} for the four-membered ring trimethylene sulfide (c-C_3H_6S) using synchrotron light at the far infrared beamline of the Canadian Light Source. The spectra are complicated by the presence of ring inversion tunneling which gives rise to dense patterns of overlapping rotation-vibration-inversion transitions in the infrared region. These are well-resolved using the Bruker IFS125HR instrument. In this talk, we will discuss the progress of the analysis of the two lowest frequency bands which correspond to the c-type ring puckering vibration at 139 cm^{-1}, and the a-type in-plane ring deformation at 529 cm^{-1}. The ideas of genetic evolution via evolutionary strategies have been used to aid in the analysis of the observed spectra. In particular, we have applied the covariance matrix adaptation evolution strategy which uses the concept of mutation and the results of previous trials to give solutions to fit the quite dense spectra.

Intermission

RJ10 4:08 – 4:23

FIRST OBSERVATION OF THE SPIN ROTATIONAL STRUCTURE OF THE HYDROXYMETHYL RADICAL (H_2COH) IN THE CH_2 ASYMMETRIC MODE

CHIH-HSUAN CHANG, FANG WANG, *JILA, The University of Colorado, Boulder, CO, USA*; DAVID NESBITT, *Department of Chemistry, JILA CU-NIST, Boulder, CO, USA.*

Rotationally-resolved direct infrared absorption spectra of hydroxymethyl radical (H_2COH) in the CH_2 asymmetric mode (ν_2) were observed for the first time using the Boulder difference frequency generation infrared spectrometer. Hydroxymethyl radical was formed with chemical selectivity via the reaction of Cl radical with CH_3OH in a discharge slit-jet supersonic expansion. As a result of sub-Doppler linewidth and low rotational temperature, the b-type rotational structure and spin-rotation splitting were fully resolved. In particular, tunneling splitting was observed due to the large-amplitude COH torsional mode. Because of the feasible permutation of hydrogens in the methylenic group, nuclear spin intensity alternation was given as: 3:1 for K_a=even, odd in the 0^+ level, and 1:3 for K_a=even, odd in the 0^- level of the ground vibrational state. The assignments were confirmed rigorously by four-line ground state combination differences, which agreed within the experimental frequency uncertainty (10 MHz). The identified transitions were fit with a Watson A-reduction Hamiltonian including the spin rotational interaction, leading to unambiguous determination of asymmetric top spectroscopic constants, as well as spin rotational constants (ϵ_{aa}, ϵ_{bb}, ϵ_{cc}) for the first time.

RJ11 4:25 – 4:40

EFFECTS OF MULTIPLE ARGON TAGGING IN ALKALI METAL $M^+H_2OAr_n$ AND $M^+D_2OAr_n$ STUDIED BY IRPD SPECTROSCOPY

CHRISTIAN VAN DER LINDE, HAOCHEN KE, JAMES M. LISY, *Department of Chemistry, University of Illinois at Urbana-Champaign, Urbana, IL, USA.*

Metal ions play important roles in biologically processes. Among these, alkali metal ions are of great importance. Their solvation is one of the fundamental processes having great effect on their activity and has been in focus of research for many years. M^+H_2O cluster are a good model for microscopic solvation. However, modeling gas-phase cluster ions has been a challenge both experimentally and theoretically due to uncertainties in the temperature (or internal energy) of these species. The temperature depends primarily on the composition and method of preparation of the cluster ion. Infrared photodissociation spectroscopy in combination with argon tagging is a well suited tool to study these species. While argon tagging itself is a well established technique for lowering the temperature and sharpening spectral features, experiments with multiple attached Ar atoms are rare.

The influence of up to three Ar atoms on cluster temperature, vibrational band origins and rotational constants will be discussed for M = Na, K, Rb and Cs. The structure, arising from rotation of the water subunit, of the asymmetric bands will be analyzed in detail. Many spectra show broadening/splitting of features and an unusual intensity distribution for features associated especially to transitions involving the |K|=1 level. The influence of a low barrier, hindering rotation of the water subunit within the cluster, will be discussed.

RJ12 4:42 – 4:57

LARGE-AMPLITUDE TUNNELING DYNAMICS IN HYDROXYMETHYL RADICAL

DAVID NESBITT, *Department of Chemistry, JILA CU-NIST, Boulder, CO, USA*; CHIH-HSUAN CHANG, FANG WANG, *JILA, The University of Colorado, Boulder, CO, USA.*

Hydroxymethyl radical (CH_2OH) is an important radical in modeling combustion flame chemistry as well as chemistry in the interstellar medium. We have recently observed high resolution jet cooled infrared spectra of CH_2OH, which reveal a rich pattern of splittings due to large-amplitude torsional tunneling of the COH moiety with respect to the methylenic framework. In order to facilitate a detailed analysis of these tunneling splittings, we have pursued high level ab initio CCSD(T) calculations of the multidimensional torsional tunneling potential energy surface using a MOLPRO package with correlated cc-pvnz-f12 basis sets and extrapolated to the complete basis set limit (CBS). Such high level potential energy surface calculations as a function of intrinsic reaction coordinate (IRC) reveal novel multidimensional tunneling dynamics and make possible preliminary estimates of the barrier height and tunneling splittings in the ground vibrational state.

RJ13 4:59 – 5:14

SYNCHROTRON RADIATION AND THE FAR-INFRARED AND MID-INFRARED SPECTRA OF NCNCS

MANFRED WINNEWISSER, BRENDA P. WINNEWISSER, FRANK C. DE LUCIA, *Department of Physics, The Ohio State University, Columbus, OH, USA*; DENNIS TOKARYK, STEPHEN CARY ROSS, *Department of Physics, University of New Brunswick, Fredericton, NB, Canada*; BRANT E BILLINGHURST, *EFD, Canadian Light Source Inc., Saskatoon, Saskatchewan, Canada.*

The large-amplitude in-plane bending vibration of NCNCS at 85 cm^{-1} has a potential energy function which includes a barrier to linearity with a height of about 285 cm^{-1}. The topology of the surface of the space defined by this two-dimensional potential function exhibits non-trivial monodromy. Therefore an energy/momentum map for a quantum system with its motion determined by such a potential takes the form of a lattice which contains a defect associated with the top of the barrier. In NCNCS, the wavenumber values of the fundamental vibrational excitation and the barrier height mean that easily accessible energy levels allow us to observe 3 bending vibrational levels below and 3 above the barrier, yet still below all of the other vibrational levels, allowing the study of all the levels in the neighborhood of the defect. In three measuring campaigns at the Canadian Light Source in May of the years 2011, 2012, and 2013 we have now obtained 8 of the 9 fundamental vibrational band systems of NCNCS in high resolution, in particular that of the large-amplitude bend in the FIR. So far only a-type spectra have been assigned. Thus we have now determined the $\Delta v_b = 1$, and $\Delta K_a = 0$ vibrational intervals (using bent molecule notation) but do not yet have experimental values for either rotational $\Delta K_a = +/- 1$ intervals nor ro-vibrational $\Delta v_b = 1$, $\Delta K_a = +/- 1$ intervals. In May of 2014 we will have our last measuring campaign and hope to observe the more elusive b-type transitions.

RJ14   5:16 – 5:31

SPECTROSCOPY OF NCNCS AT THE CANADIAN LIGHT SOURCE: THE FAR-INFRARED SPECTRUM OF THE ν_7
REGION FROM 60-140 cm^{-1}

DENNIS TOKARYK, STEPHEN CARY ROSS, *Department of Physics, University of New Brunswick, Fred-
ericton, NB, Canada*; BRENDA P. WINNEWISSER, MANFRED WINNEWISSER, FRANK C. DE LUCIA,
Department of Physics, The Ohio State University, Columbus, OH, USA; BRANT E BILLINGHURST, *EFD,
Canadian Light Source Inc., Saskatoon, Saskatchewan, Canada.*

We report on the analysis of our spectrum from 60-140 cm^{-1} of the ν_7 bending fundamental and associated hot band
sequence of NCNCS, obtained on the far-infrared beamline at the Canadian Light Source synchrotron. The data were collected
in May 2013, building upon what we learned conducting experiments in May 2011 and 2012 on this molecule. Calculations
indicated that the ν_7 system was very weak (one of the four weakest fundamental bands, all of comparable strength), but
its spectrum became evident when 30 mTorr of NCNCS was admitted into the 2-m-long sample cell, through which the
synchrotron beam passed 40 times. The best spectrum so far has been obtained with 121 mTorr of gas. Loomis-Wood plots
reveal many branches, some of which were unambiguously assignable to $\Delta\nu_7 = +1$ subbands for $\nu_7'' = 0, 1, 2, 3$ and for
$K_a = 0, 1, 2$ with $\Delta K_a = 0$ (a-type subbands) by comparison of lower-state combination differences with those obtained
from the published pure-rotational data. We will continue the analysis by assigning as many a-type subbands as possible and
by searching for b-type subbands with $\Delta K_a = \pm 1$ so that the connections between K_a-stacks can be measured. Finally, we
will simultaneously fit the infrared and rotational data with a generalized semi-rigid bender Hamiltonian.

RJ15   5:33 – 5:48

FITTING THE HIGH-RESOLUTION SPECTROSCOPIC DATA FOR NCNCS

ZBIGNIEW KISIEL, *ON2, Institute of Physics, Polish Academy of Sciences, Warszawa, Poland*; BRENDA P.
WINNEWISSER, MANFRED WINNEWISSER, FRANK C. DE LUCIA, *Department of Physics, The Ohio
State University, Columbus, OH, USA*; DENNIS TOKARYK, STEPHEN CARY ROSS, *Department of Physics,
University of New Brunswick, Fredericton, NB, Canada*; BRANT E BILLINGHURST, *EFD, Canadian Light
Source Inc., Saskatoon, Saskatchewan, Canada.*

NCNCS is a quasi-linear molecule that displays plentiful spectroscopic signatures of transition from the asymmetric top
to the linear rotor regime. The transition takes place on successive excitation of the ν_7 bending mode at *ca* 80 cm^{-1}. The
unusual spectroscopic manifestations on crossing the barrier to linearity are explained by quantum monodromy and described
quantitatively by the generalised semi-rigid bender Hamiltonian.[a,b,c]

Nevertheless, analysis to experimental accuracy of the extensive mm-wave spectrum of NCNCS recorded with the
FASSST technique has only so far been achieved with the use of separate $J(J + 1)$ expansions for each (v_7, K_a) transition sequence.[c] In addition, several selective perturbations identified between transition sequences in different vibrational
levels[c] are still unfitted. Presently we seek effective approximations to the vibration-rotation Hamiltonian that would allow
combining multiple sequences into a fit, would allow a perturbation analysis, and could use mm-wave data together with
high-resolution infrared measurements of NCNCS made at the Canadian Light Source.

The understanding of effective fits to low-K_a subsets of rotational transitions in the FASSST spectrum has already
allowed confident assignment of the ^{34}S and both ^{13}C isotopic species of NCNCS in natural abundance, as will be described.

[a] B.P.Winnewisser, et al., *Phys. Rev. Lett.* **95**. 243002 (2005).

[b] M.Winnewisser, et al., *J. Mol. Struct.* **798**, 1 (2006).

[c] B.P.Winnewisser, et al., *Phys. Chem. Chem. Phys.* **12**, 8158 (2010).

RJ16   **5:50 – 6:05**

NON COVALENT INTERACTIONS AND INTERNAL DYNAMICS IN ADDUCTS OF FREONS

WALTHER CAMINATI, QIAN GOU, LUCA EVANGELISTI, GANG FENG, LORENZO SPADA, *Dep. Chemistry 'Giacomo Ciamician', University of Bologna, Bologna, Italy*; MONTSERRAT VALLEJO-LÓPEZ, ALBERTO LESARRI, *Department Quimica Fisica y Quimica Inorganica, Universidad de Valladolid, Valladolid, Spain*; EMILIO J. COCINERO, *Physical Chemistry Department, Universidad del País Vasco, Bilbao, Spain.*

The complexation of chlorofluorocarbons (CFCs) with atmospheric water and pollutants of the atmosphere affects their reactivity and it seems to accelerate, for example, the decomposition rate of freons in the atmosphere [1]. For this reason we characterized shapes, stabilities, nature of the non-covalent interactions, structures and internal dynamics of a number of complexes of CFCs with water and of their dimers or oligomers by rotational spectroscopy. It has been found that hydrogenated CFCs form adducts with other molecules through weak hydrogen bonds (WHBs). Their C-H groups can act as proton donors, enhanced by the electron withdrawing of the halogen atoms, interacting with the electron rich regions of the partner molecules [2]. Also in adducts or oligomers of hydrogenated CFCs the monomer units are held together by nets of WHBs [3]. When CFCs are perhalogenated, the positive electrostatic region ("σ-hole") can interact electrostatically with negative sites of another, or of the same molecular entity, giving rise, according to IUPAC, to the so called halogen bond (HaB). However, it has been observed that when the perhalogenated CFCs has a Π electron system, a lone pair$\bullet\bullet\bullet\Pi$ interaction (Bürgi-Dunitz) is favoured [4]. We describe here the HaBs that CF_4 and CF_3Cl form with a variety of partner molecules such as water, ammonia, dimethyl ether, etc. Important spectroscopic features outline strong dynamics effects taking place in this kind of complex.

References

[1] V. Vaida, H. G. Kjaergaard, K. J. Feierabend, Int. Rev. Phys. Chem. 22 (2003) 203.

[2] See, for example: W. Caminati, S. Melandri, A. Maris, P. Ottaviani, Angew. Chem. Int. Ed. 45 (2006) 2438.

[3] G. Feng, L. Evangelisti, I. Cacelli, L. Carbonaro, G. Prampolini, W. Caminati, Chem. Commun. 50 (2014) 171.

[4] Q. Gou, G. Feng, L. Evangelisti, W. Caminati, Angew. Chem. Int. Ed. 52 (2013) 52 11888.

FA. Mini-symposium: Astronomical Molecular Spectroscopy in the Age of ALMA
Friday, June 20, 2014 – 8:30 AM
Room: 116 Roger Adams Lab

Chair: Susanna L. Widicus Weaver, Emory University, Atlanta, GA, USA

FA01   8:30 – 8:45

HIGH-PRECISION SUB-DOPPLER INFRARED SPECTROSCOPY OF HeH^+

ADAM J. PERRY, JAMES N. HODGES, CHARLES MARKUS, G. STEPHEN KOCHERIL, PAUL A JENK-INS II, *Department of Chemistry, University of Illinois at Urbana-Champaign, Urbana, IL, USA*; BENJAMIN J. McCALL, *Departments of Chemistry and Astronomy, University of Illinois at Urbana-Champaign, Urbana, IL, USA.*

The helium hydride ion, HeH^+, is the simplest heteronuclear diatomic, and is composed of the two most abundant elements in the universe. It is widely believed that this ion was among the first molecules to be formed; thus it has been of great interest to scientists studying the chemistry of the early universe.[ab] HeH^+ is also isoelectronic to H_2 which makes it a great target ion for theorists to include adiabatic and non-adiabatic corrections to its Born-Oppenheimer potential energy surface. The accuracy of such calculations is further improved by incorporating electron relativistic and quantum electrodynamic effects.[c]

Using the highly sensitive spectroscopic technique of Noise Immune Cavity Enhanced Optical Heterodyne Velocity Modulation Spectroscopy (NICE-OHVMS) we are able to perform sub-Doppler spectroscopy on ions of interest. When combined with frequency calibration from an optical frequency comb we fit line centers with sub-MHz precision as has previously been shown for the H_3^+, HCO^+, and CH_5^+ ions.[d] Here we report a list of the most precisely measured rovibrational transitions of HeH^+ to date. These measurements should allow theorists to continue to push the boundaries of *ab initio* calculations in order to further study this important fundamental species.

[a]S. Lepp, P. C. Stancil, A. Dalgarno *J. Phys. B* (2002), **35**, R57.

[b]S. Lepp, *Astrophys. Space Sci.* (2003), **285**, 737.

[c]K. Pachucki, J. Komasa, *J. Chem. Phys* (2012), **137**, 204314.

[d]J. N. Hodges, A. J. Perry, P. A. Jenkins II, B. M. Siller, B. J. McCall *J. Chem. Phys.* (2013), **139**, 164201.

FA02   8:47 – 9:02

HeH^+, HeD^+, AND HeT^+ POTENTIALS THAT REPRODUCE ALL MEASURED ENERGY TRANSITIONS, AND ARE ACCURATE UP TO THE INCLUSION OF RELATIVISTIC AND LEADING FOURTH ORDER QED EFFECTS IN THE LONG-RANGE REGION BEYOND AVAILABLE EXPERIMENTS

STASZEK WELSH, *Physical and Theoretical Chemistry Laboratory, Oxford University, Oxford, United Kingdom*; MARIUSZ PUCHALSKI, *Department of Chemistry, Adam Mickiewicz University, Poznan, Poland*; GRZE-GORZ LACH, *International Institute of Molecular and Cell Biology, Warsaw, Poland*; TUNG WEI-CHENG, *Department of Chemistry and Biochemistry, University of Arizona, Tucson, AZ, USA*; LUDWIK ADAMOW-ICZ, *Department of Chemistry and Astronomy, University of Arizona, Tuscon, AZ, USA*; NIKESH S. DATTANI, *Department of Chemistry, Oxford University, Oxford, United Kingdom.*

According to the big bang theory, HeH^+ was the first molecule ever produced in our universe, along with He_2^+. It is also the simplest two-electron molecular system and lightest heteronuclear molecule, with a stable ground electronic state, aside from the isotopologues of H_2. These facts make HeH^+ very interesting for experimental studies of the early universe and theoretical studies of isotope effects including Born-Oppenheimer breakdown (BOB), as well as an extremely important benchmark system for *ab initio* methods. However, no spectroscopic measurements, nor *ab initio* calculations have been reported for HeT^+, and the most accurate empirical potentials for HeH^+ and HeD^+ are 15 years old and are unreliable outside of the data range. We build the most accurate analytical empirical potentials and BOB functions for HeH^+ and HeD^+ to date. These BOB functions are then used to predict the HeT^+ potential, and are in excellent agreement with our *ab initio* potential for HeT^+. Outside the data-range, the MLR (Morse/long-range) model forces the analytic empirical potential to become the theoretically correct long-range potential based on the multipole polarizabilties of He. We include the dipole and quadrupole polarizabilities. The quadrupole value includes finite-mass and relativistic corrections, and an estimate of the leading α^3 QED corrections. The dipole value includes finite-mass, relativistic, and QED corrections up to the leading term of the α^4 effects.

FA03

ACCURATE, ANALYTIC, EMPIRICAL POTENTIALS AND BORN-OPPENHEIMER BREAKDOWN FUNCTIONS FOR THE $X(1^1\Sigma)$-STATES OF BeH, BeD, and BeT

NIKESH S. DATTANI, STASZEK WELSH, *Physical and Theoretical Chemistry Laboratory, Oxford University, Oxford, United Kingdom.*

Being the simplest neutral open shell molecule, BeH is a very important benchmark system for *ab initio* calculations. However, the most accurate empirical potentials and Born-Oppenheimer breakdown (BOB) functions for this system are nearly a decade old and are not reliable in the long-range region. Particularly, the uncertainties in their dissociation energies were about ± 200 cm^{-1}, and even the number of vibrational levels predicted was at the time very questionable, meaning that no good benchmark exists for *ab initio* calculations on neutral open shell molecules. We build new empirical potentials for BeH, BeD, and BeT that are much more reliable in the long-range. Being the second lightest heteronuclear molecule with a stable ground electronic state, BeH is also very important for the study of isotope effects, such as BOB. We extensively study isotope effects in this system, and we show that the empirical BOB functions fitted from the data of any two isotopologues, is sufficient to predict crucial properties of the third isotopologue.

FA04

RELIABLE IR LINE LISTS FOR 13 CO_2 ISOTOPOLOGUES, UP TO 1000K, E'=18,000 CM^{-1} AND WITH LINE SHAPE PARAMETERS INCLUDED

XINCHUAN HUANG, *Carl Sagan Center, SETI Institute, Moutain View, CA, USA*; ROBERT R. GAMACHE, *Department of Environmental, Earth, and Atmospheric Sciences, University of Massachusetts, Lowell, MA, USA*; RICHARD S FREEDMAN, *Carl Sagan Center, SETI Institute, Moutain View, CA, USA*; DAVID SCHWENKE, *MS T27B-1,NAS Facility, NASA Ames Research Center, Moffett Field, CA, USA*; TIMOTHY LEE, *Space Science and Astrobiology Division, NASA Ames Research Center, Moffett Field, CA, USA.*

We report both Ames-296K (1E-42 cm.molecule^{-1}) and Ames-1000K (1E-36 cm.molecule^{-1}) IR line lists up to E'=18000 cm^{-1} for 13 CO_2 isotopologues, including symmetric species 626, 636, 646, 727, 737, 828, 838, and asymmetric species 627, 628, 637, 638, 728, 738. Line shape parameters are also determined for four different temperature ranges: Mars, Earth, Venus, and higher temperatures. A size-reducing selection algorithm allows us to remove $> 90\%$ of the weak transitions while keeping all main absorption features and $> 99.9\%$ intensity sums in each cm^{-1}. The final data size for file storage and transfer is further reduced by a compact format we have developed. The total size of 13 datasets (1.15 billion transitions) is 18.7 GB , i.e. 16.2 bytes per line. Partition functions, coverage, line positions and line intensities of the Ames IR lists are compared with available data in HITRAN2012. We believe these IR lists are reliable alternatives for missing bands or transitions in existing databases. They should greatly facilitate spectroscopic analyses in future laboratory or astronomical observations. We also provide two independent, merged lists at 296K and 1000K with terrestrial isotopologue abundances which are the same as those in HITRAN and CDSD. It is hoped that these will be useful in the analysis or modeling of "natural" CO_2 studies.

FA05   9:38 – 9:53

VIBRATIONAL AVERAGING OF THE ISOTROPIC HYPERFINE COUPLING CONSTANTS FOR THE METHYL RADICAL

<u>AHMAD ADAM</u>, PER JENSEN, *Fachbereich C-Physikalische und Theoretische Chemie, Bergische Universität Wuppertal, D-42097 Wuppertal, Germany*; ANDREY YACHMENEV, SERGEI N. YURCHENKO, *Department of Physics and Astronomy, University College London, Gower Street, London WC1E 6BT, United Kingdom.*

Electronic contributions to molecular properties are often considered as the major factor and usually reported in the literature without ro-vibrational corrections. However, there are many cases where the nuclear motion contributions are significant and even larger than the electronic contribution. In order to obtain accurate theoretical predictions, nuclear motion effects on molecular properties need to be taken into account. The computed isotropic hyperfine coupling constants for the nonvibrating methyl radical CH_3 are far from the experimental values. For CH_3, we have calculated the vibrational-state-dependence of the isotropic hyperfine coupling constant in the electronic ground state. The vibrational wavefunctions used in the averaging procedure were obtained variationally with the TROVE program. Analytical representations for the potential energy surfaces and the hyperfine coupling constant surfaces are obtained in least-squares fitting procedures. Thermal averaging has been carried out for molecules in thermal equilibrium, i.e., with Boltzmann-distributed populations. The calculation methods and the results will be discussed in detail.

Intermission

FA06   10:10 – 10:25

THE TORSIONAL SPECTRUM OF DOUBLY DEUTERATED METHANOL CHD_2OH

M. NDAO, <u>L. H. COUDERT</u>, F. KWABIA TCHANA, *LISA, CNRS, Universités Paris Est Créteil et Paris Diderot, Créteil, France*; J. BARROS, *LAL, UMR 8607, CNRS-Université Paris Sud, Orsay, France*; L. MARGULÈS, *Laboratoire PhLAM, UMR 8523 CNRS - Université Lille 1, Villeneuve d'Ascq, France*; LAURENT MANCERON, *MONARIS, CNRS, UMR 7075, Paris, France*; P. ROY, *AILES beam line, Synchrotron Soleil, Gif-sur-Yvette, France.*

Although the torsional spectrum of several isotopic species of methanol with a symmetrical CH_3 or CD_3 was analyzed some time ago, it is recently,[a] and only for the monodeuterated species CH_2DOH, that such an analysis was extended to the case of an asymmetrical methyl group.

In this talk, based on a Fourier transform high-resolution spectrum recorded in the 20 to 670 cm^{-1} region, the first analysis of the torsional spectrum of doubly deuterated methanol CHD_2OH will be presented. The Q branch of many torsional subbands could be observed and their assignment was initiated using a theoretical torsion-rotation spectrum computed with an approach accounting for the torsion-rotation Coriolis coupling and for the dependence of the generalized inertia tensor on the angle of internal rotation.[b] 46 torsional subbands were thus assigned. For 28 of them, their rotational structure could be assigned and fitted using an effective Hamiltonian expressed as a $J(J+1)$ expansion; and for 2 of them microwave transitions within the lower torsional level could also be included in the analysis.[c] In several cases these analysis revealed that the torsional levels are strongly perturbed.[d]

In the talk, the torsional parameters retrieved in the analysis of the torsional subband centers will be discussed. The results of the analysis of the rotational structure of the torsional subbands will be presented and we will also try to understand the nature of the perturbations. At last, preliminary results about the analysis of the microwave spectrum will be presented.

[a]El Hilali, Coudert, Konov, and Klee, *J. Chem. Phys.* **135** (2011) 194309
[b]Lauvergnat, Coudert, Klee, and Smirnov, *J. Mol. Spectrosc.* **256** (2009) 204
[c]Quade, Liu, Mukhopadhyay, and Su, *J. Mol. Spectrosc.* **192** (1998) 378
[d]Pearson, Yu, and Drouin, *J. Mol. Spectrosc.* **280** (2012) 119

FA07   **10:27 – 10:42**

THE MICROWAVE SPECTRUM OF MONODEUTERATED ACETAMIDE $CH_2DC(=O)NH_2$

I. A. KONOV, *Department of Physics, Tomsk State University, Tomsk, Russia*; L. H. COUDERT, C. GUTLE, *LISA, CNRS, Universités Paris Est Créteil et Paris Diderot, Créteil, France*; T. R. HUET, L. MARGULÈS, R. A. MOTIYENKO, *Laboratoire PhLAM, UMR8523 CNRS - Université Lille 1, Villeneuve d'Ascq, France*; H. MØLLENDAL, *CTCC, Dept. of Chemistry, University of Oslo, Oslo, Norway*; J.-C. GUILLEMIN, *ISCR, UMR 6226, Rennes, France.*

Acetamide is an oblate asymmetric top displaying almost free internal rotation of its methyl group. The microwave spectrum of the normal species ($CH_3C(=O)NH_2$) has already been studied and a value[a] of only 25 cm^{-1} was retrieved for the height of the potential barrier hindering the internal rotation. No spectroscopic results are available about the monodeutared species with a partially deuterated CH_2D methyl group which will be the subject of the present talk.

The effects of deuteration on the hindering potential[b] will be investigated first. They lead to qualitative changes of the hindering potential no longer resembling that of the normal species and displaying several inequivalent minima.[c] A determination of the torsional potential will be attempted through an analysis of the microwave spectrum of the monodeuterated species in which torsion-rotation energies are calculated with the approach developed for monodeuterated methanol,[d] accounting for the torsion-rotation Coriolis coupling and for the dependence of the inertia tensor on the torsional angle.

A low temperature spectrum, recorded with the MB-FTMW spectrometer in Lille, has already been analyzed and 14 transitions could be assigned up to $J = 6$. Room temperature spectra have also been recorded in the 7–91 and 150–165 GHz frequency ranges and more than 100 transitions have been assigned up to $J = 16$ for the ground torsional state.

In the paper, deuteration effects will be discussed and we hope to assign a sufficient number of microwave transitions in order to obtain the first quantitative information about the hindering potential of monodeuterated acetamide.

[a] Ilyushin, Alekseev, Dyubko, Kleiner, and Hougen, *J. Molec. Spectrosc.* **227** (2004) 115

[b] Lauvergnat, Coudert, Klee, and Smirnov, *J. Molec. Spectrosc.* **256** (2009) 204

[c] Margulès, Coudert, Møllendal, Guillemin, Huet and Janečkovà, *J. Molec. Spectrosc.* **254** (2009) 55

[d] Coudert, Zemouli, Motiyenko, Margulès, and Klee, *J. Chem. Phys.* **140** (2014) 064307

FA08   **10:44 – 10:59**

TERAHERTZ SPECTROSCOPY OF DEUTERATED ACETALDEHYDE: CH_2DCHO

L. MARGULÈS, R. A. MOTIYENKO, *Laboratoire PhLAM, UMR 8523 CNRS - Université Lille 1, Villeneuve d'Ascq, France*; L. H. COUDERT, *LISA, CNRS, Universités Paris Est Créteil et Paris Diderot, Créteil, France*; J.-C. GUILLEMIN, *Institut des Sciences Chimiques de Rennes, UMR 6226 CNRS - Université de Rennes 1, Rennes, France.*

This study follows our recent investigations about deuterated methyl-top species of complex organic molecules: methanol,[a] methyl formate,[b] and dimethyl ether.[c] In particular these works led the first ISM detection of $HCOOCH_2D$ and CH_2DOCH_3. Acetaldehyde is not very abundant in the ISM, but this is a very interesting case from the spectroscopic point of view as it is an intermediate case between methyl formate and methanol. In the normal species of acetaldehyde, the barrier to internal rotation[d] V_3 is 408cm^{-1} which is close to the value in methyl formate[e]: 373 cm^{-1}. However, the value of the Coriolis coupling constant ρ is 0.33 in acetaldehyde which is a much larger value than in methyl formate, 0.08, meaning that the coupling between the torsion and the overall rotation is more important.

The sample was not a commercial one and half of its amount is the normal species which leads to a more difficult line assignment. The spectra were recorded in Lille between 75 and 950 GHz with a solid-state submillimeter-wave spectrometer. The starting point of the analysis was the centimeter-wave measurements carried out for the sym and asym- conformers.[f] A comparison between the approach developed for deuterated methyl formate ($HCOOCH_2D$), based on the water dimer formalism, and that designed recently for deuterated methanol[a] (CH_2DOH) will be presented.

This work is supported by the CNES and the Action sur Projets de l'INSU, PCMI.

[a] Coudert, L. H.; *et al. J. Chem. Phys.* **140**, (2014) 64307

[b] Coudert, L. H.; *et al.* ApJ **779**, (2013) 119

[c] Richard, C.; *et al.* A&A **552**, (2013) A117

[d] Smirnov, I. A.; *et al. J. Mol. Spectrosc.* **295** (2014) 44

[e] Ilyushin, V.; *et al. J. Mol. Spectrosc.* **255** (2009) 32

[f] Turner, P. H.; and Cox, A. P. *Chem. Phys. Lett.* **42**, (1976) 84 - Turner, P. H.; Cox, A. P.; and Hardy, J. A. *J.C.S. Farady Trans.* **2**, (1981) 1217

250

HIGH RESOLUTION SPECTROSCOPY OF THE TWO LOWEST VIBRATIONAL STATES OF QUINOLINE C_9H_7N

OLIVIER PIRALI[a], *AILES beamline, Synchrotron SOLEIL, Saint Aubin, France*; ZBIGNIEW KISIEL, *ON2, Institute of Physics, Polish Academy of Sciences, Warszawa, Poland*; MANUEL GOUBET, *Laboratoire PhLAM, Université de Lille 1, Villeneuve de Ascq, France*; SÉBASTIEN GRUET, *AILES beamline, Synchrotron SOLEIL, Saint Aubin, France*; MARIE-ALINE MARTIN-DRUMEL[b], ARNAUD CUISSET, GAEL MOURET, *Laboratoire de Physico-Chimie de l'Atmosphère, Université du Littoral Côte d'Opale, Dunkerque, France.*

PAHs molecules and derivatives have long been suspected to be present in different objects of the universe but no unambiguous detection (based on rotationally resolved spectroscopy) has been reported yet. Pure rotation transitions in the ground state (GS) of quinoline (which belongs to the C_s point group) have been analyzed in a previous work of Kisiel et al.[c] using micro-wave and sub-mm techniques. We will present in this talk a collective effort to record and assign the rotational structure of the two lowest vibrational states of quinoline (namely ν_{45} and ν_{44} centered at about 168.5 cm^{-1} and 177.6 cm^{-1} respectively). In this study, high resolution synchrotron based FT-FIR together with pure rotation spectroscopy in the sub-mm range permitted to obtain very complementary data. The spectral analysis permitted to identify relatively large perturbation patterns across a K_a difference of 2 between the two excited states (both states are A" symmetry). Successful treatment of the perturbations allowed for the accurate determination of i) the rotational constants in the excited states ii) the Fermi and Coriolis perturbation terms and iii) the relative energies of ν_{45} and ν_{44} to the GS. Since most of the PAHs molecules possess low frequency out of plane vibrations which are relatively close in energy, this study emphasizes the difficulties encountered to simulate accurately the rotational structures of such large molecules in excited vibrational states.

[a]also at : Institut des Sciences Moléculaires d'Orsay, UMR 8214 CNRS-Université Paris-Sud 91405 Orsay Cedex
[b]Present address: Harvard-Smithsonian Center for Astrophysics, Cambridge, MA 02138, USA
[c]Z. Kisiel, O. Desyatnyk, L. Pszczolkowski, S. B. Charnley and P. Ehrenfreund, J. Mol. Spectrosc. 217, 115 (2003)

CALCULATED DIPOLE MOMENTS AND DIPOLE POLARIZABILITIES OF OBSERVED AND CANDIDATE ASTRO-MOLECULES CONTAINING SILICON AND PHOSPHORUS

DAVID E. WOON, *Department of Chemistry, University of Illinois at Urbana-Champaign, Urbana, IL, USA*; HOLGER S. P. MÜLLER, *I. Physikalisches Institut, Universität zu Köln, Köln, Germany.*

Due to their role in determining the intensities of rotational transitions, accurate dipole moments are an important resource for evaluating the prospects of observing a candidate astromolecule. We have used high level ab initio calculations [mostly RCCSD(T) with aug-cc-pVQZ basis sets] to calculate the equilibrium dipole moments and dipole polarizabilities of nearly 80 compounds with up 6 atoms that contain silicon or phosphorus. To date, 17 observed astromolecules contain these elements, and more detections are likely as experimental data for their rotational transitions becomes available for interpreting observations from ALMA and other telescopes. We will compare the results from RCCSD(T) with other methods, including B3LYP and MP2. We will also describe the basis set dependence of our results, which varies considerably within the set of Si- and P-containing compounds we have studied.

FA11 $\qquad$ **11:35 – 11:50**

ANALYTIC EMPIRICAL POTENTIALS FOR ALL STABLE ISOTOPOLOGUES OF THE GROUND $X(^1\Sigma^+)$ STATE OF ZnO FROM PURELY ROTATIONAL MEASUREMENTS

<u>NIKESH S. DATTANI</u>, *Physical and Theoretical Chemistry Laboratory, Oxford University, Oxford, United Kingdom*; LINDSAY ZACK, *Department of Chemistry, University of Basel, Basel, Switzerland*; MING SUN, *Department of Chemistry and Astronomy, University of Arizona, Tuscon, AZ, USA*; ERIN R JOHNSON, *Chemistry and Chemical Biology, University of California Merced, Merced, USA*; ROBERT LE ROY, *Department of Chemistry, University of Waterloo, Waterloo, ON, Canada*; LUCY ZIURYS, *Steward Observatory, Departments of Chemistry and Physics, University of Arizona, Tucson, AZ, USA.*

We report eight new ultra-high precision (± 5 kHz) measurements of purely rotational $N(1 \leftarrow 0)$ transitions in several vibrational states of all stable isotopologues of the ground $X(1^1\Sigma^+)$-state of ZnO. Combined with previous high-resolution (± 50 kHz) measurements of purely rotational transitions between higher rotational states for the same system[a], we are able to build analytic potentials for ^{64}Zn^{16}O, ^{66}Zn^{16}O, ^{67}Zn^{16}O, ^{68}Zn^{16}O, and ^{70}Zn^{16}O, that are in full agreement with all known spectroscopic measurements of the system. Despite there being absolutely no vibrational information, our empirical potentials are able to determine the size of the vibrational spacings and the bond lengths, each with a precision of more than two orders of magnitude greater than the most precise empirical values previously known. We then use the XDM method to calculate values for the C_6, C_8, and C_{10} long-range constants for this molecule, and use these to accurately anchor the long-range regions of the potentials, where no measurements have yet been performed. In the region lying between the short-range measurements and the long-range theory on which our potentials are based, our final analytic global potentials are in very good agreement with state of the art *ab initio* potentials.

---

[a]L. N. Zack, R. L. Pulliam, L. M. Ziurys, *J. Mol. Spec.*, **256**, 186-191 (2009).

FB. Clusters/Complexes
Friday, June 20, 2014 – 8:30 AM
Room: 100 Noyes Laboratory

Chair: Michael Heaven, Emory University, Atlanta, GA, USA

FB01   **8:30 – 8:45**

DEVELOPMENT OF A CAVITY RINGDOWN SPECTROMETER FOR MEASURING ELECTRONIC STATES OF Be CLUSTERS

JACOB STEWART, MICHAEL SULLIVAN, MICHAEL HEAVEN, *Department of Chemistry, Emory University, Atlanta, GA, USA.*

Metal clusters are of interest in exploring the transition from atomic to bulk properties in metals. Spectroscopic measurements of metal clusters also provide important benchmarks for theoretical investigations of bonding in metallic systems. In particular, theoretical predictions of properties of small Be clusters are highly dependent on the level of theory employed, and experimental data would provide an important test of the accuracy of the theoretical predictions. To obtain experimental data on the electronic structure of Be clusters, we have constructed a pulsed cavity ringdown spectrometer designed to measure gas-phase metal clusters. We generate the clusters via laser ablation of a metal rod followed by supersonic expansion of the ablated material. The supersonic expansion is then probed by cavity ringdown spectroscopy using an excimer-pumped pulsed dye laser. We will present preliminary results using the spectrometer to observe small clusters of Cu and Al in preparation for observing Be clusters. We will also present any data on Be clusters that are available.

FB02   **8:47 – 9:02**

A THEORETICAL SEARCH FOR AN ELECTRONIC SPECTRUM OF THE He-BeO COMPLEX

ADRIAN GARDNER, MICHAEL HEAVEN, *Department of Chemistry, Emory University, Atlanta, GA, USA.*

The surprisingly high dissociation energy of the He-Be bond in the He-BeO complex was first reported 25 years ago.[a] Following which, a number of theoretical studies have investigated similar closed shell helium containing complexes. However, despite these investigations, a complex containing a strong He-X bond has thus far eluded experimental detection. In this work, potential energy surfaces of electronically excited states of the He-BeO complex have been calculated employing high level CASSCF+MRCI+Q methodologies and utilizing extended basis sets. Several excited states show strong interactions between helium and BeO lying in Franck-Condon accessible windows of electronic transitions arising from the vibrationless electronic ground state. It is hoped that the conclusions of this study will result in the observation an electronic spectrum of this long hypothesized strongly bound complex in the near future.

---

[a] W. Koch, J. R. Collins and G. Frenking, *Chem. Phys. Lett.* 1986, **132** 330-333.

FB03   **9:04 – 9:19**

A UNIFIED PERSPECTIVE ON THE NATURE OF PAIRWISE INTERATOMIC INTERACTIONS FROM Ar_2 TO CARBON MONOXIDE.

CHARLES K. ROSALES, LUIS A. RIVERA-RIVERA, BLAKE A. McELMURRY, ROBERT R. LUCCHESE, JOHN W. BEVAN, *Department of Chemistry, Texas A & M University, College Station, TX, USA*; JAY R. WALTON, *Department of Mathematics, Texas A & M University, College Station, TX, USA.*

A method is developed that gives a unified perspective on the nature of a wide range of pairwise interatomic interactions. The approach is applied to the diatomic molecules CO, H_2^+, H_2, HF, LiH, Li_2, O_2, and Ar_2, and one-dimensional cuts through the potentials of OC-HBr, OC-HF, OC-HCCH, OC-HCN, OC-HCl, OC-HI, OC-BrCl, and OC-Cl_2. Systems selected for investigation illustrate different bound categories varying from van der Waals, halogen bonded, hydrogen bonded to strongly bound covalently bound carbon monoxide with binding energies varying over almost three orders of magnitude, from 99.3 cm^{-1} to 90683 cm^{-1}. Accurate semi-empirically determined Rydberg-Klein-Rees or morphed interatomic potentials are used in transformations for this wide range of species to a reduced potential demonstrating commonality in their fundamental nature.

FB04 9:21 – 9:36

MID-INFRARED SPECTRUM OF THE ATMOSPHERICALLY SIGNIFICANT N_2-H_2O COMPLEX

SEAN D. SPRINGER, BLAKE A. McELMURRY, ROBERT R. LUCCHESE, <u>JOHN W. BEVAN</u>, *Department of Chemistry, Texas A & M University, College Station, TX, USA*; L. H. COUDERT, *LISA, CNRS, Universités Paris Est Créteil et Paris Diderot, Créteil, France.*

Rovibrational transitions associated with tunneling states in the vibration of the N_2–H_2O complex have been recorded using a supersonic jet quantum cascade laser spectrometer at $6.2\mu m$. Analysis of the resulting spectra is facilitated by incorporating fits of previously recorded microwave and submillimeter data accounting for Coriolis coupling to obtain the levels of the ground vibrational state. The results are then used to confirm assignment of the ν_3 vibration and explore the nature of tunneling dynamics in associated vibrationally excited states of the complex.

FB05 9:38 – 9:53

HIGH RESOLUTION INFRARED SPECTROSCOPY OF N_2O-DIACETYLENE AND $CS_2 - C_2D_2$ DIMERS

<u>MAHDI YOUSEFI</u>, S. SHEYBANI-DELOUI, JALAL NOROOZ OLIAEE, *Physics and Astronomy/Institute for Quantum Science and Technology, University of Calgary, Calgary, AB, Canada*; BOB McKELLAR, *Steacie Laboratory, National Research Council of Canada, Ottawa, ON, Canada*; NASSER MOAZZEN-AHMADI, *Physics and Astronomy/Institute for Quantum Science and Technology, University of Calgary, Calgary, AB, Canada.*

Rotationally-resolved infrared spectra of normal and bideuterated N_2O-diacetylene van der Waals complexes were observed in the N_2O ν_1 region ($\sim$2224 cm^{-1}) using a Quantum Cascade Laser to probe a supersonic slit-jet expansion. The observed rotational constants show that the structure has C_s symmetry with the constituent monomers in a slipped near-parallel configuration. *ab initio* calculations at MP2 level indicate that the observed structure is the lowest energy isomer.

Our recent investigation on complexes containing CS_2 and C_2D_2 using an Optical Parametric Oscillator has resulted in the observation of several bands, one of which is centered at 2438.16 cm^{-1}. This b-type band was assigned to $CS_2 - C_2D_2$ dimer with parallel monomers and C_{2v} symmetry, in agreement with *ab initio* calculations for the lowest energy isomer. In this talk, we discuss our observation and analysis of N_2O-diacetylene, $CS_2 - C_2D_2$ dimer and probable complexes containing CS_2 and C_2D_2 in our future observation.

FB06 9:55 – 10:10

FUNDAMENTAL AND COMBINATION BANDS OF CO_2-C_2H_2 AND CO_2-C_2D_2 IN THE MID-INFRARED REGION

MOJTABA REZAEI, JOBIN GEORGE, LUIS WELBANKS, <u>NASSER MOAZZEN-AHMADI</u>, *Department of Physics and Astronomy, University of Calgary, Calgary, AB, Canada.*

Spectra of the weakly-bound CO_2-C_2H_2 and CO_2-C_2D_2 complexes are observed in the regions of CO_2 ν_3 ($\sim$ 2340 cm^{-1}) and C_2D_2 ν_3 ($\sim$ 2440 cm^{-1}) fundamental vibrations, using an infrared Optical Parametric Oscillator (OPO) to probe a pulsed supersonic slit-jet expansion. Five bands are measured and analyzed: the fundamental asymmetric stretch of the C_2D_2 component, two combination bands involving the out-of-plane torsional vibrations for CO_2-C_2D_2, and two combination bands involving an intermolecular in-plane bending vibration for CO_2-C_2H_2 and CO_2-C_2D_2. The measured intermolecular vibrational frequencies are 61.408(1), 54.5(5), 39.9(5), and 39.961(1) cm^{-1}, respectively. Torsional vibrational frequencies are in good agreement with the value of 41 cm^{-1} estimated by Muenter[a] using a harmonic oscillator model. The intermolecular vibrations provide clear spectroscopic data against which theory can be benchmarked. These results will be discussed, along with a brief introduction of our pulsed-jet supersonic apparatus which has been retrofitted with an infrared tunable Optical Parametric Oscillator (Lockheed Martin Aculight Argos^{TM}).

---

[a]J. S. Muenter, J. Chem. Phys. **90**, 2781 (1991)

Intermission

QUANTUM MONTE CARLO SIMULATION OF VIBRATIONAL FREQUENCY SHIFTS OF CO IN SOLID para-HYDROGEN

LECHENG WANG, ROBERT LE ROY, PIERRE-NICHOLAS ROY, *Department of Chemistry, University of Waterloo, Waterloo, ON, Canada.*

Stimulated by Fajardo's remarkable study of the rovibrational spectra of CO isotopologues trapped in solid *para*-hydrogen,[a] we have performed quantum Monte Carlo simulations to predict his observed vibrational frequency shifts and inertial rotational constants using 2-body potentials based on the best available models for the pH_2-pH_2 [b] and CO-pH_2 [c] potential energy functions. We started by fitting an analytic 'Morse/Long-Range' (MLR) function[d] to the 1D "adiabaic hindered rotor" version of Hinde's 5D pH_2-pH_2 potential developed by Faruk *et al.*[e] We then modified it to take account of many-body effects by scaling it until it yielded the correct equilibrium lattice parameters for the *fcc* and *hcp* structures of pure solid *para*-hydrogen. A CO molecule was then placed at different interstitial or substitution sites in large equilibrated *fcc* or *hcp para*-hydrogen lattices, and the structural and dynamical behaviors of the micro-solvation environment around CO were simulated with a PIMC algorithm using a 2D effective pH_2-CO potential based on the 5D H_2–CO potential energy surface recently reported by Li *et al.*,[f] with a lattice sum of values of the 2D CO vibrational difference potential being use to predict the vibrational frequency shift. The effective rotational constants B_{eff} for CO in different solid *para*-hydrogen structures were also calculated and compared with the experimental observations and with predicted B_{eff} values for CO in large-sized *para*-hydrogen–CO clusters.[g]

[a] M. E. Fajardo, *J. Phys. Chem. A* **117**, 13504 (2013).

[b] R. Hinde, *J. Chem. Phys.* **128**, 154308 (2008).

[c] H. Li, X-L. Zhang, R.J. Le Roy, and P.-N. Roy, *J. Chem. Phys.* **139**, 164315 (2013).

[d] R.J. Le Roy, C.C. Haugen, J. Tao and Hui Li, *Mol. Phys.* **109**, 435 (2011).

[e] N. Faruk, R.J. Le Roy, and P.-N. Roy, *J. Chem. Phys.* (submitted December 2013).

[f] H. Li, X-L. Zhang, R. J. Le Roy, and P.-N. Roy, *J. Chem. Phys.* **139**, 164315 (2013).

[g] Y. Mizumoto and Y. Ohtsuki, *Chem. Phys. Lett.* **501**, 304 (2011).

PROGRESS IN UNDERSTANDING THE INFRARED SPECTRA OF He- AND Ne-C_2D_2

NASSER MOAZZEN-AHMADI, *Department of Physics and Astronomy, University of Calgary, Calgary, AB, Canada*; BOB McKELLAR, *Steacie Laboratory, National Research Council of Canada, Ottawa, ON, Canada.*

Infrared spectra of He-C_2H_2 were recorded around 1990 in Roger Miller's lab, but detailed rotational assignment was apparently not possible even with the help of theoretical predictions. So there were no published experimental spectra of helium-acetylene van der Waals complexes until our recent work on He-C_2D_2 in the ν_3 region ($\sim$2440 cm^{-1}).[a] The problem is that this complex lies close to the free rotor limit, so that most of the intensity in the spectrum piles up in tangles of closely spaced lines located close to the monomer rotational transitions, $R(0)$, $P(1)$, etc.

Our previous He-C_2D_2 assignments were limited to the $R(0)$ region, that is, the $j = 1 \leftarrow 0$ subband, where j represents C_2D_2 rotation. Here, we extend the analysis to $j = 0 \leftarrow 1$ and $2 \leftarrow 1$ transitions with the help of new spectra obtained using a tunable OPO laser probe and a cooled supersonic jet nozzle. These subbands are weaker, not only because of the Boltzmann factor, but also the 2:1 nuclear spin statistics of $j'' = $ even:odd C_2D_2 levels. Moreover, the $j = 0 \leftarrow 1$ subband is overlapped by strong $(C_2D_2)_2$ transitions. We use a term value approach, obtaining a self-consistent set of "experimental" energy levels which can be directly compared with theory or fitted in terms of a Coriolis model. Challenges also arise with Ne-C_2D_2, which is not quite so close to the free rotor limit, but still has many overlapping lines. Insights gained here help in assigning the tricky $R(1)$ region for Ne-C_2D_2.

[a] M. Rezaei, N. Moazzen-Ahmadi, A.R.W. McKellar, B. Fernández, and D. Farrelly, Mol. Phys. **110**, 2743 (2012).

FB09

FIRST OBSERVATION AND ANALYSIS OF OCS $-$ C_4H_2 DIMER AND $(OCS)_2$ $-$ C_4H_2 TRIMER

S. SHEYBANI-DELOUI, MAHDI YOUSEFI, JALAL NOROOZ OLIAEE, *Physics and Astronomy/Institute for Quantum Science and Technology, University of Calgary, Calgary, AB, Canada*; BOB McKELLAR, *Steacie Laboratory, National Research Council of Canada, Ottawa, ON, Canada*; NASSER MOAZZEN-AHMADI, *Physics and Astronomy/Institute for Quantum Science and Technology, University of Calgary, Calgary, AB, Canada.*

Infrared spectrum of a slipped near parallel isomer of OCS $-$ C_4H_2 was observed in the region of ν_1 fundamental band of OCS monomer ($\sim$2062 cm^{-1}) using a diode laser to probe the supersonic slit jet expansion. The *ab initio* calculations at MP2 level indicate that the observed structure is the lowest energy isomer. The OCS $-$ C_4H_2 band is composed of hybrid a/b-type transitions and was simulated by a conventional asymmetric top Hamiltonian with rotational constants of A=2892.15(10) MHz, B=1244.178(84) MHz, and C=868.692(52) MHz. The spectrum shows a relatively large red-shift of $\sim$6 cm^{-1}with respect to the OCS monomer band origin. Also, one band for $(OCS)_2$ $-$ C_4H_2 trimer is observed around 2065 cm^{-1}. This band is blue-shifted by 3 cm^{-1}relative to the ν_1 fundamental band of OCS monomer. Our analysis shows that this trimer has C_2 symmetry with rotational constants of A= 855.854(61) MHz, B=733.15(11) MHz, and C=610.10(38) MHz and c-type transitions. This structure is comparable with that of $(OCS)_2$ $-$ C_2H_2 where the OCS dimer unit within the trimer is non-polar.[a] In addition to the normal isotoplogues, OCS $-$ C_4D_2 and $(OCS)_2$ $-$ C_4D_2 were observed. In this talk, we discuss our observations and analysis on OCS $-$ C_4H_2 dimer and $(OCS)_2$ $-$ C_4H_2 trimer.

[a]Mojtaba Rezaei, A. R. W. McKellar, and N. Moazzen-Ahmadi, J. Phys. Chem. A, **115**, 10416 (2011).

FB10

CHIRPED-PULSE BROADBAND MICROWAVE SPECTRA AND STRUCTURES OF THE OCS TRIMER AND TETRAMER

LUCA EVANGELISTI, CRISTOBAL PEREZ, NATHAN A SEIFERT, BROOKS PATE, *Department of Chemistry, The University of Virginia, Charlottesville, VA, USA*; MEHDI DEHGHANY, *Department of Physics, University of Calgary, Calgary, AB, Canada*; NASSER MOAZZEN-AHMADI, *Department of Physics and Astronomy, University of Calgary, Calgary, AB, Canada*; BOB McKELLAR, *Steacie Laboratory, National Research Council of Canada, Ottawa, ON, Canada.*

Structure determination of weakly bound OCS clusters is a challenging problem due to many low energy isomers on the potential energy surface. The premier tool for studying these clusters is high-resolution infrared spectroscopy, as it can be used to analyze non-polar clusters. Following the analysis of high-resolution IR spectra of clusters formed in a molecular beam expansion of OCS there were some outstanding questions about the structures of the observed clusters. The chirped-pulse Fourier transform microwave spectrum in the 3-9 GHz frequency range was measured for a pulsed molecular beam of OCS in neon (1%). All ^{13}C, ^{18}O and ^{34}S isotopologues of the previously detected OCS trimer have been observed in natural abundance in the 3-9 GHz band using chirped-pulse Fourier transform microwave spectroscopy. The structure of this trimer features a barrel-shaped structure with two aligned and one anti-aligned OCS monomers. A new OCS trimer is also observed for the first time, and its structure is consistent with a barrel-shaped structure with 3 aligned monomers.

Using the infrared spectrum for guidance, a spectrum corresponding to a polar OCS tetramer has been assigned. This cluster has a similar barrel-like structure but with an additional tilted OCS monomer added to the top of the barrel. All ^{13}C and ^{34}S isotopologues have been assigned for the tetramer. However, due to sign ambiguities in Kraitchman's equations, and small rotational constant differences between aligned and anti-aligned combinations of OCS molecules in the trimer barrel, absolute structural assignment is indeterminate without additional constraints. Therefore a combinatoric approach was used to compute the most reasonable tetramer structure using distance and sign constraints between pairs of carbon and sulfur coordinates, assuming the experimental OCS monomer structure. Results of this approach will be presented, as well as a comparison of the experimental results with the most recent ab initio structures for the OCS tetramer.

FB11

OCS TRIMER AND TETRAMER: CALCULATED STRUCTURES AND INFRARED SPECTRA

MEHDI DEHGHANY, *Department of Physics, University of Calgary, Calgary, AB, Canada*; NASSER MOAZZEN-AHMADI, *Department of Physics and Astronomy, University of Calgary, Calgary, AB, Canada*; BOB McKELLAR, *Steacie Laboratory, National Research Council of Canada, Ottawa, ON, Canada.*

An OCS trimer was originally observed in the 1990s by microwave spectroscopy.[a] New broadband chirped-pulse microwave spectra (preceding talk) reveal an OCS tetramer and a second distinct trimer isomer. In the present talk, we discuss OCS cluster structures and infrared spectra. Our structure calculations are based on a recent *ab initio* potential energy surface[b] and assume pairwise additivity. There are also recent direct *ab initio* trimer and tetramer calculations, which are (necessarily) at a lower level of theory.[c] We find that the observed OCS trimers indeed correspond to the two lowest energy isomers in both calculations, and that there is fairly good agreement of experimental and theoretical structures. For the tetramer the global minimum is at -2773 cm^{-1} relative to dissociation, and we calculate (at least) twenty different isomers within 100 cm^{-1} of this minimum (and seven within 20 cm^{-1}). Remarkably, the observed microwave tetramer does correspond to our lowest calculated isomer. However this isomer is not included in the published direct *ab initio* calculation – it may just have been overlooked due to the large number of isomers!

In the mid-infrared region of the OCS ν_1 fundamental ($\tilde{2}060$ cm^{-1}), we observe two bands which are clearly due to the same microwave OCS tetramer. But a third band is assigned to a different tetramer not observed in the microwave spectrum. It appears to correspond to our seventh calculated isomer, located about 20 cm^{-1} above the most stable one, and it is also missing from the direct *ab initio* calculation. Neither observed tetramer has any symmetry elements.

---

[a] J.P. Connelly, A. Bauder, A. Chisholm, and B.J. Howard, Mol. Phys. **88**, 915 (1996); R.A. Peebles and R.L. Kuczkowski, J. Phys. Chem. A **103**, 6344 (1999).

[b] J. Brown, X.-G. Wang, R. Dawes, and T. Carrington, Jr., J. Chem. Phys. **136**, 134306 (2012).

[c] N. Sahu, G. Singh, and S.R. Gadre, J. Phys. Chem. A **117**, 10964 (2013).

FB12

SPECTROSCOPIC AND COMPUTATIONAL CHARACTERIZATION OF HYDRATED PYRIMIDINE ANIONS

JOHN T. KELLY, NATHAN I HAMMER, *Chemistry and Biochemistry, University of Mississippi, Oxford, MS, USA.*

Pyrimidine is known to possess a negative electron affinity. Anions created from such molecules, whose energies are higher than those of their neutral counterparts, are unstable with respect to autodetachment. The solvation of pyrimidine with just one water molecule results in a positive electron binding energy. The addition of water molecules stabilizes the excess charge and increase the binding energy. The most interesting feature is the orientation of the hydrated pyrimidine complex to help accommodate an excess electron.

FC. Theory and Computation
Friday, June 20, 2014 – 8:30 AM
Room: B102 Chemical and Life Sciences

Chair: Anne B McCoy, The Ohio State University, Columbus, OH, USA

FC01   8:30 – 8:45

THE THREE-DIMENSIONAL POTENTIAL ENERGY SURFACE OF Ar-CO

YOSHIHIRO SUMIYOSHI, *Division of Pure and Applied Science, Faculty of Science and Technology, Gunma University, Maebashi, Japan*; YASUKI ENDO, *Department of Basic Science, The University of Tokyo, Tokyo, Japan.*

A three-dimensional intermolecular potential energy surface of the Ar-CO complex has been determined by utilizing previously reported spectroscopic data, where 971 transition frequencies by microwave, millimeter-wave, submillimeter-wave, and IR spectroscopy were reproduced simultaneously within experimental accuracies. A free rotor model Hamiltonian considering all the freedom of motions for an atom-diatom system was used to calculate vibration-rotation energies. The three-dimensional potential energy surface by *ab initio* calculations at the CCSD(T)-F12b/aug-cc-pV5Z level of theory were parameterized by a model function consisting of 46 potential parameters and they were used as initial values in the least-squares analysis. In total 20 potential parameters were optimized to reproduce all the experimental data.

FC02   8:47 – 8:57

AB INITIO CALCULATIONS OF THE GROUND ELECTRONIC STATES OF THE C_3-Ar AND C_3-Ne COMPLEXES

YI-REN CHEN, YI-JEN WANG, YEN-CHU HSU, *Institute of Atomic and Molecular Sciences, Academia Sinica, Taipei, Taiwan.*

The C_3Ar and C_3Ne complexes have four large amplitude vibrations. These are the in- and out-of-plane C_3 bending motions, and the two van der Waals vibrations. Assignment of the spectra of the complexes is therefore challenging. The ab initio potential energies of their ground electronic states have been calculated at the CCSD(T)/pVQZ level. 46620 points have been computed to describe the four-dimensional potential of C_3Ar: $\angle C - C - C = 112\text{-}179.5°$, r (C-C bond length) = 1.298 Å, R(C_3-Ar) = 3.4-6.0 Å, ϕ (azimuth angle between Ar and the principal axis of C_3)=0-180°, and θ (colatitude angle) = 0-180°. For C_3Ne, 69190 points have been computed: $\angle C - C - C$=106-179.5°, r(C-C)= 1.298Å, R(C_3-Ne) = 3.0-7.0 Å, $\phi = 0\text{-}180°$, $\theta = 0\text{-}180°$. Basis set superposition errors have been corrected in the C_3-Ne potential energy calculations. Morse type functions and power series were used to fit the calculated points.

FC03   8:59 – 9:14

VIBRATIONAL LEVEL STRUCTURES OF THE GROUND ELECTRONIC STATES OF THE C_3-Ar and C_3-Ne COMPLEXES

YI-REN CHEN, YEN-CHU HSU, *Institute of Atomic and Molecular Sciences, Academia Sinica, Taipei, Taiwan.*

The Heidelberg multiconfiguration time-dependent Hartree package[a] was used to calculate the vibrational level structures of the ground electronic states of the C_3-Ar and C_3-Ne complexes. The previously reported 4-D ab initio potentials were converted to 6-D potentials by adding the potential energies of the C-C symmetric and antisymmetric stretching vibrations of C_3. They were subsequently transformed from internal coordinates to Jacobi coordinates. The kinetic-energy operators were taken from Yang and Kühn[b]. Preliminary results show that large amplitude motions occur in five coordinates: C-C-C bond angle, out-of-plane tilt angle, van der Waals stretch, van der Waals bend and one of the C-C bonds.

---

[a]G.A. Worth, M.H. Beck, A. Jäckle, H.-D. Meyer, F. Otto, M. Brill, and O. Vendrell, The MCTDH package, version 8.4, Heidelberg University, Heidelberg, Germany, 2011.
[b]Y. Yang and O. Kühn, Mol. Phys. **106, 2445(2008)**

FC04   **9:16 – 9:31**

THEORETICAL STUDY OF THE VIBRATIONAL SPECTROSCOPY OF THE ETHYL RADICAL

DANIEL P. TABOR, EDWIN SIBERT, *Department of Chemistry, The Univeristy of Wisconsin, Madison, WI, USA.*

The rich spectroscopy of the ethyl radical has attracted the attention of several experimental and theoretical investigations. Its molecular spectrum contains signatures of hyperconjugation, torsion-inversion coupling, and Fermi coupling. We present a full-dimensional theoretical treatment of this vibrational problem using a combination of Van Vleck perturbation theory and variational approaches to further explore these effects. A CCSD(T)/cc-pVTZ potential energy surface and a numerical computation of the kinetic energy operator are employed to construct the Hamiltonian. Our calculations use coordinates that exploit the system's G_{12} PI symmetry to produce compact expressions for the potential energy. This symmetry also allows for a simplified evaluation of the Hamiltonian and greatly reduces the cost of diagonalization. For this talk we illustrate the Van Vleck treatment of the vibrational modes in the ethyl radical and how such a treatment accelerates basis set convergence of the resulting eigenvalues.

FC05   **9:33 – 9:48**

SIMULATION OF ACCURATE VIBRATIONALLY RESOLVED ELECTRONIC SPECTRA: THE INTEGRATED TIME-DEPENDENT AND TIME-INDEPENDENT FRAMEWORK

ALBERTO BAIARDI, VINCENZO BARONE, *Scuola Normale Superiore, Scuola Normale Superiore, Pisa, Italy*; MALGORZATA BICZYSKO, JULIEN BLOINO, *Istituto di Chimica dei Composti OrganoMetallici (ICCOM-CNR), UOS di Pisa, Consiglio Nazionale delle Ricerche, Pisa, Italy.*

Two parallel theories including Franck–Condon, Herzberg–Teller and Duschinsky (i.e., mode mixing) effects, allowing different approximations for the description of excited state PES have been developed in order to simulate realistic, asymmetric, electronic spectra line-shapes taking into account the vibrational structure: the so-called sum-over-states or time-independent (TI) method and the alternative time-dependent (TD) approach, which exploits the properties of the Fourier transform.

The integrated TI-TD procedure included within a general purpose QM code [1,2], allows to compute one photon absorption, fluorescence, phosphorescence, electronic circular dichroism, circularly polarized luminescence and resonance Raman spectra. Combining both approaches, which use a single set of starting data, permits to profit from their respective advantages and minimize their respective limits: the time-dependent route automatically includes all vibrational states and, possibly, temperature effects, while the time-independent route allows to identify and assign single vibronic transitions. Interpretation, analysis and assignment of experimental spectra based on integrated TI-TD vibronic computations will be illustrated for challenging cases of medium-sized open-shell systems in the gas and condensed phases with inclusion of leading anharmonic effects.

1. V. Barone, A. Baiardi, M. Biczysko, J. Bloino, C. Cappelli, F. Lipparini Phys. Chem. Chem. Phys, 14, 12404, (2012)
2. A. Baiardi, V. Barone, J. Bloino J. Chem. Theory Comput., 9, 4097–4115 (2013)

FC06   **9:50 – 10:05**

HIGH ACCURACY AB INITIO CALCULATION OF METAL QUADRUPOLE-COUPLING PARAMETERS

LAN CHENG, JOHN F. STANTON, *Department of Chemistry, The University of Texas, Austin, TX, USA*; JÜRGEN GAUSS, *Institut für Physikalische Chemie, University of Mainz, Mainz, Germany.*

Accurate quantum-chemical calculations of metal quadrupole-coupling parameters are challenging due to the sensitivity of these parameters to both relativistic and electron-correlation effects. In the present study we have employed the spin-free exact two-component theory in its one-electron variant for a cost-effective treatment of scalar-relativistic effects in combination with coupled-cluster methods for a systematic incorporation of electron correlation. Spin-orbit effects have been included by means of perturbation theory at the Hartree-Fock self-consistent-field level. The accuracy and applicability of the model presented here is demonstrated with calculations of metal quadrupole-coupling parameters for a set of copper and gold compounds.

FC07 10:07 – 10:22

ELECTRONIC EXCITATIONS OF ALKALI-ALKALINE EARTH DIATOMIC MOLECULES - RESULTS FROM AB INITIO CALCULATIONS

JOHANN V. POTOTSCHNIG, GÜNTER KROIS, FLORIAN LACKNER, WOLFGANG E. ERNST, *Institute of Experimental Physics, Graz University of Technology, Graz, Austria.*

Recently interest in polar diatomic molecules with a magnetic dipole moment has been growing. An example for such molecules is the combination of an alkali metal atom and an alkaline earth metal atom. These systems are quite small, containing only three valence electrons. Nevertheless calculations of excited states are challenging. Ab initio calculations for two sample systems, LiCa[a] and RbSr, will be presented. The potential energy curves and transition dipole moments for the ground state and several excited states were determined, up to 25000 cm^{-1} for LiCa and up to 22000 cm^{-1} for RbSr. Multireference configuration interaction calculations (MRCI) based on complete active space self-consistent field wave functions (CASSCF) were used to determine the properties of the system as implemented in the MOLPRO software package.[b] Effective core potentials (ECPs) and core polarization potentials (CCPs) were applied to reduce the computational effort, while retaining accuracy. The similarities and differences of the two systems will be discussed. In both systems the accurate description of the asymptotic values of the PECs corresponding to atomic D-states proved to be difficult. The results will be compared to recent experiments, showing that a combination of theory and experiment gives a reliable description of the systems.

[a]G. Krois, J.V. Pototschnig, F. Lackner and W.E. Ernst, J. Phys. Chem. A, 117, 13719-13731 (2013)

[b]H.-J. Werner and P. J. Knowles and G. Knizia and F. R. Manby and M. Schütz et al., MOLPRO, version 2010.1, see http://www.molpro.net/

Intermission

FC08 10:39 – 10:54

CONICAL INTERSECTIONS BETWEEN VIBRATIONALLY ADIABATIC SURFACES IN METHANOL

MAHESH B. DAWADI, DAVID S. PERRY, *Department of Chemistry, The University of Akron, Akron, OH, USA.*

The discovery of a set of seven conical intersections (CI's) between vibrationally adiabatic surfaces in methanol is reported. The intersecting surfaces represent the energies of the two asymmetric CH stretch vibrations, ν_2 and ν_9, regarded as adiabatic functions of the torsional angle, γ, and COH bend angle, ρ. One conical intersection, required by symmetry, is located at the C_{3v} geometry where the COH group is linear ($\rho = 0°$); the other six are in eclipsed conformations with $\rho = 62°$ and $94°$. The three CI's at $\rho = 62°$ are close to the equilibrium geometry ($\rho = 71.4°$), within the zero-point amplitude of the COH bending vibration. CI's between electronic surfaces have long been recognized as crucial conduits for ultrafast relaxation, and recently Hamm, and Stock[a][b] have shown that vibrational CI's may also provide a mechanism for ultrafast vibrational relaxation.

The ab initio data reported here are well described by an extended Zwanziger and Grant model[c] for E $\otimes$ e Jahn-Teller systems in which Renner-Teller coupling is also active. However, in the present case, the distortion ρ from C_{3v} symmetry is much larger than is typical in the Jahn-Teller coupling of electronic surfaces and accordingly higher-order terms in ρ are required. The present results are also consistent with the two-state model of Xu et al[d]. The cusp-like features, which they found along the internal-rotation path, are explained in the context of the present work in terms of proximity to the CI's. The presence of multiple CI's near the torsional minimum energy path impacts the role of geometric phase in this three-fold internal-rotor system. When the dimensionality of the low-frequency space is extended to include the CO bond length as well as γ and ρ, the individual CI's become seams of CI's. It is shown that the CI's at $\rho = 62°$ and $94°$ lie along the same seam of CI's in this higher dimensional space.

[a]P. Hamm and G. Stock, *Phys. Rev. Lett.* **109**, 173201, (2012).

[b]P. Hamm, and G. Stock, *Mol. Phys.* **111**, 2046, (2013).

[c]J. W. Zwanziger, and E. R. Grant, *J. Chem. Phys.* **87**, 2954, (1987).

[d]L.-H. Xu, J. T. Hougen, and R. M. Lees, *J. Mol. Spectrosc.* **293-294**, 38, (2013).

260

FC09   **10:56 – 11:11**

NONADIABATIC PHOTO-PROCESS INVOLVING THE $\pi\sigma^*$ STATE IN INTRAMOLECULAR CHARGE TRANSFER: A CONCERTED SPECTROSCOPIC AND COMPUTATIONAL STUDY ON 4-(DIMETHYLAMINO)BENZETHYNE AND 4-(DIMETHYLAMINO)BENZONITRILE.

TAKASHIGE FUJIWARA, *Department of Chemistry and Biochemistry, The Ohio State University, Columbus, OH, USA*; JAVIER SEGARRA-MARTÍ, PEDRO B. COTO, *Institut für Theoretische Physik Theoretische Festkörperphysik, Friedrich-Alexander-Universität, Erlangen-Nürnberg, Germany.*

The ubiquitous nature of the low-lying $\pi\sigma^*$ state in the photo-excited aromatic molecules or biomolecules is widely recognized to play an important role in nonadiabatic photo-process such as photodissociation or intramolecular charge transfer (ICT). For instance, the O–H elimination channel in phenol is attributed to the state-cross of the repulsive $\pi\sigma^*$ state that exhibits a conical intersection with the lowest bright $\pi\pi^*$ state and with the ground state, leading to ultrafast electronic deactivation. A similar decay pathway has been found in the ICT formation of 4-(dialkylamino)benzonitriles in a polar environment, where an initially photoexcited Frank-Condon state bifurcates in the presence of a dark intermediate $\pi\sigma^*$ state that crosses the fluorescent $\pi\pi^*$ state, followed by a conical intersection with the twisted intramolecular charge transfer (TICT) state. We proposed such a two-fold decay mechanism that $\pi\sigma^*$-state highly mediates intramolecular charge transfer in 4-(dialkylamino)benzonitriles, which is supported from both our high-level *ab initio* calculations and ultrafast laser spectroscopies in the previous study.

4-(Dimethylamino)benzethyne (DMABE) is isoelectronic with 4-(dimethylamino)benzonitrile (DMABN), and the electronic structures and electronic spectra of the two molecules bear very close resemblance. However, DMABN does show the ICT formation in a polar environment, whereas DMABE does not. To probe the photophysical differences among the low-lying excited-state configurations, we performed concerted time-resolved laser spectroscopies and high level *ab initio* multireference perturbation theory quantum-chemical (CASPT2//CASSCF) computations on the two molecules. In this paper we demonstrate the importance of the bound excited-state of a $\pi\sigma^*$ configuration that induce highly $\pi\sigma^*$-state mediated intramolecular charge transfer in 4-(dialkylamino)benzonitriles.

FC10   **11:13 – 11:28**

A COMPUTATIONAL TDDFT STUDY ON INTRAMOLECULAR CHARGE TRANSFER IN DI-*TERT*-BUTYLAMINOBENZONITRILES AND 2,4,6-TRICYANOANILINES.

TAKASHIGE FUJIWARA, *Department of Chemistry and Biochemistry, The Ohio State University, Columbus, OH, USA*; MAREK Z. ZGIERSKI, *Steacie Laboratory, National Research Council of Canada, Ottawa, ON, Canada.*

We have carried out TDDFT computational studies on the low-lying excited states of di-*tert*-butylaminobenzonitrile and 2,4,6-tricyanoaniline compounds that exhibit unusual photophysical behaviors associated with the intramolecular charge transfer (ICT). For both 3- and 4-di-*tert*-butylamino)benzonitriles (*m*-DTBABN and *p*-DTBABN, respectively) show the ICT formation, and *p*-DTBABN appears to be the only *meta*-substituted aminobenzonitrile that exhibits the ICT formation. The TDDFT calculations indicate evidence that the ultrafast ICT formation in *p*-DTBABN and *m*-DTBABN is due to the sequential state switches: $\pi\pi^*(L_a) \rightarrow \pi\sigma^* \rightarrow$ ICT in the presence of conical intersections among the three closely-lying excited-states. On the other hand, 2,4,6-tricyanoaniline does not show clear evidence for the LE (locally excited) state $\rightarrow$ ICT state formation from steady-state fluorescence studies, despite the greater electron acceptor strength of tricyanobenzene as compared to monocyanobenzene, which is part of a 4-(dimethylamino)benzonitrile (*p*-DMABN) compound.

However, it is predicted that 2,4,6-tricyano-*N,N*-dimethylaniline (TCDMA), but not 2,4,6-tricyanoaniline (TCA), possesses two ICT states, which show the ICT-characterized quinoidal structures and lie below the initially photo-excited $S_1(\pi\pi^*)$ state. The CC2 calculations further predict two conformers as labeled with quinoidal (ICT–Q) and anti-quinoidal (ICT–AQ) structures are rapidly interconnecting with each other. The lower energy ICT–Q structure tends to be populated from the unstable ICT–AQ structure, which is responsible for the observed time-resolved fluorescence as well as the excited-state absorption from the mixed $S_1(\pi\pi^*)$/ICT state of TCDMA. In both cases for TCDMA and TCA, the $\pi\sigma^*$ state locates significantly higher in energy than the $S_1(\pi\pi^*)$ state (and the ICT state for TCA), thus precluding the $\pi\sigma^* \rightarrow$ ICT formation, which is believed to occur in a *p*-DMABN in polar environments.

FC11 **11:30 – 11:45**

FULL DIMENSIONAL VIBRATIONAL CALCULATIONS FOR METHANE USING AN ACCURATE NEW AB INITIO BASED POTENTIAL ENERGY SURFACE

MOUMITA MAJUMDER, RICHARD DAWES, *Department of Chemistry, Missouri University of Science and Technology, Rolla, MO, USA*; XIAO-GANG WANG, TUCKER CARRINGTON, *Department of Chemistry, Queen's University, Kingston, ON, Canada*; JUN LI, HUA GUO, *Chemistry, University of New Mexico, Albuquerque, NM, USA*; SERGEI MANZHOS, *Department of Mechanical Engineering, National University of Singapore, Singapore, China.*

New potential energy surfaces for methane were constructed, represented as analytic fits to about 100,000 individual high-level ab initio data. Explicitly-correlated multireference data (MRCI-F12(AE)/CVQZ-F12) were computed using Molpro [1] and fit using multiple strategies. Fits with small to negligible errors were obtained using adaptations of the permutation-invariant-polynomials (PIP) approach [2,3] based on neural-networks (PIP-NN) [4,5] and the interpolative moving least squares (IMLS) fitting method [6] (PIP-IMLS). The PESs were used in full-dimensional vibrational calculations with an exact kinetic energy operator by representing the Hamiltonian in a basis of products of contracted bend and stretch functions and using a symmetry adapted Lanczos method to obtain eigenvalues and eigenvectors. Very close agreement with experiment was produced from the purely ab initio PESs.

References 1- H.-J. Werner, P. J. Knowles, G. Knizia, 2012.1 ed. 2012, MOLPRO, a package of ab initio programs. see http://www.molpro.net. 2- Z. Xie and J. M. Bowman, J. Chem. Theory Comput 6, 26, 2010. 3- B. J. Braams and J. M. Bowman, Int. Rev. Phys. Chem. 28, 577, 2009. 4- J. Li, B. Jiang and Hua Guo, J. Chem. Phys. 139, 204103 (2013). 5- S Manzhos, X Wang, R Dawes and T Carrington, JPC A 110, 5295 (2006). 6- R. Dawes, X-G Wang, A.W. Jasper and T. Carrington Jr., J. Chem. Phys. 133, 134304 (2010).

FD. Chirped pulse
Friday, June 20, 2014 – 8:30 AM
Room: 112 Chemistry Annex

Chair: Steven Shipman, New College of Florida, Sarasota, FL, USA

FD01   8:30 – 8:45

MONITORING THE REACTION PRODUCTS OF PERFLUOROPROPIONIC ACID AND ALLYL PHENYL ETHER USING CHIRPED-PULSE FOURIER TRANSFORM MICROWAVE (CP-FTMW) SPECTROSCOPY

DEREK S. FRANK, DANIEL A. OBENCHAIN, *Department of Chemistry, Wesleyan University, Middletown, CT, USA*; WEI LIN, *Department of Chemistry, University of Texas, Brownsville, TX, USA*; STEWART E. NOVICK, *Department of Chemistry, Wesleyan University, Middletown, CT, USA*; S. A. COOKE, *Natural and Social Science, Purchase College SUNY, Purchase, NY, USA*; G. S. GRUBBS II, *Department of Chemistry, Missouri University of Science and Technology, Rolla, MO, USA.*

The pure rotational spectra of the reaction mixture of perfluoropropionic acid, CF_3CF_2COOH, and allyl phenyl ether, $C_6H_5OCH_2CH=CH_2$, have been studied by a pulsed nozzle, chirped-pulse Fourier transform microwave spectrometer in the frequency range of 8-14 GHz. Transitions corresponding to multiple species, two of which being starting materials allyl phenyl ether and perfluoropropionic acid, have been observed and analyzed. Determination of the reaction products was carried out by matching observed rotational constants with ab initio quantum chemical calculations of predicted products and will be discussed. Rotational constants, centrifugal distortion constants and the assignment of allyl phenyl ether and reaction products spectra will all be discussed.

FD02   8:47 – 9:02

ENANTIOMER IDENTIFICATION IN CHIRAL MIXTURES WITH BROADBAND MICROWAVE SPECTROSCOPY

V. ALVIN SHUBERT, DAVID SCHMITZ, CHRIS MEDCRAFT, *CoCoMol, Max-Planck-Institut für Struktur und Dynamik der Materie, Hamburg, Germany*; DAVID PATTERSON, JOHN M. DOYLE, *Department of Physics, Harvard University, Cambridge, MA, USA*; MELANIE SCHNELL, *CoCoMol, Max-Planck-Institut für Struktur und Dynamik der Materie, Hamburg, Germany.*

In nature and as products of chemical syntheses, chiral molecules often exist in mixtures with other chiral molecules. The analysis of these complex mixtures to identify the components, determine which enantiomers are present, and to measure the enantiomeric excesses (ee) is still one of the challenging but very important tasks of analytical chemistry. These analyses are required at every step of modern drug development, from candidate searches to production and regulation.

We present here a new method of identifying individual enantiomers in mixtures of chiral molecules in the gas phase.[a,b] It is based on broadband rotational spectroscopy and employs a sum or difference frequency generation three-wave mixing process that involves a closed cycle of three rotational transitions. The phase of the acquired signal bares the signature of the enantiomer (see figure), as it depends upon the combined quantity, $\mu_a\mu_b\mu_c$, which is of opposite sign between members of an enantiomeric pair. Furthermore, because the signal amplitude is proportional to the ee, this technique allows for both determining which enantiomer is in excess and by how much. The high resolution of our technique allows us to perform molecule specific measurements of mixtures of chiral molecules with $\mu_a\mu_b\mu_c \neq 0$, even when the molecules are very similar (e.g. conformational isomers). We introduce the technique and present results on the analysis of mixtures of the terpenes, carvone, menthone, and carvomenthenol.

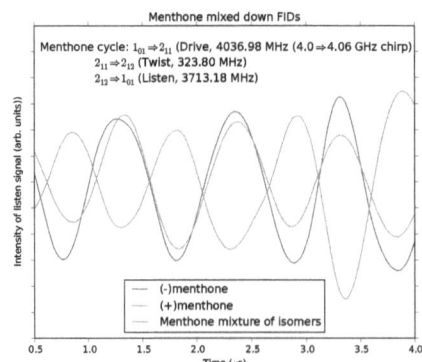

[a]D. Patterson, M. Schnell, J. M. Doyle, *Nature.* **497**, 475-477, 2013.

[b]V. A. Shubert, D. Schmitz, D. Patterson, J. M. Doyle, M. Schnell, *Ang. Chem. Int. Ed.* **53**, 1152-1155,2014.

FD03 9:04 – 9:19

BROADBAND MICROWAVE SPECTRUM AND STRUCTURE OF CYCLOPROPYL CYANOSILANE

NATHAN A SEIFERT, SIMON LOBSIGER, BROOKS PATE, *Department of Chemistry, The University of Virginia, Charlottesville, VA, USA*; GAMIL A GUIRGIS, JASON S OVERBY, *Chemistry, College of Charleston, Charleston, SC, USA*; JAMES R. DURIG, PETER GRONER, *Department of Chemistry, University of Missouri, Kansas city, MO, USA.*

The strucure of cyclopropane cyanosilane has been studied using chirped-pulse Fourier transform microwave (CP-FTMW) spectroscopy in the 6.5-18 GHz band. Two conformers of similar intensity were detected, one with a gauche orientation of the cyanosilane group with respect to the plane of the ring, and the other with a staggered conformation. The sensitivity of the CP-FTMW experiment was sufficient enough to assign spectra for all common singly-substituted heavy atom isotopologues (^{13}C, $^{29/30}Si$, ^{15}N) for each conformer, resulting in a full heavy atom Kraitchman structure of the molecule in good agreement with the predicted structure. Additionally, the hyperfine effects have been analyzed for the ^{14}N-containing parent species.

Results will also be presented on the potential tunneling spectrum arising from the symmetric double well torsional potential of the gauche conformer. Some observed transitions, especially with frequencies near the upper end of the measured band, exhibit splittings that could potentially be associated with a tunneling splitting. However, the resolution is not sufficient to provide a complete quantitative analysis of this effect.

FD04 9:21 – 9:36

BROADBAND MICROWAVE SPECTROSCOPY AND AUTOMATED ANALYSIS OF 12 CONFORMERS OF 1-HEXANAL

NATHAN A SEIFERT, CRISTOBAL PEREZ, DANIEL P. ZALESKI, JUSTIN NEILL, AMANDA STEBER, RICHARD D. SUENRAM, BROOKS PATE, *Department of Chemistry, The University of Virginia, Charlottesville, VA, USA*; STEVEN SHIPMAN, *Department of Chemistry, New College of Florida, Sarasota, FL, USA*; IAN FINNERAN, *Division of Chemistry and Chemical Engineering, California Institute of Technology, Pasadena, CA, USA*; ALBERTO LESARRI, *Department Quimica Fisica y Quimica Inorganica, Universidad de Valladolid, Valladolid, Spain.*

The rotational spectrum of 1-hexanal is used as a test case for developing automated assignment algorithms in molecular rotational spectroscopy, for the purpose of lowering the barrier to new users of rotational spectroscopy. There are two ways that the automated fitting algorithm, implemented in the AUTOFIT program, is used: 1) Assignment of the rotational spectrum of a molecule expected to be in the sample mixture (in this case, a conformer of 1-hexanal), using quantum chemistry estimates of the spectroscopic parameters to efficiently guide the search for the experimental spectrum. 2) Once a new spectrum is assigned, the algorithm is used to automatically assign isotopologue spectra (sensitivity permitting) to provide verification of the molecular structure.

Using a combination of quantum chemical calculations and automated spectral assignment, 12 conformations of 1-hexanal have been identified using chirped-pulse Fourier transform (CP-FTMW) spectroscopy in the 6.5-18, 18-26 and 26-40 GHz bands. Of these 12 conformers, the four lowest energy conformers were intense enough to resolve each of the six ^{13}C isotopologues for each conformer, and sufficient intensity was achieved to assign the ^{18}O isotopologues of the two lowest energy conformers. The full set of assignments were made using the AUTOFIT program, and a summary of results for all 38 observed species via automated assignment will be presented.

Additionally, by using all 12 conformers of 1-hexanal as a benchmark set, a discussion of dispersion-corrected density functional theory for the purpose of automated broadband spectroscopic searches will be presented, with specific results regarding rotational constant prediction. Results will also be presented on the correlation between the predicted conformational energetics predicted by multiple levels of theory and the sensitivity limits of the 1-hexanal CP-FTMW spectrum.

FD05 9:38 – 9:53

CONFORMATIONAL ANALYSIS OF IBUPROFEN USING BROADBAND MICROWAVE SPECTROSCOPY

SABRINA ZINN[a], THOMAS BETZ, MELANIE SCHNELL, *CoCoMol, Max-Planck-Institut für Struktur und Dynamik der Materie, Hamburg, Germany.*

The broadband rotational spectrum of ibuprofen ((RS)-2-(4-isobutylphenyl)-propanoic acid), a well-known drug, will be presented. As it is used to relieve pain, reduce fever, and inhibit inflammation, the knowledge of its biological activity is very interesting. Insights to the conformational flexibility of this drug might lead to a better understanding of the class of non-steroidal anti-inflammatory drugs that ibuprofen belongs to.

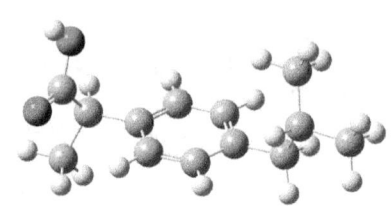

The spectrum was recorded with our broadband chirped-pulse Fourier transform microwave spectrometer in the frequency range of 2.0 - 8.3 GHz. With the obtained results, we are able to identify several conformers of ibuprofen and to determine their rotational constants. Density functional theory calculations were performed and used to support the conformational assignments. Fragments of ibuprofen could be also identified in the spectrum, which can be explained by thermal decomposition during the heating process for vaporizing it. The analysis of this fragmentation process as a function of temperature might provide us with some interesting insights into its mechanism.

---

[a]The author thanks "The Hamburg Centre for Ultrafast Imaging" for financial support.

FD06 9:55 – 10:10

ROTATIONAL SPECTRA OF HYDROGEN BONDED NETWORKS OF AMINO ALCOHOLS

DI ZHANG, TIMOTHY S. ZWIER, *Department of Chemistry, Purdue University, West Lafayette, IN, USA.*

The rotational spectra of several different amino alcohols including D/L-allo-threoninol, 2-amino-1,3-propanediol and 1,3-diamino-2-propanol over the 6.5-18.5 GHz range have been investigated under jet-cooled conditions using chirped-pulsed Fourier transform microwave spectroscopy. Despite the small size of these molecules, a great variety of conformations have been observed in the molecular expansion. While the NH_2 group is typically thought of as a H-bond acceptor, it often acts both as acceptor and donor in forming H-bonded networks. With three adjacent H-bonding substituents (a combination of OH and NH_2 groups), many different hydrogen bonding patterns are possible, including H-bonded chains and H-bonded cycles. Since many of these structures differ primarily by the relative orientation of the H-atoms, the analysis of these rotational spectra are challenging. Only through an exhaustive conformational search and the comparison with the experimental rotational constants, nuclear quadrupolar splittings, and line strengths are we able to understand the complex nature of these interactions. The ways in which the presence and number of NH_2 groups affects the relative energies, and distorts the structures will be explored.

Intermission

FD07

PREFERENCE FOR TOP- VS. SIDE-BINDING IN FLUORINATED ETHYLENE$\cdots$CO$_2$ COMPLEXES

REBECCA A. PEEBLES, ASHLEY M. ANDERTON, CORI L. CHRISTENHOLZ, RACHEL E. DORRIS, SEAN A. PEEBLES, *Department of Chemistry, Eastern Illinois University, Charleston, IL, USA.*

The weakly bound complexes between 1-fluoroethylene (FE), 1,1-difluoroethylene (DFE), and 1,1,2-trifluoroethylene (TFE) and carbon dioxide have been investigated using reduced bandwidth chirped-pulse (CP) and resonant-cavity Fourier-transform microwave (FTMW) spectroscopy. In FE$\cdots$CO$_2$, two distinct planar isomers are observed, corresponding to the CO$_2$ interacting with the CHF end of the FE (side-binding) or roughly parallel to the C=C bond (top-binding). Both structures contain a C–H$\cdots$O contact between one FE hydrogen atom and CO$_2$. In DFE$\cdots$CO$_2$, only a top-binding configuration is possible, consistent with the observed structure. Finally, although both top- and side-binding orientations are possible for TFE$\cdots$CO$_2$, only the side-binding conformation has been observed. The C–H$\cdots$O distances in the four species vary from 2.58 Å to 2.73 Å, while the observed F$\cdots$C distances are much more consistent, varying by only about 0.05 Å across the series. Ab initio calculations at the MP2/6-311++G(2d,2p) level have provided exceptionally accurate estimates of the rotational constants of these CO$_2$ complexes, although the energy ordering is, in several cases, inconsistent with the observed geometries.

FD08

CHIRPED-PULSE FOURIER-TRANSFORM MICROWAVE SPECTROSCOPY OF THE PROTOTYPICAL C-H$\cdots\pi$ INTERACTION: THE BENZENE$\cdots$ACETYLENE WEAKLY BOUND DIMER

NATHAN ULRICH, *Department of Chemistry, Eastern Illinois University, Charleston, IL, USA*; NATHAN A SEIFERT, *Department of Chemistry, The University of Virginia, Charlottesville, VA, USA*; RACHEL E. DORRIS, REBECCA A. PEEBLES, SEAN A. PEEBLES, *Department of Chemistry, Eastern Illinois University, Charleston, IL, USA*; BROOKS PATE, *Department of Chemistry, The University of Virginia, Charlottesville, VA, USA.*

The rotational spectrum of the CH$\cdots\pi$ bonded complex between benzene and acetylene has been measured in the $6-20$ GHz range using chirped-pulse Fourier-transform microwave spectroscopy. The spectra for the normal isotopologue, three unique ^{13}C substituted species, and the d_1-benzene$\cdots$HCCH species have allowed determination of the dimer structure. The spectrum is that of a symmetric top, with effective C_{6v} symmetry, and a CH$\cdots\pi$ distance of 2.4921(1) Å. The dipole moment has been measured using the Stark effect, and is 0.438(11) D. In addition to the ground state spectrum, three additional sets of transitions corresponding to similar rotational constants have been observed, likely due to excitation of the three low energy intermolecular vibrational modes of the dimer. Analysis of these excited state transitions is in progress. Comparison of the binding energy and structure of the benzene$\cdots$HCCH dimer with other H$\cdots\pi$ complexes will be presented.

FD09

EVIDENCE FROM BROADBAND ROTATIONAL SPECTROSCOPY FOR A COMPLEX BETWEEN AgCCH AND C$_6$H$_6$

DANIEL P. ZALESKI, SUSANNA LOUISE STEPHENS, NICK WALKER, *School of Chemistry, Newcastle University, Newcastle-upon-Tyne, United Kingdom*; ANTHONY LEGON, *School of Chemistry, University of Bristol, Bristol, United Kingdom.*

Last year, at the 68th International Symposium of Molecular Spectroscopy, the rotational spectrum of a complex formed between C$_2$H$_2$ and AgCCH was presented. The geometry was found to be T-shaped with the silver atom coordinated to the center of the CC bond in acetylene. Evidence for a new complex formed between AgCCH and C$_6$H$_6$ is now presented in the form of deep-averaged broadband rotational spectra. The spectra are observed only when both C$_2$H$_2$ and C$_6$H$_6$ are present in the gas sample. The relative intensities of the observed spectra are consistent with the naturally-occurring abundance ratio of the isotopes of silver. The shift on substitution of ^{107}Ag for ^{109}Ag implies a silver atom positioned close to the center of mass. The isotopic shifts observed when C$_2$D$_2$ is used as a precursor instead of C$_2$H$_2$ are also consistent with assignment to a complex formed between C$_6$H$_6$ and AgCCH/D. The geometry of the complex is yet to be precisely established.

MICROWAVE SPECTRUM AND GEOMETRY OF $CF_3I \cdots PH_3$

SUSANNA LOUISE STEPHENS, NICK WALKER, *School of Chemistry, Newcastle University, Newcastle-upon-Tyne, United Kingdom*; ANTHONY LEGON, *School of Chemistry, University of Bristol, Bristol, United Kingdom.*

A chirped-pulse Fourier transform microwave spectrometer has been used to measure the microwave spectrum of $CF_3I \cdots PH_3$ between 6.5 and 18.5 GHz. The complex was stabilised by supersonic expansion of a gas sample containing small percentages of CF_3I and PH_3 in a balance of 6 bar of argon. The observed spectrum is consistent with a C_{3v} prolate symmetric top structure which displays evidence for internal rotation of the PH_3 subunit about the principal axis. Over two hundred hyperfine transitions across eleven $J'' \leftarrow J'$ transitions have been assigned to the internal rotor A-state allowing the rotational (B_0) and centrifugal distortion (D_J and D_{JK}) constants as well as the nuclear electric quadrupole coupling constant of iodine ($\chi_{aa}(I)$) to be assigned for this state. For the E-state, the additional distortion terms D_{Jm} and D_{JKm} have been determined. The length of the halogen bond between the iodine and phosphorus atoms and the force constant of this bond have also been determined.

MICROWAVE SPECTRA AND GEOMETRIES OF $C_2H_2 \cdots AuI$ AND $C_2H_4 \cdots AuI$

SUSANNA LOUISE STEPHENS, JOHN CONNOR MULLANEY, MATT JOHN SPRAWLING, *School of Chemistry, Newcastle University, Newcastle-upon-Tyne, United Kingdom*; DAVID PETER TEW, *School of Chemistry, University of Bristol, Bristol, United Kingdom*; NICK WALKER, *School of Chemistry, Newcastle University, Newcastle-upon-Tyne, United Kingdom*; ANTHONY LEGON, *School of Chemistry, University of Bristol, Bristol, United Kingdom.*

A chirped-pulse Fourier transform microwave spectrometer has been used to measure the microwave spectra of both $C_2H_2 \cdots AuI$ and $C_2H_4 \cdots AuI$. These complexes are generated via laser ablation at 532 nm of a gold surface in the presence of CF_3I and either C_2H_2 or C_2H_4 and argon and are stabilized by a supersonic expansion. Rotational (A_0, B_0, C_0) and centrifugal distortion constants (Δ_J, Δ_{JK} and δ_J) of each molecule have been determined as well the nuclear electric quadrupole coupling constants of gold and iodine atoms (χ_{aa} (Au), $\chi_{bb} - \chi_{cc}$ (Au), χ_{aa} (I) and $\chi_{bb} - \chi_{cc}$ (I)). The spectrum of each molecule is consistent with a C_{2v} structure in which the metal atom interacts with the π-orbital of the ethene or ethyne molecule. Isotopic substitutions of atoms within the C_2H_2 or C_2H_4 subunits are in progress and in conjunction with high level *ab initio* calculations will allow for accurate determination of the geometry of each molecule.

FD12   **11:52 – 12:02**

DEUTERATED WATER HEXAMER OBSERVED BY CHIRPED-PULSE ROTATIONAL SPECTROSCOPY

LUCA EVANGELISTI, CRISTOBAL PEREZ, SIMON LOBSIGER, NATHAN A SEIFERT, DANIEL P. ZA-LESKI, BROOKS PATE, *Department of Chemistry, The University of Virginia, Charlottesville, VA, USA*; ZBIGNIEW KISIEL, *ON2, Institute of Physics, Polish Academy of Sciences, Warszawa, Poland*; BERHANE TEMELSO, GEORGE C. SHIELDS, *Dean's Office, College of Arts and Sciences, and Department of Chemistry, Bucknell University, Lewisburg, USA.*

The water hexamer is the first cluster size where three dimensional structures become the most stable. For cluster sizes up to the water decamer, the hexamer is unique because there are three low-lying isomers with distinct oxygen framework geometries that can be observed in pulsed molecular beams: the prism, cage, and book. Each structure has a different number of hydrogen bonds (prism: 9, cage: 8, book: 7). The prediction of the lowest energy form by different water potentials has become a benchmark in computational studies of water clusters. The experimental determination that the cage is the lowest energy form was presented by our group in 2012 using the population changes that occur when different carrier gases are used in the molecular beam expansion. Recently, it has been proposed by Babin and Paesani that the relative energy ordering of these isomers might be useful for testing the ability of theory to include zero-point energy effects.[a] Their calculations suggested that the prism might become the lowest energy isomer in the fully deuterated water hexamer. At the simplest level, this can explained by the fact that the prism has the most hydrogen bonds and would, therefore, experience the most energy lowering upon isotopic substitution. Broadband rotational spectroscopy in a pulsed supersonic expansion has been used to study the cage, prism and book isomers of deuterated water hexamer. These data, in conjunction with new computational advances, quantify the changes in the oxygen framework structure respect to the normal water hexamers. Moreover, by using different gases in the expansion we have established that the prism isomer becomes the minimum energy structure as suggested by Babin and Paesani.

[a] V. Babin and F. Paesani, "The curious case of the water hexamer: Cage vs. Prism", Chem. Phys. Lett. 580, 1-8 (2013).

FE. Planetary atmospheres

Friday, June 20, 2014 – 8:30 AM

Room: 274 Medical Sciences Building

Chair: Vincent Boudon, CNRS / Université de Bourgogne, Dijon, France

FE01   **8:30 – 8:45**

PHOTON AND WATER MEDIATED SULFUR OXIDE AND ACID CHEMISTRY IN THE ATMOSPHERE OF VENUS

<u>JAY A KROLL</u>, VERONICA VAIDA, *Department of Chemistry and Biochemistry, University of Colorado, Boulder, CO, USA.*

Sulfur compounds have been observed in the atmospheres of a number of planetary bodies in our solar system including Venus, Earth, Mars, Io, Europa, and Callisto. The global cloud cover on Venus located at an altitude between 50 and 80 kilometers is composed primarily of sulfuric acid (H_2SO_4) and water. Planetary photochemical models have attempted to explain observations of sulfuric acid and sulfur oxides with significant discrepancies remaining between models and observation. In particular, high SO_2 mixing ratios are observed above 90 km which exceed model predictions by orders of magnitude. Work recently done in the Vaida lab has shown red light can drive photochemistry through overtone pumping for acids like H_2SO_4 and has been successful in explaining much of the sulfur chemistry in Earth's atmosphere. Water can have a number of interesting effects such as catalysis, suppression, and anti-catalysis of thermal and photochemical processes. We investigate the role of water complexes in the hydration of sulfur oxides and dehydration of sulfur acids and present spectroscopic studies to document such effects. We investigate these reactions using FTIR and UV/Vis spectroscopy and will report on our findings.

FE02   **8:47 – 9:02**

LABORATORY SIMULATIONS OF TITAN'S SURFACE COMPOSITION AND ITS RELATION TO ATMOSPHERIC HAZE LAYERS

<u>JOSHUA A SEBREE</u>, ANGELA M SCHMITT, *Department of Chemistry and Biochemistry, University of Northern Iowa, Cedar Falls, IA, USA*; MELISSA G TRAINER, XIANG LI, VERONICA T PINNICK, STEPHANIE A GETTY, MARK LOEFFLER, CARRIE M ANDERSON, WILLIAM B BRINCKERHOFF, *Solar System Exploration Division, NASA Goddard Space Flight Center, Greenbelt, MD, USA.*

The arrival of the Cassini spacecraft in orbit around Saturn has led to the discovery of benzene at ppm levels, as well as large positive ions evocative of polycyclic aromatic hydrocarbons (PAHs) in Titan's atmosphere. Recently, the assignment of the band at 3.28 μm as observed by the Visual–Infrared Mapping Spectrometer (VIMS) to gas-phase PAHs provides further evidence that these molecules are prevalent on Titan. These observations suggest that aromatic reaction pathways play an important role in the photochemistry of Titan's atmosphere, in particular in the formation of large organic species. These aerosols eventually settle out of the atmosphere onto the surface of Titan giving rise to the different surface albedos that are observed by the VIMS instrument onboard Cassini.

We will present results from a laboratory study of the UV irradiation of ppm-level aromatic precursors to understand their influence on the observable characteristics of Titan's surface. Spectroscopic measurements of our analog aerosols compare favorably to observations of Titan's haze by VIMS and by the Composite Infrared Spectrometer (CIRS) in the far-infrared. In addition, the broad aerosol emission feature centered at approximately 145 cm^{-1} is of particular interest. From the broadness of this feature, we speculate that the emission is a blended composite of low-energy vibrations of large molecules such as polycyclic aromatic hydrocarbons (PAHs) and their nitrogen containing counterparts, polycyclic aromatic nitrogen heterocycles (PANHs). A further comparison of our aerosol spectra to the surface observations carried out by Cassini also shows a strong correlation between the aerosol makeup and the surface albedo of Titan. Using laser desorption mass spectrometry (LDMS) and collision-induced dissociation (CID) MS/MS techniques we confirm the presence of large (5+ rings) PAHs/PANHs in our aerosols and discuss possible formation pathways.

FE03 9:04–9:19

THE HIGH-RESOLUTION EXTREME-ULTRAVIOLET SPECTRUM OF N_2 BY ELECTRON IMPACT

ALAN HEAYS, *Leiden Observatory, University of Leiden, Leiden, Netherlands*; JOE M AJELLO, ALEJAN-DRO AGUILAR[a], *Jet Propulsion Laboratory, Science Division, California Institute of Technology, Pasadena, CA, USA*; BRENTON R LEWIS, STEPHEN GIBSON, *Research School of Physics and Engineering, Australian National University, Canberra, ACT, Australia.*

We have recorded high-resolution (FWHM = 0.2 Å) extreme-ultraviolet (EUV, 800–1350 Å) laboratory emission spectra of molecular nitrogen excited by 20 and 100 eV electron impact under mostly optically thin conditions. From these, emission cross sections were determined for a total of 491 features arising from N_2 electronic-vibrational transitions and atomic N I and N II multiplets. Molecular emission was observed from those excited levels which are not completely predissociative and to ground-state vibrational levels as high as $v = 17$.

The frequently-blended molecular emission bands were disentangled with the aid of a coupled-channels model of excited N_2 states that includes the strong coupling between valence and Rydberg electronic states and the effects of predissociation. The observed emission bands probe a large range of vibrational motion so that internuclear-distance-dependent electronic transition moments could be deduced experimental. The coupled-channels model could then be used to predict the emission cross sections of unobserved bands and those that are optically thick in the experimental spectra.

The electron-impact-induced fluorescence measurements and model were compared with Cassini UVIS observations of emissions from Titan's upper atmosphere.

[a]Present address: Lawrence Berkeley National Laboratory, Advanced Light Source.

FE04 9:21–9:36

FORMATION OF HYDROXYLAMINE FROM AMMONIA AND HYDROXYL RADICALS

LAHOUARI KRIM, EMILIE-LAURE ZINS, *Chemistry/ MONARIS, CNRS, UMR 8233, Sorbonne Universités, UPMC Univ Paris 06, Paris, France.*

In the interstellar medium, as well as in icy comets, ammonia may be a crucial species in the first step toward the formation of amino-acids and other prebiotic molecules such as hydroxylamine (NH_2OH). It is worth to notice that the NH_3/H_2 ratio in the ISM is 3 10^{-5} compared the H_2O/H_2 one which is only 7 10^{-5}. Using either electron-UV irradiations of water-ammonia ices or successive hydrogenation of solid nitric oxide, laboratory experiments have already shown the feasibility of reactions that may take place on the surface of ice grains in molecular clouds, and may lead to the formation of this precursor. Herein is proposed a new reaction pathway involving ammonia and hydroxyl radicals generated in a microwave discharge. Experimental studies, at 3 and 10 K, in solid phase as well as in neon matrix have shown that this reaction proceed via a hydrogen abstraction, leading to the formation of NH_2 radical, that further recombine with hydroxyl radical to form hydroxylamine, under non-energetic conditions.

FE05 9:38–9:53

THE HIGH J SPECTRUM OF THE $2\nu_2$ and ν_4 STATES OF AMMONIA

JOHN PEARSON, SHANSHAN YU, *Jet Propulsion Laboratory, California Institute of Technology, Pasadena, CA, USA.*

Modeling highly excited states of ammonia has become a priority due to its prevalence in the atmospheres of hot exo-plants. To date, the only viable approach has been to use the results of variational calculations utilizing spectroscopically fitted potential surfaces. In the laboratory, this approach has proven to be a useful guide in assigning many bands in NH_3, but most of the hot spectral data above the ν_2 state remains to be assigned in spite of the calculations. This is especially problematic for the low lying vibrational states which contribute significantly to the overall flux through hot bands. We report an extension of the assignments of the $2\nu_2$ and ν_4 states of NH_3 to high J. This exercise has clearly demonstrated that the primary challenge in utilizing theoretical line lists is the imprecise treatment of perturbations. These perturbations are wide spread in the $2\nu_2$ and ν_4 states of ammonia and effectively render large regions of the quantum number space poorly suited for assignment with the available line lists. The high J spectrum and energy levels are presented and compared with the the theoretical calculations. The capability of effective rotation-inversion Hamiltonian to model the spectrum is also discussed.

Intermission

FE06

HIGH-RESOLUTION ABSORPTION CROSS SECTIONS OF ETHANE AT LOW TEMPERATURES

ROBERT J. HARGREAVES, *Department of Chemistry and Biochemistry, Old Dominion University, Norfolk, VA, USA*; DOMINIQUE APPADOO, *800 Blackburn Road, Australian Synchrotron, Melbourne, Victoria, Australia*; PETER F. BERNATH, *Department of Chemistry and Biochemistry, Old Dominion University, Norfolk, VA, USA*.

High-resolution infrared absorption spectra have been created for the ν_9 band of ethane (C_2H_6) at 823 cm^{-1} using the Fourier transform spectrometer at the Australian Synchrotron . Infrared spectra were recorded at four different pressures for four temperatures (200, 160, 120 and 90 K) relevant to typical conditions on Titan. The THz/Far-IR beamline at the Australian Synchrotron is unique in combining a high-resolution Fourier transform spectrometer with an 'enclosive flow cooled' (EFC) cell designed to study gas phase molecules at low temperatures. The low vapor pressure of ethane at 90 K means that the EFC cell is necessary to obtain high-resolution spectra. Our cross sections and line parameters are needed to improve retrievals of ethane on Titan.

FE07

THZ SPECTROSCOPY OF 1d-ETHANE: Assignment of ν_{18}

ADAM M DALY, BRIAN DROUIN, LINDA BROWN, *Jet Propulsion Laboratory, California Institute of Technology, Pasadena, CA, USA*; PETER GRONER, *Department of Chemistry, University of Missouri, Kansas city, MO, USA*.

We have measured[a] over 130 pure rotational transitions of the lowest torsional state, ν_{18}, of C_2H_5D using a double pass 3 meter cell held at 0.2 Torr of sample pressure in the frequency ranges of 540-600, 680-800 and 940-1080 GHz. The program ERHAM[b], Effective Rotational Hamiltonian Method, was used to construct the Hamiltonian that included ρ, ϵ_1, β, 9 rotational and centrifugal distortion constants and 8 torsional constants. Fitted values of $\epsilon_1 = 1127.82(35)$ MHz, $\rho = 0.4342$ MHz and $\beta = 1.317(22)$ MHz enable predictions to experimental accuracy of both a and b-dipole allowed pure rotational transitions which have A - E splittings of 70 MHz and 1.3 GHz respectively. The data, combined with ground state data, will be useful to derive information regarding the potential barrier to internal rotation. This analysis supports our ongoing work to assign the infrared spectrum in the 700-900 cm^{-1} region to enable the first detection in outer planet atmospheres.

[a]*Research described in this paper was performed at the Jet Propulsion Laboratory, California Institute of Technology, under contracts and cooperative agreements with the National Aeronautics and Space Administration.*

[b]P. Groner *J. Mol. Spec.* 278 (2012) 52-67.

FE08

LINE POSITIONS AND INTENSITIES FOR THE ν_{12} BAND OF $^{13}C^{12}CH_6$

V. MALATHY DEVI, D. CHRIS BENNER, *Department of Physics, College of William and Mary, Williamsburg, VA, USA*; KEEYOON SUNG, *Jet Propulsion Laboratory, Science Division, California Institute of Technology, Pasadena, CA, USA*; TIMOTHY J CRAWFORD, *Jet Propulsion Laboratory, California Institute of Technology, Pasadena, CA, USA*; ARLAN MANTZ, *Department of Physics, Connecticut College, New London, CT, USA*; MARY ANN H. SMITH, *Science Directorate, NASA Langley Research Center, Hampton, VA, USA*.

High-resolution, high signal-to-noise spectra of mono-substituted ^{13}C-ethane ($^{13}C^{12}CH_6$) in the 12.2 μm region were recorded with a Bruker IFS 125HR Fourier transform spectrometer. The spectra were obtained for four sample pressures at three different temperatures between 130 and 208 K using a 99% ^{13}C-enriched ethane sample contained in a 20.38-cm long coolable absorption cell[a]. A multispectrum nonlinear least squares fitting technique[b] was used to fit the same intervals in the four spectra simultaneously to determine line positions and intensities. Similar to our previous analyses of $^{12}C_2H_6$ spectra in this same region[c], constraints were applied to accurately fit each pair of doublet components arising from torsional Coriolis interaction of the excited $\nu_{12} = 1$ state with the nearby torsional $\nu_6 = 3$ state. Line intensities corresponding to each spectrum temperature (130 K, 178 K and 208 K) are reported for 1660 ν_{12} absorption lines for which the assignments are known, and integrated intensities are estimated as the summation of the measured values. The measured line positions and intensities (re-scaled to 296 K) are compared with values in recent editions of spectroscopic databases.[d]

[a]K. Sung, A. W. Mantz, L. R. Brown, *et al.*, *J. Mol. Spectrosc.* **162** (2010) 124-134.

[b]D. C. Benner, C. P. Rinsland, V. Malathy Devi, M. A. H. Smith and D. Atkins, *JQSRT* **53** (1995) 705-721.

[c]V. Malathy Devi, C. P. Rinsland, D. Chris Benner, *et al.*, *JQSRT* **111** (2010) 1234-1251; V. Malathy Devi, D. Chris Benner, C. P. Rinsland, *et al.*, *JQSRT* **111** (2010) 2481-2504.

[d]Research described in this paper was performed at Connecticut College, the College of William and Mary, NASA Langley Research Center and the Jet Propulsion Laboratory, California Institute of Technology, under contracts and cooperative agreements with the National Aeronautics and Space Administration.

FE09

HIGH-RESOLUTION INFRARED SPECTRUM OF THE $\nu_3+\nu_8$ COMBINATION BAND OF JET-COOLED PROPYNE

DONGFENG ZHAO, HAROLD LINNARTZ, *Leiden Observatory, Sackler Laboratory for Astrophysics, Universiteit Leiden, Leiden, Netherlands*.

Propyne (CH_3-C≡CH) is an important molecule in astrophysics and planetary atmospheres, and an important constituent of fuels. Spectroscopic investigation of propyne is also of fundamental interest in intramolecular vibrational redistribution (IVR) dynamics of hydrocarbons. Although extensive spectroscopic studies on this simple organic molecule have been performed, the $\nu_3+\nu_8$ band has not been reported before. In this presentation, the high-resolution infrared spectrum of the $\nu_3+\nu_8$ combination band of propyne is presented.[a] Continuous-wave cavity ring-down spectroscopy is used to measure this weak infrared band in the 3175 cm^{-1} region using a supersonic free jet. The rotational analysis of the experimental spectrum results in accurate spectroscopic parameters for the $\nu_3+\nu_8$ combination vibrational state. Severe perturbations are found for $K = 3$ and 4 rotational levels, and are likely due to near-resonant or non-resonant interactions between the $\nu_3+\nu_8$ and other vibrational states. Moreover, three parallel-transition type subbands are observed and their analysis is presented as well.

[a]D. Zhao, H. Linnartz, Chem. Phys. Lett. (2014), DOI: 10.1016/j.cplett.2014.02.016.

FE10 11:13 – 11:28

HIGH-RESOLUTION INFRARED SPECTRA OF THE ν_1 FUNDAMENTAL BANDS OF ^{13}C MONO-SUBSTITUTED PROPYNE IN A SUPERSONIC SLIT JET

<u>DONGFENG ZHAO</u>, KIRSTIN D DONEY, HAROLD LINNARTZ, *Leiden Observatory, Sackler Laboratory for Astrophysics, Universiteit Leiden, Leiden, Netherlands.*

In the past few decades, many high-resolution spectroscopic studies have been dedicated to the C-H stretch vibrations in propyne (CH_3-C$\equiv$CH), aiming to understand the intramolecular vibrational redistribution in isolated small hydrocarbons. In this talk, we present the sensitive detection of the ν_1 (acetylenic C-H stretch) fundamental bands of the three ^{13}C mono-substituted isotopologues of propyne. The infrared absorption spectra are recorded using continuous-wave cavity ring-down spectroscopy (CRDS) in combination with a supersonic jet expansion of propyne/argon gas mixtures. A 0.05x30 mm slit nozzle is used in the present experiment to realize an effective rotational cooling to $\approx$14 K and a reduced Doppler width of $\approx$90 MHz. The high sensitivity of CRDS allows us to detect the three ^{13}C isotopologues in their 1.1% natural abundance. Different infrared band intensities of ν_1 are found for the three isotopologues. Detailed rotational analyses of the experimental spectra are performed to derive effective spectroscopic constants for the upper ν_1 vibrational state. The ^{13}C-substitution effect of the near/non-resonant perturbations to ν_1 of propyne is discussed. In addition, more accurate infrared data of ^{12}C-propyne, including the ν_1 fundamental band, are also obtained from our experimental spectra.

FE11 11:30 – 11:45

FT-IR MEASUREMENTS OF COLD CROSS SECTIONS OF BENZENE (C_6H_6) FOR CASSINI/CIRS

<u>KEEYOON SUNG</u>, *Jet Propulsion Laboratory, Science Division, California Institute of Technology, Pasadena, CA, USA*; LINDA BROWN, GEOFFREY C. TOON, *Jet Propulsion Laboratory, California Institute of Technology, Pasadena, CA, USA.*

Titan's stratosphere is abundant in hydrocarbons (C_xH_y) producing highly complicated and crowded features in the spectra of Cassini/CIRS. Among these, benzene (C_6H_6) is the heaviest hydrocarbon ever seen in the Titan and cold planets. For this reason, a series of pure and N_2-broadened C_6H_6 spectra were recorded in the 640 to 1540 cm^{-1} region at gas temperatures down to 231 K using a Fourier transform spectrometer (Bruker IFS-125HR) at the Jet Propulsion Laboratory. We report temperature dependent absorption cross sections for three strong fundamental bands (ν_4, ν_{14}, ν_{13}). We also derived pseudo-line parameters, which include mean intensities and effective lower state energies on a 0.005 cm^{-1} frequency grid, obtained by fitting all the laboratory spectra simultaneously. For the pseudoline generation, details can be found in a JPL MK-IV website, http://mark4sun.jpl.nasa.gov/data/spec/Pseudo). The resulting pseudolines of the strong bands reproduce observed cross sections to within ~3 %. These new results are compared to earlier work, including the $C_6H_6+N_2$ spectra recorded at PNNL.[a][b]

[a]S. W. Sharpe, et al., Appl Spectrosc 58, 1452-1461 (2004); C. P. Rinsland, et al. JQSRT, 109, 2511-2522 (2008).

[b]Research described in this paper was performed at the Jet Propulsion Laboratory and California Institute of Technology, under contracts and cooperative agreements with the National Aeronautics and Space Administration.

FE12

SPECTROSCOPIC INVESTIGATION OF *O-,M-,* AND *P*-CYANOSTYRENES

<u>JOSEPH A. KORN</u>, *Department of Chemistry, Purdue University, West Lafayette, IN, USA*; STEPHANIE N. KNEZZ, ROBERT J. McMAHON, *Department of Chemistry, The Univeristy of Wisconsin, Madison, WI, USA*; TIMOTHY S. ZWIER, *Department of Chemistry, Purdue University, West Lafayette, IN, USA.*

The atmosphere of Titan contains nitrogen, methane, and a rich mixture of more complex hydrocarbons and nitriles produced by photochemical processing. Data from the 2005 Cassini-Huygens mission suggests that among the more complex compounds are substituted benzenes that are themselves precursors to large polymeric tholins.[a] Nitriles are particularly prevalent in Titan's atmosphere due to the dominance of N_2 in the atmosphere. The cyanostyrenes are of particular interest, in part because they have the same molecular formula (C_9H_7N) as quinoline, a prototypical heteroaromatic, and therefore could engage in photochemical isomerization to form this molecule of significant pre-biotic relevance. As a first step in understanding the pathways leading to heteroaromatics, we have studied the isotope-selective spectroscopy of *o-, m-,* and *p*-cyanostyrene under jet-cooled conditions relevant to Titan's atmosphere. In this talk, the excitation and emission spectra for the three isomers will be presented. Using a combination of resonant two-photon ionization, LIF excitation, and dispersed fluorescence spectroscopies, the vibronic spectroscopy of the three isomers were recorded and compared. The *meta* isomer has two conformational isomers, which have been distinguished and studied using hole-burning methods. The talk will compare and contrast the UV spectral signatures of the set of structural and conformational isomers of the cyanostyrenes, using the ethynylstyrene counterparts as points of comparison.[b]

[a] Sebree, J. A.; Kidwell, N. M.; Selby, T. M.; Amberger, B. K.; McMahon, R. J.; Zwier, T. S., Photochemistry of Benzylallene: Ring-Closing Reactions to Form Naphthalene. *Journal of the American Chemical Society* **2012**, 134 (2), 1153-1163.

[b] Selby, T. M.; Clarkson, J. R.; Mitchell, D.; Fitzpatrick, J. A. J.; Lee, H. D.; Pratt, D. W.; Zwier, T. S., Isomer-Specific Spectroscopy and Conformational Isomerization Energetics of *o-, m-,* and *p*-Ethynylstyrenes. *The Journal of Physical Chemistry A* **2005**, 109 (20), 4484-4496.

AUTHOR INDEX

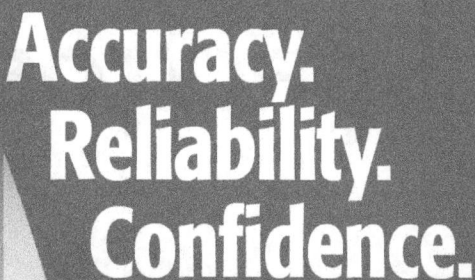

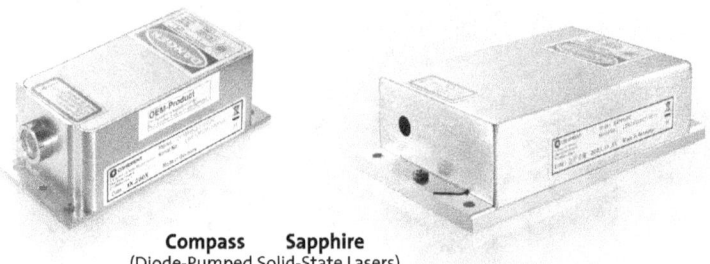
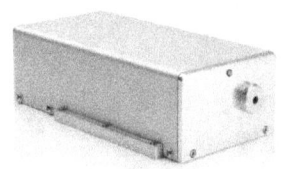

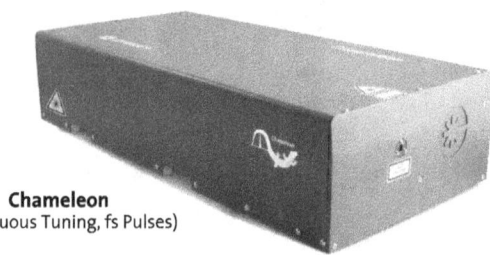

Journal of Molecular Spectroscopy

Give your research the advantage it deserves

The *Journal of Molecular Spectroscopy* presents experimental and theoretical articles on all subjects relevant to molecular spectroscopy and its modern applications. An international medium for the publication of some of the most significant research in the field, the Journal of Molecular Spectroscopy is an invaluable resource for astrophysicists, chemists, physicists, engineers, and others involved in molecular spectroscopy research and practice.

Six reasons to publish in the *Journal of Molecular Spectroscopy*

- Fast editorial times (total time from submission until final decision < 5 weeks)
- High refereeing standards
- Fast journal publication times (final corrected article online < 8 weeks)
- Wide dissemination: over 7,000 institutes worldwide have access to the journal
- High exposure of your article online; >130,000 article downloads per year on www.sciencedirect.com
- High archival value; cited half-life >10.0 years

Feature articles

The *Journal of Molecular Spectroscopy* publishes peer reviewed Feature articles - these articles will give an overview of areas of particular significance in molecular spectroscopy. They may review and consolidate an area of theoretical development or a collection of experimental data, in each case offering some new insights. The articles may also summarize the present status of a rapidly developing and/or evolving field. All the articles should serve as introductions to areas of spectroscopy other than one's specialty and should be particularly valuable to students entering the field.

Feature articles will be solicited by invitation of the Editor. However, the Editor invites you to suggest, with a reasonable level of detail, a topics that could be of interest. Self-nominations by potential authors are particularly encouraged.

For more information
www.elsevier.com/locate/jms

The Symposium thanks *The Journal of Physical Chemistry A* for subsidizing the cost of the picnic for the students

THE JOURNAL OF
PHYSICAL
CHEMISTRY

The leader in physical chemistry »»

EDITOR-IN-CHIEF:
George C. Schatz
Northwestern University

JPC A DEPUTY EDITOR:
Anne B. McCoy
The Ohio State University

JPC A: Isolated Molecules, Clusters, Radicals, and Ions; Environmental Chemistry, Geochemistry, and Astrochemistry; Theory

WHY AUTHORS CHOOSE *JPC*:

> Rapid Publication

> Reputation for Excellence

> Broad, Worldwide Dissemination

> Fast and Informed Peer Review by Editors Who Are Distinguished, Active Researchers

> Exciting, New Open Access Initiatives: go to **www.acsopenaccess.org**

View presentations of *JPC Letters* research narrated by the authors at **pubs.acs.org/page/jpclcd/acsliveslides.html**

pubs.acs.org/JPCA

ACS Publications
Most Trusted. Most Cited. Most Read.

Newport.
The Brands of Innovation.

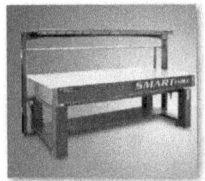

Newport
Optical Tables

Newport
XMS Linear Motor Stages

NewFocus
Velocity™ Widely Tunable Laser

Spectra-Physics
Sirah Pulsed Dye Lasers

Spectra-Physics
Quanta-Ray Pulsed Nd:YAG Lasers

At Newport our mission is to continuously evolve our knowledge and experience in order to deliver innovative products and solutions that advance our customer's technologies. To deliver upon our mission, we feel it is vital to possess expertise and experience across a broad spectrum of technologies and interconnected products.

For over 40 years, Newport has continued to grow and today is built upon world-class brands such as Corion®, New Focus™, Oriel® Instruments, Richardson Gratings™ and Spectra-Physics®. Alone, each of these brands has a rich history of product innovation and expertise. Together, we provide a synergy of knowledge across a broad spectrum of products along with the ability to deliver unsurpassed solutions and integration.

Today, Newport is one company comprised of industry leading brands, stronger and more responsive to the full breadth of your product needs. We thank you for over 40 years of support, and we look forward to many more years of partnership.

Visit **www.Newport.com/brands-5** or call **800-222-6440** for more information.

**Enjoy FREE donuts,
sponsored by Newport**

©2014 Newport Corporation

◁▷ Newport.
Experience | Solutions

Quantel Welcomes You to the
International Symposium on Molecular Spectroscopy
69th Meeting – June 16-20, 2014

Sponsor of the Women's Lunch – Wednesday June 18th

Quantel
2 bis, avenue du Pacifique
Z.A. de Courtaboeuf – BP 23
91941 Les Ulis Cedex – France
33 (0)1 69 29 16 45

Quantel USA
601 Haggerty Lane
Bozeman, MT 59715
1-877-QUANTEL

www.quantel-laser.com

Transmit and Receive Systems
Covering the 70GHz-3THz Spectrum

VDI offers a wide variety of transmit and receive systems covering the 70GHz-3THz spectrum. These systems incorporate VDI's frequency extension and mixer components coupled with commercially available microwave oscillators and amplifiers.

For transmit systems, VDI can configure them with or without a drive oscillator. A VDI Amplifier / Multiplier Chain (AMC) requires a customer supplied low frequency source (typically <20GHz, 10dBm nominal). A VDI Transmitter (Tx) integrates a source (oscillator or synthesizer) with the VDI AMC. A VDI Mixer / Amplifier / Multiplier Chain (MixAMC) requires a customer low frequency local oscillator. A VDI Receiver (Rx) integrates the LO drive oscillator with the Mixer and LO Chain for turn-key operation.

Standard AMCs and MixAMCs have been developed to provide high performance RF drive multiplication and downconversion for full waveguide band coverage. These systems can be used to extend traditional spectrum analyzers and signal generators into the THz and mm-wave ranges. VDI's standard AMC and MixAMCs offer various modes of operation. VDI AMCs can be operated in standard frequency mode (<20GHz, 10dBm nominal) or high frequency RF drive mode (<45GHz, 0dBm nominal). VDI MixAMCs can also operate in standard and high frequency LO drive modes. Customers also have the option to operate MixAMCs for block-downconversion (<20GHz IF) or as a spectrum analyzer extender. Standard AMCs and MixAMCs are available from WR15 (50-75GHz) to WR1.0 (750-1,100GHz).

VDI offers both narrow-band high-power and broadband low-power systems. High power systems use VDI's D-series X2 multipliers to achieve maximum multiplier efficiency and power handling. VDI has developed many high power systems for special customer applications, such as a novel multiplier based source with output power of 160mW at 200GHz.

Reconfigurable / modular AMCs are also available upon request.

Call for Award Nominations

Visit **www.coblentz.org** for more information

ABB Bomem-Michelson Award: ABB sponsors the Bomem-Michelson Award to honor scientists whom have advanced the technique(s) of vibrational, molecular, Raman, or electronic spectroscopy. Contributions may be theoretical, experimental, or both. The recipient must be actively working and at least 37 years of age. The nomination should include a resume of the candidate's career as well as a synopsis of the special research achievements that make the candidate an eligible nominee for the ABB sponsored Bomem-Michelson Award. Nominations for the award are open between February 1st and **May 1st** each year. Further information regarding the ABB Bomem-Michelson Award can be found at www.coblentz.org/awards/the-bomem-michelson-award.

Coblentz Award: The Coblentz Award is presented annually to an outstanding young molecular spectroscopist under the age of 40. The candidate must be under the age of 40 on January 1st of the year of the award. Nominations should include a detailed description of the nominee's accomplishments, a curriculum vitae and as many supporting letters as possible. Annual updates of files of nominated candidates are encouraged. Nominations for the Coblentz Award are open between January 3rd and **July 15th** each year. Further information regarding the Coblentz Award is available at www.coblentz.org/awards/the-coblentz-award.

Craver Award: The Craver Award is presented annually to an outstanding young molecular spectroscopist whose efforts are in the area of applied analytical vibrational spectroscopy. The candidate must be under the age of 45 on January 1st of the year of the award. The work may include any aspect of (near-, mid-, or far-infrared) IR, THz, or Raman spectroscopy in applied analytical vibrational spectroscopy. Nominees are welcome from academic, government, or industrial research. Nominations must include a detailed description of the nominee's accomplishments, curriculum vitae or resume, and a minimum of three supporting letters. Nominations for the Craver Award are open between March 30th and **August 30th** each year. Further information about the Craver Award is available at www.coblentz.org/awards/the-craver-award.

Ellis R. Lippincott Award: The Ellis R. Lippincott Award is presented annually in recognition of significant contributions and notable achievements in the field of vibrational spectroscopy. The medal is jointly sponsored by the Coblentz Society, the Optical Society of America and the Society for Applied Spectroscopy. Recipients must have made significant contributions to vibrational spectroscopy as judged by their influence on other scientists. Because innovation was a hallmark of the work of Ellis R. Lippincott, this quality in the contributions of candidates will be carefully appraised. Nominations for the award are open between January 1st and **October 1st** each year. Nominations should be submitted to: Lippincott Award Chairperson, awards@osa.org. Further information regarding the Ellis R. Lippincott Award is available at www.coblentz.org/awards/the-lippincott-award.

Honorary Membership: The Coblentz Society awards honorary memberships in the Society to people who have made outstanding contributions to the field of vibrational spectroscopy or any other field related to the purposes of the Society. Nominations close on **February 1st** each year, with awards announced at the Annual Members Meeting at Pittcon and presented at FACSS. Send your nomination for 2015 to Dr. James Rydzak, Coblentz Society President at James.W.Rydzak@gsk.com

BOLOGNA2014

The 23rd International Conference on High Resolution Molecular Spectroscopy
Bologna, Italy, September 2-6, 2014

EXECUTIVE COMMITTEE

Walther Caminati, chairman
Jens-Uwe Grabow
Per Jensen
Štěpán Urban
Monica Sanna, scientific secretary

LOCAL ORGANIZING COMMITTEE

Małgorzata Biczysko
Camilla Calabrese
Claudio degli Esposti
Luca Dore
Luca Evangelisti
Laura B. Favero

Qian Gou
Assimo Maris
Sonia Melandri
Cristina Puzzarini
Lorenzo Spada

INTERNATIONAL STEERING COMMITTEE

Jens-Uwe Grabow, chairman
Walther Caminati
Per Jensen
Zbigniew Kisiel
Juan Carlos López
Laurence S. Rothman
Keiichi Tanaka
Štěpán Urban
Andrei Vigasin
Yunjie Xu

The subjects covered by the meeting include all experimental and theoretical aspects of high resolution rotational, vibrational, or electronic spectroscopy of molecules (radicals, ions, complexes, clusters, ...) in the gas phase or in matrices, together with the application of high-resolution molecular spectroscopy in related fields such as the physics and chemistry of the atmospheres of planets and cool stars, the physics and chemistry of the interstellar medium, chemical kinetics, remote sensing, ...

Further information is available from the local organizing committee

Prof. Walther Caminati, chairman
Dipartimento di Chimica
"Giacomo Ciamician" dell'Università

Via Selmi, 2
I-40126 Bologna
Italy

Tel: +39 051 20 9 9480
Fax: +39 051 20 9 9456
E-mail: walther.caminati@unibo.it

Monica Sanna, scientific secretary
Theoretical and Computational
Chemistry Group
Scuola Normale Superiore
Piazza dei Cavalieri 7
I-56126 Pisa
Italy

Tel: +39 050 50 97 63
Fax: +39 050 09 430
E-mail: monica.sanna@sns.it

WWW: http://www.chem.uni-wuppertal.de/conference/

ISMS MEETING VENUE INFORMATION

All contributed talks will be held in the Chemistry complex (and immediately adjoining buildings). The plenary talks will be held across the quad (about 600') in the newly renovated Lincoln Hall, with intermissions held on Foellinger Forecourt (just North of Foellinger Auditorium), with a backup rain location inside Lincoln Hall.

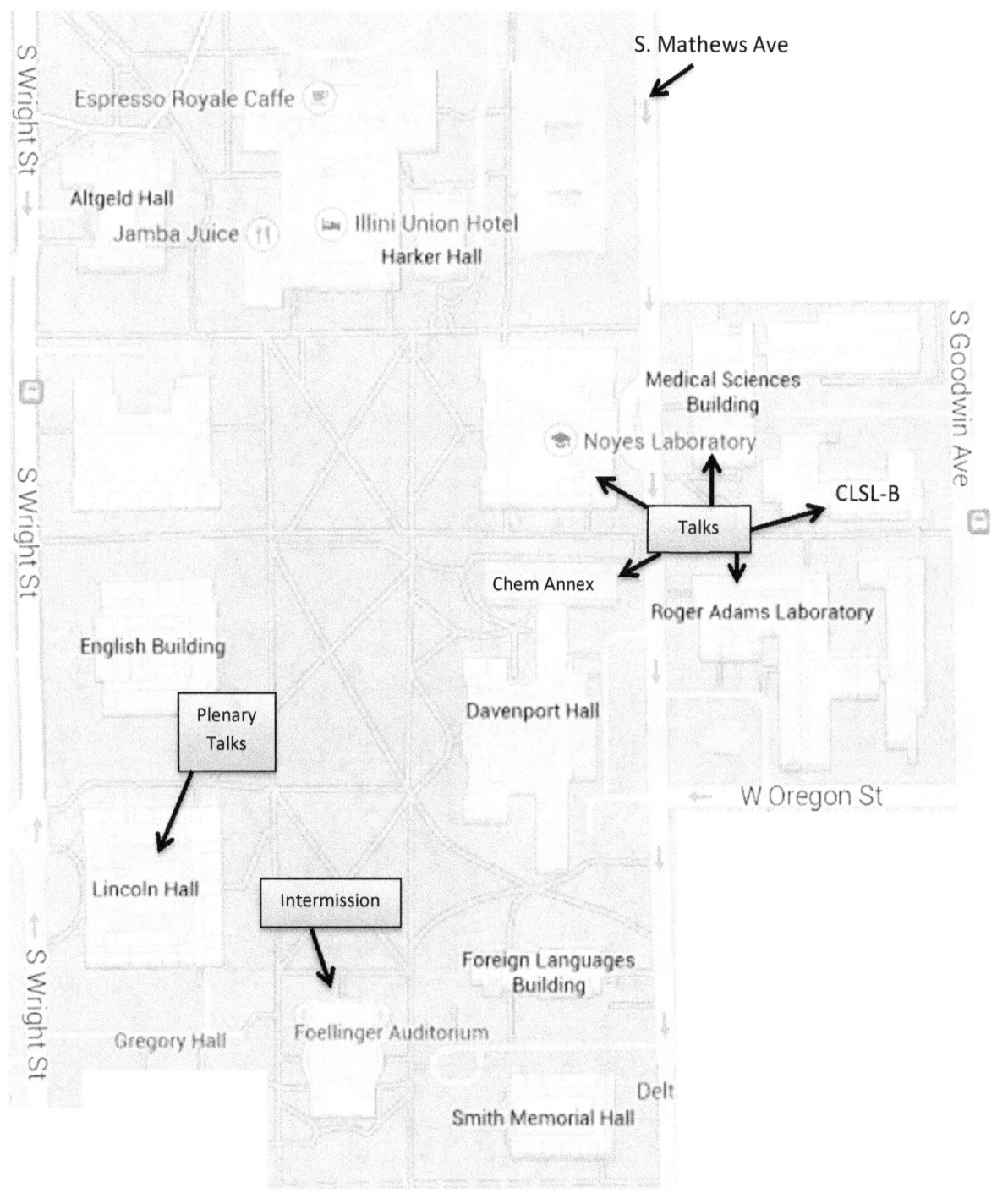

NOYES LABORATORY (NL)

Noyes Laboratory houses our Registration and Exhibitor/Refreshment Rooms (163/165), the Computer Lab (151), two lecture halls (NL 100 and NL 217), and the Chemistry Library.

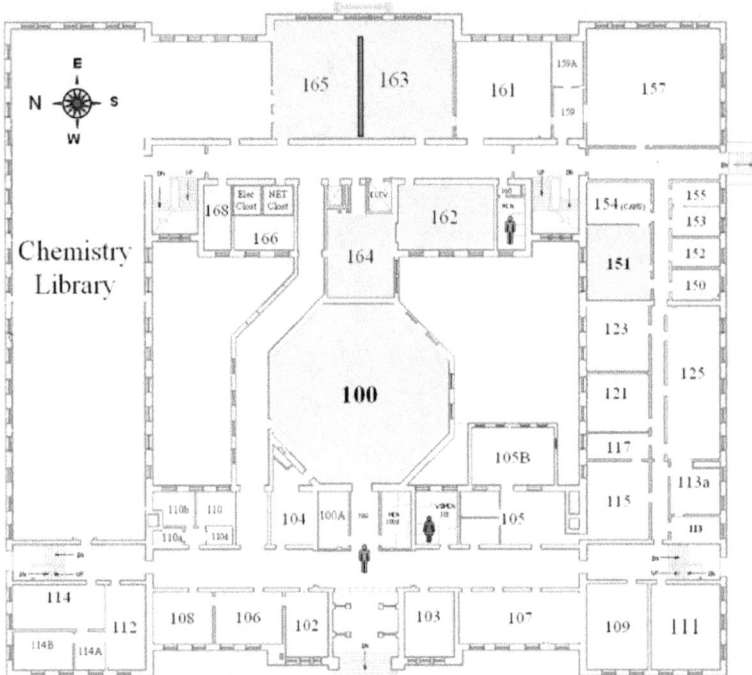

Noyes Laboratory - 1st Floor

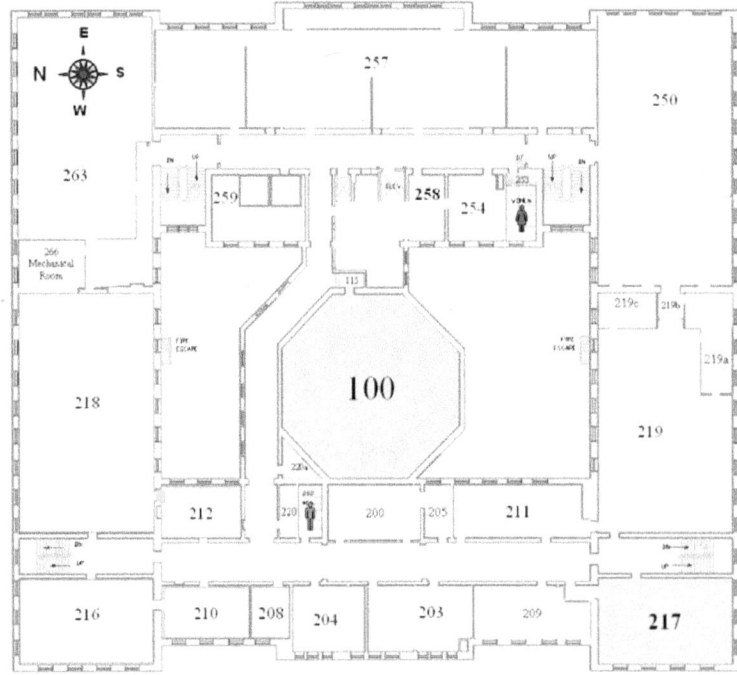

Noyes Laboratory - 2nd Floor

CHEMISTRY ANNEX (CA)

Chemistry Annex is immediately to the south of Noyes Laboratory across a pedestrian walkway. It has one lecture hall (CA 112).

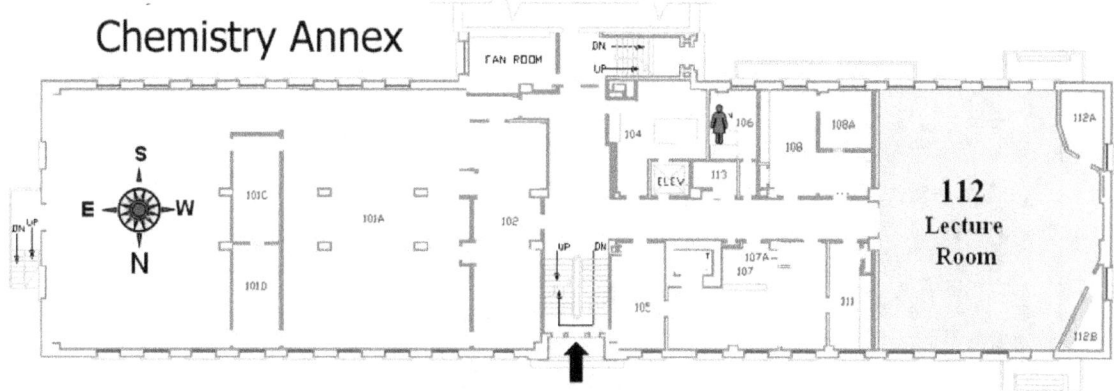

ROGER ADAMS LABORATORY (RAL)

Roger Adams Laboratory is across the street to the east of Chemistry Annex. It has one lecture hall (RAL 116). Please note that in Roger Adams Lab, the ground level is called "Ground" and the First Floor is equivalent to the Second Floor in the other buildings.

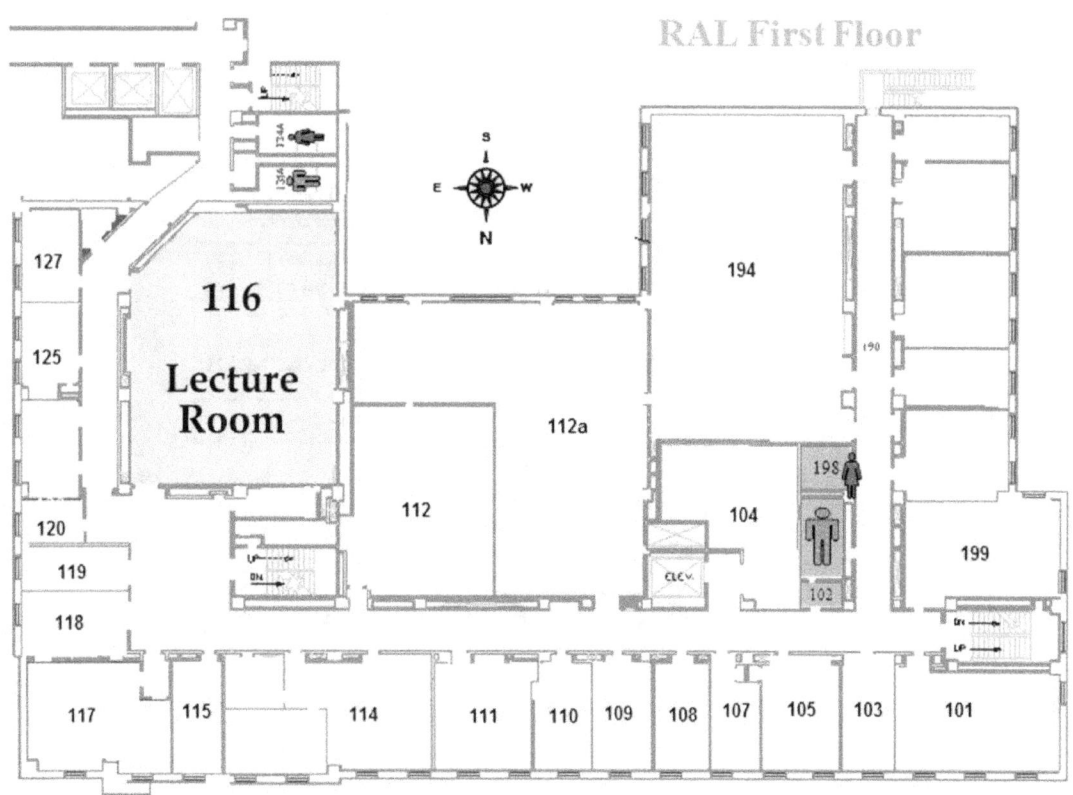

CHEMICAL AND LIFE SCIENCES (CLSL)

CLSL is a multi-wing building located across the street to the east of Noyes Laboratory. The lecture hall (CLSL B102) is in the B wing across the pedestrian walkway to the northeast of Roger Adams.

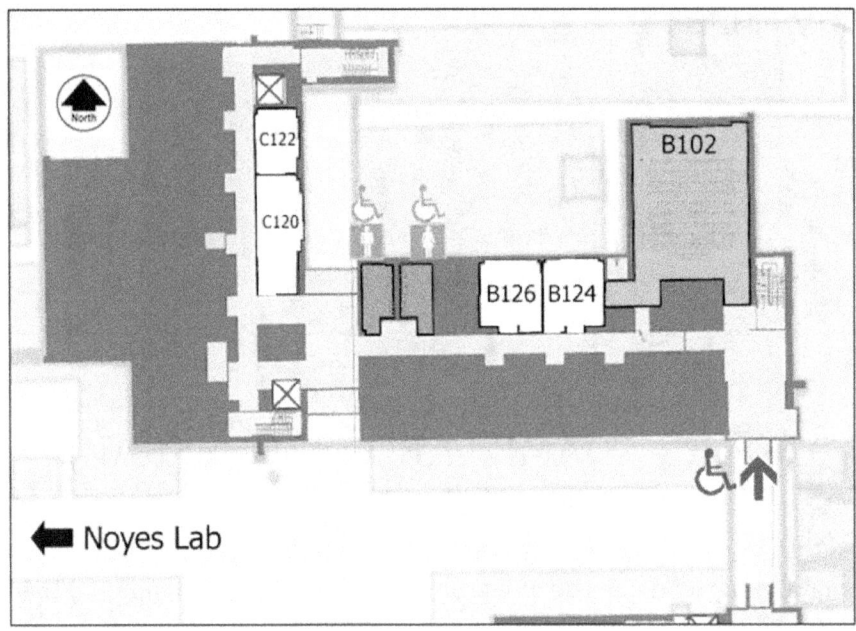

MEDICAL SCIENCES BUILDING (MSB)

Medical Sciences is across the pedestrian walkway to the north of Roger Adams. It has one lecture hall (MSB 274).

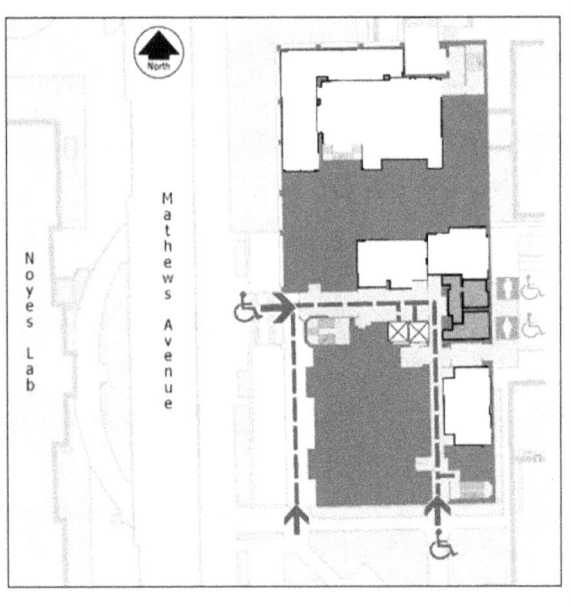
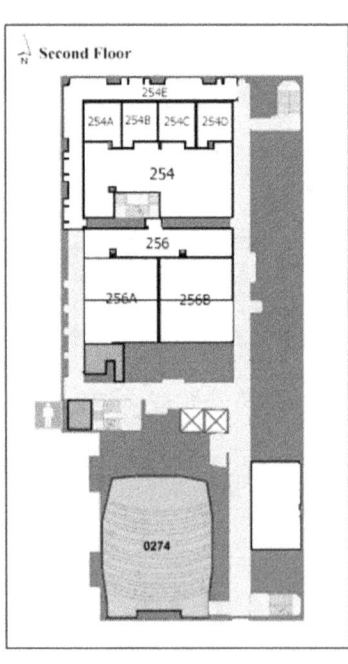

PARKING (E14) TO BOUSFIELD DORM

If you purchased a parking permit and are staying at the dorm, you will park in lot E14 (any spot). E14 is nearly due south of Bousfield Hall Dorm.

Parking enforcement begins at 6:00 AM on Monday, so you will need to have your car in lot E14 with your permit displayed before then. There are many parking meters on E. Peabody Drive (and in the lot across from Bousfield) if you wish to park closer for short periods (25 cents/15 minutes – generally between 6 AM and 6 PM, but check the meter because some go until 9 PM).

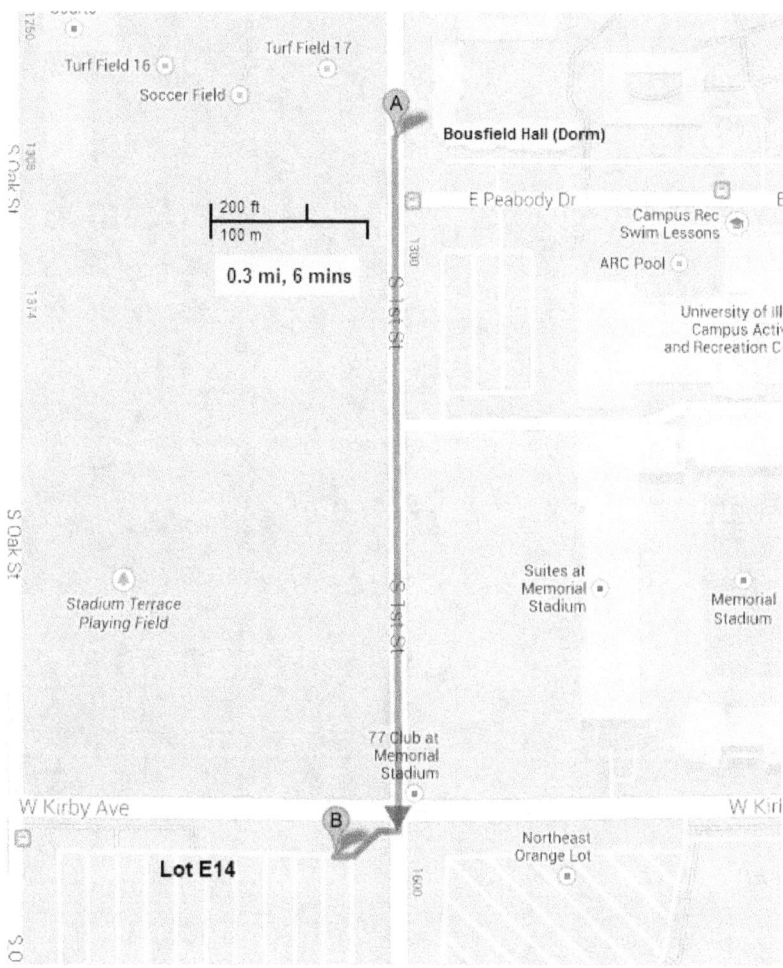

BOUSFIELD/NUGENT DORM to MEETING VENUE (walking)

Bousfield & Nugent Halls are just under a mile (15-20 minute walk) from the main symposium buildings

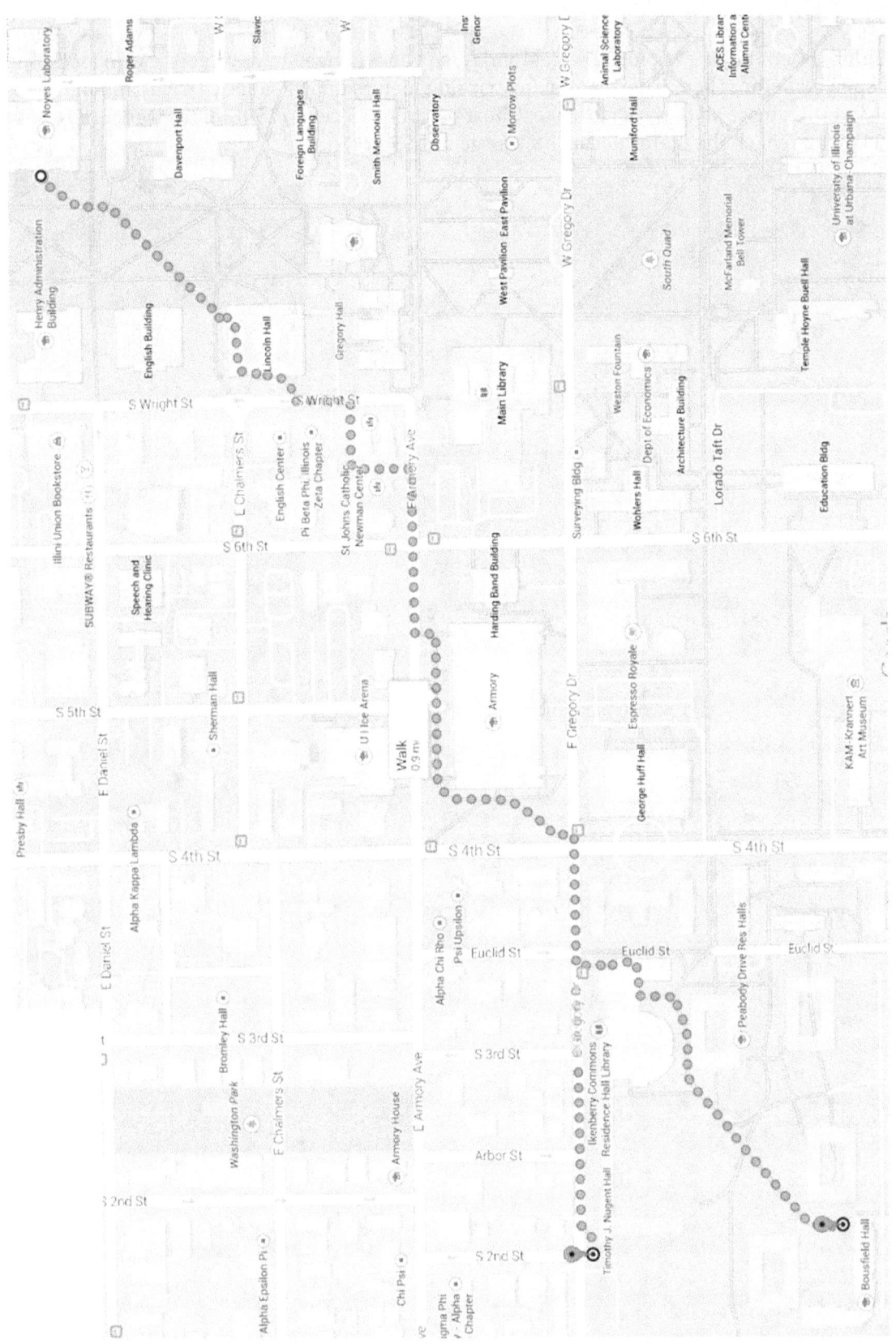

BOUSFIELD/NUGENT DORM to MEETING VENUE (bus)

There is convenient and free bus service between Bousfield/Nugent Dorms and 1 block from the meeting venue. The Yellow Line picks up on the corner of First and Peabody (Bousfield), and also on Gregory Drive (Nugent) in front of Ikenberry Commons, and drops off at the Wright Street Terminal (just outside of the Henry Administration Building). Return locations are the same but across the street. The Yellow Line will also take you to downtown Champaign, but you will need to pay for your return (only iStops are free). Approximately every 10 minutes during the day.

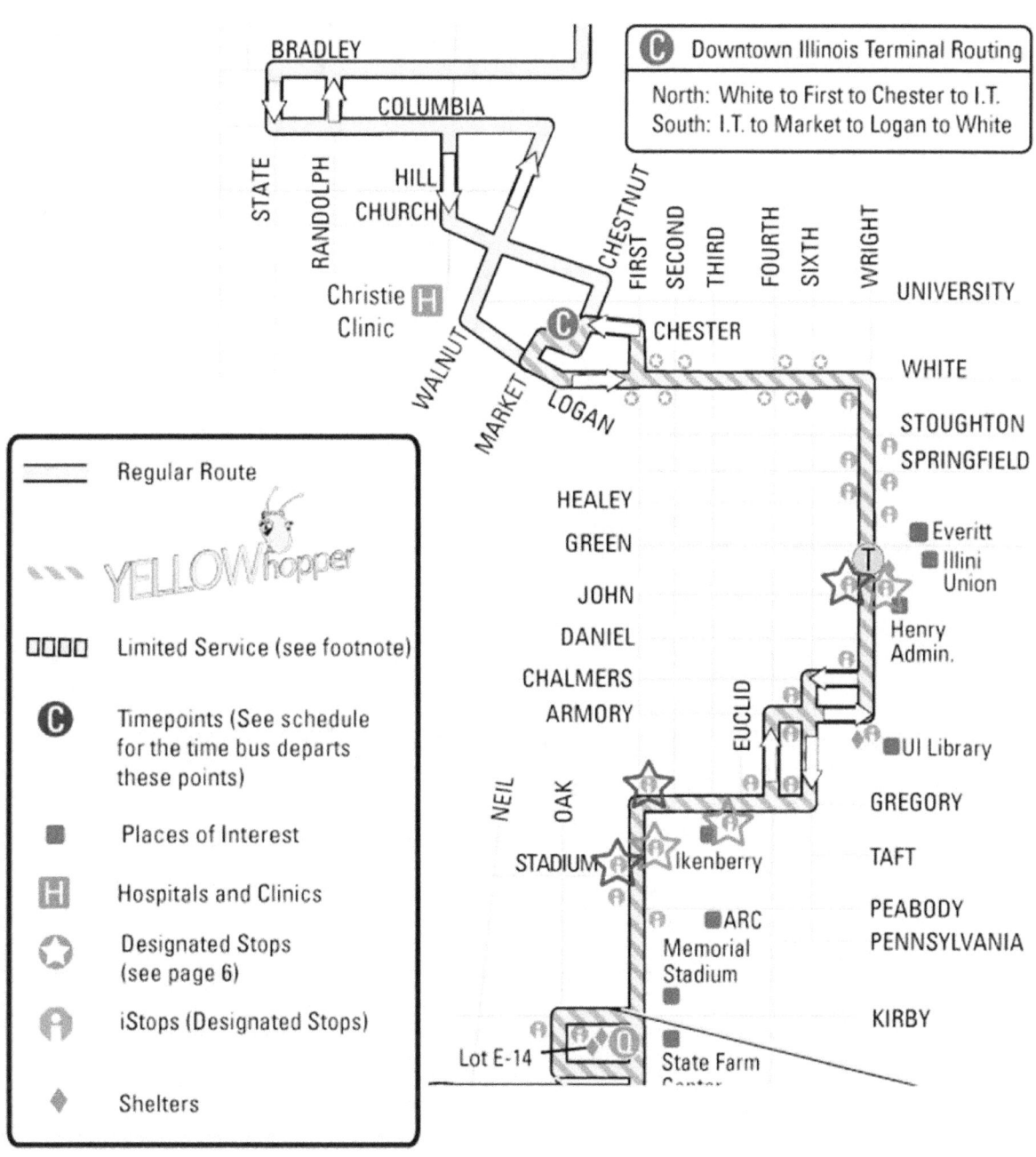

The Gold Line picks up on the corner of First and Peabody, and also on Gregory Drive in front of Ikenberry Commons and drops off at the Krannert Center (across the street from CLSL-B). Return locations are the same but across the street. Runs approximately every 10 minutes during the day (offset from the Yellow Line by 5 minutes).

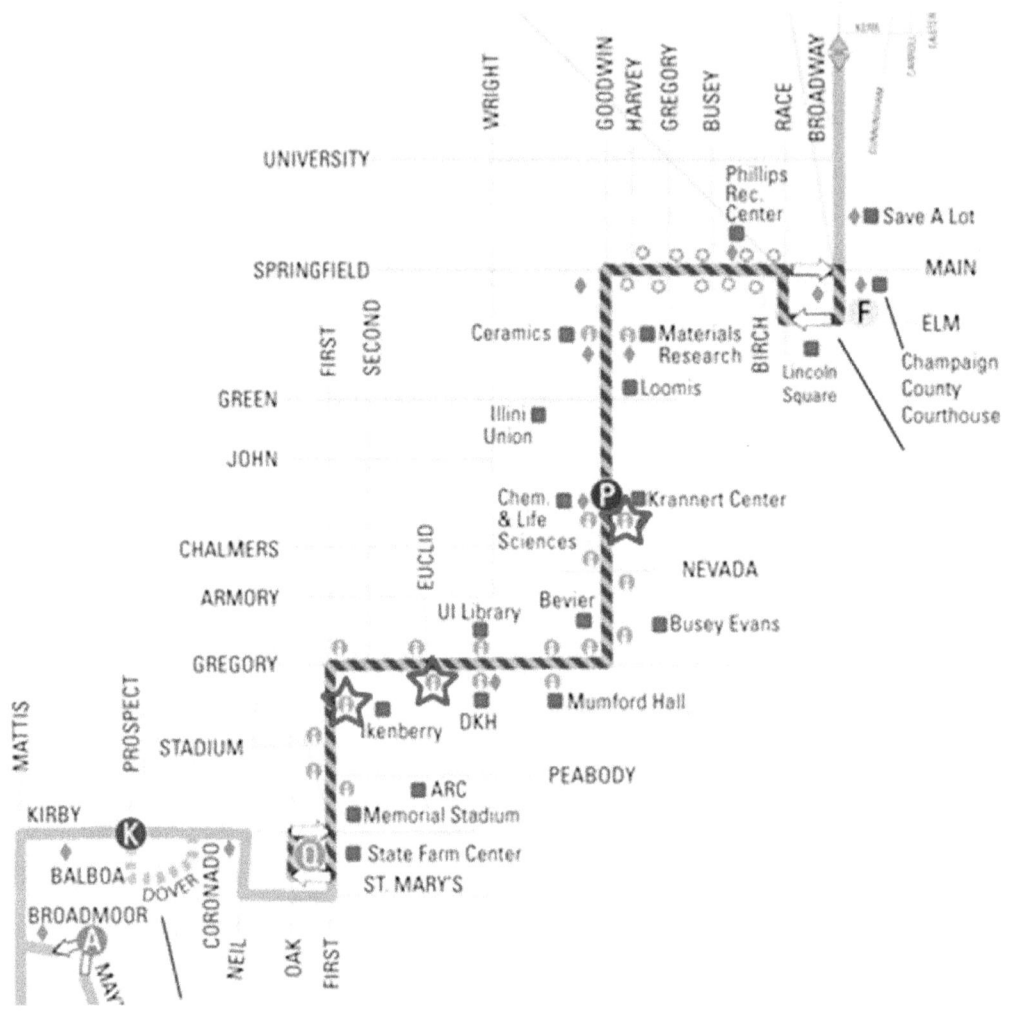

Bus Stops (Yellow Line = Left Arrow, Gold Line = Right Arrow, Lincoln Hall (Plenary) and Noyes Lab = Stars)

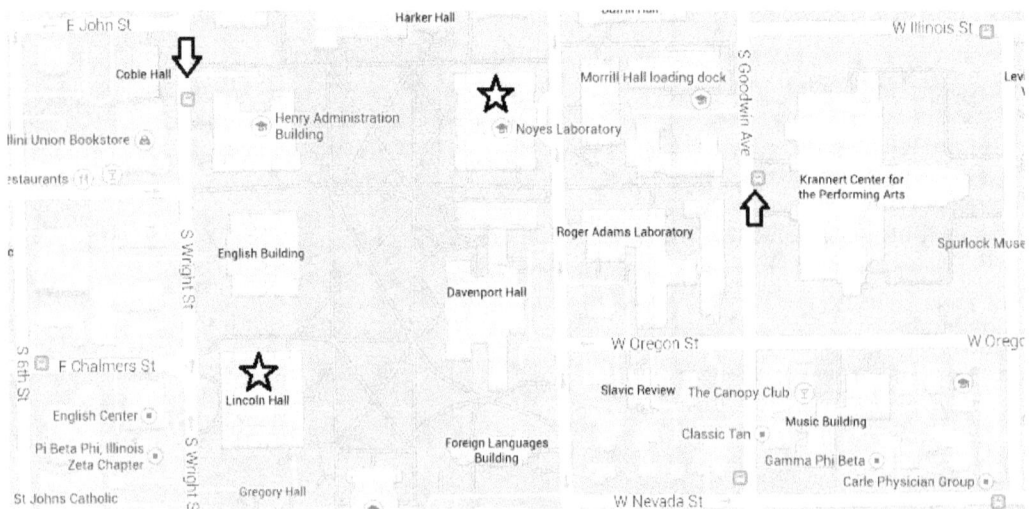

A: Alice Campbell Alumni Center
B: Bousfield Hall (Dorm)
C: Chem Annex (Talks)
D: Nugent Hall (Dorm)
G: Green Street (Restaurants)
H: Hampton Inn
I: Ikenberry Commons (Picnic)
L: Lincoln Hall (Plenary)
M: Medical Sciences Building (Talks)
N: Noyes Lab (Talks/Donuts/Coffee)
P: Parking Lot (E14) for Paid Permits
R: Roger Adams Lab (Talks)
S: Chem Life Sciences B (Talks)
U: Illini Union (Hotel, Restaurants)
Z: iHotel

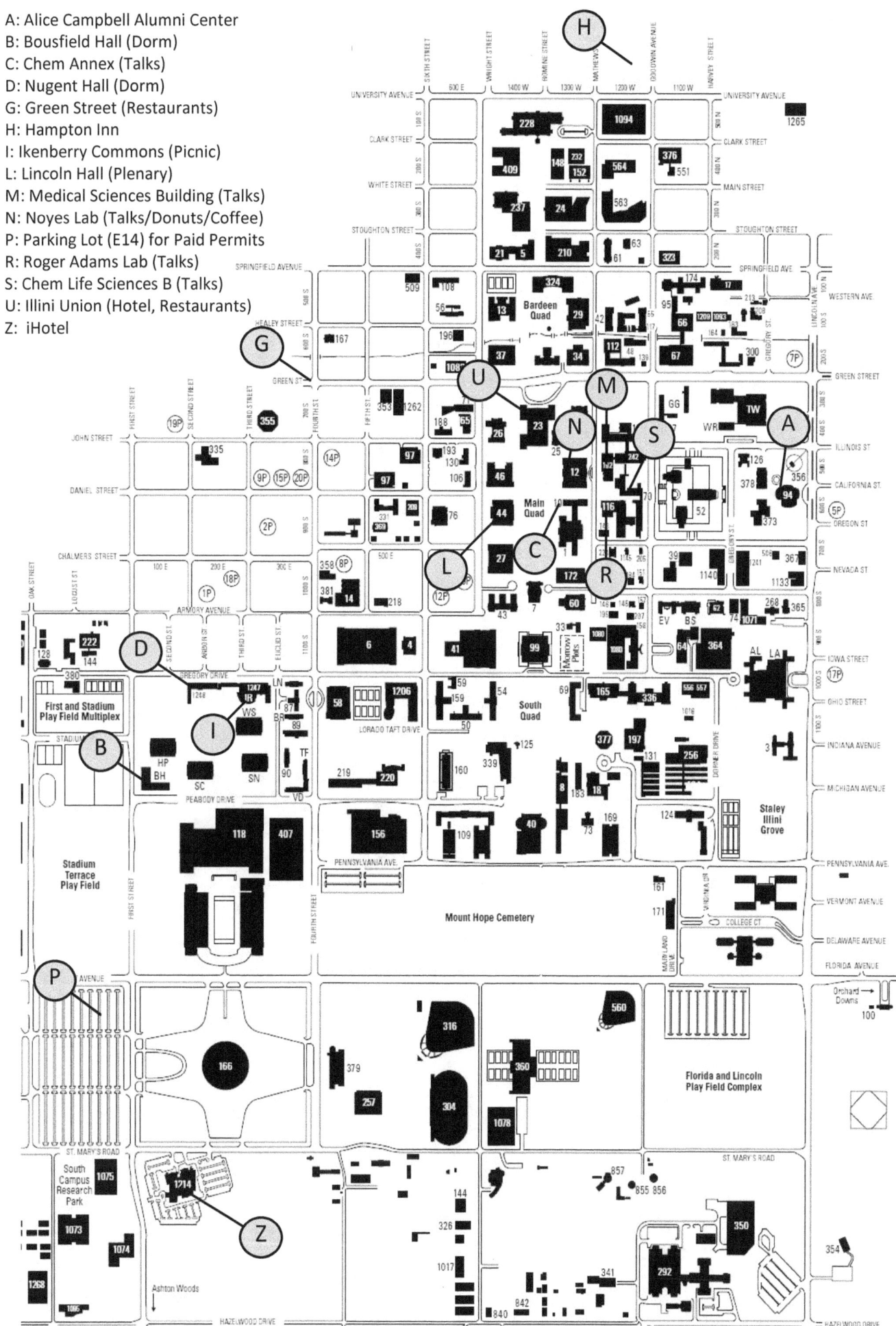

NOTES

NOTES

NOTES